Chemistry and the Sense
of Smell

Chemistry and the Sense of Smell

Charles S. Sell

WILEY

Published by John Wiley & Sons, Inc., Hoboken, New Jersey.
Published simultaneously in Canada.

For general information on our other products and services or for technical support, please contact our Customer Care Department within the United States at (800) 762-2974, outside the United States at (317) 572-3993 or fax (317) 572-4002.

Wiley also publishes its books in a variety of electronic formats. Some content that appears in print may not be available in electronic formats. For more information about Wiley products, visit our web site at www.wiley.com.

Library of Congress Cataloging-in-Publication Data:

Sell, Charles S., author.
 Chemistry and the sense of smell / by Charles S. Sell.
   pages cm
 Includes bibliographical references and index.
 ISBN 978-0-470-55130-1 (hardback)
1. Chemical senses. I. Title.
 QP458.S45 2014
 612.8′6–dc23

                                        2013034276

Printed in the United States of America.

10 9 8 7 6 5 4 3 2 1

Contents

Preface

At the very outset, I must make it clear that this book is a personal perspective on olfaction and the perfume industry. The views expressed in it are mine and not necessarily those of my colleagues, academic contacts or companies or institutions with which I have been associated. The views are those of a chemist and are admittedly biased in favour of fragrance chemists and their art.

At the start of my school life, chemistry was not my favourite subject. However, when I reached the sixth form, I was introduced to organic chemistry and immediately fell in love with the subject. I still have a vivid memory of adding a solution of adipoyl chloride in carbon tetrachloride to one of hexamethylene diamine in water, seeing a film of nylon forming at the interface and then finding that, as I pulled the film out of the mixture, more seemed to grow by magic and, as I drew the film out, it produced a long string of nylon. My interest in the living world drew me to natural products chemistry and the excitement of relating the chemicals I could synthesise in the laboratory to those in living organisms. My time at the Australian National University in Canberra with the late Professor Arthur Birch introduced me to the chemistry of terpenoids, and one of my synthetic targets was a termite trail pheromone, giving rise to my interest in chemical communication. Whilst a post-doctoral researcher at Warwick University working with Professor Bernard Golding, I deepened my understanding of enzymes. My experience in terpenoid chemistry was instrumental in my joining PPL and thus starting a career in fragrance chemistry. Since then, I have worked on analysis of perfume and perfume ingredients, chemical process development and optimisation and also on the discovery of novel fragrance ingredients. The last of these activities led me to speculation about structure/odour relationships and a fascination with the unpredictability of the odour that would be elicited by any new molecular structure. Having spent years struggling with structure/odour relationships in an attempt to understand the sense of smell, I came to the conclusion that I was asking the wrong questions. So I looked to biology to seek the right questions to ask. I was very fortunate to become part of Givaudan and to be involved in TecnoScent, Givaudan's joint venture with ChemCom to explore the olfactory receptors. The study of olfaction has made enormous advances over the last few decades, and the subject of olfactory receptors is a large part of this. We now know the primary structures of all of the human olfactory receptors and the basic principles of how they function in olfactory sensory neurons, two huge steps forward in our understanding which have both been recognised by the awarding of Nobel Prizes. The olfactory receptors are a vital first stage in the process of olfaction and the key point in the chemistry of the process, before neuroprocessing begins. For this reason, the chapter describing the receptors (Chapter 2) is the largest in the book and considerable space is devoted to providing the context of class A G-protein coupled receptors (GPCRs) in general.

CHARLES S. SELL
January 2014

Acknowledgments

I would like to thank all of my former colleagues and friends in Givaudan (including PPL, PPF and Quest) and ChemCom and in universities (my teachers, friends and consultants) for their support and encouragement and for their role in developing my interest and thinking in chemistry and fragrance.

My thanks go to Dr. Ton van der Weerdt, Dr. Philip Kraft and Stuart Reader for helpful comments on the manuscript, each in his area of expertise. I would also like to thank Dr. Sebastien Patiny for help in producing figures 2.14 and 2.15 and Dr. Philip Kraft for providing figure 8.14.

My wife, Hilary, deserves very special mention and thanks for her patience and tolerance with me during the many hours which I have spent in my study to write this book.

Introduction

René Descartes said 'I think, therefore I am'. The knowledge of one's own existence is the only certainty which each human has, the rest of what we understand about the universe is comprised of mental models based on input from our senses. Smell is often described as the most mysterious or the least understood of our senses. In the light of the very significant advances in our understanding over the last two decades, I would argue that the latter is not the case. Smell is certainly the oldest of our senses since it is present in even the most primitive living organisms and, throughout evolution, has played a crucial role in survival and development of species. Our understanding of the chemical mechanisms of odour detection in the nose has advanced enormously since Buck and Axel's discovery in 1991 of the gene family coding for the olfactory receptor proteins. The mysteries of smell revolve around the complexity of the combinatorial detection system and the neuroprocessing that converts the physical input into the mental image which we call smell. Unlike vision where we have three primary colours, each corresponding to a specific wavelength of the electromagnetic spectrum, there are no fixed reference points in odour. When we describe a smell, it is always in relation to other things that elicit a similar mental impression. We might describe one sample as smelling like roses and another as smelling like rotten eggs but neither of these is a fixed reference point. Odour exists as a continuum in a multi-dimensional mental space, and all we can do in describing a new odour is to relate it to known points in that 'odour space'. Odour classification is merely an attempt to map out regions within that space. Many parts of the brain are involved in converting the chemical stimulus in the nose into the mental odour percept, and some of these brain regions are strongly linked to memory and emotion. Thus an odour can trigger memories or influence emotional states before the subject is consciously aware of smelling it. In humans, the role of smell has extended from a survival tool, giving us information about changes in the chemical environment, through a desire to mask unpleasant odours, into a source of pleasure and artistic expression in the form of perfumery.

The chemistry of fragrance is a fascinating subject because of its breadth and the diversity of other disciplines that impinges upon it. The fragrance industry has

Chemistry and the Sense of Smell, First Edition. Charles S. Sell.
© 2014 John Wiley & Sons, Inc. Published 2014 by John Wiley & Sons, Inc.

ancient roots. For example, a perfume factory discovered on Crete dates back to 2000 B.C., and Egyptian tomb paintings often portray scenes involving the use of perfumes. In those days, the ingredients of perfumery were extracted from plant and animal sources, and plant extracts still provide many of the key notes in perfumery. Our understanding of how nature produces such an array of intricate chemical structures has grown over the last century and the natural products chemist now works alongside botanists, biochemists and molecular biologists in seeking to further our knowledge of biosynthesis. I never cease to be amazed by the variety of terpenoids that nature makes from a single precursor, isopentenyl pyrophosphate. The modern perfumery industry relies heavily on ingredients synthesised by chemists. The feedstocks include natural extracts such as pinene and petrochemicals such as isobutylene. The complexity of fragrance molecules, the performance and cost constraints of perfumes for household applications and the need to use synthetic routes that do minimal harm to the environment all combine to present a significant challenge for the process chemist. Success in this undertaking requires close collaboration with the chemical engineers who will design the process plant used in manufacture. The first generation of synthetic fragrance ingredients were exact copies of natural counterparts, such as the coumarin, vanillin and heliotropin used in 'Jicky' (1889), but non-nature-identical materials were given a boost in 1921 with the success of Chanel 5 which used small amounts of novel aldehydes to add a unique top note to the rose and jasmine oils in the heart of the fragrance. Designing novel fragrance ingredients is another very significant intellectual challenge and there are many parameters that must be taken into account. It is not sufficient just to produce a pleasing odour, the price must also be acceptable and the substance must be stable to the components of the consumer goods into which perfume is incorporated. These include acids as strong as hydrochloric, bases such as sodium hydroxide and oxidants like sodium hypochlorite and peracetic acid. The material should also be safe to use and should biodegrade easily in sewage treatment plants. Structure/activity relationships are important tools, and these bring the fragrance chemist into contact with mathematicians such as statisticians and computer modellers. Attempts to understand the relationship between molecular structure and odour brings us to the forefront of current scientific research. At least nine Nobel Chemistry Prize winners have mentioned fragrance chemistry in their Nobel Lectures and eight Nobel Prizes have gone to scientists working on the biochemistry and molecular biology of the class of receptor proteins to which the olfactory receptors belong. In 2004, the Nobel Prize for physiology/medicine went to Richard Axel and Linda Buck for their work on identifying the genes responsible for the olfactory receptor proteins which are the basis of our sense of smell. Linda Buck used this discovery to confirm that smell is a combinatorial sense, with each receptor responding to a range of odorants and each odorant stimulating a range of receptors. The 2012 Nobel Prize for chemistry was awarded jointly to Robert Lefkowitz and Brian Kobilka for their work in elucidating the structure and mechanism of action of G-protein coupled receptors, the class to which olfactory receptors belong. Chemists trying to understand the implications of these two great breakthroughs in our understanding of olfaction must be prepared to work at the frontiers between chemistry, molecular biology, neuroscience and psychology. Albert Einstein said: 'The most beautiful thing we can experience is the mysterious. It is the source of all true art and science'. We have come a long way in our understanding of fragrance but there is still plenty of mystery to provide us with intellectual challenge and beauty.

The object of this book is to review our current state of knowledge of the chemical aspects of the sense of smell, from the volatile compounds of nature to our man-made odorants that complement them; our understanding of how the nose detects odorants and produces an electrochemical signal which is translated into a mental image; and to touch on the role of this chemical sense in living organisms and in particular in humans and its contribution to our way of life and our well-being. Throughout the book, the emphasis will be on the human sense of smell, but the sense in other species will be included in order to clarify the subject or to provide the context.

Chapter 1

Why Do We Have a Sense of Smell?

THE EVOLUTION OF OLFACTION

Smell and taste are undoubtedly the oldest of our five senses since even the simplest single-celled organisms possess receptors for detection of small molecules in their environment. For example, Nijland and Burgess have shown that *Bacillus licheniformis* can detect and respond to volatile secretions (ammonia) from other members of the same species (1). One striking example of odour detection by single cells is the human sperm which possesses smell receptors identical to one of those found in the nose, a receptor known as OR1D2, and sperm will actively swim towards the source of any of the odorous molecules, such as Bourgeonal (**1.1**), that activate this receptor (2). It is presumed that the ovum releases some chemical signal which OR1D2 detects and thus the sperm is led to its target. However, the identity of this chemical signal remains unknown. Even simple organisms, such as the nematode worm *Caenorhabditis elegans*, use the sense of smell for various purposes. For example, they respond to odours by chemotaxis as a way of helping them find food (3) and they also use odorants to control population density (4).

It is easy to imagine how early living cells would gain a survival advantage by developing a mechanism to detect food sources in the primeval environment and to move towards them just as spermatozoa swim toward a source of Bourgeonal (**1.1**). Having developed such a detection mechanism, the genes coding for the proteins involved would become an important feature of the genome and would undergo development, diversification and sophistication over the course of evolution. Probably because of their evolutionary importance, the genes coding for olfactory receptor (OR) proteins are one of the fastest evolving groups of genes and form the largest gene family in the genome. An interesting recent discovery is that diet and eating habits affect the evolution of taste receptor genes (5). For example, animals such as cats, which are purely carnivorous, have lost functional variants of the sweet receptor. Sea lions and bottle-nosed dolphins were once land animals

Chemistry and the Sense of Smell, First Edition. Charles S. Sell.
© 2014 John Wiley & Sons, Inc. Published 2014 by John Wiley & Sons, Inc.

but have returned to a marine environment, and members of both species swallow their food whole without tasting it. Sea lions have lost their functional receptors for sweet and umami tastes, and the dolphins have lost these and the bitter receptors also. In all of the examples, the loss is due to mutations in the genes that have made them pseudo-genes. In other words, the genes were there in the ancestors of the species but have been lost owing to changes in diet and habit.

Smell receptors essentially recognise molecules from the environment and thus provide the organism with information about the chemistry of its environment and, more importantly, about changes in that chemistry. In single-celled organisms, the smell/taste receptors are located in the cell wall, in contact with the external environment. As animals became more complex over the course of evolution, specialized taste and smell cells developed and became located in specialised regions of the organisms. Fish have receptors on their skin, therefore in contact with the water which constitutes their environment. In air-breathing animals, the smell organs are located in the nasal cavity. Therefore, odorant molecules reach the olfactory tissue primarily through inhaled air and so must be volatile. For example, in humans the olfactory epithelium (OE) is located at the top of the nasal cavity towards its rear and, thus, under normal conditions, is accessible only to volatile substances. In some species, mice for example, the nose is sometimes placed in physical contact with the scent source (e.g. the murine urine posts which will be described later) and the animal sniffs in such a way that non-volatile materials can be drawn into contact with the sensory neurons. Much of what is commonly considered 'taste' is actually smell. The taste receptors on the tongue sense only sweet (e.g. sucrose), sour (e.g. citric acid), salt (e.g. sodium chloride), bitter (e.g. quinine) and umami (e.g. glutamate); the rest is smell. When odorants are sniffed through the nose, this is referred to as *ortho-nasal olfaction*, whereas the smell of material taken into the mouth and reaching the nose via the airways behind the mouth is known as *retro-nasal olfaction*.

Smell is the most important sense for most animals, the main exceptions being aquatic animals which rely heavily on sound, and diurnal birds and five primates for which vision is the dominant sense. Asian elephants, mice, rats and dogs all have similar olfactory acuity and outperform primates and fur seals (6). Amongst the mammals, only rhesus macaques, chimpanzees, orang-outangs, gorillas and humans rely more on sight than smell. These primates use only about half the number of OR types that other mammals do and are the only mammals with colour vision. Consequently, speculation arose that an evolutionary trade-off between odour and trichromatic vision had occurred. However, an examination and comparison of the olfactory gene repertoires of hominids, old-world monkeys and new-world monkeys led Matsui et al. to conclude that this was not the case (7).

On the other hand, there are many examples of evolutionary pressure affecting the genes for the chemical senses (taste and smell) in the animal kingdom and a few of these will suffice to illustrate this. Viviparous sea snakes do not rely on a terrestrial environment, unlike their oviparous counterparts who lay their eggs on land. The viviparous sea snakes have lost many of their OR genes, whereas the oviparous species have retained theirs (8). About 4.2 million years ago, giant pandas changed from being carnivores to being herbivores and, at about the same time, lost their umami taste receptors (9). Umami taste is due to glutamate and some nucleotides and is therefore associated with a carnivorous diet. There is therefore speculation that the two phenomena are related, but the fact that the gene is present in herbivores such as the cow and the horse suggests that the loss of the gene might have played a reinforcing role rather than a causative one. A possible alternative explanation for the change of diet has been proposed following an analysis of the panda genome in the context of other species (10).

The mosquito species *Aedes aegypti* and *Anopheles gambiae* belong to the *Culicinae* and *Anophelinae* mosquito clades, respectively. These clades diverged about 150 million years ago, yet there are OR genes that are highly conserved between the two species. Heterologous expression of the genes from both species produced receptors that respond strongly to indole, thus providing evidence of an ancient adaptation that has been preserved because of its life cycle importance (11).

Another interesting example of adaptation involves the response of a local fruit fly to the fruit of the Tahitian tree *Morinda citrifolia*. The fruit of this tree is known as *noni fruit*. It is good for humans but it contains octanoic acid which is toxic to all but one species of fruit flies of the *Drosophila* family. However, *Drosophila sechellia* flies do feed on noni fruit and choose it as a site for egg laying. Fruit flies of the *Drosophila* family have taste organs on their legs and mouthparts. It has been shown that variants in an odour-binding protein (OBP57e) are responsible for this change in food preference and also in courtship behaviour and in determination of whether the OBPs are expressed on the legs or around the mouth. The genes for this OBP are highly variable and allow for rapid evolution and adaptation as evidenced by the altered response of *D. sechellia* to octanoic acid (12).

Mice convey social signals using proteins of the lipocalin family, known as *major urinary proteins* or MUPs. Originally they were restricted in MUP types. But the development of agriculture 20,000 years ago and the resultant closer association of mice with humans, as well as the consequent increased density of murine communities, led to the need for more precise social communication and so the pool of MUP genes has increased. Mice are capable of reproduction at the age of 6 weeks, and so 20,000 years therefore represents a large number of murine generations and easily allows for such evolutionary adaptation (P. Brennan, Personal communication.).

Estimates of the number of olfactory genes per species vary slightly, a typical example (based on the analysis of Zhang and Firestein (13)) is shown in Table 1.1. In vertebrate species, the lowest number of OR genes (14) is found in the puffer

Table 1.1   Number of Intact Olfactory Genes in Different Species

Species	Chicken	Opossum	Rat	Mouse	Dog	Chimp	Human
Intact genes	554	899	1278	1194	713	353	384

fish (15) and the highest in the cow (2129) (16) (115). For rats and mice, the olfactory genes represent 4.5% of the total genome; for humans the figure is 2%.

Based on the figures in Table 1.1, it is tempting to speculate that the human sense of smell is inferior to that of rats and dogs. However, on examination of the amino acid sequences of OR proteins, we find that the human repertoire of 382 ORs covers all of the chemical space covered by the 1278 receptors of rats. The initial olfactory signal is therefore somewhat less finely tuned in humans but we have an enormous advantage in signal processing because of our very much more powerful brains. So perhaps we do not need the fine detail of input that rodents do because we can make better use of the incoming information and can therefore dispense with an unnecessarily large array of receptor types. Therefore, our sense of smell might be better than we tend to think.

The sense of smell gives organisms (from amoeba to humans) information about the changing chemistry of their environment and thus can alert them to either danger or opportunity. Just as single-celled organisms might use smell/taste to detect amino acids or sugars in their aqueous environment, highly evolved animals use smell to detect the smell of food. For example, lions use smell to detect antelopes in the savannah, monkeys use smell to detect ripe fruit in the rainforest canopy and humans use smell to find the bakery counter at the back of the supermarket. The sense of smell also warns us against the dangers of spoiled food. We quickly learn that the smell of hydrogen sulfide warns us to avoid rotten eggs or meat that has gone bad as a result of bacterial activity. Just as the lion locates the antelope using its sense of smell, the sense of smell can warn the antelope of the approach of the lion. The smell of smoke is a universal warning signal to all mammalian species. It therefore follows from this role in continuously analysing the chemistry of the environment that the sense of smell must be time-based, capable of dealing with complex mixtures of molecules (since natural odours are almost invariably mixtures) and capable of recognising previously unknown molecules. Thus the sense of smell cannot depend on a simple mechanism. The complexity of the sense will be made clear in Chapter 2.

GOOD FOOD

Taste is used to evaluate food both for its nutritious content and the possible presence of poisons. There are five tastes: sweet identifies carbohydrates for energy; umami identifies essential amino acids; salt ensures the correct electrolyte balance; sour warns against fermentation; and bitter warns against poisons such as alkaloids. The receptors for sweet, bitter and umami are G-protein coupled receptors (GPCRs), as are the ORs. Those for salt and sour are ion channels. In the mouth, there are also neurons containing receptors known as *transient receptor potential channels* (TRPs) which judge temperature, pressure and also poisons. However, much of what is normally referred to by lay people as 'taste' or 'flavour' is actually smell, and the diversity of odour signals is such that smell has to be sensitive to a much greater range of stimuli than these other senses. For instance, smell is used to judge quality of food, such as ripeness of fruit by its ester content, and the presence of poisons and bacterial contamination by the presence of amines and thiols. When we smell by sniffing ambient air, the process is known as *ortho-nasal olfaction*, whereas smelling food in the mouth involves air travelling up through the back of mouth and into the rear of the nasal cavity and is thus known as *retro-nasal olfaction*. In his book *Neurogastronomy*, Gordon Shepherd, one of the greatest figures in olfactory neuroscience, suggests that the importance of retro-nasal olfaction helped to shape human evolution (17). This view is supported by the finding that *Homo sapiens* have a larger olfactory bulb and a larger olfactory cortex than did *Homo neanderthalensis*, the only other species to have such a large brain in proportion to overall body size (18). Since neanderthals lost out in competition with *H. sapiens*, we must have had some advantage over them and perhaps the answer does lie in our superior sense of smell compared to theirs.

We all know how the smell of food attracts us. Shoppers are drawn to the smell of freshly baked bread coming from the bakery counter at the back of the supermarket, and it has been shown that blindfolded students can follow a chocolate trail in the same way that a bloodhound will follow a scent trail (19). We also know that the smell of food makes an important contribution to our enjoyment of food, and it also can control our appetite. For example, it has been shown that a complex strawberry flavour gives more feeling of satiety than a simple flavour (20). A line of ants following a food trail is a common sight, and other insects also lay trails between the nest and a food source. For example, the Australian termite species *Nasutitermes exitiosus* lays a trail of the diterpene hydrocarbon neocembrene-A to lead other members of the colony to a newly discovered food source (21) (116). Neocembrene-A (**1.2**) is virtually odourless to humans but the termites are phenomenally sensitive to it. The European grapevine moth *Lobesia botrana* is attracted to grapevines (*Vitis vinifera*) by volatiles produced by the plant. Although it is attracted to individual chemical components such as 1-hexanol (**1.3**), 1-octen-3-ol (**1.4**), (Z)-3-hexenyl acetate (**1.5**) and (*E*)- β-caryophyllene (**1.6**), the attraction is much more potent when these are present in the ratio found in the plant (22) (Figure 1.1). Similarly, blowflies are attracted to corpses by dimethyl disulfide and 1-butanol (23).

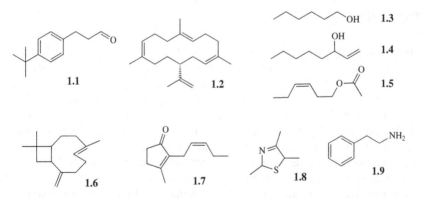

Figure 1.1   Some chemical signals.

The important role of olfaction in food selection is nicely illustrated by the following example of alteration in odour perception. After mating, the females of the cotton leafworm moth (*Spodoptera littoralis*) change their food preference from lilac flowers (*Syringa vulgaris*) to the leaves of the cotton plant (*Gossypium hirsutum*) which is the best food source for the larvae. This behaviour, which clearly gives the larvae the best survival chance, has been shown to be due to changes in the processing of the olfactory signals in the antennal lobe which is the primary olfactory centre of the insect (24).

Of course, humans represent food for some other species. Smallegange et al. investigated the relative attractiveness to the malarial mosquito *A. gambiae* of fresh human sweat, matured human sweat, used socks and some chemical components of human body odours including ammonia, lactic acid and a blend of these with various fatty acids (25). The skin residues on socks proved the most potent attractant of these. Carlson et al. showed that *A. gambiae* and *D. melanogaster* (a fruit fly) have evolved OR genes covering different parts of odour space. The narrowly tuned receptors of *A. gambiae* respond to volatiles in human sweat, whereas those of *D. melanogaster* respond to volatiles emitted by fruit (26). Cloning the gene for the mosquito's AgOr1 receptor into fruit fly neurons that had been engineered to be otherwise free of ORs resulted in the fruit fly neuron responding to p-cresol, a ligand of AgOr1 and a component of human sweat (27). The silkworm *Bombyx mori* feeds exclusively on mulberry leaves. Tanaka et al. found that the insects were guided to the mulberry by chemotaxis and identified *cis*-jasmone (**1.7**) as the volatile responsible (28). The insects' detection threshold for *cis*-jasmone is 3 pg/l. Tanaka et al. isolated 66 OR genes from the insects, cloned then into *Xenoopus* oocytes and showed that one of these receptors, BmOR56, was very selectively tuned to *cis*-jasmone. Of course, it is possible that one species could detect the trail pheromone of another and use it in controlling social behaviour. Thus one species of stingless bee, *Trigona hyalinata*, will avoid food trails left by members of the related species *Trigona spinipes* and thus prevent conflict in competition for food sources (29).

Food source identification can reach subtle levels. For example, the tick *Ixodes hexagonus* is attracted to the smell of sick hedgehogs (*Erinaceus europaeus*) in preference to that of healthy animals (30), and the predatory mite *Neoseiulus baraki* is attracted to those parts of a coconut tree that are infested by the pest *Aceria guerreronis* which is its food source (31). The ladybird, *Coccinella septempunctata*, preys on aphids and will not only detect and respond to the smell of aphids but can also learn to distinguish between the smells of two different cultivars of the same plant and will respond to one that it has already experienced to have been aphid-infested, irrespective of the smell of aphids (32).

BAD FOOD

The chemical senses provide warnings of dangers. For example, bitter taste in food warns against the possible presence of toxic alkaloids. Bacterial contamination of food is a clear danger and so something that our senses need to protect us against. Bacterial decomposition of proteins generates a number of characteristic by-products such as ammonia, hydrogen sulfide, methanethiol and dimethyl sulfide. Trimethylamine is responsible for the well-known odour of rotten fish. Lipid oxidation products are another product of bacterial action on food, and so, for example, butyric acid is an indication that milk has gone bad. Since all of these degradation products are volatile, the sense of smell offers an ideal mechanism for their detection and we quickly learn that their odours signal danger. Not only are our detection thresholds for them very low, but the resultant signals are processed faster and more accurately than those of other odours (33).

NAVIGATION

Smell is also used in navigation by animals. It is well known that salmon return to their natal stream to spawn and that they locate it by smell. Using functional magnetic resonance imaging (fMRI), it is now possible to trace the neural pathway through which this recognition occurs (34). Pigeons also use smell in finding their way back to their home and it has been demonstrated that blocking one nostril results in them taking longer and making more exploratory excursions en route. Interestingly, the effect is greater if it is the right nostril that is blocked (35).

DANGER SIGNALS

The use of smell to alert animals to danger is well known to humans. In the past, town gas was produced from coal and contained various potently malodorous thiols which soon became known as a *warning signal* of a leak of highly flammable gas. This association is so strong that cocktails of similar thiols are now added to propane and butane to serve as warnings of leaks. The smell of fire seems to be a strong warning signal for all mammals and it is obvious why it should be so. As will be discussed later, the response of an animal to the odour of a predator is an example of a kairomone, an interspecies semiochemical benefitting the receiver of the signal.

Damage to the skin of one fish has been shown to release a mixture of odorants that trigger the fear reaction in other members of the shoal and therefore drives them to flee from potential predators (36). Madagascan mouse lemurs (*Microcebus murinus* and *M. ravelobensis*) have been shown to distinguish between odours of native predators and other animals and to avoid the former (37). Similarly, rats show innate fear reaction to predator urine but not herbivore urine (38). 3,4-Dehydro-2,4,5-trimethylthiazoline (**1.8**) (also known as 2,5-dihydro-2,4,5-trimethylthiazoline or TMT) is the component in fox urine that elicits the innate fear response of 'freezing' in rodents (39). It is detected by a number of receptors in the mouse OE, but only those in certain regions elicit the fear response (40). Deactivation of those receptors prevents the fear response in mice, but these 'fearless' mice can still be trained to recognise and respond to the odour of TMT. This suggests that signals from different regions of the OE of the mouse are processed differently by the brain. The crucial factor in this recognition and response to TMT is that of the pattern of glomerular innervation in the olfactory bulb, as demonstrated by the decreased avoidance behaviour when the targeting of axons is disrupted (41). It has been found that some other odours (even if previously unknown to the rodent) can also disrupt processing of the TMT signal in some (but not all) brain regions (14, 42).

One group of receptors that are involved in detection of nitrogen-containing molecules is the trace amine activated receptors or TAARs. The role of TAAR4 (which responds to TMT) in predator detection has been studied by Liberles et al. (43). They studied the response of TAAR4 to the urine of various species and found that it responded to that of the bobcat and the mountain lion but not to others (including human). The active component was identified as 2-phenylethylamine (**1.9**) which is known to activate a variety of olfactory sensory neurons (OSNs) in mice, both in the OE and vomeronasal organ (VNO). They established that this is present not only in the urine of bobcats and mountain lions but also of lions, jaguars and servals. They confirmed its absence from the urine of humans, cows, pigs, giraffes, moose, squirrels, rats, rabbits and horses. Using the technique of Fendt (44), they found that mice showed a fear response to lion urine and 2-phenylethylamine (**1.9**). When the lion urine was treated with mono-amine oxygenase, the fear response was reduced but not totally eliminated, which led to the conclusion that there are other components in the lion urine that also elicit the fear response in mice.

CHEMICAL COMMUNICATION

Recognition of the intrinsic smells of food or danger is only part of the story as far as use of olfactory information by animals is concerned. Having developed a means of detecting odorant molecules, plants and animals then evolved the means of communicating with each other through the use of odour. Chemical communication can be used in sexual attraction and behaviour, in social organisation and in defence. When chemical communication is mentioned, the first word that springs to mind is usually 'pheromone'. However, pheromones are only part of the array of chemical messengers, and their exact role is a matter of debate in current scientific circles. Many apparently conflicting results from past experiments on chemical communication have been explained by later work, revealing the unexpected complexity of signalling systems. The chemical signals used by plants and animals are sometimes single chemical entities and sometimes mixtures, either of unrelated substances or of isomeric ratios. In some cases, the exact ratio of components in signal mixtures is crucial, and even relatively small differences from optimum result in failure of the signal to be recognised.

Chemicals used in communication between different organisms are known as *semiochemicals*. Semiochemicals can be used between different members of the same species or between members of different species. Sometimes they benefit

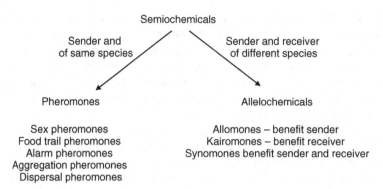

Figure 1.2 Semiochemical definitions.

the sender of the signal, sometimes its receiver, and sometimes both. Figure 1.2 shows the terms commonly used to describe these various different types of semiochemicals.

The great debate that rages in the field of chemical communication is that of learnt versus innate response to chemical signals. The argument is most intense on the subject of pheromones. Evidence for innate stereotypical response to chemical signals is strongest in insects and other invertebrates. For example, genetic variation in one of the receptors (OR47a) of the fruit fly *D. melanogaster* directly affects the fly's response to the odour of ethyl hexanoate, which is an agonist of that receptor (45). Similarly, 'hard-wired' pheromone-induced behaviour can be found in the common shore crab *Carcinus maenas*, though the structure of the pheromone remains unknown. Male crabs will attempt to mate with stones that have been treated with odours taken from a female, showing that the behaviour is independent of context and input from other senses (46). There are few such clear examples of pheromone-induced behaviour in the case of mammals where learning and context would seem to be much more significant. However, the fact that mice that have been bred in captivity for generations and never exposed to a fox or any other predator will still show the fear response to TMT suggests an innate reaction to that odour.

Part of this discussion, though often not recognised as such, is the question of whether chemicals are produced purely for communication or whether they are produced for other reasons and then a learnt response results in their being adapted for communication by the receiver of the signal. In some cases the answer is obvious, in others it is not so clear, and indeed the real situation could be somewhere between the two. Co-evolution could also contribute to the development of a signalling system in which both sender and receiver adapt so that a chemical that was originally produced for another purpose or merely as a metabolic by-product becomes part of a signalling system. Examples (such as those described below) of a damaged plant 'summoning' help in the form of predators could be considered to be examples of allomones, but the history of how such interplay between species came about is more difficult to define. Bacterial metabolism produces amines and thiols from proteins and carboxylic acids from lipids. Thus, becoming ill after eating spoiled food would clearly lead to a learnt reaction to smells associated with bacterial contamination, the odour of butyric acid giving warning of sour milk for instance. Markers for good and bad food would therefore fall into the category of kairomones and are probably largely learnt. On the other hand, the trail pheromone laid by *Nasutitermes exitiosus* as described above is clearly an example of an intentional signal. The active component, neocembrene, is not found in the food source and is only produced by the termite when it has identified one. To determine whether the response to the signal is innate or learnt would require careful experimentation with naïve insects.

Karlson and Lüscher defined a pheromone as 'a substance which is excreted to the outside by an individual and received by a second individual of the same species, in which it releases a specific reaction, for example, a definite behaviour or developmental process (47)'. Wilson and Bossert then suggested classifying pheromones into primer and releaser pheromones, primer pheromones producing neuroendocrine or developmental changes and releaser pheromones eliciting specific behaviour (48). Primer pheromones therefore would tend to fall back into the category of what were originally named *ectohormones* by Bethe There is evidence that the smell of pups induces changes in the brain of female mice that would lead to the onset of maternal behaviour (49). Such an effect would seem more hormonal than the result of communication.

It is also important to distinguish between pheromones and signature scents. Pheromones are anonymous signals, for which the detector system is hard-wired and no learning is required, the response being innate. For variable signals such as signature scents, the composition is usually complex, pattern recognition is key to interpretation, there is no hard wiring and learning is required. A pheromone is either a single chemical entity or a simple mixture of defined composition and the response to it is innate, whereas signature scents are variable mixtures characteristic of an individual or colony (50). An account of pheromone-induced behaviour will be found in the book by Wyatt (51).

Insect Pheromones

Examples of compounds that show pheromone activity in the strict sense (innate, stereptypical response with no learning having been involved) are found in insects. Perhaps the best known and most studied is bombykol (**1.10**), the sex attractant of the silkmoth *Bombyx mori*. It is released by the female and is a powerful attractant for the male (52). Other sex attractants include grandisol (**1.11**), which is a sex attractant for the male boll weevil *Anthonomus grandis*, and 2,6-dichlorophenol (**1.12**), which is a sex attractant of the Lone Star Tick *Amblyomma americanum* and also a component of disinfectants such as Dettol and TCP. Lineatin (**1.13**) is the aggregation pheromone of the striped ambrosia beetle *Trypodendron lineatum*. This beetle attacks dead and felled Douglas fir trees and uses lineatin to summon others to a newly discovered food source (Figure 1.3).

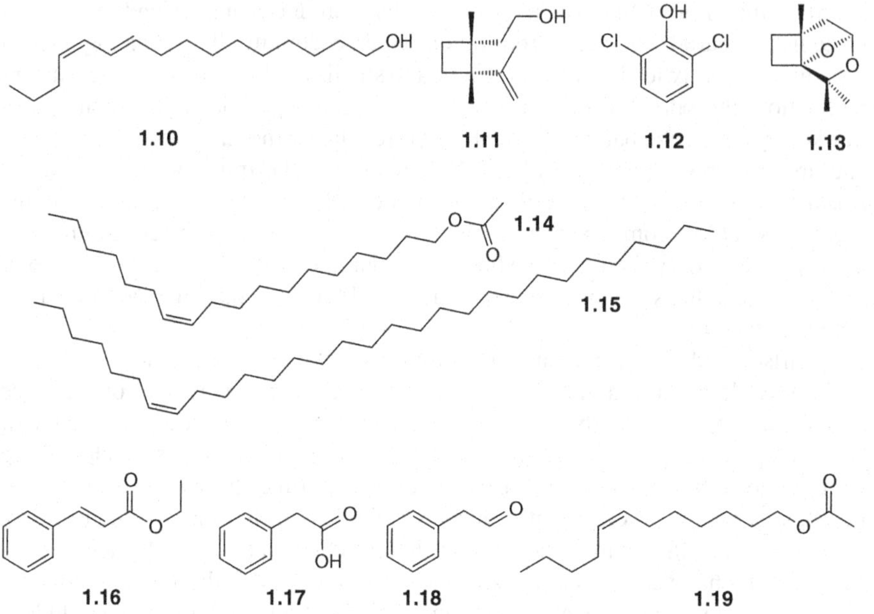

Figure 1.3   Some chemical signals used by insects.

11-(Z)-Vaccenyl acetate (**1.14**) is a pheromone that induces male–male aggression in *D. melanogaster* and is detected by the fly's olfactory system. However, another aggression-inducing pheromone, 7-(Z)-tricosane (**1.15**), is detected by their gustatory system. It was found that sensitivity to the latter was required for the former to be effective, but not vice versa, indicating a hierarchical regulation (53).

Insects often synthesise the pheromones themselves, but sometimes they obtain them from food. For example, males of the Oriental fruit moth, *Grapholita molesta*, acquire ethyl cinnamate (**1.16**) from the leaves they feed on whilst larvae, and later use it as a sex pheromone (54).

It would appear that insects process pheromone and food signals differently. After mating, male *Agrotis ipsilon* moths become less sensitive to the female sex pheromone and more sensitive to food-related odours, presumably to enable them to forage more efficiently (55).

However, it has also been found that male *D. melanogaster* flies increase their courtship behaviour when they detect phenylacetic acid (**1.17**) or phenylacetaldehyde (**1.18**), which are food signals. It is possible that this mechanism encourages the insects to breed on good food sources (56). The plasticity of response to pheromones by *D. melanogaster* is illustrated by the fact that males detect rivals by a combination of signals, including olfactory ones, and, on encountering a rival, they increase their mating activity in order to compete more effectively with the rival (57).

In many cases, two sets of signals are used together and achieve a synergistic effect; in other words, the combined signal gives a stronger response than would be expected if the signals were merely additive. An example of this is the combination of the aggregation pheromone of the American palm weevil *Rhynchophorus palmarum* with plant volatiles, which serve as kairomones (58).

(*Z*)-7-Dodecen-1-yl acetate (**1.19**) is a sex pheromone for several species of moths and butterflies and also plays a role in sexual communication in the Asian elephant. However, because insects and elephants use the same compound, this does not mean that they use it in the same way.

In social insects, the recognition odour of a colony is made up from contributions from every individual. For example, in bees, every member of the hive contributes to the comb odour, and it is this composite odour that is used for distinction between colony members and outsiders. This is a clever trick that allows genetic variation between individuals without destroying the social structure of the colony.

Vertebrate Pheromones?

Even in non-mammalian vertebrates, the role of 'pheromones' becomes less clear than in insects. In journal publications, the term *pheromone* is often used, particularly in the titles of papers, even when the role of the odorant in question is not understood, and therefore caution should always be exercised when reading.

Male budgerigars, *Melopsittacus undulates*, produce higher levels of octadecanol, nonadecanol and eicosanol in their uropygial glands than do females, and females are attracted to a mixture of these three alkanols when they are present in the right proportions (59). In another example of avian chemical communication, petrels (*Halobaena caerulea* and *H. desolata*) uropygial gland secretions contain range of fatty acid derivatives (including relative alcohols and esters) and their variation is such that they can be used by the birds to determine species, sex and identity of different birds (60). Similarly, the femoral gland secretions of male Spanish rock lizards, *Iberolacerta cyreni*, contain steroids and lipids and females are more attracted to males with high oleic acid content (61). However, this does not necessarily mean that the volatiles produced by males in either example constitute pheromones in the strict sense as defined by Karlson and Lüscher. The odours could merely be signatures that are recognised by the females whose response to them is learnt.

Amphibians such as frogs and newts have four noses, as opposed to the two of most other vertebrates. On each side, they have a 'wet' nose and a 'dry' nose. The former is used when submerged and the latter when breathing air. They can therefore use water-soluble chemicals such as proteins for communication under water (62) and volatile chemicals for communication through the air. An example of such volatile chemicals is the mixture of (*R*)-8-methyl-2-nonanol (**1.20**) and (*S*)-phoracantholide (**1.21**) produced by males of the Madagascan frog *Mantidactylus multiplicatus*, though their role in communication is not known at present (Figure 1.4) (63).

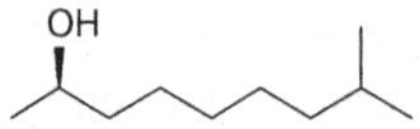

1.20 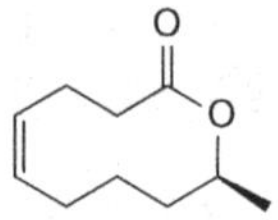   **1.21**

Figure 1.4   Semiochemicals of Mantidactylus multiplicatus.

Mammalian Pheromones?

In his rigorous analysis of the most significant studies claiming to have identified mammalian pheromones, Dick Doty proposes that mammalian pheromones do not exist (64). In many of the cases concerned, the real situation is complex and many different factors contribute to the behaviour. In other examples, the actual effect is not clear, or control experiments were found to give similar results to those claiming pheromone activity. In some cases, such as the alleged induction of menstrual synchrony in women living closely together (e.g. in hostels), the results are judged by some to be more likely to be a result of aberrations/flaws/omissions in experimental design or in statistical treatment of results. In many instances, the 'pheromone' might simply be a signature scent and the response to it a learnt one, analogous to the response of Pavlov's dogs to the sound of a bell.

A couple of examples of the best known alleged pheromones will serve to illustrate Doty's thesis.

Perhaps the best known of all is the effect of androstenone (**1.22**) on sows, reported by Melrose et al. in 1971 (65). This steroid is produced by boars and is found in their saliva. When they chomp their jaws, an aerosol containing androstenone is released into the air. The scent of androstenone either from boars or produced synthetically and administered to a sow as an aerosol will cause it to adopt the mating stance (lordosis). So, at first sight, there appears to be evidence for a pheromone effect. However, it is only effective for sexually experienced sows and gives a variable response even within the positive group. Androstenone is not necessary for induction of lordosis, and the sound of the boar grunting can also have the same effect. The activity therefore would seem to be a conditioned response in which the odour cue is learnt and is reinforced by context. This in turn raises the question of whether the androstenone is synthesised for chemical communication at all or whether it is simply a by-product of steroid metabolism that happens to be used in this way (Figure 1.5).

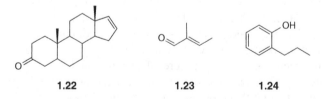

1.22   **1.23**   **1.24**

Figure 1.5   Some chemicals resulting in behavioural responses in mammals.

Response to the rabbit nipple search 'pheromone' was originally thought to be innate because it is displayed by newly born rabbit kits (66). However, the candidate 'pheromone', 2-methyl-2-butenal, (**1.23**) (67) is present in the amniotic fluid and it is now known that mammalian embryos do learn to recognise odours *in utero* and even birds can learn odours *in ovo* and the learning does affect behaviour in later life (68). It is therefore likely that the kits have learnt the odour *in utero* and seek it because of familiarity (69).

As mentioned earlier, (Z)-7-dodecen-1-yl acetate (**1.19**) is a pheromone for over 120 insect species, mostly from the order *Lepidoptera*, and was also found to be used as a sex attractant by female Indian elephants (*Elephas maximus*) who produce it in their urine (70). Male Indian elephants living in the absence of con-specifics in American zoos responded to (Z)-7-dodecen-1-yl acetate (**1.19**), thus giving further credence to the idea that it is a pheromone. However, the degree of response was lower than that to intact urine and a control substance, *o*-propylphenol (**1.24**), elicited the same response. When tested on working elephants in Burma (hence elephants living in close proximity to others), the responses of dominant and subordinate males were different, showing that the response is context-dependent and therefore not a pheromone in the sense defined by Karlson and Lüscher.

In humans, the areas of the body, other than the head, where hair growth is greatest are the armpits and groin regions. The role of hair there is to prevent chafing as the limbs move relative to the torso. The hair is lubricated by secretions containing water, fats and various other chemicals. This provides an ideal location for bacterial action and, consequently, the formation of volatile metabolites. The production of these body odours could therefore be entirely coincidental but their formation does give rise to signature combinations of odorants that can be used for identification and communication.

There are many reports of human beings able to recognise the signature odours of other humans, for example, by the ability to pick out from a variety of T-shirts those that were worn by themselves, those that were worn by a close friend and those that were worn by a stranger. It has also been reported that humans tend to prefer T-shirts that have been worn by people with the most different major histocompatability complex (MHC) (71). This tends to suggest the role of sweat as a pheromone. Differences in body odour generation and in olfactory sensitivity between the sexes could possibly be used in mate selection and sexual behaviour. Consequently, there has been much research and even more speculation on the subject. Differences in olfactory acuity between the sexes has been studied extensively, and the results are often contradictory. However, more researchers find that women outperform men than vice versa. This could be due to the effect of hormones, or it could be related to many other parameters such as social conditioning. In an attempt to better understand the phenomenon, Doty and Cameron carried out an extensive review of the subject. They concluded that there is no simple relationship between reproductive hormones and olfactory capability and that the interplay of the two is very complex (72).

Most social signals in higher animals are mixtures rather than single chemicals as can be found in insects. Human sweat, for example, contains hundreds, if

not thousands, of individual chemical components and it is the different proportions of these components that allow us to recognise different individuals or to
pick our own T-shirt out of a selection of otherwise identical ones that have each
been worn by a different person. Fresh human sweat is odourless, and it is bacterial
action that produces the characteristic smell. Of course, the exact composition of
the complex mixture of odorants that results is the result of both the nature of the
human metabolic substrate and the blend of flora on that individual's skin (73, 74).
For example, most humans show distinct patterns of composition of axillary sweat
components and these can be distinguished by smell, whereas those of monozygotic
twins are very similar and not readily distinguishable (75). Furthermore, humans
have also been shown to be capable of distinguishing between the body odours of
different Western lowland gorillas (*Gorilla gorilla gorilla*) (76). Humans can also
distinguish between male and female mouse urine because of differences in the
volatile components.

These findings are clear evidence that the human sense of smell is better than
Freud would have had us believe. However, as will be described in the next chapter,
humans lack the physical organs and brain structures that are involved in detection of putative pheromones in other mammals. Taking this and all of the above
into account, it would seem more likely that human odours are signature odours
and social markers with a learnt response rather than pheromones in the sense of
Karlson and Lüscher.

Of course, the complexity of mammalian odours allows for almost infinite
variation from species to species and individual to individual. The exact balance
between odorous materials produced directly by the mammal and those produced
by microbial action on mammalian substrates enables the resultant signature odour
to be used for such purposes as recognition of conspecifics, members of the same
or of different social groups, recognition of individuals and determination of sex,
reproductive status and social hierarchy. A simple example is the well-known
recognition of its own lamb by a mother sheep. After lambing, a ewe's hormones
cause it to lick her lamb and, in doing so, she learns the odour of the lamb. Of
course, as with most such effects, the odour cue is supported by input from other
senses. In this case, it would be learning the visual appearance of the lamb. Female
sea lions can also identify their own pups by their smell (77).

As will be further discussed in the next chapter, rodents have four different
systems for detection of environmental chemicals. Their VNO contains receptors
known as *vomeronasal receptors*, falling into the VR1 and VR2 subtypes.
Their ORs are found in the OE. The VR1 receptors are highly sensitive and
selective, and the VR2 receptors are highly specific, whereas the ORs are
broadly tuned. This makes the VN receptors much more suitable for pheromone
detection. Pheromone signals and odours are interpreted in different parts of the
brain in hamsters and mice, as are signals from conspecific and heterospecific
animals.

An important contribution to the mammalian pheromone debate came from the
team of Hurst at Liverpool University. Previous work on murine sex pheromones
had given confusing results and left the question of whether such chemicals existed.

The explanation is now clear and the reason for previous confusion apparent. Male mice build urine posts at strategic points around their territory and will drive off any competing males. Thus the mark is characteristic of an individual mouse and is used for territorial and status identification. If a mouse adds to the urine post of another, this is taken as a hostile action and the owner of the post will find and attack the mouse responsible. The dominant male mouse will also attack any other intact male entering his territory and, if the urine of an intact male is painted onto the back of a castrated mouse, it will also be attacked and driven off. In any area therefore, the predominant odour is that of the urine of the dominant male. Some of the compounds found as odour markers in the signatures of male mice are the thiazole derivative (**1.25**), the bicyclic acetal (**1.26**), the hydroxy ketone (**1.27**) and the farnesene isomers (**1.28**) and (**1.29**). Mice excrete vast amounts of protein in their urine in the form of MUPs. These proteins are lipocalins, similar to the odour-binding proteins of other mammals and are in the 18–20 kDa range. Each mouse produces a large number of distinct MUPs and the patterns have a genetic basis. The MUPs are detected by the VNO, which is designed to detect proteins, and therefore physical contact is necessary since the proteins are non-volatile. Signals originating in the VNO are processed in the accessory olfactory bulb (AOB). The volatile odorants of the urine are trapped in and slowly released from the MUPS. The urine also contains a protein that Hurst has named *darcin*, after the character created by Jane Austen. Like the MUPs, darcin, a non-volatile protein, is detected by female mice rubbing their noses on the urine posts and sniffing it into the nose. Darcin is the real attractant but the females learn to associate it with the volatile odour of the dominant male and therefore will be attracted to his scent (78). Exposure to darcin also leads the mice into developing a preference for areas where they have detected it, even if the scent mark is no longer present (Figure 1.6) (79).

There is no inherent attraction to the volatile urine components, the response of the female is learnt and the signals are specific to specific males. Females are more attracted to the scent of a male they know than to that of a stranger, even if the marks are 24 h old. However, if one male has over-marked the mark of another, the female's preference will be for the new male presumably because he has shown himself to be more dominant and hence better material for producing offspring. The MUP genes are found on chromosome 4, whereas the MHC is on chromosome 16. It has been shown that it is the MUP/odorant combination, rather than the MHC, that will control mate choice by females. However, laboratory mice are

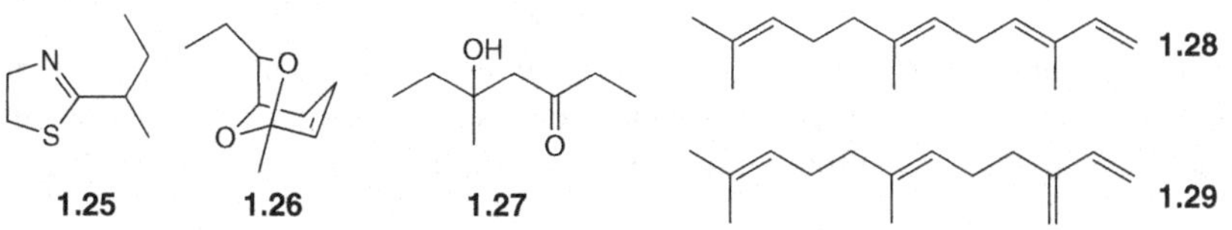

Figure 1.6   Some odorants found in mouse urine.

heavily inbred. Mitochondrial DNA shows that they are descended from only three lineages, and the Liverpool group has shown that they come in only two MUP types. Therefore, results on laboratory mice do not necessarily reflect the situation with wild-type mice. Mice originally were restricted in MUP types. The development of agriculture 20,000 years ago, the resultant closer association of mice with humans and the consequent increased density of murine communities led to the need for more precise social communication and so the pool of MUP genes has expanded. Mice are capable of reproduction at the age of 6 weeks, and so 20,000 years represents a large number of murine generations (P. Brennan, Personal communication.).

If a drop of the urine of a strange mouse is applied to the nose of a pregnant female mouse, 80% of them will abort their litter. This does not work if the stranger's urine is replaced by water or by the urine of the father of the litter. The male signal works by inhibiting prolactin release and by removing luteotropic support. Signals from the VNO of the female cause release of hormones and dopamine in the brain, and this blocks the pregnancy hormone patterns. Increased local inhibition in the AOB at memory recall is hypothesised to disrupt transmission of the pregnancy blocking signal (P. Brennan, Personal communication.).

Their specificity depends on certain anchor residues. Brennan has shown that these could be the factors that determine specificity of pregnancy markers. However, the peptides do not. Exposure to male murine urine accelerates puberty in prepubertal females (together with other effects). The signals act via VNRs (TRP2C) and so are probably due to non-volatile components. MUPs from strange males do not block pregnancy, whereas lower molecular weight (MW) proteins show more effect. The hypothesis is that these proteins (possibly nonapeptides) are related to the MHC and bind to the MHC proteins of the female, therefore carrying match/no-match messages. Leinders-Zufall et al. have shown that the VNO responds to the nonapeptides (80). They work in isolation and are not testosterone dependant and the MHC proteins are absent from urine. So, the nonapeptides are involved but it is not known how (P. Brennan, Personal communication.).

Urine signals are not necessarily restricted to rodents. Some primates deposit urine on their hands and then rub them over the rest of their body. It is thought that this might play some role in social communication. Support for this hypothesis includes the fact that fMRI showed that the brains of female capuchin monkeys processed odour signals from the urine of mature males differently from that of immature males (81).

Caveat

A danger for animals using chemicals to communicate with conspecifics is that predators can eavesdrop on their signals and use these to find them (82). Of course, the predators will also be the source of odorous substances and so the potential prey must learn to be able to distinguish the signals from its conspecific and those from the predator so that it can adopt appropriate behaviour (83).

Communication in Plants

Plant Volatiles as Attractants

Flowers use volatile scents to attract insects, and even primitive plants such as mosses have developed complex chemical signalling systems to influence the behaviour of insects (84). The use of volatile chemicals by plants to attract pollinators and as a means of seed dispersal by attracting fruit eaters is so well known that it requires no further discussion here. However, not all attractants serve the plant well. For example, the apple blossom boll weevil, *Anthonomus pomorum*, is attracted to the volatiles released from developing fruit buds. It then lays its eggs in the bud and the larvae that result feed on the apple (85). This volatile signal would therefore be considered to be a kairomone, that is, one that benefits the receiver.

Plant Volatiles for Defence (Repellents and Anti-Feedants)

d-Limonene (**1.30**) is an example of an allomone (an allelochemical that benefits the sender) in that it is produced by the Australian tree *Araucaria bidwilli* and repels termites that would otherwise attack it (86). d-Limonene is an alarm pheromone of the termites and so the tree essentially uses the insect's own communication system to deter it.

Nepetalactone is produced by catmint (*Nepeta cataria*) and is a mixture of two isomers, (**1.31**) and (**1.32**), the former being the major (87). It is insect-repellent, thus serving to deter unwanted herbivorous insects from the catmint. Interestingly, it also induces grooming and rolling behaviour in all felines, from domestic cats to lions and tigers (Figure 1.7).

1.30 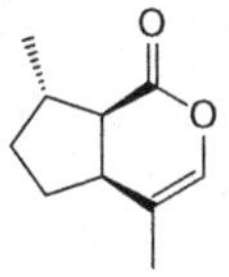   **1.31** 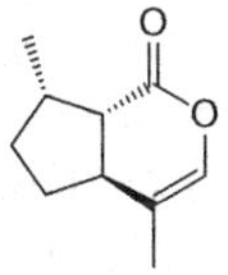   **1.32**

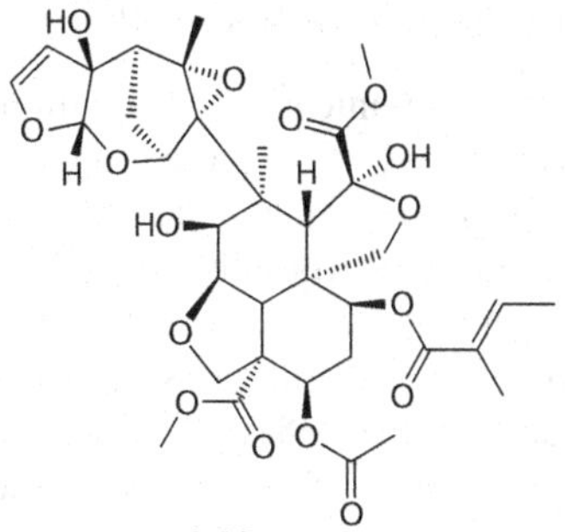

1.33

Figure 1.7   Some semiochemicals produced by plants.

An anti-feedant is a substance that a plant produces to prevent herbivorous insects from eating it. Arguably, the best known anti-feedant is azadirachtin (**1.33**), a product of the neem tree *Azadirachta indica*. It was found during a locust plague in 1959 that desert locusts (*Schistocerca gregaria*) left neem trees untouched whilst devouring everything else. The structure of the active principle was not elucidated until 1968 because of its complexity (88). Azadirachtin is not volatile and so the insect has to taste it to detect its presence.

Pest Predator Attraction

Producing chemicals that attract desired animal species or repel unwanted ones is a fairly straightforward way for plants to look after their interests. However, more complex mechanisms also exist, as discovered by Turlings and his co-workers. They showed that damage to maize (*Zea mays*) roots by the beetle *Diabrotica virifega virifega* causes attraction of the nematode *Heterorhabditis megedis*, which is a predator of the beetle. (*E*)-β-Caryophyllene (**1.34**) is the active agent released by the maize plant (89). Even more complex is the reaction of maize to attack by the beet army worm (*Spodoptera exigua*). Volicitin (**1.35**) is produced by the beet army worm and is present in its saliva. When these caterpillars browse on maize plants (*Zea mays*), some volicitin is transferred to the plant. This triggers a chemical change in the plant, and it starts to produce a variety of odorous chemicals such as α-*trans*-bergamotene (**1.36**), (*E*)-β-farnesene (**1.29**) and (*E*)-nerolidol (**1.37**). These attract a parasitic wasp, *Cotesia marginoventris*, which preys on the beet army worm. It has been shown that other damage to the maize leaves, such as cutting with a knife, does not induce this change in the plant's chemistry. Thus maize plants have developed a clever system of summoning predators to fight off attack by the army worm, and this defensive mechanism is only called into play when necessitated by the onslaught of beet army worms (Figure 1.8) (90).

Communication Between Plants

Chemical communication is not limited to animals but can also occur between plants as can be seen from the following examples. Methyl jasmonate (**1.38**) potentiates defence mechanisms in tomatoes and other members of the *Solanaceae* and *Fabaceae* families. Initially this effect was discovered by direct application to the leaves; then it was found that it could be achieved by keeping the tomato plant in a closed space with sagebrush (*Artemisia tridentata*) which is known to contain methyl jasmonate (**1.38**) and allowing the jasmonate to diffuse through the air (91). A further example of one species eavesdropping on the signals of another is that of the native tobacco (*Nicotiana attenuata*) which also picks up the signals from clipped sagebrush to prime its defence system into increasing its resistance to predation by caterpillars of the moth *Manduca sexta* (92, 93). This can be used in pest control since clipping of sagebrush plants in the field stimulates the release of methyl jasmonate (**1.38**) and this affects any neighbouring tobacco plants (94).

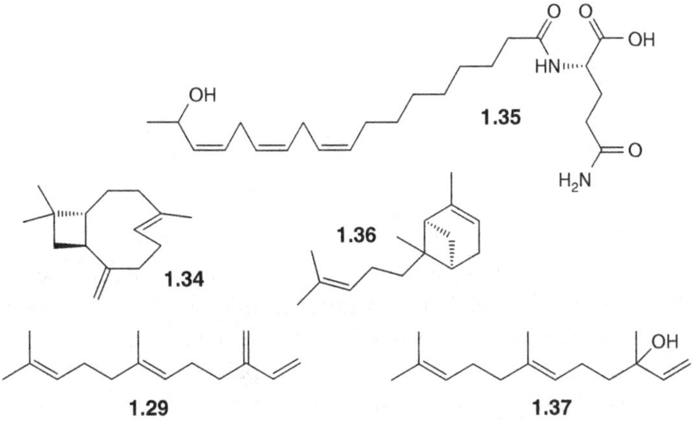

Figure 1.8   Some plant semiochemicals.

The natural cocktail of volatiles released by the clipped sagebrush includes not only methyl jasmonate (**1.38**) but also methacrolein (**1.39**), some terpenoids and various other chemicals (95).

Micro-organism- and Parasite-Induced Communication

The protozoan parasite *Toxoplasma gondii* infects the brain of rats and alters their reaction to the odour of cats from one of fear and avoidance to one of sexual attraction, and thus the infected rats are more likely to pass on the infection to cats (96).

Viruses also use chemical signals to their advantage by causing their host organisms to produce signals that work in favour of the virus. For example, cucumber mosaic virus affects the squash *Curcubita pepo* and is spread by aphids. Transmission is most effective if the aphids move rapidly from one plant to another. The virus causes the plant to produce aphid-repellent chemicals to ensure that aphids move quickly to another plant from the plant it has just infected (97). Another example is that of the mouse mammary adenovirus which is passed from a female to her pups in her milk. Infected pups then produce mammary tumours at the reproductive stage of their lives. The virus also causes increased production of 3,4-dehydro-*exo*-brevicomin (DHB) (**1.40**) in the urine of infected females. Since DHB is an attractant for males, the virus ensures its success by making infected females more attractive to males (Figure 1.9) (98).

HUMAN OLFACTION IN CONTEXT

Much of the above discussion relates to species other than humans. It does have relevance to humans but we must always be careful when making interspecies

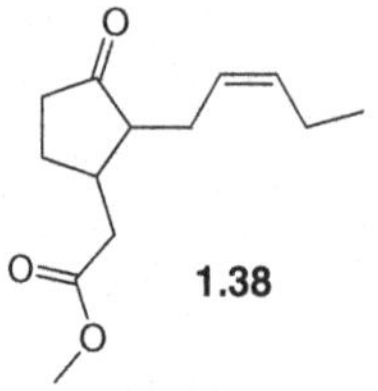

Figure 1.9 Some semiochemicals produced in response to injury or infection.

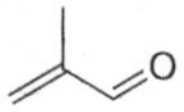

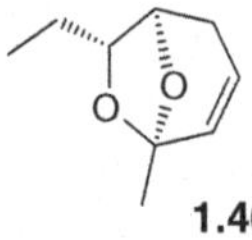

comparisons and more will be said on this subject in Chapter 2. Certain species, such as the fruit fly *D. melanogaster*, mice or rats, are often selected for study because they are easier to work with than humans, and, generally, the simpler a species is, the easier it is to study one facet relatively independently of others. The much greater complexity of humans means that conclusions drawn from studies in simpler species might bear little relevance to us. Comparison with insects is particularly risky because of some significant differences between vertebrate and invertebrate olfaction. Similarities and differences between insects and vertebrates have been nicely reviewed by Kaupp (99), and will be discussed in more detail in Chapter 2. Evolutionary pressure has generated complex and sophisticated systems including our own, and we can learn much by studying smell in other species but there are always caveats in extrapolation to human olfaction. The rest of this book is devoted to human olfaction. Reference will be made to findings in other species, but I will try to indicate the relevance of these and point out necessary caveats.

Our sense of smell has evolved to give us information about chemical changes in the environment and to enable us to select good food and avoid ingestion of harmful substances. It must be able to detect and, both accurately and reliably, identify the odours of those chemicals of importance for survival. What is more, we must be able to detect these odours against a complex odour background. The animal that fails to detect and recognise the odour of the approaching lion because it is surrounded by the odour of flowers or trees will not leave descendants to preserve its genes. Similarly, the sense of smell must be time-based because we need to know immediately that the lion is approaching and how close it is. These simple evolutionary guiding principles based on macroscopic considerations must give us strong clues about how the sense of smell operates at the microscopic level. Evolution tends to adapt and refine systems that work rather than to discard them and look for something better. Therefore we can learn about our sense of smell by studying that of other animals, including much simpler ones. However, we must do so with caution because of that very process of adaptation and refinement.

OLFACTION IN THE CONTEXT OF THE SENSES

René Descartes made the now famous observation '*Cogito ergo sum*', 'I think therefore I am'. The only certainty for any of us is that of our own existence. Beyond that, everything we know of the universe comes through our five senses; olfaction (smell), gustation (taste), vision (sight), audition (hearing) and somatosensation (touch, though the term somatosensation also covers heating, cooling, tingling and the detection of irritants). We use the input from these senses to build models of the universe around us. However, in his *Principles of Psychology*, written in 1890, William James gives the following warning. 'The general law of perception: Whilst part of what we perceive comes through our senses from the object before us, another part (and it may be the greater part) comes from inside our heads.' So, whilst most of us believe that we have a good idea of how the universe is, I am reminded of the song from Gershwin's opera 'Porgy and Bess' that 'It ain't necessarily so.' Our brains use all the senses together in order to build these models, and this is a mechanism that normally improves accuracy. For example, the interaction between olfaction and audition has been shown to improve reaction times when subjects try to locate a stimulus by sound (100).

However, such cross-modal effects can allow for tricks to be played. The classic example is the red wine/white wine experiment in which addition of a tasteless red dye prevents wine experts from giving accurate descriptions of it because the red colour signal coming from the visual sense alters the way in which the olfactory and gustatory signals are interpreted (101). Consumer goods manufacturers and fragrance marketers know very well how smell can affect judgements of softness of freshly laundered clothes or the creaminess and cleaning ability of soap. However, expectation also plays a part in forming olfactory percepts, and it has been shown that beliefs about flavour of chocolates can outweigh either the colour or taste that is actually perceived (102).

THE CHEMICAL BASIS OF ALL THE SENSES

Smell and taste are normally referred to as the *chemical senses* though, in fact, all five senses rely on chemistry in the form of transmembrane proteins. These are proteins that sit in cell membranes with one face exposed to the world outside the cell and the opposite to the cell interior. Touch (103) and hearing (104) rely on pressure-sensitive ion channels that alter their ability to allow ions to pass across the membrane depending on pressure applied to the membrane. Of the five tastes (sweet, sour, bitter, salt and umami) two, salt and sour, also rely on ion channels. The salt taste receptor is a variant of the vanilloid receptor (105), and the sour receptors which are sensitive to proton concentration are the ion channels PKD2L1 and PKD1L3 (106, 107). Vision, olfaction and the other three tastes (sweet, bitter and umami) use a family of membrane proteins known as *7-trans-membrane G-protein coupled receptors* or GPCRs for short. Vision, olfaction and bitter taste use class A GPCRs, whereas sweet and umami tastes rely on class C GPCRs (108). Whilst sweet and umami tastes are dependent on a single receptor system, bitter taste is closer to olfaction in that it uses a combinatorial mechanism, allowing a wide variety of diverse molecules to be recognized and identified as 'bitter' (109). Much more detail about GPCRs can be found in the next chapter.

DISTINGUISHING FEATURES OF SMELL AS A SENSE

Vision and smell receptors send signals directly to the cortex, whereas signals from the other senses (audition, taste and somatosensation) pass through the brain stem before reaching the cortex. The olfactory route is the most direct and therefore fastest of our senses. It interacts closely with the brain centres involved in memory and emotion, thus accounting for the well-known effects of smell on them. Smell is a crucial part of flavour and hence of great importance for nutrition, and thus the neuroscientist Gordon Shepherd argues that its role in human evolution and development has been much more significant than it has been given credit for.

Touch is located widely throughout the body whereas taste is found only in the tongue. The other three senses all have two centres for detecting incoming signals. We have two eyes for vision, two ears for hearing and two noses for smell. Having two eyes and two ears enables us to have stereoscopic vision and stereophonic hearing. However, the ability to locate the direction from which a smell originates is not due to olfaction but to the trigeminal nerves in the nose (110). The reason for having two separate noses is rather different. The air flow is always different in each nostril and so the temporal pattern of activation of the receptor sheet is different and this almost certainly gives the brain additional information (111). Another interesting difference between the two eyes and two noses is that visual processing is contra-lateral, that is signals from the right eye are processed in the left visual cortex and those from the left eye in the right visual cortex. Olfaction is ipsilateral; that is because the initial olfactory processing region, the olfactory bulb, sits directly above the epithelium from which it receives input and thus signals from the right OE are processed by the right olfactory bulb and the left by the left.

For those in the fragrance industry, especially chemists involved in the design of novel fragrance ingredients, there is one distinguishing feature of smell that is extremely important. Sight, hearing and touch all have simple physical parameters that can be used to measure their inputs; wavelength and intensity of light for sight; frequency and amplitude of sound waves for hearing; and pressure for touch. Olfaction has no such references and this leads to significant difficulties in measuring smell as will be discussed in Chapter 3. Taste is in between. Salt and sour tastes correlate with Na^+ and H^+ ion concentrations, respectively, whilst sweet, umami and bitter are usually measured by sensory comparison with known concentrations of standards, usually sucrose, monosodium glutamate and quinine, respectively.

ODOUR IS NOT A MOLECULAR PROPERTY

A dominant theme of this book is the assertion that odour is not a molecular property. This seems to be a very difficult concept for physical scientists to accept. However, until we realise that odour is a mental percept and not a fundamental property of a molecule in the way that vapour pressure, log P and so on are, our ability to understand odour is severely impaired. This misunderstanding has led to an enormous amount of futile and at times quite acrimonious debate as will be seen in Chapter 8. In Chapter 2, I hope to give a clear and detailed account of how recognition of an odorant molecule by ORs is translated into a mental percept and why the connection between the two is not straightforward.

Smell is created in the brain based on inputs from the nose and elsewhere. The law of specific nerve energies, also known as *Müller's law*, was first postulated by Müller in 1835. A modern statement of the law would read something like, 'Irrespective of how it is stimulated, each type of sensory nerve gives rise to a particular sensation which depends, not on the nerve but on the part of the brain in which it terminates'. So, for example, pressing on the eye gives an impression of a flash of light even though pressure rather than light was involved in stimulating the nerve. In other words, we thus see pressure. Similarly, nowadays using optogenetics, as will be seen in Chapter 2, mice can be made to smell light. Smell is therefore shown clearly to be a mental percept and not a molecular property since, in optogenetics, there are no molecules to smell.

Going back to the basic principles through which our sense of smell evolved, it is clearly nonsense to think that smell is geared to analyse components of a mixture let alone to analyse the structural features of the molecules comprising it. An animal does not need to know whether it is smelling a ketone or an ester, a terpenoid or a shikimate, it needs to know the survival implications of the total odour which it senses, in other words, food or poison, prey or predator.

The leading neuroscientist Gordon Shepherd concludes that 'Smell is not present in the molecules that stimulate the smell receptors'. (112) and he goes on to point out that the poet T. S. Eliot had also grasped the truth that sensory images exist in the mind and are only our personal interpretations of reality when he wrote in his poem 'The Dry Salvages' ' ... you are the music whilst the music lasts'. (113) Gordon then paraphrases this as ' ... you are the flavour whilst the flavour lasts'. This is similar to my conclusion on smell which is that 'The odour elicited upon recognition of a volatile substance by the receptors in the OE is a property of the person perceiving it and not of the molecules being perceived'.

REFERENCES

1. R. Nijland and J. G. Burgess, Bacterial olfaction. *Biotechnol. J.*, **2010**, *5*, 974–977.
2. M. Spehr, G. Gisselmann, A. Poplawski, J. A. Riffell, C. H. Wetzel, R. K. Zimmer, and H. Hatt, *Science* **2003**, *299*, 2054.
3. A. Kauffman, L. Parsons, G. Stein, A. Wills, R. Kaletsky, and C. Murphy, *J. Vis. Exp.*, **2011** doi: 10.3791/2490.
4. K. Yamada, T. Hirotsu, M. Matsuki, R. A. Butcher, M. Tomioka, T. Ishihara, J. Clardy, H. Kunitomo, and Y. Iino, *Science*, **2010**, *329*, 1647–1650.
5. P. Jiang, J. Losue, X. Li, D. Glaser, W. Li, J. G. Brand, R. F. Margolskee, D. R. Reed, and G. K. Beauchamp, *PNAS*, **2012**, *109(13)*, 4956–4961 doi: 10.1073/pnas.1118360109.
6. J. Arvidsson, M. Amundin, and M. Laska, *Physiol. Behav.*, **2012**, *105*, 809–814.
7. A. Matsui, T. Go, and Y. Nimura, *Mol. Biol. Evol.*, **2010**, *27*, 1192–1200.
8. T. Kishida and T. Hikida, *J. Evol. Biol.*, **2010**, *23*, 302–310.
9. H. Zhao, J. R. Yang, H. Xu, and J. Zhang, *Mol. Biol. Evol.*, **2010** doi: 10.1093/molbev/msq153.
10. K. Jin, C. Xue, X. Wu, J. Qian, Y. Zhu, Z. Yang, T. Yonezawa, M. J. Crabbe, Y. Cao, M. Hasegawa, Y. Zhong, and Y. Zheng, *PLoS One 6*, **2011**, e22602.

11. J. D. Bohbot, P. L. Jones, G. Wang, R. J. Pitts, G. M. Pask, and L. J. Zwiebel, *Chem. Senses*, **2011**, *36(2)*, 149–160 doi: 10.1093/chemse/bjq105.
12. T. Matsuo, S. Sugaya, J. Yasukawa, T. Aigaki, and Y. Fuyama, *PLoS Biology*, **2007**, *5(5)*, 985–996.
13. X. Zhang and S. Firestein, *Results Probl. Cell Differ.*, **2008** doi: 10.1007/400_2008_28.
14. M. Matsukawa, M. Imada, T. Murakami, S. Aizawa, and T. Sato, *Brain Res.*, **2011**, *1381*, 117–123.
15. Y. Nimura and M. Nei, *J. Hum. Genet.*, **2006**, *51*, 505–517.
16. M. Nei, Y. Nimura, and M. Nozawa, *Nat. Rev. Genet.*, **2008**, *9*, 951–963.
17. G. M. Shepherd, *Neurogastronomy: How the Brain Creates Flavour and Why It Matters*, Columbia University Press, New York, 2012, ISBN: 978-0-231-15010-4, 98-0-231-53031-6 (e-book).
18. M. Bastir, A. Rosas, P. Gunz, A. Peña-Melian, G. Manzi, K. Harvati, R. Kruszynski, C. Stringer, and J.-J. Hublin, *Nat. Commun.*, **2011**, *2*, 588 doi: 10.1038/ncomms1593.
19. J. Porter, B. Craven, R. M. Khan, S.-J. Chang, I. Kang, B. Judkewitz, J. Volpe, G. Settles and N. Sobel, *Nat. Neurosci.*, **2007**, *10(1)*, 27–29.
20. R. M. A. J. Ruijschop, A. E. M. Boelrijk, M. J. M. Burgering, C. de Graaf, M. S. Westerterp-Plantenga, *Chem. Senses*, **2009**, *35(2)*, 91–100.
21. B. P. Moore, *Nature*, **1966**, *211*, 746–747.
22. M. von Arx, D. Schmidt-Busser, and P. Guerin, *J. Insect Physiol.*, **2011**, *57(10)*, 1323–1331 doi: 10.1016/j.jinsphys.2011.06.010.
23. C. Frederickx, J. Dekeirsschieter, F. J. Verheggen, and E. Haubruge, *J. Forensic Sci.*, **2011**. *57(2)*, 386–90 doi: 10.1111/j.1556-4029.2011.02010.x.
24. A. M. Saveer, S. H. Kromann, G. Birgersson, M. Bengtsson, T. Lindblom, A. Balkenius, B. S. Hansson, P. Witzgall, P. G. Becher, and R. Ignell, *Proc. R. Soc. B*, **2012**, *279(1737)*, 2314–2322 doi: 10.1098/rspb.2011.2710.
25. R. C. Smallegange, B. G. Knols, and W. Takken, *J. Med. Entomol.*, **2010**, *47*, 338–344.
26. A. F. Carey, G. Wang, C-Y Su, L. J. Zwiebel, and J. R. Carlson, *Nature*, **2010**, *464*, 66–72 doi: 10.1038/nature08834.
27. E. A. Hallem, A. N. Fox, L. J. Zwiebel, and J. R. Carlson, *Nature*, **2004**, *427*, 212–213 doi: 10.1038/427212a.
28. K. Tanaka, Y. Uda, Y. Ono, T. Nakagawa, M. Suwa, R. Yamaoka, and K. Touhara, *Curr. Biol.*, **2009**, *19(11)*, 881–890.
29. E. M. Lichtenberg, M. Hrncir, I. C. Turatti, and J.C. Nieh, *Behav. Ecol. Sociobiol.*, **2011**, *65*, 763–774.
30. T. Bunnell, K. Hanisch, J. D. Hardege, and T. Breithaupt, *J. Chem. Ecol.*, **2011**, *37(4)*, 340–347.
31. J. W. Melo, D. B. Lima, A. Pallini, J. E. Oliveira, and M. G. Gondim Jr., *Exp. Appl. Acarol.*, **2011** doi: 10.1007/s10493-011-9465-1.
32. R. Glinwood, E. Ahmed, E. Qvarfordt, and V. Ninkovic, *Oecologia*, **2011**, *166(3)*, 637–647 doi: 10.1007/s00442-010-1892-x.
33. S. Boesveldt, J. Frasnelli, A. R. Gordon, and J. N. Lundstrom, *Biol. Psychol.*, **2010**, *84*, 313–317.
34. H. Bandoh, I. Kida, and H. Ueda, **2011**, *PLoS One 6(1)*, e16051.
35. A. Gagliardo, C. Filannino, P. Ioale, T. Pecchia, M. Wikelski, and G. Vallortigara, *J. Exp. Biol.*, **2011**, *214*, 593–598.
36. A. S. Mathuru, C. Kibat, W. F. Cheong, G. Shui, M. R. Wenk, R. W. Friedrich, and S. Jesuthasan, *Curr. Biol.*, **2012**, *22*, 538–544.
37. P. Kappel, S. Hohenbrink, and U. Radespiel, *Am. J. Primatol.*, **2011**, *73*, 928–938. doi: 10.1002/ajp.20963.
38. M. Fendt, *J. Chem. Ecol.*, **2006**, *32(12)*, 2617–2627 doi: 10.1007/s10886-006-9186-9.
39. T. Endres, R. Apfelbach, and M. Fendt, *Behav. Neurosci*, **2005**, *119(4)*, 1004–1010.
40. K. Kobayakawa, R. Kobayakawa, H. Matsumoto, Y. Oka, T. Imai, M. Ikawa, M. Okabe, T. Ikeda, S. Itohara, T. Kikusui, K. Mori, and H. Sakano, *Nature*, **2007**, *450*, 503–508.
41. J. H. Cho, J. E. Prince, C. T. Cutforth, and J. F. Cloutier, *J. Neurosci.*, **2011**, *31*, 7920–7926.
42. Y. Nikaido, S. Miyata, and T. Nakashima, *Physiol. Behav.* **2011**, *103(5)*, 547–556.

43. D. M. Ferrero, J. K. Lemon, D. Fluegge, S. L. Pashkovski, W. J. Korzan, S. R. Datta, M. Spehr, M. Fendt, and S. D. Liberles, *PNAS*, **2011**, *108*, 11235–11240 doi: 10.1073/pnas.1103317108).
44. M. Fendt, *J. Chem. Ecol.*, **2006**, *32*, 2617.
45. P. K. Richgels and S. M. Rollmann, *Chem. Senses*, **2012**, *37(3)*, 229–240 doi: 10.1093/chemse/bjr097.
46. J. D. Hardege, A. Jennings, D. Hayden, C. T. Müller, D. Pascoe, M. G. Bentley, and A. S. Clare, *Mar. Ecol. Prog. Ser.*, **2002**, *244*, 179–189.
47. P. Karlson and M. Lüscher, *Nature*, **1959**, *183*, 55–56.
48. E. O. Wilson and W. H. Bossert, *Recent Prog. Horm. Res.*, **1963**, *19*, 673–710.
49. S. V. Canavan, L. C. Mayes, and H. B. Treloar, *Front Psychiatry*, **2011**, *2*, 40.
50. T. D. Wyatt, *J. Comp. Physiol. A Neuroethol. Sens. Neural Behav. Physiol.*, **2010**, *196*, 685–700.
51. T. D Wyatt, *Pheromones and Animal Behaviour*, Cambridge University Press, England, 2003, ISBN: 0521485266.
52. A. Butenandt, R. Beckmann, D. Stamm, and E. Hecker, *Z. Naturforsch. B*, **1959**, *14*, 284–284.
53. L. Wang, X. Han, J. Mehren, M. Hiroi, J. C. Billeter, T. Miyamoto, H. Amrein, J. D. Levine, and D. J. Anderson, *Nat. Neurosci.*, **2011**, *14*, 757–762.
54. P. J. Landolt and T. W. Phillips, *Annu. Rev. Entomol.*, **1997**, *42*, 371–391 doi: 10.1146/annurev.ento.42.1.371.
55. R. B. Barrozo, D. Jarriault, N. Deisig, C. Gemeno, C. Monsempes, P. Lucas, C. Gadenne and S. Anton, *Eur. J. Neurosci.*, **2011**, *33*, 1841–1850.
56. Y. Grosjean, R. Rytz, J-P. Farine, L. Abuin, J. Cortot, G. X. S. E. Jefferis, and R. Benton, *Nature*, **2011**, *478*, 236–240 doi: 10.1038/nature10428.
57. A. Bretman, J. D. Westmancoat, M. J. Gage, and T. Chapman, *Biol. Lett.*, **2011**, *21(7)*, 617–622 doi: 10.1098/rsbl.2011.0544.
58. I. Said, B. Kaabi, and D. Rochat, *Chem. Cent. J.*, **2011**, *5*, 14 doi: 10.1186/1752-153X-5-14.
59. J-X. Zhang, W. Wei, J-H. Zhang, and W-H. Yang, *Chem. Senses*, **2010**, *35*, 375–382 doi: 10.1093/chemse/bjq025.
60. J. Mardon, S. M. Saunders, M. J. Anderson, C. Couchoux, and F. Bonadonna, *Chem. Senses*, **2010**, *35*, 309–321.
61. J. Martin and P. Lopez, *Chem. Senses*, **2010**, *35*, 253–262.
62. R. M. Belanger and L. D. Corkum, *J. Herpetol.*, **2009**, *43*, 184–191.
63. D. Poth, K. C. Wollenburg, M. Vences, and S. Schulz, *Angew. Chem. Int. Ed.*, **2012**, *51*, 2187–2190 doi: 10.1002/anie.201106592.
64. R. L. Doty, *The Great Pheromone Myth*, Johns Hopkins University Press, Baltimore, 2010, ISBN-13: 978-0-8018-9347-6, ISBN-10: 0-8018-9347-X.
65. D. R. Melrose, H. C. Reed, and R. L. Patterson, *Brit. Vet. J.*, **1971**, *127*, 497–502.
66. R. Hudson, and H. Distel, *Behaviour*, **1983**, *85*, 260–275.
67. B. Schaal, G. Coureaud, D. Langlois, C. Ginies, E. Semon, and G. Perrier, *Nature*, **2003**, *424*, 68–72.
68. A. Bertin, L. Calandreau, C. Arnoud, and F. Lévy, *Chem. Senses*, **2012**, *37(3)*, 253–261 doi: 10.1093/chemse/bjr101.
69. R. Hudson, *J. Comp. Physiol. A*, **1999**, *185*, 297–304.
70. L. E. Rasmussen, T. D. Lee, W. L. Ropelofs, A. Zhang, and G. D. Daves Jr., *Nature*, **1996**, *379*, 684.
71. (a) C. Wedekind, T. Seebeck, F. Betters, and A. J. Poepke, *Proc. R. Soc. London, Ser. B* **1995**, *260*, 245. (b) C. Wedekind and S. Furi, *Proc. R. Soc. London, Ser. B*, **1997**, *264*, 1471.
72. R. L. Doty and E. L. Cameron, *Physiol. Behav.*, **2009**, *97*, 213–228 doi: 10.1016/j.physbeh.2009.02.032.
73. A. Natsch, J. Schmid, and F. Flachsmann, *Chem. Biodiv.*, **2001**, *1*, 1058–1072.
74. A. Natsch, S. Derrer, F. Flachsmann, and J. Schmid, *Chem. Biodiv.*, **2006**, *3*, 1–20.
75. F. Kuhn and A. Natsch, *J. R. Soc. Interface*, 2008, doi: 10.1098/rsif2008.0223.
76. P. G. Hepper and D. L. Wells, *Chem. Senses*, **2010**, *35*, 263–268 doi: 10.1093/chemse/bjq015.
77. B. J. Pitcher, R. G. Harcourt, B. Schaal, and I. Charrier, *Biol. Lett.*, **2011**, *7(1)*, 60–62.

78. S. A. Roberts, D. M. Simpson, S. D. Armstrong, A. J. Davidson, D. H. Robertson, L. McLean, R. J. Beynon, and J. L Hurst. *BMC Biology*, **2010**; *8*:75 doi: 10.1186/1741-7007-8-75.
79. S. A. Roberts, A. J. Davidson, L. McLean, R. J. Beynon, and J. L. Hurst, *Science*, **2012**, *338*, 1462–1465 doi: 10.1126/science.1225638.
80. T. Leinders-Zufall, P. Brennan, P. Widmayer, P. Chadramani, A. Maul-Pavicic, M. Jäger, X.-H. Li, H. Breer, F. Zufall, and T. Boehm, *Science*, **2004**, *306(5698)*, 1033–1037.
81. K. A. Phillips, C. A. Buzzell, N. Holder, and C. C. Sherwood, *Am. J. Primatol.*, **2011**, *73*, 578–584.
82. N. K. Hughes, C. J. Price, and P. B. Banks, *PLoS One 5(9)*: e13114. doi: 10.1371/journal.pone.0013114.
83. R. Hamer, F. L. Lemckert, and P. B. Banks, *Biol. Lett.*, **2011**, *7*, 361–363.
84. T. N. Rosenstiel, E. E. Shortlidge, A. N. Melnychenko, J. F. Pankow, and S. M. Eppley, *Nature*, **2012**, *489*, 431–433 doi: 10.1038/nature11330.
85. R. Piskorski and S. Dorn, *Chem. Biodiv.*, **2010**, *7(9)*, 2254–2260 doi: 10.1002/cbdv.201000221.
86. A. J. Birch, *J. Proc. R. Soc. New South Wales*, **1938**, *71*, 259.
87. S. M. McElvain, R. D. Bright, and P. R. Johnson, *J. Am. Chem. Soc.*, **1941**, *63(6)*, 1558–1563 doi: 10.1021/ja01851a019.
88. J. H. Butterworth and E. D. Morgan, *Chem. Commun.*, **1968**, *1*, 23–24. doi: 10.1039/C19680000023.
89. S. Rasmann, T. G. Köllner, J. Degenhardt, I. Hildtpold, S. Toepfer, U. Kuhlmann, J. Gershenzon, and T. C. J. Turlings, *Nature*, **2005**, *434*, 732–737.
90. (a) T. C. J. Turlings, J. H. Tumlinson, and W. J. Lewis, *Science*, **1990**, *250(1)*, 1251–1253. (b) H. T. Alborn, T. C. J. Turlings, T. H. Jones, G. Stenhagen, J. H. Loughrin, and J. H. Tumlinson, *Science*, **1997**, *276*, 945 doi: 10.1126/science.276.5314.945.
91. F. E. Farmer and C. A. Ryan, *PNAS*, **1990**, *87(19)*, 7713–7716.
92. C. A. Preston, G. Laue, and I. T. Baldwin, *J. Chem. Ecol.*, **2004**, *30(11)*, 2193–2214.
93. A. Kessler, R. Halitsche, C. Diezel, and I. T. Baldwin, *Oecologia*, **2006**, *148(2)*, 280–292.
94. R. Karban, I. T. Baldwin, K. J. Baxter, G. Laue, and G. W. Felton, *Oecologia*, **2000**, *125(1)*, 66–71.
95. R. Karban, K. Shiojiri, M. Huntzinger, and A. C. McCall, *Ecology*, **2006**, *87(4)*, 922–930.
96. P. K. House, A. Vyas, and R. Sapolsky, *PLoS One*, **2011**, *6* e23277.
97. K. E. Mauck, C. M. De Moraes, and M. C. Mescher, *PNAS*, **2010**, *107(8)*, 3600–3605 doi: 10.1073/pnas.0907191107.
98. K. Matsumura, M. Opiekun, K. Mori, T. Tashiro, H. Oka, K. Yamazaki, and G. Beauchamp, *Chem. Senses*, **2009**, *34*, A1–121, abstract 30 doi: 10.1093/chemse/bjp032.
99. U. B. Kaupp, *Nat. Rev. Neurosci.*, **2010**, *11*, 188–200 doi: 10.1038/nrn2789.
100. V. La Buissonniere-Ariza, J. Frasnelli, O. Collignon, and F. Lepore, *Neurosci. Lett.* **2012**, *506*, 188–192.
101. G. Morrot, F. Brochet, and D. Dubourdieu, *Brain Lang.*, **2001**, *79*, 309–320.
102. C. A. Levitan, M. Zampini, R. Li, and C. Spence, *Chem. Senses*, **2008**, *33*, 415–423 doi: 10.1093/chemse/bjn008.
103. R. O'Hagan, M. Chalfia, and M. B. Goodman, *Nat. Neurosci.*, **2004**, *8*, 43–50 doi: 10.1038/nn1362.
104. I. J. Russell, *Nature*, **1983**, *301*, 334–336 doi: 10.1038/301334a0.
105. V. Lyall, G. L. Heck, A. K. Vinnikova, S. Ghosh, T.-H. T. Phan, R. I. Alam, O. F. Russell, S. A. Malik, J. W. Bigbee, and J. A. DeSimone, *J. Physiol.*, **2004**, *558(1)*, 147–159.
106. R. B. Chang, H. Waters, and E. R. Liman, *PNAS*, **2010**, *103(33)*, 12569–12574 doi: 10.1073/pnas.1013664107.
107. Y. Ishimaru, H. Inada, M. Kubota, H. Zhuang, M. Tominaga, and H. Matsunami, *PNAS*, **2006**, *107(51)*, 22320–22325.
108. M. Behrens, W. Meyerhof, C. Hellfritsch, and T. Hoffmann, *Angew. Chem. Int. Ed.*, **2011**, *50*, 2220–2242 doi: 10.1002/anie201002094.
109. W. Meyerhof, C. Batram, C. Kuhn, A. Brockhoff, E. Chudoba, B. Bufe, G. Appendino, and M. Behrens, *Chem. Senses*, **2010**, *35*, 157–170 doi: 10.1093/chemse/bjp092.

110. J. Frasnelli, T. Hummel, J. Berg, G. Huang, and R. L. Doty, *Chem. Senses*, **2011**, *36(4)*, 405–410 doi: 10.1093/chemse/bjr001.
111. N. Sobel, R. M. Khan, A. Saltman, E. V. Sullivan, and J. D. S. Gabrieli, *Nature*, **1999**, *402*, 35.
112. G. M. Shepherd, *Neurogastronomy: How the Brain Creates Flavour and Why It Matters*, Columbia University Press, New York, 2012, p. 86, ISBN 978-0-231-15910-.
113. T. S. Eliot, *The Dry Salvages*, Collected Poems, Faber, London, 1963.
114. E. Roura, B. Humphrey, G. Tedó, and I. Ipharraguerre, *Can. J. Animal Sci.*, **2008**, *88(4)*, 535–558 doi: 10.4141/CJAS08014.
115. A. J. Birch, W. V. Brown, J. E. T. Corrie, and B. P. Moore, *J. Chem. Soc., Perkin Trans. 1*, **1972**, 2653–2658 doi: 10.1039/P19720002653.
116. A. Bethe, *Naturwissenschaften*, **1932**, *11*, 177–181.

Chapter 2

The Mechanism of Olfaction

The most beautiful thing we can experience is the mysterious. It is the source of all true art and science. *Albert Einstein*

OVERVIEW

In humans, the process of olfaction begins when volatile molecules enter the nasal cavity either in inhaled air (orthonasal olfaction) or by diffusion from the mouth and throat (retronasal olfaction) (Figure 2.1). In the nose, these molecules activate receptors in the olfactory epithelium (OE). Odorant molecules are also exposed to odorant binding proteins (the role of which is unclear) and enzymes such as cytochrome P450s and esterases, both of which are known to metabolise odorants rapidly enough to affect olfaction. The olfactory receptor (OR) proteins are located in hair-like projections (called *cilia*) of the olfactory sensory neurons (OSNs). Activation of the receptor proteins sets in chain a complex sequence of biochemical reactions which eventually generate an electrical signal in synapses at the other end of the neuron. Odour is not dependent on any single receptor type but relies on recognition of a combinatorial pattern, with each odorant activating a range of receptors and each receptor responding to a range of odorants. This combinatorial mechanism was proposed by Polak (1) and confirmed by Malnic et al. (2). Its validation represents one of the great advances in our understanding of olfaction. The signals from the OR cells responding to odorants and their metabolites are picked up by the olfactory bulb (OB) where mapping of receptor types onto glomeruli occurs. The signal patterns leaving the OB then travel to the piriform cortex (PC) which, together with the entorhinal cortex and amygdala, is known as the *primary olfactory cortex*. Odour objects are formed in the PC and compared against previously stored odour objects. Output from the PC goes to various other brain regions and, combined with other inputs, is eventually interpreted as an odour percept in the orbitofrontal cortex (OFC). It is important to note that the conscious percept

Chemistry and the Sense of Smell, First Edition. Charles S. Sell.
© 2014 John Wiley & Sons, Inc. Published 2014 by John Wiley & Sons, Inc.

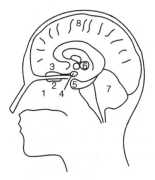

1 Nasal airspace
2 Olfactory epithelium
3 Olfactory bulb
4 Piriform cortex
5 Amygdala
6 Thalamus
7 Cerebellum
8 Neocortex

Figure 2.1   The location of some of the key olfactory organs in humans.

of odour only exists at this stage, and that it is not present in the molecular stimulus. Olfactory signals are generated in the epithelium within tens of milliseconds of presentation of an odorant to the nose, and cortical activity is detectable within 100 ms. This is a very much simplified overview of the process, and at every stage there are complex interactions with other brain regions, as will be described later. The entire process in the brain has been reviewed well by a number of experts, and anyone wishing to know more detail than is present in this book is recommended to read those by Wilson and Stevenson (3), Hawkes and Doty (4) and Shepherd (5) as well as the review by Delano and Sobel (6). This chapter will describe the olfaction process starting with the nasal air and following through, as closely as possible, in the sequence just described.

COMPARISON WITH OTHER SPECIES

Before launching into a review of what is known about olfaction, it is perhaps worth sounding one small note of caution. Much of our knowledge has been gleaned from work with diverse species. Titles of research papers and even their abstracts often make statements about discoveries in olfaction, but careful reading of the whole publication is necessary in order to determine which species was studied and how relevant the results are for human olfaction.

The animals most used to study olfaction are the nematode worm *Caenorhabditis elegans*, the fruit fly *Drosophila melanogaster*, the moth *Manduca sexta*, the zebrafish *Danio rero*, the frog *Xenopus laevis* and, for mammals, various strains of mouse and rat. These are selected for practical reasons such as knowing the full sequence of the genome and/or because they are easy to work with, and not because they are the best models for human olfaction. Therefore, we need to be careful in translating results from these other species into humans. Flies and mice have much shorter life cycles than humans and so genetic modification is easier. For example, 'knock-out' variants in which a specific gene is absent can be used to see whether a key protein is important for a given function. In human studies, subjects with missing or defective genes must be sought by screening, a practice which is so far confined mostly to investigation of specific anosmias, as will be described later in the chapter. If olfactory deficits are associated with a specific illness, then investigation of the illness might produce information about olfaction.

Olfaction is a key sense for survival of all animals, and there are common features suggesting that what we learn about one species might apply to others also. Olfactory signals leaving the OBs travel by several routes to the higher centres of the brain where the phenomenon of odour eventually comes into being. The architecture of the olfactory parts of the brain is remarkably consistent across all mammalian species and so research on other mammals throws considerable light on the function in humans. A comparison of the nematode worm *Caenorhabditis elegans*, the fruit fly *D. melanogaster* and the mouse *Mus musculus* shows similarities in the function of the olfactory processing systems in their brains. For example, the antennal lobe of *D. melanogaster* performs a similar role to that of the OB of *M. musculus*; the mushroom body in insects to the PC in mammals; and the Kenyon cells of the mushroom body to the pyramidal cells of the PC.

However, there are also differences: some features of the rodent brain, whilst similar to those of insects, are different from those of primates (7). There are even structural differences between rats and mice in the brain regions involved in olfaction (8). The platypus (*Ornithorhyncus anatinus*) and the echidnas (*Tachyglossus aculeatus, Zaglossus bruijni, Z. attenboroughi and Z. bartoni*) are the only surviving examples of monotremes (egg-laying mammals) and are genetically closer to each other than to any other mammals. However, their olfactory systems are quite different from each other and evidence suggests that they diverged a long time ago in evolutionary history. One thing they do have in common and which is different from other mammals is that their olfactory organs are not fully developed on hatching and therefore neonates do not have a sense of smell (9).

Differences Between Insect and Human Olfaction

In terms of olfaction, there are considerable differences between insects and humans, particularly so at the receptor level; an account of these can be found in the reviews by Kaupp (10) and by Nei et al. (11). Great care must therefore be exercised when making comparisons between insects and humans, otherwise totally nonsensical conclusions could be reached. For instance, the fruit fly *D. melanogaster* detects carbon dioxide using specialised ORs (12), whereas humans use taste receptors in combination with a carbonic anhydrase enzyme (13).

The three main differences between insects and mammals at receptor level are the role of the odour binding proteins (OBPs) and the ratio between these and receptors, the direction of the protein sequence and the heterodimeric nature of insect receptors.

As already mentioned in Chapter 1, OBPs have significant roles to play in insect olfaction, and these have been reviewed by Fan et al. (14). The number of receptor types used by any mammalian species is many times that of the number of different OBPs used by that species. Insects show a very different ratio.

Table 2.1 Number of OBP Types Used by Representative Insects

Fruit fly	*Drosophila melanogaster*	51
Mosquito	*Anopheles gambiae*	57
Silk moth	*Bombyx mori*	44
Honey bee	*Apis mellifera*	21

For example, humans use only one OBP (OBP2A) but between 350 and 400 different types of OR, whereas the fruit fly *D. melanogaster* expresses 51 different types of OBP but has only 60 types of OR. Table 2.1 shows the number of different OBPs used by some typical insects.

In insects, the OBPs are not just passive transport proteins but are part of the coding of the signal. For example, the silkmoths *Bombyx mori* and *Antheraea polyphemus* use both OBPs and ORs in detecting bombykol (**2.1**) and bombykal (**2.2**) (Figure 2.2) (15). The response of the receptors depends on both the structure of the odorant (pheromone) and the binding protein (16). Similarly, suppression of OBP genes in *D. melanogaster* alters the insect's behavioural response to odorants because of altered recognition by the receptors (17). A specific OBP, known as *LUSH*, is required for detection of the *Drosophila* pheromone (Z)-11-octadecenyl acetate, also known as *vaccenyl acetate* (**2.3**), by the insect's receptor OR67d (18, 19). Various mutants of LUSH exist, and LUSH^{D118A} is the dominant active one. It has been shown that the conformation of LUSH changes on binding

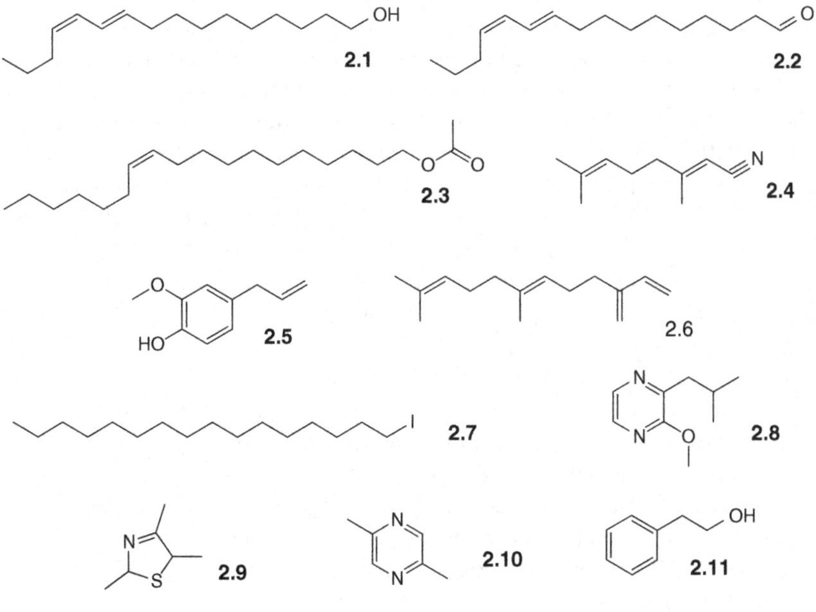

Figure 2.2 Some insect pheromones and other odorants.

vaccenyl acetate and that it is this new complex that activates the receptor (20). In fact, if LUSH is absent, not only does the OR fail to respond to 11-*cis*-vaccenyl acetate but the basal rate of firing is also reduced (19). Binding of 11-*cis*-vaccenyl acetate to LUSH is known to affect the conformation of the C-terminus and it was found that a modified LUSH, namely LUSH^{D118A}, in which the terminus resembles more closely that of the 11-*cis*-vaccenyl acetate/LUSH complex, activated the receptor in the absence of 11-*cis*-vaccenyl acetate (20).

Like ORs, OBPs show a range of binding strengths to ligands, and this variability, combined with genetic variation, adds to complexity and hence the adaptability of olfaction to improve survival capability.

The crystal structure of AmelOBP14, one of the 21 OBPs of the honeybee (*Apis mellifera*), has been determined. This OBP binds to a variety of ligands, geranyl nitrile (**2.4**) and eugenol (**2.5**) having the highest binding affinities (21). The role of the honeybee OBP genes and their evolution has been described by Forêt and Maleszka (22). Study of the crystal structure of OBP7 from the malarial vector *Anopheles gambiae* shows a very broad specificity as regards ligand binding (23), which is in contrast to the very narrow selectivity of an OBP that is highly conserved across numerous aphid species such as *Sitobion avenae* and is highly specific for (*E*)-farnesene (**2.6**) (24). Electrophysiological studies showed that the *S. avenae* OBPs, SaveOBP2 and SaveOBP3, are effective at binding volatiles from green leaves. However, SaveOBP7 bound (*E*)-farnesene (**2.6**) strongly and is therefore a candidate for an alarm pheromone detector (25).

The crystal structure of another OBP from *A. gambiae*, namely AgamOBP47, shows it to be quite different in structure from classical OBPs and thus implies that it could have a different functional role (26). The pheromone binding protein of the silkmoth *B. mori* has been shown to bind 1-iodohexadecane (**2.7**) and 2-isobutyl-3-methoxypyrazine (**2.8**) in addition to the pheromone bombykol, and X-ray crystal structures of the OBP in the unliganded state as well as bound to each of these ligands show how the binding pocket can accommodate the different structures. The OBP might therefore also play a part in recognition of non-pheromone odorants by the insect (27). Since only female mosquitoes prey on humans, any OBP that is female-specific might be involved in prey detection. OBP2 shows increased expression in female *A. gambiae* than in males and could therefore be a target for mosquito deterrence in the fight against malaria (28). Similarly, some OBPs and ORs are present at higher levels in another mosquito species *Culex quinquefasciatus* (29).

Insect ORs do not show homology with mammalian ORs (30). Moreover, in terms of location in the cell membrane, the amino acid chain runs in the opposite direction in insects from that in mammals, as shown in Figure 2.3. In mammals, the amino terminus is outside the cell and the carboxyl terminus inside, whereas in insects the carboxyl terminus is outside and the amino terminus inside the cell (31).

Sato et al. showed that insect ORs are heterodimeric, comprising one G-protein coupled receptor (GPCR), which recognises the odorant, and an ion channel, originally named OR83B but which is now called *ORCO*, which opens to start the signal transduction process thus avoiding the need for the G-protein/cyclic nucleotide

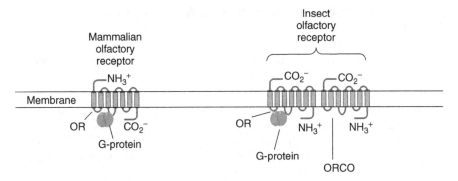

Figure 2.3 Differences between mammalian and insect olfactory receptors.

gated (CNG) cascade (32). However, Wicher et al. found that the GPCR component can also act via the G-protein mechanism as in mammals. The direct coupling of the receptor to ORCO gives a rapid but short-lived response to the odorant, whereas the G-protein mechanism gives a slower but more sustained response (33). Nichols et al. showed that ORCO does not contribute to the binding site of the heterodimer and identified agonists, partial agonists and antagonists for an OR/ORCO dimer (34).

Unlike mammals, insects do not have temperature control, and it has been shown that the signals at receptor level are temperature dependent (35). The insect brain must therefore be able to compensate for this, so that the significance of the signal intensity can be interpreted appropriately.

Differences Between Fish and Mammalian Olfaction

Since they live entirely in an aqueous medium, fish have rather different requirements of olfaction from those of air-breathing animals. Mammals, including humans, have some receptors that resemble those of fish in homology, and these are known as the *fish-like receptors*. Both fish and mammals also use another group of receptors, known as *trace amine associated receptors (TAARs)* which are sensitive, as their name indicates, to various amines. Mammals use a much smaller range of TAARs than fish. For example, humans have 6 different types of TAAR, mice 15 and rats 17, but the zebrafish has no fewer than 112 different TAAR genes in their genome. Fish therefore are more adapted to detection of water-soluble substances, whereas air-breathing animals are more adapted to volatile, and therefore less hydrophilic, molecules. Human OSNs are ciliated, whereas fish also have microvillous and crypt neurons. There is a general rule that each OSN expresses only one type of OR protein. The crypt neurons in fish are slightly different in that all crypt neurons express the same receptor, the V1R-like ora4 receptor (36).

Differences Between Reptile and Mammalian Olfaction

Snakes present an interesting and dramatic example of inter-species differences in olfaction. In snakes, the vomeronasal organ (VNO) is located on the roof of the mouth rather than in the nasal cavity as in mammals. The tip of the snake's tongue collects odorants from the air and tastants from food and delivers them to the VNO. In studying the colubrid snake *Nerodia fasciata*, Daghfous et al. showed that oscillatory tongue-flicks were used to sample the air for volatile odorants whilst simple downward extensions of the tongue were used to pick up non-volatile substances from sources such as potential food (37).

Differences Between Human Olfaction and that of Other Mammals

Figure 2.1 shows a simplified cross section of the human head indicating where the main olfactory organs are located. Comparison with the equivalent drawing for a rodent in Figure 2.4 shows a number of significant differences. First, the OE of rodents and their olfactory brain regions are much larger in proportion to the size of the remainder of their brains than is the case with humans. Dogs are more similar to rodents than to humans in that respect. Second, it can be seen that rodents have four separate organs capable of detecting volatile chemicals: the major olfactory epithelium (MOE), the VNO (also known as *Jacobsen's organ*), the septal organ (SO) and the Grueneberg ganglion. A number of reviews of the roles of these various organs have been published (38–41).

Olfactory Epithelium

The human OE and the MOE of rodents are the most important sensors for odorants in the respective species and are the main sites for expression of OR proteins. Figure 2.4 shows how the MOE of rodents and the major OB, to which signals from the OE travel, are close to the roof of the skull and are therefore relatively easily studied by techniques such as functional magnetic resonance imaging (fMRI) and optogenetics. On the other hand, Figure 2.1 shows that the corresponding organs in

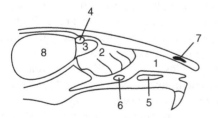

1 Nasal airspace
2 Major olfactory epithelium
3 Major olfactory bulb
4 Accessory olfactory bulb
5 Vomeronasal organ
6 Septal organ
7 Groeneberg ganglion
8 Remainder of brain

Figure 2.4   Some features of the rodent brain and olfactory organs.

humans are located on the roof of the nasal cavity and the base of the brain, respectively. Thus, ethical considerations apart, they are much less accessible in humans and consequently much more difficult to study.

In addition to OSNs expressing OR proteins, the OE contains cells expressing other sensory receptors. TAARs are another family of GPCRs and were first found by Liberles and Buck who showed their presence in the OE (42). Humans have 6 different types of TAARs, mice 15, rats 17 and zebrafish 112. Of the murine TAARS, 14 evolved in olfactory tissue. The TAARs detect various small amines and for rodents, one in particular is the predator odour 2,4,5-trimethylthiazoline (TMT) (**2.9**), which is present in fox urine.

Lin et al. found that some sensory neurons in the MOE of rodents express the transient receptor potential channel M5 (TRPM5) and that those which project axons to the ventral area of the main OB are involved in processing information from semiochemicals (43–45).

The dog is normally considered to have the keenest sense of smell of all mammals. They have 10 times the number of OSNs that humans do (100 million as opposed to 10 million) and twice as many OR types (800 vs 350–400) yet humans have twice as many glomeruli as dogs (6000 vs 3000). In humans, this means that each glomerulus receives input from about 1700 OSNs, whereas in the dog the figure is only 850. Furthermore, humans have a much larger cortex for processing of the olfactory signals, which compensates for the poorer input from the epithelium. Approximate statistics for four species are shown in Table 2.2.

Vomeronasal Organ

The VNO provides an interesting and dramatic example of inter-species differences. In snakes, the VNO is on the roof of the mouth and odorants collected from the air by adsorption or tastants from food by touching are delivered to it by the tip of the tongue. In contrast, the VNO of rodents is located on the floor of the nasal cavity and responds to semiochemicals.

There are two main types of receptor in the VNO, vomeronasal receptors (VNRs) and formyl peptide receptors (FPRs). Mice have two classes of VNRs, V1R and V2R receptors, and five members of the FPR family (46). VNO neurons project to the accessory OB and signal processing in the accessory OB is different from that of the main OB (47).

Table 2.2 Numbers of OSNs, OR Types and Glomeruli Used by Four Mammalian Species

Species	OSNs	OR types	Glomeruli
Rat	10 M	1300	
Rabbit	50 M		2000
Dog	100 M	800	3000
Human	10 M	350–400	6000

As with ORs, G-proteins are also essential for functioning of the VNRs of the VNO (48, 49). However, unlike the ORs, the VNRs give a response that is not dependent on concentration (50). Their agonists include sulfated steroids which are products of steroidal hormone catabolism and which are significant ligands in female mouse urine. These sulfated steroids are non-volatile and therefore cannot reach the VNO through the vapour phase and must be taken in physically by, for example, rubbing the nose on the odour source and sniffing particles up into the nasal cavity. Isogai et al. investigated VNRs in mice and showed that they detect a wide range of socially relevant chemicals. The selectivity profile enabled them to distinguish between the sex of conspecifics and to recognise but not to respond to mice of related species and to detect and distinguish between the odours of various predators such as ferrets, owls and snakes. More VNRs were found to be devoted to predator odorants than to sex-related odours (51).

The FPRs are GPCRs, and other members of the family function in the immune system. However, those in the VNO are more similar in sequence to the VNRs than are those in the immune system. Rivière et al. found evidence that the olfactory FPRs have a role in detecting pathogens or pathogenic states (52).

The human genome contains five remnant genes for vomeronasal type-1 receptors and these have been found to respond to various odorants when expressed *in vitro* in HeLa/Olf cells (53). One of these genes, hVN1R3, is a pseudogene. In humans, the VNO has a role in the development of sex hormone secretion but then regresses. The vomeronasal cavities can be observed in some adults, but there are no neural connections in adults; the vomeronasal genes are mutated and non-functional and there is no accessory OB (54). Furthermore, there is no other related neuroprocessing (4). It must therefore be concluded that the VNO is not functional in adult humans.

Septal Organ

This organ is sometimes referred to as the *septal organ of Masera*, after its discoverer (55). Although the sensory neurons in it have some morphological differences compared to those of the MOE, its ORs are the same as some of those in the MOE and their axons project to the glomeruli in the main OB. Over half of the OSNs in the SO express the MOR256-3 receptor. The breadth of tuning of its receptors and the details of subsequent neuroprocessing suggest that it has a role as a general odorant detection and alerting system (38).

Grueneberg Ganglion

The sensory neurons in the Grüneberg ganglion, or the SO of Grüneberg (SOG), express olfactory marker protein (OMP) and their axons terminate in a certain subset of glomeruli in the major OB. They are therefore part of the olfactory system. TAARs have also been identified in the Grüneberg ganglion (56). The Grüneberg ganglion has been shown to respond to 2,5-dimethylpyrazine (**2.10**) and related odorants and also to low temperature, and so it is possibly a dual-function sensory organ (57). Both the chemosensory and thermosensory functions appear to depend on the cyclic guanosine monophosphate (cGMP) cascade, but since the chemosensory response is subject to adaptation whereas the thermosensory response is not, it would seem that the two systems only partially overlap (58). In view of all the above and the physical location of the SOG, its neurons might serve as an early warning system for olfaction (59).

THE HUMAN SENSE OF SMELL

Freud described humans as microsmatic, in other words, having a poor sense of smell. Vision is certainly our dominant sense: we have a smaller number of ORs then rodents or dogs, we have fewer types of ORs than rodents or dogs and, through evolution, we are losing OR genes more rapidly than rodents and dogs. So, does modern science support Freud's assertion? At least one leading contemporary olfactory neuroscientist would question that belief. Gordon Shepherd of Yale University points out that we have very much greater processing power than any other species and so need less input data to enable our brains to create meaningful odour percepts (60, 61). This theme is developed further in his book on neurogastronomy (5). Behavioural experiments with two different monkey species (the squirrel monkey *Saimiri sciureus* and the pigtail macaque *Macaca nemestrina*) concluded that neither the number of functional OR genes nor the relative size of their OBs affected their ability to discriminate between the odours of nine different pairs of enantiomers (62). This would tend to support Shepherd's viewpoint. My belief is that Shepherd is closer to the truth than Freud and that we tend to undervalue the wonderful sense of smell that we have.

BASIC ANATOMY OF THE HUMAN NOSE

The Olfactory Epithelium and the Olfactory Receptors

We tend to talk about the nose as a single entity, yet there is a septum that divides it into two physically separate cavities. On the roof of each of these, and extending down onto the septum, is a patch of tissue known as the *olfactory epithelium*. Odorants can reach the epithelium either from front, by inhalation of air from the environment (referred to as the *orthonasal route*), or by diffusion from the mouth and respiratory and gustatory tracts (referred to as the *retronasal route*). The latter is vital in flavor detection, as the tongue contains receptors only for sweet, sour, salt, bitter and umami tastes, and the rest of what is normally referred to as *flavour* is actually retronasal smell. On the sides of the nasal cavities are found bony plates called *turbinates*, which cause turbulence in the air flow and hence help to ensure that odorants reach the epithelia.

The OE is a patch of greenish-yellow tissue several square centimetres in area and 100–200 µm thick. The receptor cells that it contains run from the nasal cavity through the base of the skull (the cribriform plate of the ethmoid bone) into the OB. On the side of the nasal cavity, the receptor cells have hairs or cilia, which are 20–200 µm long. These cilia are bathed in a mucus layer which is 35 µm thick and flows backwards continually at a rate of 1–6 cm/min. The receptor proteins are expressed in the cilia of the receptor cells. Individual receptor cells fire spontaneously at a rate of 3–60 impulses/s and this rate of firing is increased when the cells are stimulated by an odorant. There are about 6 million OSNs per nostril with an average of 25 cilia per neuron (63, 64). In dogs, there can be hundreds of cilia per OSN.

Body Position

Lundström et al. showed that body position affects odour perception. Subjects had a lower detection threshold for 2-phenylethanol (**2.11**) when sitting upright than when lying down (65), though a later report by the same team suggested that the effect is observed only at perithreshold levels of the odorant (66). There are various possible explanations, such as effects on nasal airflow or on the cardiovascular system, but so far the mechanism for this effect remains unclear. One suggestion is that the effect stems from activity in the locus coeruleus region of the brain (67).

Two Noses

Different degrees of swelling of veins in the nasal turbinates results in the air flow through each nostril being different, with air flow through one nostril being faster than through the other. The faster flow alternates from one nostril to the other in line with changes in swelling of the blood vessels. The periodicity of cycling varies from about 25 min to 200 min (68). The cycling between the two noses is regulated by the sympathetic and parasympathetic branches of the autonomic nervous system and is coupled with a similar cycling of cortical dominance in the brain (69). However, simple theories about regular cycling are challenged by various other observations, reviewed by Hawkes and Doty (4). For example, Gilbert and Rosenwasser found that a regular cycle, alternating between fast flow in one nostril with slow in the other, was only present in 13% of the subjects they observed (70).

Notwithstanding the variability between individuals and even within one individual, the difference in flow rate between the two noses does provide a mechanism for increased sensitivity and discrimination based on transport phenomena because the timing and pattern of spread of signal across the epithelium will depend, in part, on the air flow. This would be expected because the ratio of adsorption of hydrophilic versus hydrophobic odorants is more in favour of the former at high flow rates and the latter at lower flow rates (71). Indeed, it has been shown that the differences in flow rate do result in differences in perception between the two noses (72). Intriguingly, detection thresholds are lower when the left nostril is occluded relative to the right than when the occlusion is reversed (73). Signals from both noses are integrated at the level of the PC and higher. Similarly, information on recognition features is known to be exchanged at a higher level in the brain because a molecule whose odour has been learnt using one nostril only will be recognized by the other (74). Neither the handedness nor the sex of subjects affects their judgement of odours when using only one nose for assessment (75). Evidence from mice suggests that occluding one naris affects not only the occluded side but also the response from the unaffected side (76). When different odours are presented simultaneously to the two noses, a phenomenon known as *binaural rivalry* results in alternating odour percepts, similar to the alternating visual images when different pictures are shown simultaneously to the two eyes (77).

Other Receptors in the Nose

The motile cilia lining the airways serve to move mucus from its source to the gastrointestinal tract and hence remove irritants and other potentially harmful substances from the lungs and airways. Irritants are detected by T2R receptors, and it has been shown that the motile cilia also have bitter taste receptors, stimulation of which causes the cilia to beat faster and hence improve removal of airborne contaminants (78, 79).

Nasal irritants produce sensations such as burning, tingling, cooling and heating, and these are grouped together as nasal pungency. The sensations arise from stimulation of receptors known as *transient receptor potential channels (TRPs)* which are located in the membrane of the trigeminal nerves enervating the nasal cavity. The family of TRP channels has been shown to contain about 30 members (80). It has also been shown that solitary chemosensory cells (SCSs) in the respiratory mucosa of rodent and human noses contain sweet, bitter and umami taste receptors and that these play a role in mediating trigeminal reflexes (81). For example, Gulbransen et al. showed that some SCSs in mice (and probably the same is the case in humans) detect bitter substances via the T2R bitter taste receptor and that these cells mediate the response of the nearby trigeminal nerve (82). Other murine SCSs have been found to express the TRPM5 receptor which responds to irritants (45). Nasal pungency has been reviewed by Doty and Cometto-Muñiz (83). The trigeminal nerve response is modulated by adenosine via the A2A receptor, and it has been shown that some odorants affect the adenosine-based mechanism and hence the trigeminal response to irritants (84).

About 70% of odorants also activate the trigeminal nerve in the nasal cavity. For most compounds, olfactory detection thresholds are lower than pungency thresholds. In a homologous series, small changes in structure have a small and relatively predictable effect on the pungency threshold, whereas small structural changes can have large and unpredictable effects on odour thresholds. The relationship between olfactory and pungency thresholds has been studied by Abraham et al. (85).

Comparison of congenitally anosmic subjects with normosmic controls showed that the ability to detect, recognise and discriminate between odorants that activate the trigeminal nerve was similar in the two groups. This suggests that the trigeminal sense plays a significant role in odour recognition (86). Localisation of odours (that is, the ability to tell the direction from which an odour is coming) is dependent on the trigeminal nerves rather than the OSNs. Sniffing improves the ability to localise an odour but only if the odorant stimulates the trigeminal nerve (87). Thus, just as having two eyes gives us stereoscopic vision and having two ears allows us to tell the direction from which sound is coming, having two noses allows us to determine the location of the source of a smell. However, this is done via the trigeminal nerve system rather than the olfactory system (87). fMRI has been used to show that sniffing, that is, activating the trigeminal nerves by the change of gas flow in the nose, and smelling, that is, activation of the ORs by odorants, are processed differently in the brain (88). Taking all of these observations into consideration, we must conclude that the percept which we call 'odour' is a synthesis of inputs including at least those from both the olfactory and trigeminal systems.

TRANSPORT TO THE OLFACTORY RECEPTORS

It is self-evident that transport properties must be of importance in olfaction since, if odorants cannot reach the OE, they will not be detected by the ORs and no odor will result. Volatility is the most obvious requirement, and for organic compounds this results in a cutoff point at about 18–20 carbon atoms in the molecule equivalent to a molecular weight of about 300 Da. Larger molecules are simply not volatile enough to reach the OE in sufficient concentration to be perceived. Solubility is also important. This is partly due to the fact that water solubility implies a polar molecular structure and this, in turn, implies a low vapour pressure relative to the molecular weight, because of intermolecular hydrogen bonding. However, solubility properties *per se* also seem to be important, perhaps related to the ability of molecules to cross the aqueous mucus layer to reach the receptor proteins. A computational model of airflow in the nose of the dog showed that the pattern of odorant deposition across the OE depended on the water solubility of the odorants (89). Fragrance molecules generally have log $P_{oct/water}$ in the region of 2–5. This constraint in turn limits the number and nature of heteroatoms in the molecular structure of odorants.

Rodríguez et al. produced a linear regression analysis of published odour detection thresholds with respect to vapour pressure, water solubility and water/octanol partition coefficient. The resultant equation had an r^2 of 0.769, indicating that these three properties account for 77% of variance in threshold. Thus, these simple physical chemical properties would seem to dominate in transport of the odorant from nasal air to the receptor. If OBPs are involved in transport across the mucus, their interaction with the odorants must be relatively non-specific; that is, in humans, where there is only one type of OBP, they do not improve transport of high log P odorants relative to other odorants (90). As will be seen later, displacement of water from around the ligand and from around the ligand-binding site is known to contribute significantly to binding energy and so this could be an alternative or additional explanation for the dependence of olfactory threshold on solubility parameters.

Nasal Airflow

Between 5% and 15% of orthonasal airflow is directed to the OE. The sensitivity of ORs has been shown to be greater *in vivo* than in cell culture (91) and this has been shown to be due to improved delivery of the odorants to the receptor because of the airflow in the nose (92). It might be expected that a greater amount of air flowing through the nose would lead to greater sensitivity, and it has been shown that surgically induced increases in nasal volume result in an increase in nasal patency (93).

Sniffing

Sniffing results in an increased flow of air through the nose and it is not surprising, therefore, that it is an important feature in olfaction. Taking in an increased volume of air might be expected to increase sensitivity. Also, sniffing could affect the distribution of odorants to different regions of the OE and, so, if olfactory neurons expressing a given receptor type are not uniformly distributed, different glomerular activation patterns could result. For practical reasons, most of the studies on the role of sniffing have been done using rats and mice. So, although the following observations could well apply to humans, we must be mindful of the possibility that they might not.

Sniffing behaviour in mice is a dynamic process and is dependent on olfactory cues and other cues and the behavioural context (94). The normal breathing frequency of mice is about 2 Hz, and this increases to about 10 Hz when sniffing. Murine OSNs respond reliably to oscillating stimuli with a frequency of 2 Hz, but when this is increased to 5 Hz, the reliability is reduced considerably. This is considered to be an adaptive filter, which allows the animal to adjust the output of its sensory neurons by altering the breathing rate (95). Using optogenetics (see the section on neuroprocessing for details of this technique), it was found that mice could determine accurately the point of activation relative to sniff phase as their ability to determine timing was in the order of 25 ms with some individuals having an accuracy of 10 ms. The average accuracy is therefore about 1/10 of a sniff cycle. Interestingly, mice were not able to time auditory clicks to the same level of accuracy. This temporal accuracy would therefore seem to be unique to the olfactory system and could be used as part of odour coding (96).

The glomerular activation pattern is altered by sniffing. Greater nasal airflow gives more sensitivity for some glomeruli and less for others, showing that the effect is dependent on the chemical structure of odorant rather than the receptor (92). There is a faster response due to increased flow rate during sniffing, but both flow rate and frequency of sniffing affect total output from the OB (97). Similarly, the rate of air flow affects the electrical activity of the OB in rats (98). In rats, high-frequency sniffing increases the power of the gamma oscillations (which reflect the synchronous activity of mitral/tufted (M/T) cells) in the OB and increase sensitivity to odorants and is thought to play a role in forming odour objects relating to novel odours (99). With faster sniffing, the spike bursts in the mitral cells of the OB of mice are shorter and sharper, allowing the post-synaptic network to maintain robust temporal coding irrespective of the sniffing behaviour of the animal (100).

It is therefore clear that, in rodents at least, sniffing affects sensitivity to odorants, the nature of the output from the OB and the ability to form and retain odour images.

In humans, it has been found that reduced sniffing abilities in Parkinson's disease result in decreased olfactory acuity (101). However, in healthy subjects, nasal patency (102, 103) and sniff volume (104, 105) do not affect smelling. In contrast, odour perception, or imagination of odour perception, does affect the respiratory behaviour of humans (106) and the magnitude of the sniff is modulated by odour quality and intensity as well as by cognitive factors (107). In humans, it has been shown by fMRI that sniffing causes activation of the PC even when no odorant is present in the inhaled air. It is considered that this is a somatosensory response due to increased airflow. This is probably a way of indicating that active air sampling is in progress and, as would be expected, the pattern of brain activation is different from that when an odorant is present (88).

The Olfactory Mucus

Mammalian airways are coated with a protective layer of mucus which serves to maintain the correct level of hydration and to remove unwanted airborne contaminants by carrying them out of the airways and into the gastrointestinal tract. Nasal mucus contains a vast array of chemicals including glycoproteins (108), carbohydrates, sugars and amino sugars. Any of these could help transport odorants through the mucus by clathration/binding and/or formation of transient species. For example β-lactoglobulin has been shown to bind various odorants (109).

OSNs are primary neurons and are exposed to the external environment. They are therefore a potential route to the brain through which pathogens and toxins could bypass the protection offered to the brain by the blood/brain barrier. Unlike the motile cilia of the respiratory mucosa, the cilia of the OSNs have no motor control and simply sway with movement of the fluid around them. Therefore in order to protect the brain, there is high metabolic activity in the mucus of the OE, including a range of cytochrome P450 oxidative enzymes and various hydrolytic enzymes. Proteomic analysis of the olfactory mucus proteins of 16 healthy adult volunteers, using two-dimensional gel electrophoresis, matrix-assisted laser dissociation/ionisation time-of-flight (MALDI-ToF) mass spectrometry, reverse-phase liquid chromatography (RPLC) and Edman sequencing, identified 83 proteins with functions such as anti-inflammatory, antimicrobial, protease inhibition, antioxidant, transport, transcription, transduction, cytoskeletal, regulation, binding, and metabolism of odorant molecules (110).

The olfactory mucosal fluid is produced in Bowman's glands which are distributed throughout the OE (111). Aquaporin pathways in Bowman's glands help to prevent dehydration of the mucosa and to keep ion concentrations in the correct range (112). The proton-secreting enzyme V-ATPase is also present and probably plays a role in maintaining the optimal pH and possibly also has a role in carbon dioxide detection via hydration (113).

Odour-Binding Proteins

OBPs belong to the lipocalin family. They are produced in cells other than the sensory neurons and are excreted into the olfactory mucus. They were first identified as being involved in binding of pyrazines and so were originally named *pyrazine-binding proteins* (114). However, it soon became clear that they are capable of binding a very wide range of odorant molecules and the name 'odorant-binding proteins' was coined (115–117). As discussed earlier, OBPs are more important to insects than to mammals although rodents use OBPs extensively, as discussed in Chapter 1. The crystal structure of a bovine OBP has been published (118) and the expression of a human OBP, with affinity for a wide range of odorants, in the OE has been demonstrated (119).

Possible roles of OBPs in humans include transport of odorants across the mucus to the receptor, signal attenuation and removal of excess odorant. Since there are no lipocalins present in *in vitro* studies such as those of Spehr et al. (120), where both cloned ORs in cell culture and behavioural studies with intact spermatozoa were used, it is clear that odorants can be detected by receptor proteins in the absence of OBPs. In insects, OBPs are crucial in recognition, but the possibility that the receptor proteins distinguish between free and liganded lipocalins rather than detecting free odorants cannot be the case in mammals because the observed receptive ranges are the same for cloned receptors in cell culture as for intact neurons in whole animals, as will be seen later in this chapter.

Yabuki et al. introduced five odorants (Figure 2.5) into the gas phase over a film of rat OBP3 in water. They found a significant uptake leading to a 50,000-fold increase in the concentration of the odorants in the aqueous phase. The rate of dissociation was affected by the affinity of the OBP for the odorant and by the airflow. Competition between odorants was determined by their relative affinities for the OBP. It is therefore plausible that OBPs help to increase the concentration of odorants in the vicinity of the ORs. These authors point out that mass transfer in the nose is poorly understood. Various factors such as log P, vapour pressure, odorant

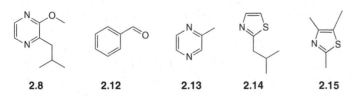

2.8   **2.12**   **2.13**   **2.14**   **2.15**

Figure 2.5 OBP ligands of Yabuki et al.

affinity for OBP and airflow are involved and interact to give a complex dynamic situation (121). Similarly, an OBP from the blowfly (*Phormia regina*) was found to trap airborne odorants and take them into the aqueous medium (122). Site-directed mutagenesis showed that a single lysine residue, Lys112, is responsible for binding of the human OBP, hOBP-2A, to aldehydes and acids. This result hints at the possibility of reactions such as imine formation and salt formation, respectively, occurring between the ligand and the OBP (123). OBPs are dimeric in nature. Binding studies on the rat OBP, rOBP-1 F, showed that only one odorant is bound per dimeric structure. Binding of the odorant to the binding pocket of one monomer induces conformational changes in the other monomer which prevent it from binding to a second molecule of odorant (124).

Nasal Metabolism

Metabolic enzymes have been shown to occur in the nasal mucosa (125, 126) and these include about 12 cytochrome P450s, the principal one of which being CYP2A13 (127–129). The earliest evidence that such enzymes are involved in nasal metabolism involved the bullfrog *Rana catesbeiana*. When ^{14}C-labelled octane was introduced to the frog's nose, the exhaled air contained water-soluble radiolabelled compounds, indicating that metabolism had taken place (130). It has also been shown that glucuronase enzymes (specifically uridine 5'-diphospho (UDP)-glucuronosyltransferase) are present in the nose and, together with the cytochromes, play a role in removal of xenobiotics from the mucus (131). Oxidation by cytochromes increases the water solubility of odorants and also paves the way for glucuronylation, as illustrated in Figure 2.6, further increasing water solubility and eventually leading to excretion by the kidneys. In the example, coumarin (**2.16**) is oxidised to 7-hydroxycoumarin (**2.17**) from which the glucuronate (**2.18**) is produced.

In addition to their role as protection from pathogens and toxins, the most obvious function of these enzymes would be to remove odorants from the mucus, thereby terminating signal generation and resetting the detection system. In other words, they would contribute to provision of the time dimension of olfaction, which is an important feature in survival. For example, when a male insect follows the sex pheromone trail of a female, it is important that the pheromone signal is quickly eliminated and the detection system reset. Females of the scarab beetle *Phyllopertha diversa* produce 1,3-dimethyl-2,4-(1*H*,3*H*)-quinazlinedione (DMQ) (**2.19**) as a sex attractant, and the half-life of DMQ in the male insect's antenna is normally 15 ms. When males are treated with an inhibitor of the cytochrome CYP4AW1, which degrades DMQ, they become unable to follow the pheromone trail (132).

The rate at which nasal metabolism occurs suggests that perception of odorants might be affected because their metabolites will also be present in the mucus. Nagashima and Touhara (133) investigated metabolism in the murine nose and showed that aldehydes are oxidised to acids and reduced to alcohols in the nose. For example, they demonstrated that the glomerular activation patterns are different if

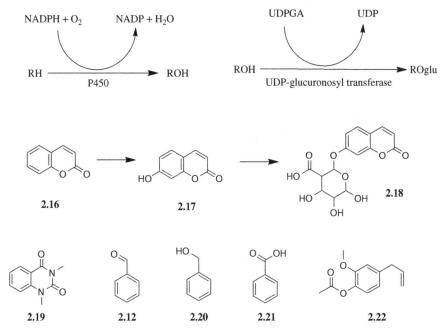

Figure 2.6   Nasal metabolism.

enzyme inhibitors are added to the mucosa. In a more definitive experiment, they added benzaldehyde (**2.12**) to the mucosa and then extracted the mucosa to prove that 80% of the benzaldehyde (**2.12**) had been metabolised to benzyl alcohol (**2.20**) and benzoic acid (**2.21**) (in the ratio 1 : 4). Addition of eugenyl acetate (**2.22**) to the mucosa and subsequent extraction showed >90% conversion to eugenol (**2.5**). The rates of metabolism of a range of substrates were in line with what would be expected on the basis of chemical reactivity of the substrates. Behavioural tests showed that, when treated with enzyme inhibitors, the behaviour of the mice, and hence presumably their olfactory perception, was affected only in response to those odorants that had been shown to be subject to nasal metabolism. They did not observe conversion of alcohols, thiols or lactones.

The work of Schilling et al. gives clear confirmation that these effects do indeed occur in humans and that they do affect human perception of odorants (134–138). Figure 2.7 shows some of the reactions that have been demonstrated to take place in the human nose, and within the space of one breath cycle. The 7-hydroxylation of coumarin (**2.16**), as proposed in Figure 2.6, has been confirmed and 3-carene (**2.23**) was found to be epoxidised. 2-Methoxyacetophenone (**2.25**) and *N*-methyl methyl anthranilate (**2.27**) are both demethylated, presumably via hydroxylation of the methyl group and subsequent hydrolysis. Pulegone (**2.29**) is oxidised initially to menthofuran (**2.30**) and this, in turn, is oxidised to the lactone (**2.31**). Of particular interest is the hydroxylation of the nor-terpenoid (**2.32**) to the allylic alcohol (**2.33**).

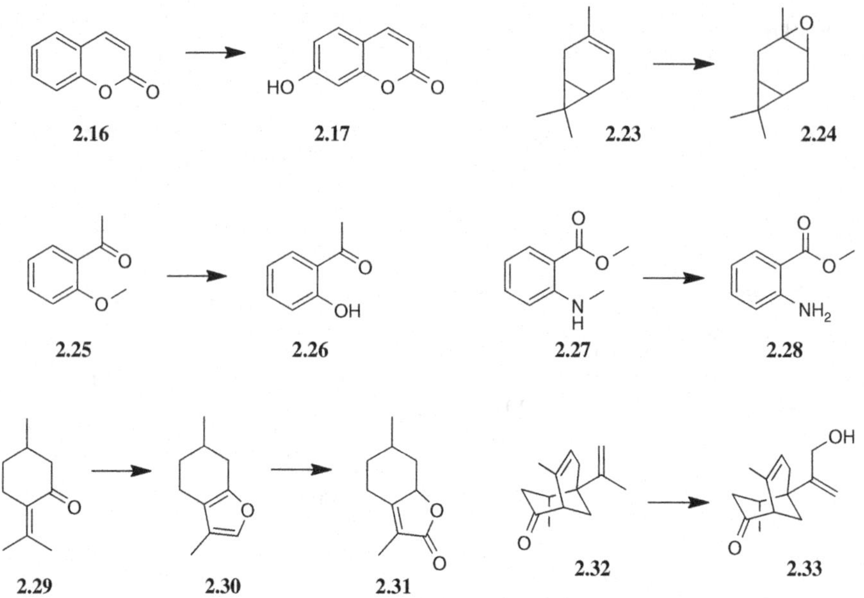

2.16   **2.17**   **2.23**   **2.24**

2.25   **2.26**   **2.27**   **2.28**

2.29   **2.30**   **2.31**   **2.32**   **2.33**

Figure 2.7   CYP2A13 transformations shown to occur in the human nose.

Some people describe (**2.32**) as smelling woody, whilst others perceive a raspberry quality. The difference does not lie in the individual subject's OR repertoire but in whether they express CYP2A13 or not. The raspberry character is elicited by the alcohol (**2.33**), so those who do express the enzyme perceive raspberry, whereas those who do not, perceive the woody character of (**2.32**). If the alcohol (**2.33**) is given to the latter group they then perceive raspberry, proving that it is not their receptor profile that distinguishes them from the former group.

The possibility that odorants could react with the ORs themselves will be discussed later in this chapter.

Solubility Effects and Delivery of Odorants to Receptors

Several studies have shown that odour detection thresholds are related to water solubility of members of a homologous series. The thresholds generally become lower as the hydrophobicity increases (139–141). This could be due to transport through the mucus or to the effects of hydrophobic bonding with the receptor. In a computer model of ligand approach to GPCRs, Dror et al. found that a major energetic barrier was likely to arise from removal of the hydration sheaths around the approaching ligand and the extracellular surface (ECS) of the receptor (142). Since odorant molecules are mostly quite hydrophobic, this raises interesting questions about their first contact with the receptor and their entry into the binding pocket. Because of the repulsive interactions between the odorant and the surrounding aqueous medium, entry into the receptor might represent a significant energy gain in the case of odorant/OR interaction.

DISTRIBUTION OF OLFACTORY RECEPTOR TYPES ACROSS THE OLFACTORY EPITHELIUM

The air drawn into the nose travels across the OE, and so it would be expected that an almost chromatographic effect might be observed with the distribution of odorants being determined by their volatility and hydrophobicity. In orthonasal olfaction, the more hydrophilic odorants would be expected to be adsorbed more at the front of the epithelium than would hydrophobic odorants, and these latter would, conversely, be expected to have a greater chance of reaching the back of the epithelium. Changes in nasal airflow would then produce a different adsorption profile. It is therefore important to know whether neurons expressing a given type of receptor are spread uniformly across the epithelium or only in localised regions. This is also important because, as mentioned in Chapter 1, it has been shown that in mice, signals generated by the same odorant are interpreted differently depending on which area of the epithelium the signals are generated (143).

Using probes for the RNA associated with a given receptor, it has proved possible to investigate OR expression patterns in the olfactory epithelia of rodents. There do not appear to be any results on humans as yet. In mice, the receptor OR3 is expressed in a subset of cells that are distributed symmetrically across the epithelium (144). Similarly, in rats, neurons expressing the OR37 receptor are found in broad zones (145). RNA probes for 11 different receptor types in rats showed each of them to be localised to distinct zones (146). A broad survey of murine receptors also found that each is localised with bilateral symmetry but that receptor types from the same subfamily are not clustered (147). In a review of such publications, Ma concluded that, although all OSNs expressing the same type of OR project their axons to only one or two glomeruli, it would seem that these neurons are spread in broad, overlapping groups across the MOE (38). Transplantation of cilia-rich membranes from various regions of the murine epithelium into *Xenopus* oocytes and subsequent measurement of their response to odorants showed that the location of ORs across the epithelium does not depend on physical and chemical properties of their agonists (148). The OR37 group of receptors is highly conserved across mammalian species (149), and neurons expressing them are confined to one area of the ventral OB, unlike many other receptors that are spread more broadly (150). This led Bautze et al. to suggest that they responded to odorants of significance and that they did respond to long-chain fatty aldehydes associated with fur, each member of the receptor family showing a preference for certain aldehydes (151).

It will be interesting to see whether similar patterns exist in humans. Since an odorant elicits the same percept whether it is sampled by the orthonasal or retronasal route, the brain must be capable of compensating for any 'chromatographic' effect on odorant adsorption.

OLFACTORY RECEPTORS

As mentioned previously, the percept of odour depends on input from various groups of receptors, namely ORs, TAARs, the receptors of SCSs, the receptors of the trigeminal nerves and, in non-human mammals, the VNRs and FPRs. With the exception of the ion channels of the trigeminal nerve, these receptors are all members of the family of 7-TM GPCRs. The most important, especially in terms of human olfaction, are the ORs and so they are the major focus of the remainder of this chapter. GPCRs are notoriously difficult to work with experimentally and the ORs are the most difficult subset. No OR has yet been isolated in crystal form and consequently much of our understanding of their mechanism of action is extrapolated from what we have learnt about other members of the same class of GPCRs. Therefore, this chapter contains considerable detail on these other Class A GPCRs. Examination of what we know about these other Class A GPCRs makes it abundantly clear that all members of the Class A GPCR family function in a similar way and so there can be no doubt that the ORs also operate in the same way.

There are two nomenclature systems in common use for ORs, and both will be used in this chapter depending on which was used in the source material cited. The older system relates to the location of the gene encoding the receptor. Thus, OR17-4 indicates a gene at location 4 on chromosome 17. Lower-case prefixes indicate the species: h for human, m for mouse, r for rat and so on. The more recent system refers to the sequence homology of the protein and takes the form of the letters OR, then one number followed by a capital letter and then a second number, OR1D2 for example.

The OR proteins belong to the family of 7-transmembrane (TM) GPCRs. The gene family coding for the OR proteins is the largest in the genome. It was discovered by Linda Buck and Richard Axel, who were awarded the 2004 Nobel Prize in medicine/physiology for their work. They estimated that there are codes for about 1000 OR proteins (152). A later review reports estimates of 853 for humans, 1490 for mice, 1493 for rats, 971 for dogs and 1091 for chimpanzees (153). In humans, many of the genes suffer from mutations, making them pseudogenes and so humans use a much reduced set containing only 350–400. Variation between humans is such that, statistically, it is unlikely that any two humans use the same set of ORs (154). Events following receptor activation involve the normal train of G-protein (G-olf, in the case of olfaction), second messenger (cyclic adenosine monophosphate (cAMP) or IP3, in the case of olfaction) and ion channel chemistry leading to polarization of the cell and hence generation of a discharge at the synapse with neurons of the OB (155). Each OSN expresses only one type of OR, and all OSNs expressing the same OR converge on the same glomerulus in the OB. The biochemical mechanisms controlling this selectivity remain unknown.

There has been a considerable volume of research into the genetics of the human ORs, the principal research group in the field being that of Doron Lancet at the Weizman Institute. In excellent reviews of the subject (156, 157), they trace the evolution of the approximately 900 human OR genes. About 63% of these have undergone pseudogenisation. Some pseudogenes have been found to be functional, for example OR5P3 (158), and it is also suggested that they might play a role in mRNA regulation and therefore also exert a different type of influence on olfaction (159).

The genes are divided into 17 families spread across the genome, the heaviest clustering being on chromosome 11. The Y-chromosome is the only chromosome not to contain OR genes. About 10% of human ORs belong to the subfamily known as *class 1* or *fish-like* ORs and these are the oldest in evolutionary terms. Details of all human OR genes can be found in the human olfactory receptor database (HORDE) (160). Using a DNA microarray, Zhang et al. showed that 437 (76%) ORs were expressed in the human OE (161).

Olfactory Receptors in Organs Other Than the Olfactory Epithelium

ORs have been found to be expressed in various organs other than the OE. The receptor OR1D2 (formerly known as *hOR 17–4*) is expressed in human sperm as well as in the OE (120, 162). The receptor responds to Bourgeonal (**2.34**), and human sperm will swim towards a source of Bourgeonal (**2.34**). Similarly, the murine receptor mOR23 occurs in both the OE and sperm of mice (163, 164) and responds to Lyral (**2.35**) but not Bourgeonal (**2.34**) (165) (Figure 2.8). As is the case with the above human example, murine sperm move towards a source of the agonist Lyral (**2.35**) (163). Taken together, these results suggest that directional control of sperm swimming is modulated by chemical signals which are received by ORs, a phenomenon known as *chemotaxis*. The human receptor OR51E2, also known as the *prostate-specific GPCR* or PSGR, is over expressed in prostate cancer cells (166) and ORs have also been found in the human kidney (167). One of these, OR78, has been found to respond to short-chain fatty acids and, via coupling with GPR41, another GPCR, to play a role in control of blood pressure (168). Indeed, a study of gene expression showed that ORs are found in many different organs in both humans and mice (153). An analysis of the evolutionary aspects of ectopically expressed genes suggested that they do not have alternative functions (161). However, a more recent study found that in mice the OR mOR23 occurs in muscle tissue where it plays a role in tissue regeneration (169).

7-Transmembrane G-Protein Coupled Receptors (GPCRs)

The OR proteins belong to the family of 7-TM GPCRs (170) and clearly belong to a stable phylogenetic cluster which is a subset of the rhodopsin family (171).

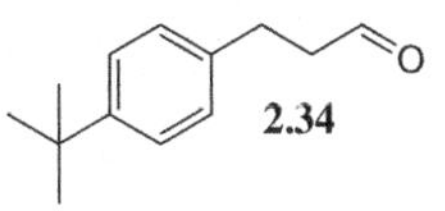

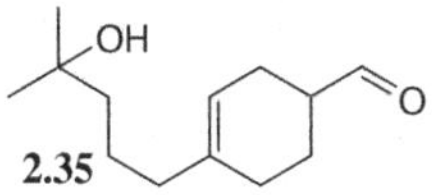

Figure 2.8 Odorants causing sperm chemotaxis.

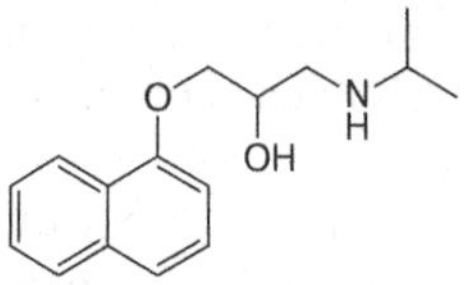

2.36 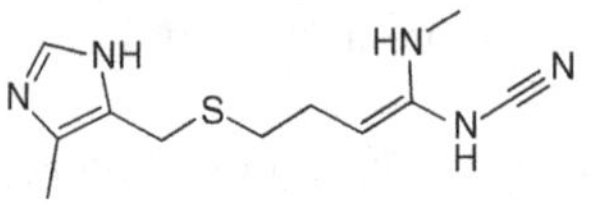   **2.37**

Figure 2.9 Examples of prescription drugs acting on GPCRs.

There are probably more than 900 OR-like sequences in the human genome (172) and as many as 460 of these could code for unique functional ORs (173). The genes coding for the GPCR family of proteins is the largest in the genome and, indeed, the OR genes alone would still be the largest family in the genome and represent about 2% of all human genes. The GPCR family of receptors includes those that detect adrenalin (**2.45**), adenosine (**2.57**), histamine, prostaglandins, dopamine (**2.278**) and so on. GPCRs are vital for recognition of many hormones and neurotransmitters, and about 60% of prescription drugs act on GPCRs (Figure 2.9). Our understanding of GPCRs is used in drug design, successful examples including propranolol (**2.36**) (a β-blocker) and cimetidine (**2.37**) (an antiulcer drug). Our knowledge of GPCRs has advanced enormously over the first decade of the twenty-first century, and the contribution of two of the leading researchers in the field, Robert Lefkowitz and Brian Kobilka, was recognised by their being awarded the Nobel Prize for chemistry in 2012.

The GPCR proteins are characterised by having a structure that crosses the cell membrane seven times. In mammalian GPCRs, the amino terminus is outside the cell and the carboxyl terminus inside. The sections that are found in the cell membrane fold into α-helices, and these sit together in the membrane providing a columnar structure. The helices are labelled TM 1 to TM7, TM 1 being that nearest the N-terminus. The presence of proline (**2.38**) (Figure 2.10) in any of these TM helices produces a kink in the helix because of the fact that the nitrogen atom of proline (**2.38**) is tied back in a five-membered ring. There are three extra-cellular and three intra-cellular loops and these are labelled EL1 to EL3 and IL1 to IL3, respectively, again with the numbering starting from the amino terminus (extracellular tail). A simple schematic is shown in Figure 2.11.

Various classifications have been proposed and used. The commonest system uses three main classes. Class A (also known as *class 1* or *rhodopsin-like GPCRs*) includes the ORs; class B (also known as *class 2* or *secretin-like receptors*) and class C (also known as *class 3* or *metabotropic glutamate-like receptors*) include the sweet and umami taste receptors. As far as ligand binding is concerned, the rhodopsin family is distinct in having an agonist binding site only

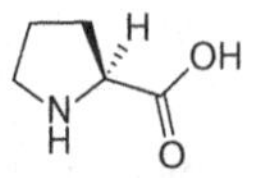

2.38   **Figure 2.10** Proline.

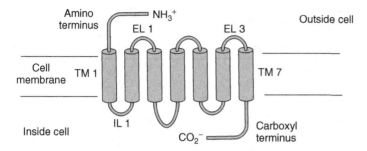

Figure 2.11   Schematic diagram of a GPCR.

in the TM section, whereas both secretin and metabotropic type receptors also have binding sites in their large extracellular tails and this has significance in terms of ligand design (174). In a comprehensive review of GPCR types, Fredriksson and co-workers have proposed a system for human GPCRs based on protein homology which they call the GRAFS system. In this acronym, G stands for metabotropic glutamate-like receptors, R for rhodopsin-like receptors, A for adhesion receptors, F for frizzled/taste 2 receptors and S for secretin-like receptors (175). The ORs are most closely related to rhodopsin, and all classification systems clearly place them together accordingly. In addition to rhodopsin and the ORs, other members of the family include the adrenergic receptors which are sensitive to adrenalin (**2.45**), the muscarinic receptors which respond to acetylcholine (**2.280**) and the A_{2A} adenosine receptors (A_{2A}ARs) which respond to adenosine (**2.57**).

There are two systems of nomenclature for ORs. The earlier one identifies them by their location in the animal's DNA. A small letter at the start of the name indicates the species (h for human, m for mouse, r for rat etc.); then OR indicates that it is an olfactory receptor. There follow two numbers. The first indicates the chromosome on which the DNA coding for the receptor is found, and, the second, its location on that chromosome. So, for example, hOR 17−4 is a human OR, the DNA coding sequence for which is found on chromosome 17. Sometimes, instead of using the chromosome numbers, the receptor is identified by a ligand. For example, the first murine receptor found to respond to eugenol (**2.5**) was dubbed mouse eugenol olfactory receptor (mOREG). The more recent system characterises ORs by their sequence homology using an initial OR followed by a number then a letter then a second number. So, for example, in the new system, hOR 17−4 is known as *OR1D2*.

Before launching into a description of GPCR structure, it is probably worthwhile describing some of the techniques, terminology and abbreviations used by biochemists and molecular biologists.

For convenience, shorthand codes are used for amino acids so that the primary sequences of proteins can be written in a concise way. The three-letter codes were the first to be used but, with the enormous growth in data available on protein structures, the single letter codes tend to be used mostly nowadays. These symbols are shown in Table 2.3, and the structures of the amino acids are shown in Figure 2.12.

Table 2.3   Amino Acids and Their Three and One Letter Symbols

Name	Three letter symbol	One letter symbol
Alanine	Ala	A
Cysteine	Cys	C
Aspartic acid	Asp	D
Glutamic acid	Glu	E
Phenylalanine	Phe	F
Glycine	Gly	G
Histidine	His	H
Isoleucine	Ile	I
Lysine	Lys	K
Leucine	Leu	L
Methionine	Met	M
Asparagine	Asn	N
Proline	Pro	P
Glutamine	Gln	Q
Arginine	Arg	R
Serine	Ser	S
Threonine	Thr	T
Valine	Val	V
Tryptophane	Trp	W
Tyrosine	Tyr	Y

The primary sequence of a protein can be obtained from the sequence of bases in the gene coding for it, each three base 'letter' of the DNA code representing one specific amino acid. The protein's primary sequence can then be written as a string of letters. For example, a fragment of the sequence of a GPCR might be written as:

CNLEGFFATLGGEIALWSLVVLAIERYVVVCKPMSNF.

This describes a sequence starting from the amino terminal end with cysteine followed by asparagine then leucine, aspartic acid, glycine, two phenylalanines, then alanine and so on. In fact, the above example is the sequence of TM 3 of bovine rhodopsin starting at the extracellular end and running on into IL2. It is now possible to compare this sequence with that of other proteins, as shown in Figure 2.13. The shaded area of the figure is that part of the proteins which constitutes TM 3, and the subsequent fragment is the start of IL2.

In Figure 2.13, we see how the sequence of bovine rhodopsin compares with those of three hypothetical ORs. Such figures are known as *alignments*. In this case, we can see that the first amino acid in all of the sequences is cysteine. In such cases, where the same amino acid is persistently found at the same position, it is said to be *conserved*. Thus it would be said that the cysteine in question is conserved across all four proteins in the example. If, in a larger sample of proteins

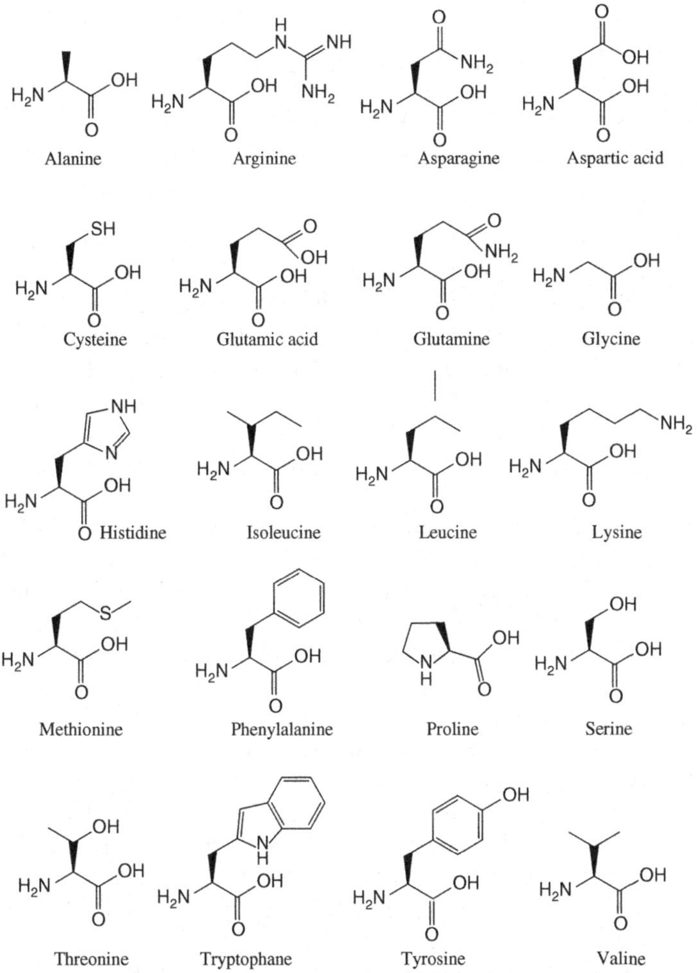

Figure 2.12 The amino acids of which proteins are constituted.

it was found that 90% of them had cysteine at this position, then it would be said that cysteine is 90% conserved at this position. The overall similarity between different protein sequences is known as the *homology* between them. It can be seen from the figure that there is a high degree of homology between rhodopsin and the OR proteins and an even higher level between the three ORs. Of particular interest here is the sequence at the very bottom of TM 3 and the start of IL2. Rhodopsin has an ERY (glutamic acid-arginine-tyrosine) motif and the ORs all have DRY (aspartic acid-arginine-tyrosine). Aspartic and glutamic acids are close homologues and therefore fulfil similar roles in protein structure and function. This DRY or ERY (abbreviated to (E)DRY) motif is very highly conserved across all GPCRs and DRY across all ORs and is therefore considered to be crucial for coupling to

```
C N L E G F F A T L G G E I A L W S L V V L A I E R Y V V V C K P M S N F

C L T Q M Y F M I A L A K A D S Y I L A A M A Y D R A V A I S C P L H Y T
C L M Q M Y F L V S L G N T D S Y T L A V M A Y D R Y V A I S H P L H Y T
C L T Q L Y F M I A L V A L D N L I L A A M A Y D R Y V A I C C P L H Y T
```

Figure 2.13   Comparison of part of the bovine rhodopsin sequence (top line) with those of some hypothetical olfactory receptors.

the G-protein. This conclusion is confirmed by all relevant X-ray crystal structure data. In the ORs, the three amino acids ahead of the DRY motif are also highly conserved and so MAYDRY is recognised as a conserved feature of ORs. In fact, Buck and Axel used conserved sequences such as PMYF in TM 2, MAYDRYVAIC in TM 3, KAFSTCA in TM 6 and PMLNPFIYSLRN in TM 7 in order to search the genome for the genes encoding the ORs (152).

When comparing large numbers of proteins, the level of homology can be visualised using a chart of the type shown in Figure 2.14. In this chart, the chance of finding a particular amino acid at a particular position in the sequence is shown by giving the code letter for that amino acid the corresponding percentage of the height of the chart. The chart in Figure 2.14 is for the three hypothetical OR fragments shown in Figure 2.13. At the left-hand side, the first position is fully occupied by a letter C, indicating that there is a 100% chance of finding a cysteine at that position. Two positions further along, we see that there is a 67% chance of finding threonine and a 33% chance of finding methionine. One advantage of such charts is that conserved features such as the MAYDRY motif stand out clearly and we can also see that in conserved regions, if the commonest amino acid is not there, the substitute is often a close analogue: for example, glutamic acid instead of aspartic acid. Colour coding of the letters can be used to indicate amino acid residue type (e.g. acidic, neutral or basic). A figure for a larger selection of proteins is shown in Figure 2.15 for a number of real mouse ORs. In this figure, the conservation of the MAYDRY motif is clearly visible just to left of centre of the fourth line.

There are two numbering systems in common use for GPCRs. The first simply starts at the amino terminus and numbers the amino acids sequentially along the peptide chain to the carboxyl terminus. The second is called the *Ballesteros–Weinstein system* after its inventors (176). They devised a system of numbering the amino acids in each of the helices, with the number 50 being assigned to the most highly conserved position and the others being numbered

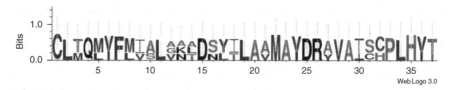

Figure 2.14   Homology of sample receptor sequences.

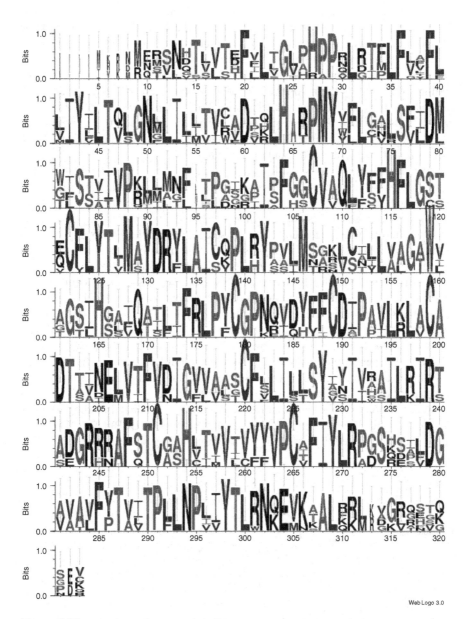

Figure 2.15 Homology of some murine olfactory receptor sequences.

relative to it. Thus $130^{3.50}$ would indicate the 130th amino acid from the terminus and show that it lies in the third TM helix and is the most highly conserved amino acid in that helix across all GPCRs. An amino acid numbered $130^{3.48}$ would be the 130th amino acid from the terminus, would lie in the third TM helix and would be two positions away from the most highly conserved amino acid in that helix.

When there are variations in the amino acid found at a given position, the mutation is indicated by a shorthand code which gives the letter for one of the possibilities followed by the number of the position in the overall sequence and then the code for the other alternative amino acid. For example, D123E would indicate two variants in which one has aspartic acid at position 123 and the other has glutamic acid instead of aspartic. It is usual to place the amino acid of the functional variant first if one is functional and the other not.

We can learn quite a lot just from studying the amino acid sequences of ORs. Those residues that are highly conserved are likely to be important in forming the basic structural features of the receptor and in its link to the G-protein, whereas the highly variable residues are more likely to be involved in forming the binding pocket and thus giving each OR a unique selectivity profile. For example, based on homology, Man et al. (177) suggested that the main binding site is between helices 3, 4, 5 and 6 and in the upper part (i.e. towards the cell surface) of the TM section. This prediction has been borne out by more recent X-ray, nuclear magnetic resonance (NMR) and infrared (IR) studies. A schematic view, looking from the top of the OR down into the TM barrel, of the predicted binding site is shown in Figure 2.16. Similarly, combining analysis of the sequences of 197 different ORs with models based on homology with rhodopsin led to the prediction that 17 specific hyper-variable amino acid positions would be found to be those that form the ligand binding site of ORs. Support for this prediction comes from the fact that 12 of the 17 positions are known to be involved in the ligand binding of a number of other GPCRs and that the majority of the predicted binding residues do point inwards into the proposed ligand binding pocket (LBP) (178). The same authors also suggested the possibility of a minor binding pocket, and this has also

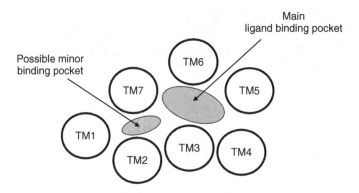

Figure 2.16   Predicted binding site in olfactory receptor proteins.

been proposed by another research team, as will be discussed below (179). Matsui et al. allowed a radiolabelled substrate to bind to the α_2-adrenergic receptor and then used peptidases to break the complex into smaller units (180). Knowing the primary sequence of the receptor (181), they were able to show that the ligand had bound to TM4 of the receptor in the proposed binding region. Similarly, TM2 was found to be involved in the binding site of the β_2-adrenergic receptor (182).

The free fatty acid receptor GPR40 (formerly known as *FFAR1*) plays a role in controlling the secretion of insulin and hence in glucose homeostasis. In an investigation of GPR40, Tikhonova et al. (183) used homology modelling based on rhodopsin supported by site-directed mutagenesis to study the LBP. GPR40 responds to blood plasma acids, such as oleic (**2.39**) and linoleic acids and (**2.40**), and is also known to bind with low nanomolar potency to the synthetic compound GW9508 (**2.41**) (184) (Figure 2.17). Their model, based on rhodopsin homology, showed the latter to fit very neatly into the predicted LBP. This illustrates an important point about the binding of ligands in that the ligand is not necessarily bound in its lowest energy configuration. The authors also showed how the shape of the hydrophobic part of the binding pocket is radically affected by the exact conformation of the amino acids in the TM region. One can therefore speculate that such conformational flexibility in both ligand and receptor might account for the relatively broad tuning of the ORs. Thus, for ORs, the 'lock and key' analogy has been replaced by what is often referred to as a *flexible key in a flexible lock*.

The conformational changes around the LBP are only a small part of the total flexibility of a GPCR. These proteins move continually between thousands of different conformations, some of which are capable of interacting with the corresponding G-protein, as will be described later. Those conformations that are able to bind to the G-protein are known as *active configurations*. Ligands that stabilise the GPCR in an active configuration are known as *agonists*, and those that stabilise

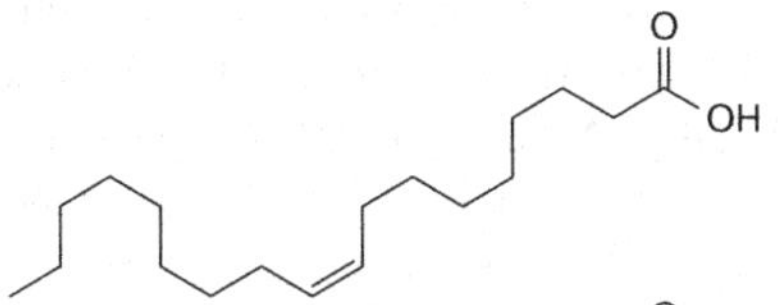

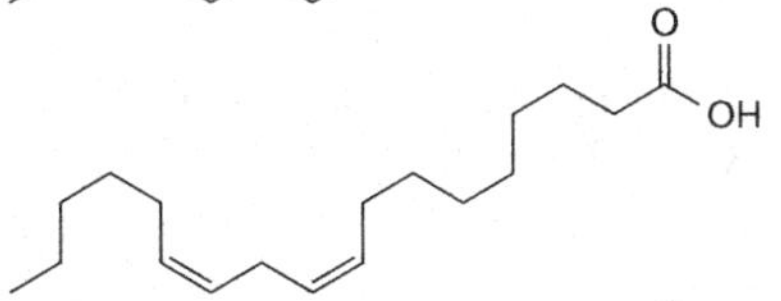

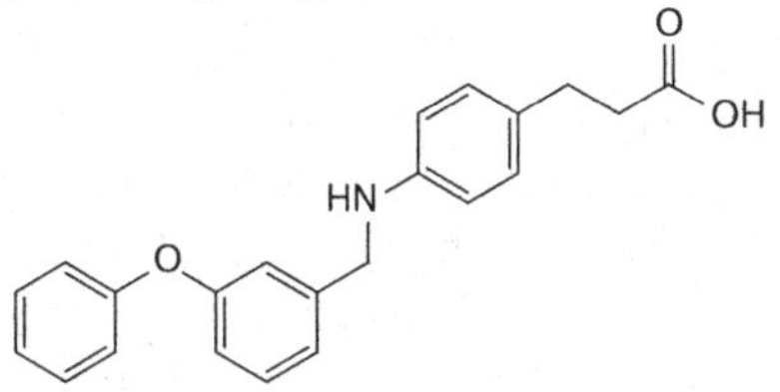

Figure 2.17 GPR40 agonists.

it in a configuration which cannot bind the G-protein (inactive conformations) are known as *inverse agonists* or *antagonists*. The fact that more than one active conformation may be possible could perhaps explain the broad but selective tuning which will be seen later for the OR proteins.

X-ray Crystallography in Structural Determination of GPCRs

GPCRs are difficult to crystallise because in their native state they are located in the cell membrane and their lipid environment around them serves to support the structure. ORs are even more difficult to crystallise, presumably because of a more flexible structure. Production of sufficient quantities of ORs for crystallisation is also the topic of current research, and some recent progress will be discussed in the next chapter under the topic of electronic noses. There is a great deal of effort being put into the search for new methods of obtaining GPCR crystals: an example is the development of a protein-based detergent which has enabled purification of some mammalian ORs (185). Although crystal structures were not obtained, circular dichroism was used to confirm the presence of α-helices, and thermophoresis proved that the receptors recognised and bound their ligands. The key physiological roles of some other GPCRs have made them primary targets of research in the pharmaceutical industry, and so much more time and money have been spent in research into them than into ORs. Taking into consideration the difficulty of crystallising ORs and the higher interest in GPCRs of pharmaceutical importance, it is not surprising that much of what we know about the detail of the three-dimensional structure of ORs has been learnt from studies on other class A GPCRs. The significance of research into defining the tertiary structures and mechanism of action of GPCRs was recognised by the award of the 2012 Nobel Prize to Robert J. Lefkowitz and Brian K. Kobilka for their enormous contributions to the field.

(Rhod)opsin. Because of its more ready availability, rhodopsin has been a primary focus of GPCR crystallography from the earliest days of investigation into GPCR structure and function; since it is a prototypical class A GPCR, it has proved particularly useful for the study of the entire family. All the earliest X-ray crystal structures of GPCRs are of rhodopsin, opsin and various intermediates between the two, and most GPCR modelling is based on the rhodopsin structure.

It must be remembered, of course, that the rhodopsin/opsin mechanism is different from that of other class A GPCRs in that the ligand (11-*cis*-retinal) (**2.42**) (Figure 2.18) is covalently bound to rhodopsin via an imine bond. Absorption of a photon causes isomerisation of the ligand to the all-trans form (**2.43**). The resultant complex is unstable, the imine bond hydrolyses and the all-trans retinal (**2.43**) dissociates to give the free protein, opsin. Thus, rhodopsin represents the inactive state of the receptor, and metarhodopsin II, the intermediate immediately before dissociation of the ligand, represents the active state. Because all the intermediates in this process have strong ultraviolet absorption spectra, the photochemical aspects of the mechanism can be studied spectroscopically, and this has been nicely reviewed by Kandori (186). Since it would be undesirable for the optical neurons to be activated

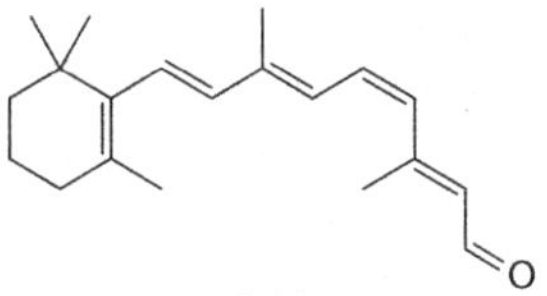

2.42 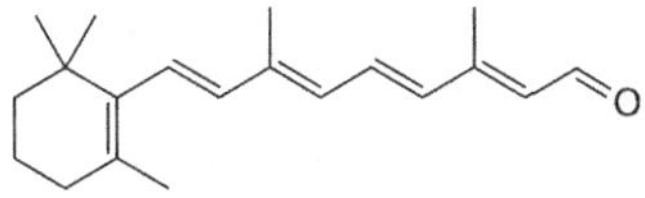   **2.43**

Figure 2.18 Retinals.

when there is no light stimulation, the visual system has evolved in such a way that this unique feature of the rhodopsin/opsin system means that rhodopsin is unusual in another aspect: unlike all other GPCRs, it has no level of background activation. However, despite all this, the rhodopsin system still provides important clues as to the structure of other members of the whole class A family, and the key intermediate in the process of vision, metarhodopsin II, is a good model for the active state of other GPCRs.

In 1990 Henderson et al. (187) reported the structural determination of bacteriorhodopsin, and in 2000 a structure was published with a resolution of 2.8 Å (188). The structure of bovine rhodopsin was solved in 2004 (189), and in the same year the resolution was improved to 2.2 Å (190). Another major breakthrough came in 2008 when Park et al. (191) published the structure of free opsin at 2.9 Å resolution showing how opsin differs from rhodopsin through conformational changes around the highly conserved (E)DRY and $NPxxY(x)_{5,6}F$ regions at the bases of TMs 3 and 7, respectively. In opsin, TM 6 is tilted out by 6–7 Å relative to its position in rhodopsin and the LBP is reorganised, opening channels as possible routes for retinal (**2.42**)/ (**2.43**) to diffuse in or out of the LBP. Scheerer et al. (192) then added further to this by obtaining a structure of opsin bound to a synthetic peptide, built to resemble the part of the G-protein that binds to opsin. This 3.2 Å resolution structure shows exactly how the G-protein binds to the receptor in, amongst others, the (E)DRY and $NPxxY(x)_{5,6}F$ regions. Choe et al. (193) introduced the all-trans retinal (**2.43**) into the active site of opsin to produce a form of metarhodopsin II which was stable enough to be used to obtain a crystal structure. Their findings concerning the binding of metarhodopsin to retinal (**2.43**) and transducin (the G-protein associated with (rhod)opsin) are in agreement with those of the earlier crystal structures. Similarly Standfuss et al. (194) produced a 3 Å resolution of transducin coupled rhodopsin in which the receptor was modified to make it constitutively active, and showed the same movement of TM6 that had earlier been identified as crucial to activation of the receptor.

β₁-Adrenergic Receptor. The adrenergic receptors (sometimes abbreviated to adrenoceptors) are activated by adrenalin (**2.45**) in the body. Warne et al. have published X-ray crystal structures of the human $β_1$-adrenergic receptor bound to an antagonist, cyanopindolol (**2.46**), (at 2.7 Å resolution) (195) and also to the full agonists carmoterol (**2.47**) and isoprenaline (**2.48**) and to the partial agonists salbutamol (**2.49**) and dobutamine (**2.50**) (196) (Figure 2.19). The conformational

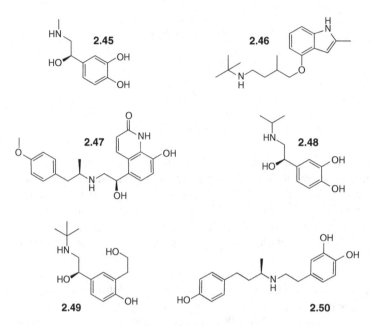

Figure 2.19 Adrenergic receptor agonists and inverse agonists.

changes in the protein structure on moving from the inactive (antagonist bound) to the active (agonist bound) states are similar to those observed on going from rhodopsin to opsin.

As mentioned earlier, whereas rhodopsin has no basal activity, hormone receptors (like ORs) do have a level of background firing. It was postulated that the salt bridge between an arginine at the base of TM3 and glutamic acid at the base of TM6 of rhodopsin was responsible for the stability of the inactive form. The distance between these two amino acids is greater in the adrenergic receptors and so it was suspected that this weakening of the link was responsible for their higher rate of basal activity. However, in a comparison of eight different crystal structures using three different agonists, Moukhametzianov et al. (197) showed that the bridge was still in place in the β_1-adrenergic receptor. The suggestion is therefore that variations in the strength of this salt bridge have evolved to allow for different rates of basal activity depending on the physiological requirement.

β_2-Adrenergic Receptor. A crystal structure of the β_2-adrenergic receptor bound to the inverse agonist carazolol (**2.51**) was published in 2007, showing that this receptor also binds its ligands in the same region as retinal (**2.42**) is bound in rhodopsin (198). In the same year, another published X-ray structure used cholesterol (**2.52**) to facilitate crystallisation and obtained a structure at 2.4 Å resolution, also with carazolol (**2.51**) bound in the LBP, and showed that the receptor exists in monomeric form (199). A third paper that year reported a structure with a modified third intracellular loop to aid crystallisation, and showed

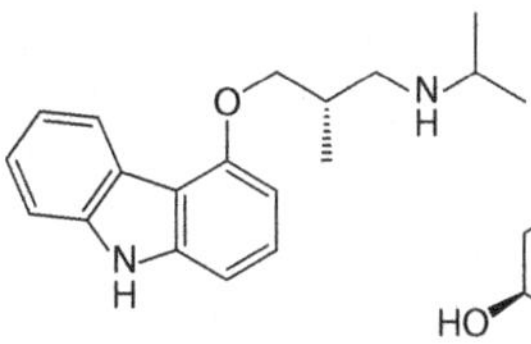

2.51 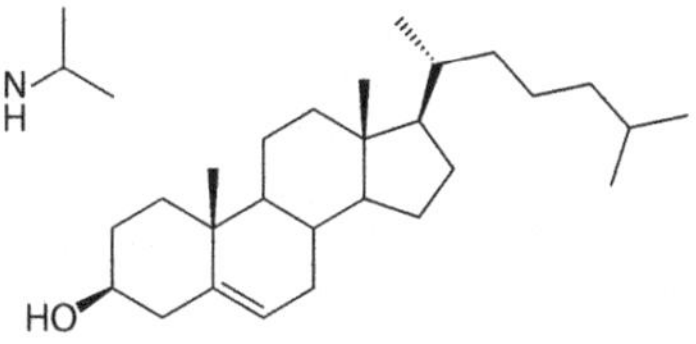   **2.52**

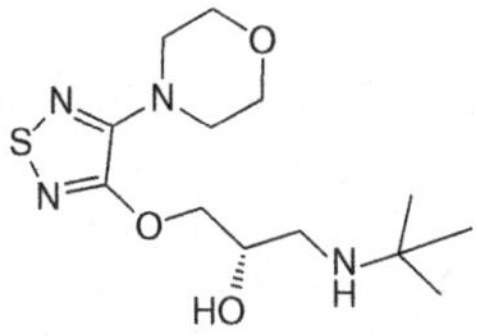

2.53

Figure 2.20 Compounds binding to the β_2-adrenergic receptor.

that neither the binding nor the pharmacologic properties were affected by more than small amounts (200) (Figure 2.20).

The following year, another study investigated the role of cholesterol (**2.52**) further, using the receptor bound to the partial inverse agonist timolol (**2.53**), and found that cholesterol (**2.52**) actually binds to the receptor and supports its structure. Two cholesterol (**2.52**) molecules were attached to each receptor and were not in the packing interface but in a pocket between TMs 1, 2, 3 and 4. From homology, the authors proposed that about half of all class A GPCRs would bind to cholesterol (**2.52**) and use the rigidity of the latter to strengthen the TM part of its structure (201). However, no further evidence has yet been found to support this suggestion.

Another study of the β_2-adrenergic receptor used bioluminescence resonance energy transfer (BRET) to further investigate the role of cholesterol (**2.52**) in the membrane on the activity of the receptor (202). They found that cholesterol (**2.52**) in the membrane had an effect of separating the receptor from the G-protein and the adenylyl cyclase. When the membrane was depleted of cholesterol (**2.52**), the three proteins tended to associate and this had the effect of increasing activity and the basal rate of firing.

An interesting series of structures of the receptor bound to three different ligands (ICI 118,551 (**2.54**), a benzofuran analogue (**2.55**) and alprenolol (**2.56**)) show how the binding site can accommodate different structures by making only minor adjustments to the conformation of the amino acid residues in it (203) (Figure 2.21). The existence of different binding conformations of the β_2-adrenergic receptor was also confirmed by plasmon resonance studies (204).

All of the above β_2-adrenergic receptor structures were obtained with inverse agonists, that is, in the inactive state of the receptor. By constructing a peptide mimic of the G-protein, Rasmussen et al. were able to obtain a structure of the receptor in its active state. As in other cases where this has been studied, TM helix

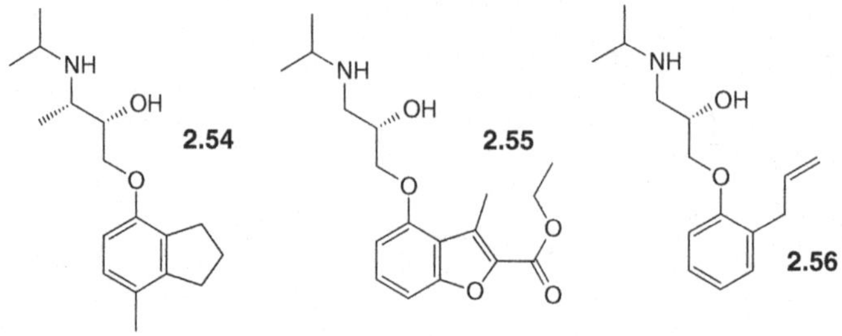

Figure 2.21 Three β₂-adrenergic receptor ligands showing flexibility of the binding pocket.

6 swings outwards, in this instance by 11 Å. Other changes, such as those to TMs 5 and 7, are also remarkably similar to those in other GPCRs (205).

One of the obstacles to obtaining X-ray crystal structures of agonist-bound active states of GPCRs is the relatively low affinity of most of the agonists and the high rate of dissociation of the complex. In order to overcome this, Rosenbaum et al. developed a ligand that could be covalently tethered to the β₂-adrenergic receptor. This work (206) (with 3.5 Å resolution), together with its companion paper (205), demonstrated how the extracellular and intracellular surfaces both cooperate in binding the ligand into the LBP and how, in the absence of the G-protein or a substitute for it, the ligand-bound active state spontaneously relaxes into the inactive state.

A₂ₐ Adenosine Receptor (A₂ₐAR). Another pair of GPCR crystal structures showing the difference between agonist and antagonist bound configurations is that of the A₂ₐAR, which is also a class A GPCR. The natural ligand for the receptor is adenosine (**2.57**), and caffeine (**2.58**) is a known antagonist (Figure 2.22). The antagonist-bound structure was determined at 2.6 Å resolution using another antagonist, namely, ZM241385 (**2.59**) (207). This receptor has four disulfide bridges in the extracellular domain which contribute to the shape of the binding pocket in the TM section. This allows ZM241385 (**2.59**) to dock into the binding pocket in a plane at right angles to the plane of the membrane and in an extended form. Using the former drug UK-432097 (**2.60**) as an agonist, it proved possible to obtain a crystal structure of A₂ₐAR in its active configuration (208). UK-432097 (**2.60**) is a substituted adenosine (**2.57**) and is similar in structure to many known agonists of A₂ₐAR. It is a highly potent and selective agonist and was developed as a drug candidate for chronic obstructive pulmonary disease (209). The crystal structure shows clearly the various hydrogen bonds and hydrophobic interactions between the ligand and the receptor. Six of the nine amino acid residues involved in hydrogen-bond formation had already been demonstrated to be of importance to binding, using site-directed mutagenesis. The changes observed on moving from the inactive (antagonist bound) form to the active (agonist bound), were mostly very similar to those involved in moving from rhodopsin (inactive) to opsin (active). Relative to the inactive (antagonist bound) state, in the active (agonist bound) state the TM helix 3 moves upwards away from the cell interior, helix 5 swings sideways

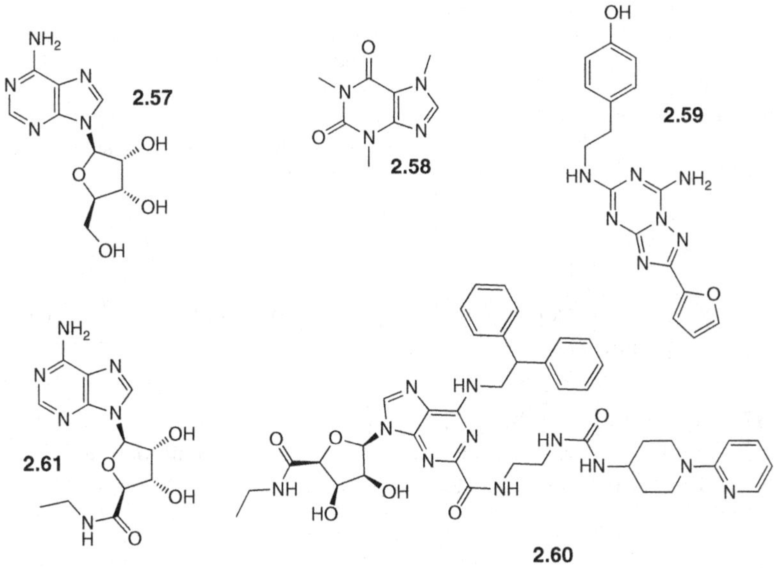

Figure 2.22 Agonists and antagonists of the $A_{2A}AR$ receptor.

and helix 6 tilts outwards. These movements break the ionic lock between helices 3 and 6. Helix 7 moves in a manner characteristic of the $A_{2A}AR$ receptor but the other movements are similar to those of rhodopsin and therefore possibly characteristic of all class A GPCRs including the ORs. The presence of a proline (**2.38**) disrupts the α-helical structure because of the fact that the amino group of proline is tied back to the side chain by being part of a cyclic system and therefore creates a bend in the helix. Xu et al. (208) show how these bends in helices 5, 6 and 7 serve to amplify the small structural changes induced by the agonist and lead to a greater change in configuration at the intracellular site at which the GPCR couples to the G_α fragment of the G-protein.

Lebon et al. determined crystal structures of a thermo-stabilised $A_{2A}AR$ bound to adenosine (**2.57**) and also to the synthetic agonist 5′-N-ethylcarboxamidoadenosine (NECA) (**2.61**) (210). By superimposing their structures of the $A_{2A}AR$ on those of the $β_1$-adrenergic receptor, they were able to show that the positions of the ligands (adenosine (**2.57**) in the $A_{2A}AR$ and isoprenaline (**2.48**) in the $β_1$-adrenergic receptor) were very similar and that similar 5–6 Å wide clefts open up on the intracellular face to allow binding to the G-protein. These features are all similar to those observed in (rhod)opsin. In the $A_{2A}AR$, the agonists have ribose groups that bind deep in the LBP forming polar interactions with Ser277 and His278. The inverse agonist ZM241385 (**2.59**) does not have the ribose moiety and therefore cannot form these bonds. This prevents TM5 from adopting the active state configuration. For both $A_{2A}AR$ and the $β_1$-adrenergic receptor, a bulge in TM5 seems necessary

2.62

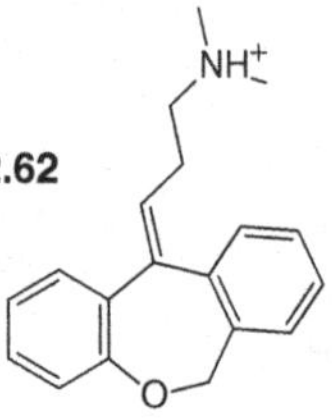

2.63

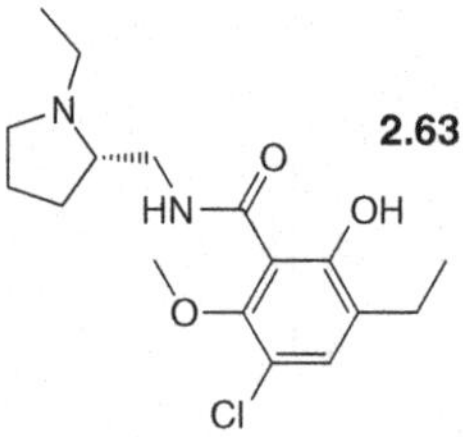

Figure 2.23   Antagonists of the H1 histamine and D3 dopamine receptors.

for activation and inverse agonists prevent its formation. Thus activation of the receptors involves pulling the extracellular ends of TMs 3, 5 and 7 closer together.

Human H1 Histamine Receptor.   The crystal structure of this receptor, bound to the antagonist doxepin (**2.62**) (Figure 2.23), shows a similar pattern to all of the other class A GPCR structures and also clearly shows EL2 lying across the top of the LBP (211).

D3 Dopamine Receptor.   The crystal structure of the D3 dopamine receptor bound with the antagonist eticlopride (**2.63**) shows a locked conformation of the ionic lock. The binding of ligands can extend towards the ECS to accommodate larger ligands (212).

CXCR4 Chemokine Receptor.   Wu et al. determined the crystal structures of the CXCR4 chemokine receptor bound both to a small molecule and to a peptide. The binding site is closer to the ECS in this receptor than for other class A GPCRs (213).

M2 Muscarinic Acetylcholine Receptor.   Haga et al. determined the structure of the human M2 muscarinic receptor which is involved in controlling the cardio-vascular function. The basic structure of the protein is similar to that of all the others described above (214). The crystal structures were of the receptor bound to an antagonist QNB (**2.64**), and show how this ligand is held in the binding pocket by a salt bridge from the carboxylate anion of an aspartate residue, hydrogen bonds to the carbonyl and hydroxyl oxygen atoms and hydrophobic bonds to other parts of the molecule. This structure is also interesting in that it shows many highly ordered water molecules covering the ECS and ICS and also lining the channel between the two, except for a region in the centre of the TMS where hydrophobic residues prevent the ingress of water. This supports ideas (discussed in more detail later in the chapter) concerning the role of water in determining the structure of GPCRs and that removal of water from the LBP constitutes a significant part of the energetics of binding. This receptor is known as being susceptible to allosteric modulation, and possible binding sites are indicated at the ECS, their role being to lock the ligand in the binding pocket and therefore increase response by preventing egress of the agonist. This concept will also be discussed in more detail later.

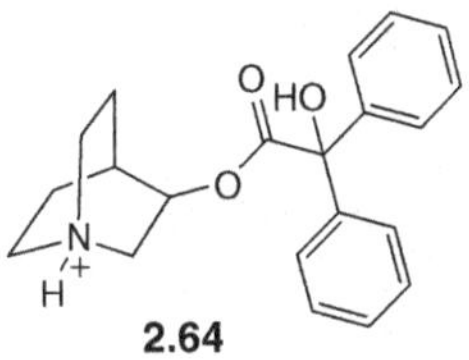

Figure 2.24 Antagonists of muscarinic acetylcholine receptors.

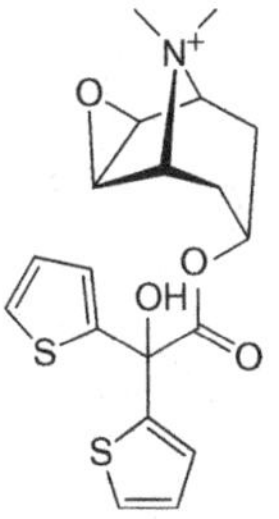

M3 Muscarinic Acetylcholine Receptor. The structure of the M3 muscarinic acetylcholine receptor bound to an antagonist, the bronchodilator drug Tiotropium (**2.65**) has been determined (215) (Figure 2.24). It is similar to that of the M2 muscarinic acetylcholine receptor, although they bind to different G-proteins. Molecular dynamics simulations suggest that Tiotropium (**2.65**) binds to an allosteric site on the ECS of both the M2 and M3 receptors (215).

The Opioid Receptors. The opioid receptors are important physiologically and respond to opioid drugs such as morphine (**2.66**) and also to endogenous ligands such as the endorphins. There are three classical opioid receptors, the δ-, κ- and μ-opioid receptors, and one other closely related receptor, the nociceptin/orphanin FQ receptor (NOP). All are class A GPCRs and bind to their ligands by a combination of polar and hydrophobic interactions in the LBP, just as described for all the other GPCRs.

Wu et al. determined the structure of the human κ-opioid receptor by X-ray crystallography and supported the findings with site-directed mutagenesis experiments (216). The crystallography was carried out on the receptor bound to an antagonist JDTic (**2.67**) which has a very high affinity for it. The team also investigated a range of other ligands and found some binding features common to all of them and some which accounted for differences in affinity of different ligands. Comparison with the chemokine and the β-adrenergic receptors showed that the binding pockets of all three are broadly similar.

A crystal structure of the μ-opioid receptor bound to the antagonist β-FNA (**2.68**) shows that its binding pocket is also typical of class A GPCRs, although it, like rhodopsin, forms a covalent bond with this ligand to give the Schiff's base (**2.69**). In the case of the μ-opioid receptor, a lysine residue in the protein undergoes a Michael addition to the fumarate fragment of the ligand, as shown in Figure 2.25 (217). In the μ-opioid receptor, the bound ligand is more exposed to the extracellular medium than is the ligand of the M3-muscarinic receptor. This is reflected in the dissociation rates of the two classes of receptors. Whereas the half-lives of the ligand/muscarinic receptor complexes is of the orders of tens of hours, those of μ-opioid receptors are of the order of tens of minutes or even a few minutes. The tight stacking of the μ-opioid receptors in the crystal suggests that they might function as dimers or oligomers *in vivo*. The TM5 and TM6 helices of one μ-opioid receptor will recognise those of another, and this would increase the

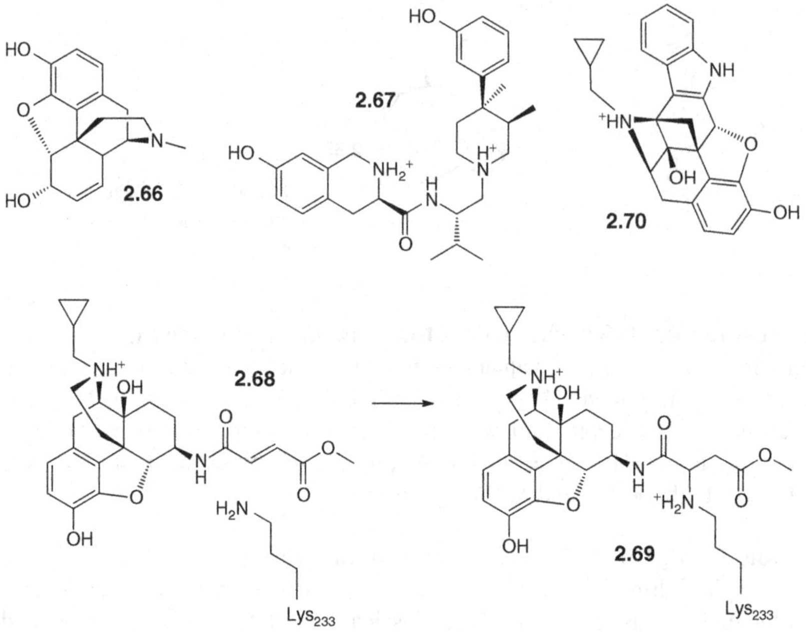

Figure 2.25   Ligands of the opioid receptors.

propensity to form homodimers. The only other class A GPCR with this property is the CXCR-4 chemokine receptor.

The crystal structure of the murine δ-opioid receptor bound to an antagonist Naltrindole (**2.70**) was solved by Granier et al. (218). They identified an allosteric binding site across the top of the main LBP. This is roughly consistent with observations described below on allosteric modulators.

The NOP has some features in common with the three classical opioid receptors, but some differences in amino acid sequence lead to a slightly different structure of the binding pocket and different ligand affinities. The crystal structure of the human receptor was solved using the receptor bound to a peptide which mimics one of its natural ligands (219).

It was postulated that the ligands of the opioid receptors contain a structural fragment that represents an 'address' which enables them to recognise the receptor family and a 'message' fragment that serves to produce the different responses to them. Taken together, these four crystal structures add weight to this theory of address and message recognition. The deepest region of the binding pockets is highly conserved across the family and binds to the common 'address' features of the ligand, whereas the 'message' part of the ligand structure binds to a more variable region of the LBP.

NTSR1 Neurotensin Receptor.   The NTSR1 receptor responds to the oligopeptide neurotensin, which is a neurotransmitter and hormone involved in some key functions in the brain. The structure of the rat variant of the receptor bound to an agonist has been determined by White et al., and their report shows how its main structural features closely resemble those of other class A GPCRs (220).

Nuclear Magnetic Resonance Spectroscopy (NMR) in Structural Determination of GPCRs

An advantage of NMR for the study of GPCRs is that, by using isotopic labels in the protein and/or ligand, those labelled positions can be made to stand out from the rest of the molecule without significantly affecting their binding properties. There-fore, GPCRs can be studied in lipid membranes and in a dynamic situation where the protein moves from an inactive to an active state. In a crude analogy, therefore, whilst X-ray crystal structures give snapshots before and after activation, NMR can provide a movie of the activation process. A good example of this is the work of Ahuja et al., who used NMR to study the binding of retinal (**2.42**) in rhodopsin (the inactive state of the protein) and in the corresponding active state metarhodopsin II (221). They labelled with ^{13}C various sites in retinal (**2.42**) and in the LBP of rhodopsin and studied the interactions between the protein and its ligand as the complex moved from rhodopsin to metarhodopsin II following activation by light. In the ligand-bound inactive state, they were able to show that EL2 closes over the LBP once the ligand is in place. They showed that, in the inactive state, there was no interaction between the methyl group of methionine 207 (which is located in TM5) and either the methyl group on the double bond of the cyclohexene ring of retinal (**2.42**) or the olefinic carbon atom to which it is attached. Upon activation by light, these two fragments come closer together, showing a movement of the cyclo-hexene ring towards TM5. In metarhodopsin II, the Schiff's base of all-trans retinal (**2.43**) serves to hold EL2 and the remainder of the ECS in the active configuration. On investigating a number of changes between the two states, they could show that activation involved changes in the conformations of TMs 5, 6 and 7 and rearrange-ment of the hydrogen-bonding network around the conserved NPxxY motif and the electrostatic bonding around the conserved ERY motif. These changes alter the configuration of the cytosolic face and expose the binding sites for the G-protein. They also conclude that the conformational changes right across the protein, from the ECS through the TM section to the cytosolic face, all work together in binding the ligand and moving from the inactive to the active state.

Bokoch et al. used NMR to study the ECS of the β_2-adrenergic receptor bound to the inverse agonist carazolol (**2.51**) (222). They showed that small molecules binding in the LBP affected the ECS and suggested the ECS as a possible site for allosteric modulation. They modified lysine residues in the receptor by methyla-tion with ^{13}C. The tertiary amines thus produced can participate in salt bridges in a manner similar to the primary amines of the original but give a distinct and sepa-rate NMR signal. They found that binding of a ligand in the LBP weakens the salt bridge from EL2 to the EL3/TM7 junction and that the ECS changes depending on which ligand is bound in the LBP. They also showed that phenylalanine 193 adopts a trans configuration pointing towards TM5 when the receptor is bound to carazolol (**2.51**), but can assume several different configurations when the receptor is bound to alprenolol (**2.56**). This latter finding is reminiscent of that of Tikhonova et al. (183) that the hydrophobic residues can adopt different configurations, albeit that Bokoch et al. were looking at the ECS whereas Tikhonova et al. were studying the LBP. The adaptability of hydrophobic residues has, of course, significant implications for ORs since the hydrophobic part of odorant molecules is known to be important in affecting the ultimate odour percept. The fact that the hydrophobic residues comprising the LBP can move to accommodate different ligands would help to explain some of the breadth of OR tuning. Another finding of Bokoch et al. is that agonists and inverse agonists induce different changes in the ECS, and they predict that the extracellular ends of TMs 6 and 7 move on activation. This also supports the concept discussed above of the whole receptor structure moving together on activation.

Shi et al. (223) have shown that, using a combination of ^{13}C and ^{15}N labels and various NMR techniques (such as magic-angle spinning (MAS), solid-state nuclear magnetic resonance (SSNMR) and dipolar assisted rotational resonance (DARR)), it is possible to determine the structure of a GPCR without recourse to crystallography; their work took only 15 days of experimental time. The system they studied was that of sensory rhodopsin II from *Anabaena* species incorporated into a lipid membrane. Using H/D exchange, they were able to determine which hydrogen atoms are exchangeable in the receptor. Incubation of the micelles in deuterium oxide resulted in exchange of the hydrogen atoms around both the ECS and ICS but not in the TM section and, significantly, not in EL3 where X-ray crystallography indicates a strong hydrogen-bonding network. The pattern of exchange suggests that the protein is located with a greater proportion of its bulk in the cytosol than outside the membrane surface.

The β_2-adrenergic receptor can activate either G-protein or β-arrestin second messenger pathways. Some ligands stabilise the receptor in the conformation that activates one pathway whilst other ligands stabilise the other pathway. Liu et al. used ^{19}F-NMR on ^{19}F-labelled receptors to show how different ligands induced the two different appropriate conformations at the intracellular ends of helices 6 and 7 (224).

The most difficult regions of GPCR structure, as far as X-ray crystallography is concerned, are the loops and tails where the exposed loops are either dynamic, and thus not visible, or else involved in crystal contacts. Higman et al. used MAS solid-state (SS) NMR to define more closely the configuration of the loops in bacteriorhodopsin (225).

Sensory rhodopsin and proteorhodopsin (a pigment found in marine bacteria and involved in photohaply) are not GPCRs but are related 7-TM proteins. They are even more difficult to crystallise then GPCRs, and so the solution of their full tertiary structures in liquid phase using NMR is a major breakthrough, which augers well for the potential of NMR for OR structural determination (226–228). Indeed, Park et al. have since used ^{15}N- and ^{13}C NMR to obtain a structure of the chemokine receptor CXCR1 in a phospholipid bilayer under physiological conditions and without modification of the receptor sequence (229). This, together with a previous publication of the same group (230), showed that the ECS is responsible for binding to the ligand (a peptide called *interleukin-8*), the intracellular face of the receptor binds to the G-protein and the lipid bilayer plays a role in helping the receptor adopt a configuration which allows ligand binding to result in G-protein activation. The structural determination also shows two disulfide bridges on the ECS, one from the amino terminal to the top of TM7 and the other from the middle of EL2 to the top of TM3.

Infrared Spectroscopy (IR) in Structural Determination of GPCRs

As with NMR spectroscopy, IR can be used to study GPCRs in the membrane environment and also to do so in a dynamic situation. Ye et al. incorporated *p*-azido-L-phenylalanine (**2.70**) into rhodopsin, which they then expressed in human embryonic kidney (HEK) cells (Figure 2.26). Using Fourier transform infrared (FTIR) spectroscopy, they were able to follow the conformational changes in the receptor as it moved from the inactive rhodopsin to opsin by monitoring the azido frequencies in the spectrum (231). Katayama et al. used IR to study the hydrogen-bonding networks in red and green rhodopsins and observe the changes in protein conformation upon activation (232). The results of both of these studies produced evidence in support of the basic conclusions from X-ray crystallography.

Other Techniques for Investigating GPCR Structure

Altenbach et al. used double electron–electron resonance (DEER) spectroscopy to show that, upon activation, TM6 of rhodopsin moves outwards by 5 Å, thus adding further support to the evidence from X-ray structures of the active and inactive forms (233). Wade et al. studied the olfactory receptor OR17-40 expressed in yeast cells using BRET. Their experiments suggest that ORs can homodimerise and that activation involves a conformational change in the receptor resulting from binding of the ligand (234). In view of the findings concerning cholesterol (**2.52**) (discussed above in the section on the β_2-adrenergic receptor) and since yeast cells have ergosterol rather than cholesterol (**2.52**) in their cell walls, further work would be necessary to determine the relevance to mammals of the findings regarding homodimerisation. Modified GPCRs containing photoactive fragments in their structures

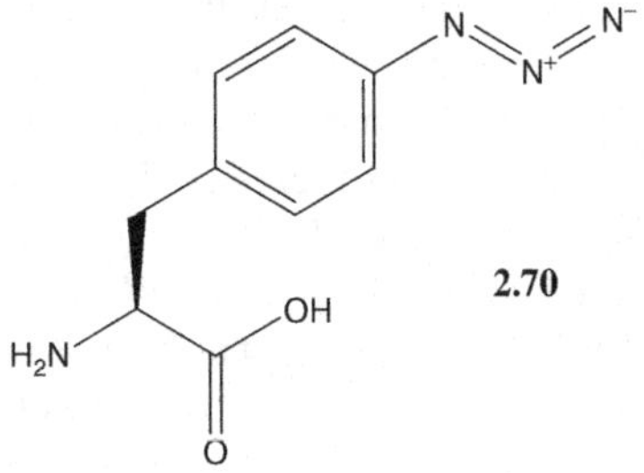

2.70

Figure 2.26   Azidophenylalanine.

have also been used to investigate mechanistic aspects. This avenue of research has been reviewed by Beck-Sickinger and Budisa (235).

Having reviewed the overall structure of GPCRs, it is perhaps appropriate at this point to look specifically at each of the three main regions, the ECS, TM region and ICS.

The Extracellular Surface

The role of the extracellular loops of class A GPCRs is less well understood than that of the intracellular loops and tail which are known to be involved in binding to the G-protein and hence in initiating the second messenger transduction cascade.

Receptors responding to peptides and proteins can recognise their ligands at the extracellular face. Hawtin et al. (236) showed that three specific amino acids, all on the ECS – one on EL1, one on EL2 and one at the top of TM3 – are key for ligand recognition in the V_{1a} vasopressin receptor (V_{1a} R). They contrast this with the amine receptors in which the ligands must enter the LBP in the TM section of the receptor. However, the oligopeptide ligands of the NTSR1 receptor do unfold from their tertiary structure and extend the peptide chain into the normal LBP (220).

A number of polar amino acid residues in EL2 are highly conserved across all type 1 GPCRs, which is strongly indicative of some role in structure and/or function, and, based on their work with mutations, Avlani et al. have proposed that this role is that of a gatekeeper over the top of the binding pocket to assist in binding and/or activation of the receptor (237). This work is discussed in more detail below because of its implications for allosteric modulation. In their above-mentioned investigation of the free fatty acid receptor GPR40, Tikhonova et al. (183) looked at the extracellular face of the receptor and postulated that EL2 opens to allow the ligand to enter the binding site and then closes over again when the ligand is bound. Their conclusions are therefore very similar to those of Avlani et al.

The importance of the disulfide bridge(s) in the extracellular loops of GPCRs was shown by Dohlman et al. who found, for the β_2-adrenergic receptor, that breaking of the bridge or replacing the cysteines by serine resulted in loss of activity and also reduced the efficacy of transport of the receptor to the cell surface (238).

In an experiment to study movement at the ECS of β_2-adrenergic receptor, Elling et al. (239) introduced histidine and cysteine residues at the tops of TMs 3, 6 and 7 in place of neutral amino acids as found in the native receptor, and then used metal ions to pull these together in a mimic of the proposed closing as the receptor binds a ligand. Of various metals tested, Cu and Zn were found to bind and to form a bridge between TM3 and TM6, thus pulling these together. They found that the metal ions did indeed cause activation of the receptor experimentally. The TM helices of class A GPCRs are not all straight; instead, some have kinks introduced by the presence of proline. The prolines are conserved across all class A GPCRs as follows: TM2 (80%), TM5 (77%), TM6 (100%), TM7 (96%). Elling et al. built 3-D models of the receptor and showed that inward movement of the tops of helices 6 and 7, because of the highly conserved prolines, caused an outward movement of

TMs 6 and 7, and TM 6 in particular. They therefore proposed that a global toggle switch mechanism operates (240). In this, closure of the ECS around the top of the LBP causes simultaneous outward movement of TMs 6 and 7, thus opening up the ICS to reveal the cleft for G-protein binding. As discussed above, we now have X-ray crystal structures of three different class A GPCRs, namely (rhod)opsin, the β_1-adrenergic receptor and the A_{2A}AR, all before and after activation and all three pairs of structures support the global toggle switch mechanism of Elling et al.

The Transmembrane Region (TM) and the Ligand Binding Pocket (LBP)

One primary function of the TM section of the receptor is to provide the overall structure, and this is achieved through polar interactions between the helices and water molecules that are located in that region. Angel et al. analysed published crystal structures of five different GPCRs to investigate the role of water molecules in the TM region (241). They found that the presence and location of water molecules is highly conserved across all of the receptors studied and suggested that this implies a role for them in defining the structure of the receptors. The work of Shi et al. (discussed above in the section on NMR spectroscopy) showed that the hydrogen atoms of these water molecules do not exchange readily, and this would support the proposition that they are fixed in place as a basic part of the GPCR structure through hydrogen bonding to the amide groups that form the helices.

The other major role of the TM section of the receptor is to provide the LBP.

As discussed previously, the first approaches to determining the nature and location of the LBP involved homology. It was argued that the sites of greatest variation in amino acid identity were likely to be those involved in binding of the ligand. The differences in the LBPs must be the key differences between different GPCRs since the G-proteins (e.g. transducin in vision, G_{olf} in olfaction and gustducin in taste) are all similar in their binding to the GPCRs and the overall mechanism is similar across the entire family.

Thus Lancet and coworkers identified likely sites for ligand binding as described above. The use of homology modelling is vindicated in a report by Kurland et al. in which they used homology modelling based on the rhodopsin structure to build a computational model of the rat I7 receptor (242). From this, they measured the calculated binding energies of those aliphatic aldehydes to which the receptor is known to respond. They then measured the actual binding energies using the receptor expressed in rat neurons and found an excellent correlation between calculated and measured binding energies. Thus they built a clear picture of the LBP. This is shown schematically in Figure 2.27. There is a salt bridge between lysine 164 in TM4 and aspartic acid 204 in TM5, and the carbonyl oxygen atom of the aldehyde function of the ligand forms a polar interaction with a hydrogen atom of the amino group of the lysine. Tyrosine 107 on TM3 also forms part of this binding. The hydrophobic part of the binding pocket is lined

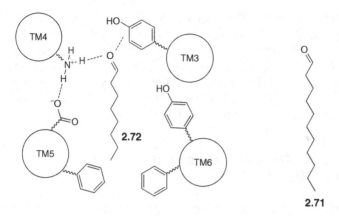

Figure 2.27 Binding of octanal to rat OR-I7.

by tyrosine 264 on TM6 and phenylalanines 205 on TM5 and 262 on TM6. The dimensions of the hydrophobic pocket account for the observed receptive range, for example, by showing that aldehydes larger than undecanal (**2.71**) cannot be accommodated inside it. Figure 2.27 shows octanal (**2.72**) bound into the LBP. Others such as Goddard approached the same question by modelling, as will be discussed below.

In another approach to identifying those amino acid residues that are most important in ligand binding, Gloriam et al. used all the published X-ray structures of rhodopsin class receptors to identify which amino acid residues are directed towards the bound ligands and which are key to ligand binding in each of the various subgroups studied. Unfortunately, this could not include the ORs since no X-ray structures exist for them (174). It would seem that TM3 is most important for discrimination of the LBP architecture. Gloriam et al. have published a review of the use of our knowledge of GPCR binding sites for drug design (174).

When mutagenesis is used to identify amino acid residues of importance, loss of activity could mean that the deleted and replaced amino acid is vital for binding, but it could also mean that a change in structure was responsible for the change in activity. For example, Kato et al. showed that mutagenesis of some residues in the intracellular loops of the mOREG destroyed its activity by preventing coupling to the G-protein (243). On the other hand, some changes were found to increase its activity. None of the changes affected the binding of eugenol (**2.5**) (Figure 2.28) to the receptor. In an elegant experiment to prove that a specific amino acid is involved in ligand binding, Strader et al. (244) proposed that replacement of Asp113 in the β_2-adrenergic receptor by serine would reduce its affinity for adrenalin (**2.45**), the amino function of which forms an ion pair with the aspartate side chain, and, instead, make it bind to compounds that can form a hydrogen bond to the OH group of serine. This is exactly what happened in practice: the D113S mutant was activated by catechol esters and ketones rather than adrenalin (**2.45**). This confirms that Asp113 is involved in binding to adrenalin (**2.45**).

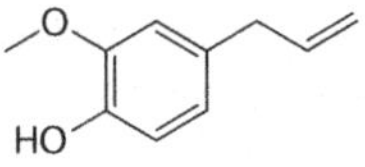

Figure 2.28   Eugenol and BI-167107.

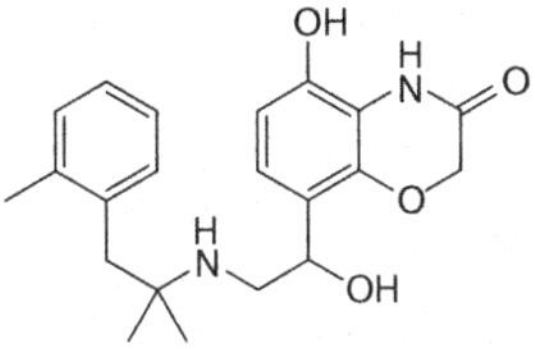

From the rapidly growing volume of data on receptors where X-ray structures exist, there can be no doubt about the location of the LBP and the residues of which it is comprised. Similarly, as also described above, NMR, IR and other techniques also give hard evidence of how and where the ligands are bound. All of this solid physical evidence is in basic support of the binding site in the upper part of the TM section (i.e. towards the ECS) of the protein, as proposed by Lancet and his coworkers and also shows clearly that the interaction between ligand and receptor is comprised of hydrogen bonding, other polar bonding, hydrophobic bonding and other non-polar interactions. This is certainly the case for all class A GPCRs for which definitive experimental evidence exists, and there is no reason to doubt that exactly the same applies to the ORs and their ligands. What experimental evidence there is for odorant binding, such as the examples discussed above and below, offers further support for this picture of odorant/OR binding. The increase in entropy that results from displacement of water molecules from the binding pocket is often the largest factor in ligand/receptor affinity (245). Since odorants are almost all hydrophobic, this factor might be even more significant for them.

The Intracellular Surface

The most highly conserved regions of all class A GPCRs are found at the ICS. It therefore seemed most likely to early researchers that these are involved in binding to the G-protein since those parts of the G-proteins involved in binding to the receptors are all similar in structure. This basic hypothesis has now been confirmed by various crystal structures and other experimental data, some of which are discussed above.

For example, comparison of the X-ray structures of rhodopsin and opsin showed how opsin differs from rhodopsin through conformational changes around the highly conserved (E)DRY and $NPxxY(x)_{5,6}F$ regions at the bases of TMs 3 and 7, respectively. In opsin, TM 6 is tilted outwards by 6–7 Å relative to its position in rhodopsin and the LBP is reorganised, opening channels as possible routes for retinal (**2.42**)/(**2.43**) to diffuse in or out of the LBP (192).

Kato, Katada and Touhara used site-specific mutagenesis to show that for the murine OR mOREG the conserved motifs of MAYDRY at the lower end of TM3, KAFSTCK at bottom of TM6 and LRNK at the start of the intracellular helix are key for binding of $G_{\alpha olf}$ (243). As would be expected, these regions of ORs are equivalent to those of (rhod)opsin which Scheerer et al. showed by X-ray crystal-lography to bind to transducin (the G-protein involved in the visual system) (192). Similarly, in agreement with the X-ray data on (rhod)opsin, Kato et al. suggested that the conformation of TM6 in mOREG changes when the ligand is bound and that serine-240 is vital for activating $G_{\alpha olf}$. It is a highly conserved part of a highly conserved motif KAFSTC which is present in most ORs. There is a H-bonding net-work at Ser-240 and Tyr-241 in KAFSTC. Kato et al. propose that, on an odorant binding to mOREG, TM6 moves in a way that unlocks a network of ionic inter-actions between the cytoplasmic ends of TMs 3 and 6 which leads to activation of the G-protein via specific interactions with newly exposed amino acids in the intracellular loops and in the C-terminal domain.

This last mentioned proposal of Kato et al. (243) is also in perfect agreement with known data on other GPCRs. The existence of the ionic lock between TMs 3 and 6 was proposed by Ballosteros et al. (246) on the basis of their work with the β_2-adrenergic receptor. In that receptor, the amino acids involved are D130 and R131 in TM3 and E268 in TM6. In rhodopsin, the ionic lock involves E134 and R135 in TM3 and E247 in TM6 and a similar ionic lock has been demonstrated in the $A_{2A}AR$ as described above. The D(E)RY(W) motif is 72% conserved in all class A GPCRs. Activation of the receptor involves a 'proton switch' in this region whereby a subtle change in the hydrogen-bonding pattern results in breaking of the ionic lock (247).

In a further confirmation of this mechanism, Rasmussen et al. have reported the crystal structure of the β_2-adrenergic receptor bound to its associated G-protein, Gs, and to the agonist BI-167107 (**2.73**) which was selected from a library of 50 ago-nists as the optimum ligand for crystallisation with Gs (248). This crystal structure of the receptor in its active state shows clearly the outward movement of helices 6 and 7 and the binding of the termini of Gs to the cleft in the intracellular face of the receptor. The conformational changes induced in the G-protein by this binding then destabilise its binding to guanosine diphosphate (GDP), opening the route for binding to guanosine triphosphate (GTP) and hence activation of ACIII.

It would appear that the entire process of ligand binding, conformational change in the receptor and binding of the G-protein occurs in a concerted manner. For example, the crystal structure of the $A_{2A}AR$ complexed to an antibody shows how this second protein attaches to the intracellular face of the receptor and allows it to bind only to antagonists and not agonists (249). Similarly, Shirokova et al. have shown that use of a different G-protein can make a ligand an antagonist rather than an agonist and also change receptive range of the receptor (250). Of course, this is an observation resulting from *in vitro* experiments with different G-proteins and does not affect the actual outcome in the nose since G_{olf} is the only G-protein expressed in the OSNs.

Modulation of GPCR Signalling by Other Cell Proteins

Ritter and Hall (251) have reviewed a variety of mechanisms by which other pro-teins present in the cell can modulate signalling by GPCRs. One example in the area of ORs is described by Li and Matsunami (252), who found that the M3 muscarinic acetylcholine receptor can positively modulate OR activity. Similarly, it has been found that some microvillous cells in the OE express TRPM5 receptors which, when activated by recognition of chemical signals, release acetylcholine (**2.280**) which then modulates the activity of the surrounding OSNs (253). Because the OSNs also express adrenergic and muscarinic receptors which respond to neurotransmitters released by the autonomic nervous system, it is clear that signals generated by odorants can be modulated by the autonomic nervous system (254).

Summary Description of Olfactory Receptor Structure and Activation

Clearly ligand/GPCR interaction is a very complex and mobile system. Water molecules, ions such as sodium, and lipids and cholesterol (**2.52**) from the membrane all play a role in stabilising the receptor structure in both the inactive and active forms and in moving from one of these states to the other, all of which are clearly demonstrated in a 1.8 Å resolution structure of a chimeric version of the $A_{2A}AR$ (255). Ligands continuously move into and out of the binding pocket, and the whole GPCR structure is in a state of continual movement between conformations. Thus, ligand binding is not binary in nature, that is, it does not operate as a simple on/off switch and there are grades of activity. In its free state, the receptor moves from one conformation to another and inevitably from time to time will spontaneously arrive at an active conformation and so there is always a background level of firing. The presence of ligands then serves to modulate the firing rate (256).

Fortunately, we have snapshots of the active and inactive states in the form of X-ray crystal structures of GPCRs bound to agonists and inverse agonists, respectively. All that we see from these confirms the proposed mechanism as outlined above and these crystal structures have moved us out of the realms of speculation about GPCR, hence OR, structure and mechanism and into the certainty of the physical reality. In view of the high degree of structural similarity across the whole family of class A GPCRs, there can be very little doubt that the ORs function in the same way. The similarities between various class A GPCRs have been reviewed by Kolb and Klebe (257). The general features of GPCRs and the mechanism by which they activate G-proteins have also been reviewed by Audet and Bouvier (258).

Evolutionary considerations readily explain one of the main differences between ORs and other members of the class A GPCR family. In the case of hormone receptors, it is important that the receptor is activated only when appropriate and therefore it is not surprising that such receptors have evolved to be highly selective, responding only to one agonist such as adrenalin (**2.45**). Synthetic agonists for such receptors are always few in number because of the constraints around the LBP. However, in olfaction, animals are exposed to a wide variety of odorants in the environment and most natural smells are complex mixtures. It therefore makes sense that the ORs are more broadly tuned and that there is a large number of them allowing for a combinatorial approach to recognising odours.

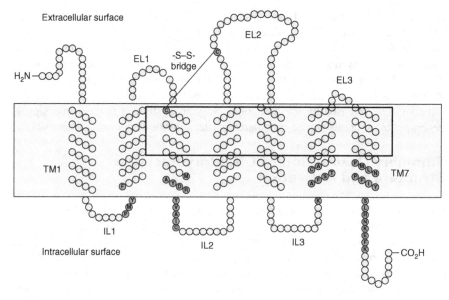

Figure 2.29   Snake plot of a typical olfactory receptor showing some key conserved residues.

Thus, based on all of the above, we can now form an overall picture of how ORs function. The general structural features of a typical OR are shown in the 'snake plot' diagram of Figure 2.29. In the figure, the shaded area represents the cell membrane and the boxed area is that comprising the binding pocket. The highly conserved motifs of PMYF, MAYDRYVAIC, KAFSTCA and PML-NPFIYSLRNKEFK are shown at the intracellular ends of helices 2, 3, 6 and 7, respectively. There is a highly conserved disulfide bridge running from the top of TM3 to a cysteine in EL2, the largest of the three extracellular loops. ORs, like all class A GPCRs, have small loops and short tails. This is in stark contrast to the sweet and umami taste receptors, as will be further discussed below.

Another interesting feature of ORs is that a disulfide bridge between a cysteine at the top of TM3 and one in EL2 helps to hold the assembly together and restricts conformational mobility. There are usually one or two other cysteines in EL2 also. It is well known that thiols usually have intense odours and that their odour character often changes from pleasant at low concentration to unpleasant at high concentration. It is therefore tempting to speculate that thiols might undergo some chemistry at the ECS of the receptors. For instance, they could undergo a metathetical interaction with the disulfide bridge leading to a more conformationally flexible receptor structure and thus possibly increase the basal firing rate by increasing the chances of an active conformation of the receptor interacting with the G-protein. Support for such a hypothesis comes from the fact that reductive enzymes have been shown to activate transient receptor potential (TRP) sensory channels in *Drosophila neurons*, by reducing the disulfide bridges on the ECS (259). Alternatively, an incoming thiol could form a disulfide bridge with one of the other cysteines in EL2 and thus form a hydrophobic cap over the LBP, hence slowing the rate of ingress or egress of agonists. Such a reaction would be promoted by oxidants, such as suitable metal ions, for example Cu^{2+}, in the mucus. It could be that at low concentrations we see only binding of the thiol to a cognate receptor but at higher concentrations such chemical interactions result in effects across a wide range of receptors resulting in a signal which the brain interprets as unpleasant.

Possible support for the involvement of copper(II) in the reaction of thiol odorants comes from a finding of Matsunami's group (260). They found that the *in vitro* response of the murine receptor m-OR244-3 to the murine semiochemical methylthiomethanethiol (MTMT) (**2.74**) was enhanced by the presence of copper(II) in the culture medium and attributed this to complexation between the thiol, the cupric ion and a histidine residue in the receptor. However, the possibility of oxidation of the thiol by the copper(II) to give the disulfide (**2.75**) cannot be ruled out (Figure 2.30). This is a well known reaction of thiols, the disulfide (**2.75**) is also an agonist of m-OR244-3 and, since it is less volatile,* it will suffer lower loss through vaporisation during incubation than would the thiol, leading to a higher observed response.

Thiols are notorious for their intense and unpleasant odours and also for the fact that the odour quality of many of them changes with concentration. Might it be that at low concentration the response of selective receptors is seen whereas at higher concentration reaction with all of the receptor types occurs and a different signal pattern is obtained? Similarly, aldehydes generally have intense odours and they are also capable of rapid reaction with functional groups on the receptor protein. For example, they could form Schiff's bases with lysine residues and at least some ORs have a lysine on the ECS. Peptide reactivity studies showed that the interaction between lysine and aldehydes at pH 7 is very slow (261). However, a recent publication reported that lysine residues in TM proteins were capable of reacting with ester functions in the lipid membrane to form amide bonds (262). This is a much more difficult reaction than Schiff's base formation (which certainly occurs in the opsin/rhodopsin system), and so the possibility of Schiff's base formation between odorants and ORs cannot be ruled out.

The process of odorant recognition and G-protein activation is shown schematically in Figures 2.31–2.33. These figures show a highly simplified and stylised cross section of the receptor through the TM assembly at the level of TMs 3 and 6. Figure 2.31 shows the receptor in its inactive state. The bottom (intracellular) ends of TMs 3 and 6 are held together by the ionic lock between

2.74   **2.75**

Figure 2.30 Oxidation of MTMT by copper(II).

---

*The b.p. of (**2.74**) is equal to 40 °C at 20 mmHg, and ~130 °C at 760 mmHg; the b.p. (**2.75**) is equal to 240 °C at 760 mmHg. Both figures are according to ChemSpider.

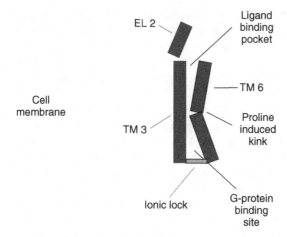

Figure 2.31   Simplified cross section of GPCR structure.

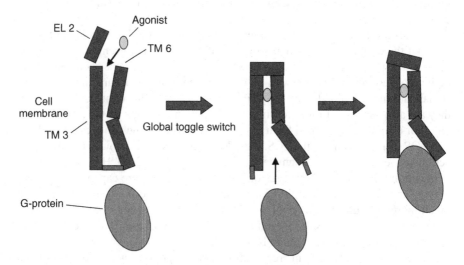

Figure 2.32   Agonist recognition and G-protein binding.

them. The proline-induced kink in TM6 creates two pockets. The LBP is located in the upper one at the top of the TM section, and the G-protein binding site is at the bottom, just above and around the ionic lock.

Figure 2.32 shows the ligand (an odorant molecule in the case of olfaction) entering the LBP. As it does so, the tops of TMs 3 and 6 are pulled together to bind to it and EL2 closes over the top of the LBP to complete the binding. This brings the global toggle switch into operation, breaking the ionic lock and widening the gap between the bottom ends of TMs 3 and 6 and moving TM3 slightly upwards towards the extracellular face. Because of the geometry of TM6, the movement of the bottom end is magnified relative to that at the top. The effect is to open the

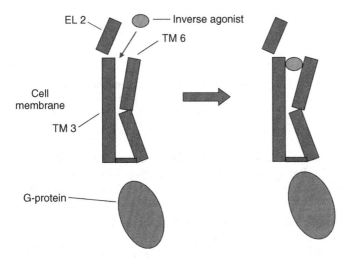

Figure 2.33 Binding of an inverse agonist.

cleft on the intracellular face of the receptor and to allow the G-protein to bind in the gap.

Figure 2.33 shows the binding of an inverse agonist. In this case, the ligand binds into the binding pocket and stabilises the inactive state of the receptor. The ionic lock remains in place and the G-protein cannot bind since the binding site for it remains too small and is obscured by the ionic lock.

The determination of the structure of the chemokine receptor CXCR1 by Park et al. using ^{15}N- and ^{13}C NMR mentioned earlier also served to show how the entire protein structure (extracellular and intracellular loops and TM section) work together with the agonist (interleukin-8), the phospholipids of the bilayer and the G-protein, in order to activate the G-protein (229).

Multiple Binding Sites

The above descriptions of GPCRs focus on one agonist binding site within the LBP. However, there is evidence that multiple binding sites might exist and this could, of course, explain some of the breadth of tuning of various ORs. Kahsai et al. used MS techniques, labelling free cysteine and lysine residues with *N*-ethylmaleiimide (**2.76**) or succinic anhydride (**2.77**) (with and without deuteration), to study the conformational effects of ligand binding to the β_2-adrenergic receptor (Figure 2.34). They investigated the effect of binding of agonists, partial agonists and inverse agonists on the conformation of the protein through their effects on the accessibility of the cysteine and lysine residues and concluded that there is conformational variability in the way that different ligands bind (263).

DEET (*N*,*N*-diethyl-*m*-toluamide) (**2.78**) is known to 'repel' a variety of insects including fruit flies and mosquitoes. It was thought that it might act by blocking

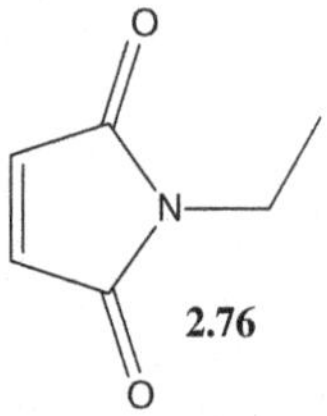

Figure 2.34   Two compounds used to study conformational effects of ligand binding to olfactory receptors.

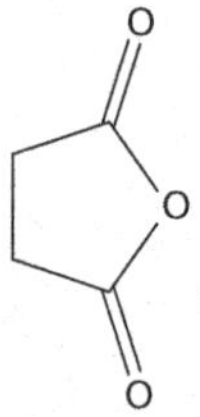

detection of odorant molecules (264, 265) or by being detected and eliciting an avoidance reaction (266). Pellegrino et al. found that it was more complex than either of these hypotheses (267). As described earlier, insect ORs function as heterodimers, with one receptor serving to recognise the odorant and the second acting as an ion channel, avoiding the need for a second messenger cascade as in mammalian olfaction. The ion channel was originally named OR83B but is now called *ORCO* and is common to all insect olfactory heterodimers. Since the action of DEET (**2.78**) was found to be dependent on specific odorant receptor couples, it was clear that it does not act on ORCO but rather on the receptor that detects the odorant. In some cases, DEET (**2.78**) was found to inhibit activation of the receptor by its ligand. Oct-1-ene-3-ol (**2.79**) activates OR59B in *D. melanogaster* when present at high concentrations, but at low concentrations it has the opposite effect and reduces the basal level of receptor firing (Figure 2.35). It was found that DEET (**2.78**) prevented this inhibition of the basal firing rate. In one strain of *D. melanogaster* (called *Boa esperança*), this latter effect of DEET (**2.78**) does not occur. Pellegrino et al. postulated that there are two binding sites in OR59B, one which has a high affinity for oct-1-ene-3-ol (**2.79**) and therefore responds to very low concentrations of it, and one which responds to higher concentrations. The first stabilises the inactive state of the receptor, and the second stabilises the active state. Therefore, it would seem that DEET (**2.79**) binds to the former. Study of the amino acid sequence of OR59B from the *B. esperança* strain showed that it contained a single nucleotide polymorphism (SNP) (V191A) which occurs at the top of TM2. This alanine-for-valine substitution was shown to prevent DEET (**2.78**) from disrupting the oct-1-ene-3-ol (**2.79**) activated inhibition of basal firing and so provides support for the concept of a second binding site.

It remains to be seen whether either of these findings regarding multiple binding sites will have a parallel in human ORs. However, the fact that the SNP lies at the top of TM2 would be consistent with the minor binding pocket first proposed by Pilpel and Lancet as discussed above.

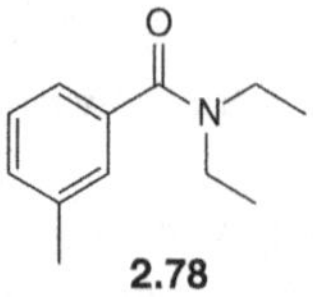

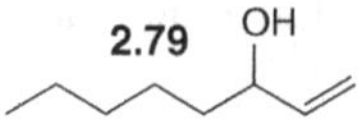

Figure 2.35   Confusing insect receptors.

Olfactory Receptor Modelling

Katada et al. have published a review of techniques used for OR modelling and also provide interesting comparisons between several ORs and some other class A GPCRs (268). In all cases, the binding site is predicted to be towards the cytosolic face of the TM bundle and between TMs 3 and 7, with TMs 3 and 5 being prominent in forming interactions with the respective ligands. This is very much in keeping with the findings from homology modelling, such as those of Pilpel and Lancet (178). Vaidehi has also reviewed the computational techniques used to study GPCR conformations (269).

Molecular modelling of receptor proteins is a useful tool, but it must be remembered that the models are only models and will represent reality only to some degrees. In order to support hypotheses based on models, many research groups test their predictions by comparison with actual binding studies and, in some cases, also with site-directed mutagenesis. In this latter technique, an amino acid which the model suggests to be key to ligand binding is replaced by one that would not have the same properties. Comparison of the binding affinity of the synthetic protein containing the modified structure with that of the native protein will then provide experimental evidence to confirm the prediction from the model.

In a survey of the modelling community (270), 29 groups submitted 206 models of the $A_{2A}AR$, and their results were compared with those from the X-ray crystal structure (207). It was found that there was some variability in the accuracy of prediction of the models. They predicted 4 ± 7 correct contacts between the ligand and the receptor, and the accuracy of the structure was 9.5 ± 3.8 Å

Most OR models are based on the crystal structure of rhodopsin. There are a number of assumptions made when building such models. For example, it is assumed that the tertiary structure adopted by rhodopsin in the crystalline form is similar to that it adopts when in the cell membrane. Recent studies using NMR and IR spectroscopy on receptors in membranes, as described above, suggest that this is a reasonable assumption. The assumption that OR proteins adopt a similar tertiary structure to that of rhodopsin is also reasonable in view of the high degree of homology across all class A GPCRs. Similarly, the assumption that ligand docking in ORs is similar to co-factor docking in rhodopsin is justified by the enormous weight of evidence typified by the examples discussed above.

One of the first exercises of this type was the work of Singer and Shepherd on the rat receptor OR5 which is known to respond to Lyral® (**2.35**) (271) (Figure 2.36). They proposed a model for the binding site of Lyral® (**2.35**) which involved TM helices 3–7. However, a later model identified a somewhat different site using only helices 3, 4 and 5 (272).

In an evaluation of the scope and limitations of protein modelling for design of new odorants, Bajgrowicz and Broger also considered OR5 and investigated the binding of Lilial (**2.80**) to it (273). (Lilial (**2.80**) and Lyral (**2.35**) both elicit similar muguet odour percepts in humans.) They developed two models. The first proposed a binding site consisting of 11 residues in helices 3, 4, 6, and 7. Threonine-152 in helix 4 forms a H-bond with the aldehyde group. The ligand's phenyl ring is stacked

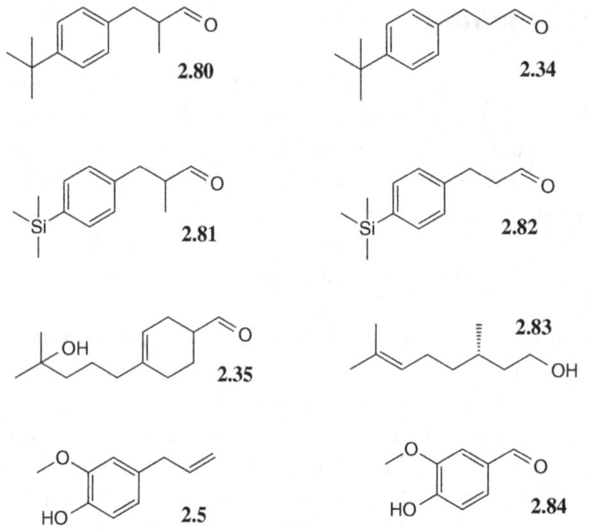

Figure 2.36 Some odorants used in OR model studies.

between phenyl rings of phenylalanine-104 (helix 3) and tyrosine-252 (helix 6), and the *t*-butyl group finds a match to residues in helices 3, 6 and 7. In the second model, the orientation of the ligand is inverted, with the aldehyde function forming a hydrogen bond to residues in helix 7.

Doszczak et al. prepared the sila derivatives of Lilial (**2.80**) and Bourgeonal (**2.34**), that is, (**2.81**) and (**2.82**), respectively, in which the tertiary carbon atom was replaced by silicon. They then predicted the binding affinities of these to OR1D2 based on a rhodopsin homology model and compared them with the experimental affinities of the receptor expressed in HEK cells. There was good agreement between the model and the experimental results (274).

Schmiedeberg et al. postulated that certain amino acid positions that interact with specific odorants would be conserved in orthologs but not in paralogs, and therefore investigated the binding properties of two paralogous human receptors, OR1A1 and OR1A2, and the mouse ortholog, Olfr43, of OR1A1, using protein models based on rhodopsin homology supported by site-directed mutagenesis and heterologous expression (275). They found that two conserved amino acid positions in the orthologs OR1A1 and Olfr43 were necessary for their sensitivity to (S)-citronellol (**2.83**) and that changes at these positions were responsible for the different sensitivity of the paralog OR1A2 to (S)-citronellol (**2.83**). The predictions from the models of the binding site were borne out by the experimental results on the native receptors and the mutations derived from them.

Katada et al. studied the responsiveness of the eugenol-sensitive mouse receptor mOREG using 22 shikimate derivatives related to eugenol (**2.5**) or vanillin (**2.84**) (276). This receptor also has a broad but selective range. The oxygen of the hydroxy group of eugenol (**2.5**) (and the corresponding oxygen atom of the other agonists) is hydrogen-bonded to serine-113 and eight other amino acid residues form the general shape of the binding pocket. They showed that recognition occurs through electrostatic (hydrogen-bonding), van der Waals and hydrophobic interactions, and their results from modelling were confirmed by *in vivo* testing of the receptors.

Lai et al. built a dynamic model of the rat olfactory receptor ORI7 (277). They incorporated 10 potential aldehydic ligands into the binding site and then set the whole assembly into normal motion. Some of the test materials remained in the binding site whilst others migrated out. Correlation with *in vivo* results was 100%. Those molecules that the model predicted would remain in the binding site were found to be agonists, whilst those that migrated out failed to activate the receptor *in vivo*. Moreover, the model elucidated the route into the binding site from the extracellular side of the protein. Similar to the results of Katada et al. on the binding of eugenol (**2.5**) to mOREG (276), they found that the oxygen atom of the agonists was tethered by electrostatic forces, in this instance to lysine-164, and the steric fit between the residues of the binding pocket and the agonist determined its stability in the binding site.

Gelis et al. used a model in which both the ligand and the amino acid residues of the binding pocket move to accommodate others and find the best fit (278). They used their dynamic homology model to predict binding and confirmed their findings by site-directed mutagenesis. They thus established that the human receptor OR2AG1 responds strongly to amyl butyrate (**2.85**), phenethyl acetate (**2.86**) and phenoxyethyl isobutyrate (**2.87**) and more weakly to isopentyl acetate (**2.88**), isoamyl benzoate (**2.89**) and prenyl acetate (**2.90**) (Figure 2.37).

Another example of dynamic simulation of ligand/GPCR interaction is that of Dror et al., who explored the binding of an agonist, isoproterenol (**2.91**), and three antagonists, alprenolol (**2.56**), dihydroalprenolol (**2.92**) and propranolol (**2.36**), to the β-1 and β_2-adrenergic receptors (142) (Figure 2.38). Their model suggests that the ligand first makes contact with the receptor via sites on the ECS which they have dubbed 'vestibule sites' and then follows a defined path down into the LBP. The ligands were initially placed at least 30 Å from the LBP and hence about 12 Å from the outer edge of the ECS. They were then allowed to move freely and without interference from the computer operator. The ligand initially docked into one of the vestibule sites and then followed a path down into the binding pocket. In most

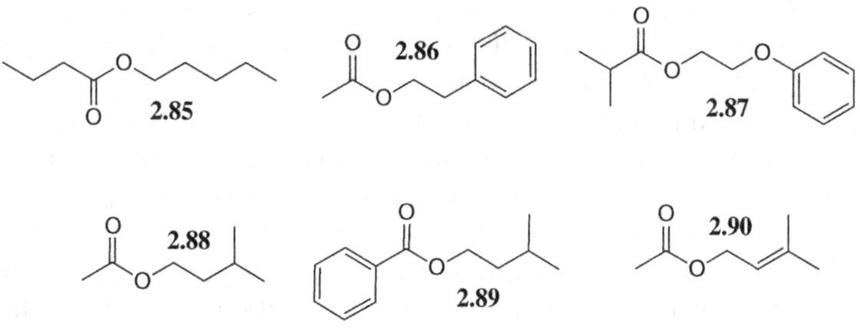

Figure 2.37   Agonists of the human olfactory receptor OR2AG1.

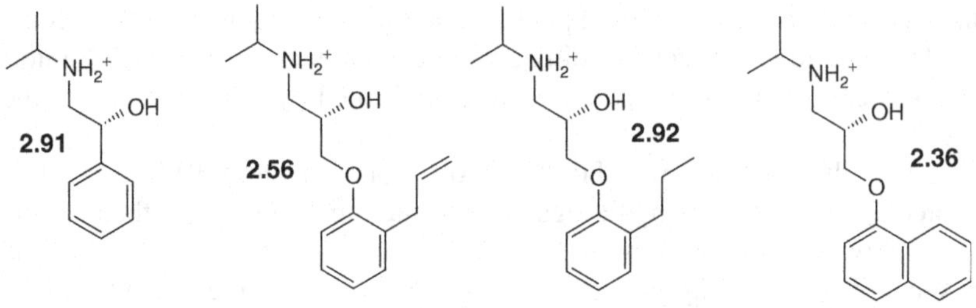

Figure 2.38   Some agonists and antagonists of the adrenergic receptors.

simulations, the ligand followed the same path and, once past the rim of the ECS, the ligand was much more likely to penetrate into the binding pocket than to return to the extracellular medium. In order to allow the ligand to pass through the narrow gap at the top of the binding pocket, the receptor must be deformed, thus widening the gap. On entering the binding pocket, alprenolol (**2.56**) first formed a salt bridge between its protonated amine function and the carboxylate anion of Asp-113. In most cases, the crystallographic configuration (pose) was adopted immediately but sometimes an alternative pose occurred initially, though this relaxed rapidly into the crystallographic pose, in which the hydroxyl group of alprenonolol (**2.56**) forms a hydrogen bond to Asn-293. The final configuration was found to replicate that of the crystal structure to a precision of <1 Å. This is perhaps not too surprising since basic receptor models are based on crystal structures. Once bound in the pocket, the ligands tended to remain there and the rate of egress was low. The calculated rate of binding was similar to the experimental value. The agonist isoproterenol (**2.91**) entered the binding pocket by the same route as alprenolol (**2.56**) and formed the same salt bridge. However, it was bound much more loosely, probably because the receptor model was based on the inactive (i.e. antagonist bound) state.

Calculations on the energetics of the process revealed an interesting observation. There was an expected energy barrier to passage of the ligand through the narrow gap at the top of the binding pocket, but there was a greater one on initial binding to the vestibule site. This was largely due to removal of water molecules solvating the ligand and lining the binding site. As alprenolol (**2.56**) enters the vestibule, about 15 water molecules are forced out of it and 500 Å^2 of hydrophobic surface is buried. The relevance of this to ORs is uncertain since odorants are hydrophobic, unlike the ionic ligands in this case. However, algorithms, such as that of Abraham et al. (279), predicting odour threshold do tend to indicate that hydrophobicity is a determining parameter. This could be due to transport to the receptor but would also be consistent with the above findings on energy of binding.

The team of Goddard uses two proprietary programs for modelling GPCRs in general as well as ORs. The first is the MembStruk program which builds the protein structure mostly from the primary sequence and relies much less on the crystal structure of rhodopsin than do other modelling techniques. The complementary program HierDock predicts the binding site for a ligand. The results from this approach are generally in fairly good basic agreement with those of rhodopsin-based models.

Vaidehi et al. have used MembStruk and HierDock software to show that there is consistency between GPCRs of different types (mouse and rat I7 ORs, the human sweet receptor, endothelial differential gene 6 and the β-adrenergic receptor) (280). They also showed that their modelling techniques work across this range of receptors and used them to predict both the tertiary structure and binding site of rhodopsin with a reasonable degree of accuracy.

Floriano et al. investigated the mouse receptor ORS25 (281). Using the Hier-Dock software, they identified binding sites and calculated the binding energies for 24 potential agonists. The binding site was found to involve 10 amino acids from TM helices III–VI, and the energies indicated that hexanol (**2.93**) and heptanol (**2.94**) should bind most strongly (Figure 2.39). Experiments with the receptor in cells showed that, indeed, of the 24 test materials, only these two alcohols elicited a response. They then went on to screen a further six mouse receptors (S6, S18, S19, S25, S46 and S50) against the same 24 odorants (C4–C9 alcohols, acids, bromoacids and dicarboxylic acids) (282). As before, they found good agreement between the HierDock-calculated binding energies and experimental receptor activation. They confirmed that the crucial TM helices are 3–6 and that extracellular loops 2 and 3 also contribute to binding. Six of the amino acid positions are key to binding and, in the examples studied, the selectivity of the receptors were determined largely by two of these. This latter result is also supported by mutation data. Extrapolating these results to all 869 ORs in the murine genome, they suggested that 34 are involved in perception of acids and 36 for alcohols. The original six under study were also found to respond to aldehydes and esters. They have also screened 56 odorants against mouse and rat OR I7 and again found good agreement between the predictions from their model and *in vivo* results (283).

The binding threshold hypothesis (BTH) postulates that a ligand should bind with a binding energy of more than a certain value in order to activate the receptor and the resultant signal cascade. In order to test this, Hummel et al. built models of the mouse and human orthologs of OR 912-93. The model suggested that ketone ligands formed a hydrogen bond with Ser-105 of the mouse ortholog. The human ortholog has 66% homology with the mouse receptor but one difference is that, in the human version, position 105 is occupied by glycine rather than serine and thus such a hydrogen bond is not possible. It was found that ketones binding to the mOR912-93 model with a predicted binding energy of 26 kcal/mol or more did activate the receptor experimentally and that no ketones activated hOR912-93 (284).

There is a consistent feature in all of these models, both rhodopsin-based and those derived from the primary structure, in that, for a good odorant/receptor fit, each requires a polar group in the odorant which can form a hydrogen bond or similar electrostatic interaction with a donor site in the receptor and the rest of the fit is determined by a spatial match with the shape of the (largely hydrophobic) binding pocket. Saturated hydrocarbons presumably lack the polar interaction and, in some cases at least, it would seem that there are also weaker non-bonded interactions, such as π-stacking, in the hydrophobic pocket.

2.93   2.94   **Figure 2.39**   Agonists of mORS25.

Most modelling studies primarily involve the LBP. However, some do consider the extracellular and intracellular loops. An example, in addition to that cited above, is the work of Goldfeld et al., who have developed a technique which they named the Protein Local Optimization Program (PLOP) and which they have used to model the extracellular loops of various GPCRs. Their results compare favourably with the corresponding X-ray crystal structures (285).

Use of Receptor Models in Ligand Design

Clearly, one objective of building receptor models is to aid in the design of new ligands and hence biologically active materials. In the case of olfaction, the biological activity is the ability to elicit an odour percept. Defining the LBP is equivalent to defining a pharmacophore for the receptor. Such pharmacophores can be used as inspiration for synthetic targets or as a means of screening candidates *in silico*, thus eliminating probable failures and reducing the number of materials to be prepared in reality. To date, the examples of this approach are from the pharmaceutical industry but it is only a matter of time before it is extended to fragrance applications.

Bhattacharya and Vaidehi used X-ray crystal structures to model binding conformations of full agonists, partial agonists and inverse agonists and to calculate energy profiles of moving the β_2-adrenergic receptor from inactive to active configurations (286). They were able to check their predictions against experimental results and thus support their model. They can therefore use the model for virtual screening to find new agonists and inverse agonists. Klabunde et al. described a method that they used to build pharmacophore models of the LBP of GPCRs based on analysis of the sequences of 270 receptor proteins (287). They were then able to use their pharmacophores for virtual screening of potential ligands. Phatak et al. developed ligand-steered homology models to enable virtual screening (288). Basically, this involves taking a crude 3D binding pocket model based on known X-ray structures and docking ligands into it, thus refining the binding site model. The refined model can then be used for virtual screening. Sanders et al. have shown that it is even possible to build a model of the receptor (they used the β_2-adrenergic receptor for their work) from its primary sequence and rhodopsin homology, and then, from the parameters of the binding pocket, predict ligands that will serve as agonists or inverse agonists (289).

Receptive Ranges of Olfactory Receptors

There are two main techniques that are used to explore the receptive ranges of ORs. Since each OSN expresses only one type of OR, it is possible to measure the response of the neuron and thus determine the range of odorants to which its receptor responds. The other approach is to clone the receptors into heterologous cells. These techniques have been reviewed by Veithen et al. (290) and by Reisert and Restrepo (291).

As always, care is needed when comparing results of research across different species. Obviously, the closer the species are, the more likely the findings on one will apply to the other. Thus research on other mammalian receptors is likely to be of more relevance to humans than those from studies on fish. The largest divide is between vertebrates and insects. As discussed earlier in this chapter, there are many differences between insect and human olfaction, and comparisons between receptive ranges of insect receptors and those of humans are not necessarily meaningful. For example, detection of 11-*cis*-vaccenyl acetate (**2.3**) by *D. melanogaster* requires both the olfactory receptor OR67d and the OBP LUSH since it is the complex between the two that is detected rather than the odorant *per se* (18, 19). In fact, as also mentioned previously, modification of the extracellular tail of OR67d to make it resemble the 11-*cis*-vaccenyl acetate/LUSH complex causes activation of the receptor in the absence of 11-*cis*-vaccenyl acetate (**2.3**). Thus, the receptor in effect activates itself (20). In contrast, comparison of the receptive ranges of mammalian ORs in their natural environment with those of the receptors when heterologously expressed in cell cultures shows that the two are similar and therefore unaffected by the presence or absence of OBPs.

Use of Neurons and Bulb Maps to Determine Receptive Range of Olfactory Receptors

Olfactory neurons can be separated out from olfactory epithelial tissue, and their responses can be measured using, for example, a dye that responds to increasing calcium levels in the cell when its receptor is activated. In this way, Araneda et al. identified an array of 59 aldehyde-responding receptors and showed how the pattern of activation of them can be used to discriminate between nine different aldehydes (292).

Another example of this approach was reported by Bieri et al., who looked for receptors responding to sandalwood odorants (293). They investigated the response of nearly 8000 rat neurons to sandalwood oil, the major odiferous components of which are α- and β-santalol ((**2.95**) and (**2.96**), respectively), Javanol (**2.97**), Radjanol (**2.98**), Sandalore (**2.99**) and Ebanol (**2.100**) (Figure 2.40). They also included 5α-androst-16-en-3α-ol (**2.101**) and octanal (**2.72**) in the study: the former because it is often reported to elicit a sandalwood odour and the latter as a blank since both its structure and the odour it elicits are quite different from those of sandalwood chemicals. They found very few neurons that responded to sandalwood compounds, although the percentage of neurons responding to octanal (**2.72**) was the same as that reported by Araneda et al. (292). None of the receptors that responded to the sandalwood compounds responded to either 5α-androst-16-en-3α-ol (**2.101**) or octanal (**2.72**). The responses to the sandalwood compounds formed a pattern as shown in Figure 2.41. This is as would be expected based on the combinatorial nature of olfaction, although the lack of receptors responding to Sandalore (**2.99**) is rather puzzling. It must also be remembered that the odour descriptions were supplied by humans, and no behavioural tests were carried out to investigate the perception of the compounds by rats.

Results such as these are valuable in that they show how the different sandalwood chemicals are recognised by different combinations of receptors. However,

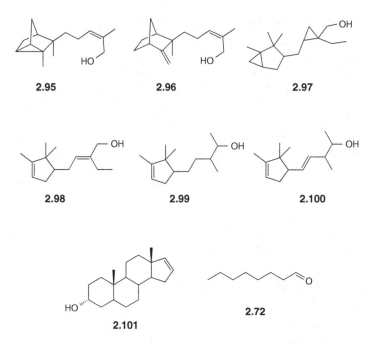

Figure 2.40 Odorants tested on sandalwood receptors by Bieri et al.

Neuron		1	2	3	4	5	6	7	8	9	10	11	12	13	14	15	16
Sandalwood oil																	
Javanol																	
Radjanol																	
Ebanol																	

Figure 2.41 Response of 16 neurons to 4 different sandalwood odorants.

they do not tell us which receptors are involved. For example, in Figure 2.41 it is possible that neurons 1, 2, 3, 5, 6 and 9 all express the same receptor, or it could be that they express different receptors with the same receptive ranges. Thanks to recent advances in molecular biology, techniques have now been developed to overcome this problem, as will be seen later.

The largest study of this type was reported by Nara et al. (294). They tested 3000 mouse neurons with 125 odorants. They first tested mixtures of odorants. These mixtures were grouped according to their chemical functional groups (such as alcohols or esters) and contained between 3 and 14 odorants each. The exception was a group of musks which comprised three macrocylic esters, two macrocyclic ketones and three nitroaromatics. The authors did point out that the odour descriptions relate to human perception and that we cannot tell what percepts are elicited in mice. Of the 3000 neurons, 217 (7%) responded to one or more of the mixtures. Of those neurons responding to the mixes, 197 were then tested with the components of the mixture to which they responded. Of these neurons, 28 failed to respond to any of the components of the corresponding mixture. A possible explanation that the authors offerred for this is as follows. Since each OSN contains many ORs, it is possible that activation by a number of ligands might be too weak to give individual responses but the cumulative effect of a mixture would result in a signal. Some of the 169 neurons that did respond to individual odorants were found to be broadly tuned and responded to a wide variety of odorants. The majority were more narrowly tuned and responded to groups of odorants with related structures. Some neurons responded to only one of the library of 125 odorants. Aldehydes tended to stimulate a wide range of neurons whereas musks stimulated few. They found that 80% of the neurons that responded to murine pheromones also responded to other odorants. This last observation is consistent with the findings of Kobayakawa et al. that some murine neurons respond to an odorant and elicit an innate response whereas others respond to the same odorant but do not elicit the innate response (143).

One method to enable identification of which receptor is expressed in a neuron is to clone a dye, such as the green fluorescent protein (GFP), into the cells expressing a specific OR. The neurons containing the dye, and hence also containing the receptor under investigation, can then be identified under a microscope and either the calcium response or the electrical signals generated by them can be measured. In this way, Touhara et al. found that the murine receptor mOR23 responded to Lyral (**2.35**) (165). In fact, the tuning appears to be quite narrow as five odorants, namely myrac aldehyde (**2.102**), hydroxycitronellal (**2.103**), hydroxycitronellal dimethyl acetal (**2.104**), dihydromyrcenol (**2.105**) and tetrahydromyrcenol (**2.106**), related to Lyral (**2.35**) in structure and (to human) odour all failed to activate it (Figure 2.42).

Araneda et al. tested 90 odorants for activity with the rat receptor I7 (295). They used an adenovirus to increase the level of I7 in the epithelium and to

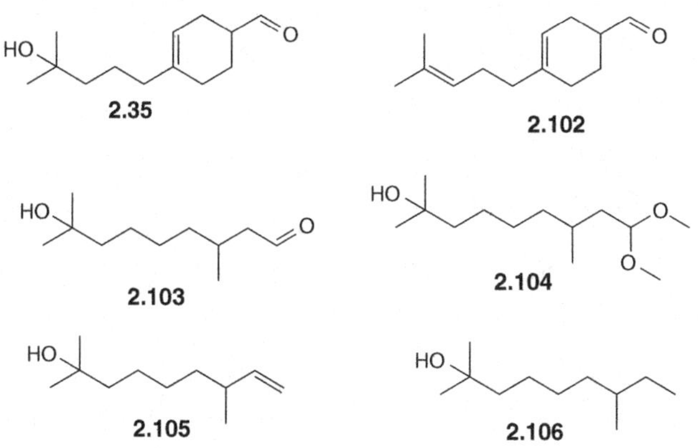

Figure 2.42   Tuning of mOR23.

simultaneously introduce GFP as a way of detecting activation. I7 was found to be very selective towards aldehydes and with a preference for eight carbon atoms in the chain. Both saturated and olefinic aldehydes were found to be active, and there was some tolerance for substituents, mostly methyl groups, on the chain. The strongest agonist was found to be (*E,E*)-2,4-octadienal (**2.107**). Other strong agonists were dihydrocitronellal (**2.108**), octanal (**2.72**), citronellal (**2.109**), 7-methyloctanal (**2.110**), (*E*)-2-octenal (**2.111**) and 3-(4′-methylcyclohex-1-yl)butanal (**2.112**). 2-Octynal (**2.113**) was a weak agonist, citral (**2.114**) very weak and 2,5,7-trimethyloct-2-enal (**2.115**) was inactive (Figure 2.43). With this knowledge and using models, they were able to define more precisely the steric constraints on the binding site.

Krautwurst et al. screened 26 odorants against 80 chimeric ORs and identified three receptors that responded to (−)-carvone (**2.116**), citronellal (**2.109**), and (+)-limonene (**2.117**), respectively. They also found that a single valine-to-isoleucine substitution caused the mouse I7 receptor to favour heptanal (**2.118**) rather than octanal (**2.72**) (296). One is tempted to speculate that this one

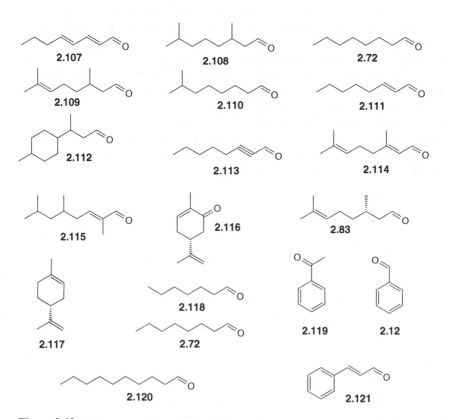

Figure 2.43   Odorants tested on mORI7 and other murine receptors.

carbon enlargement of the receptor causes a corresponding one carbon reduction in the volume of the hydrophobic part of the LBP.

Bozza et al. altered the mouse M71 gene locus to also express GFP, rather than using viral infection to introduce genes (297). They found acetophenone (**2.119**) (EC_{50} 20 µM) (an important semiochemical for mice) and benzaldehyde (**2.12**) (EC_{50} 100 µM) to be agonists of mouse M71 and for I7 and confirmed octanal (**2.72**) as an agonist (in agreement with Araneda et al.) but found decanal (**2.120**) to be weak and cinnamaldehyde (**2.121**) and citral (**2.114**) to be agonists which results are in contrast to those of Araneda et al. In all cases, the DR curve was steep, saturation occurring within 2 log units of concentration.

Grosmaitre et al. found that the receptors in the SO of mice tended to be more broadly tuned than those in the MOE (298). For example, hexanoic acid (**2.122**) elicited a response from almost 90% of the SO sensory neurons and limonene (**2.117**) 50% of them. They decided to investigate one of these in order to see whether the breadth of tuning is due to the receptor or to some other feature of the SO. For their study, they selected mSR1, which is found in both the SO and the MOE, and tested it with five odorants, namely (+)-camphor (**2.123**), amyl acetate (**2.124**), benzaldehyde (**2.12**), octanoic acid (**2.125**) and heptanal (**2.118**). The receptor responded to all of them, indicating that it is indeed broadly tuned and responds to odorants with quite different structures and functional groups (Figure 2.44).

Since all OSNs that express the same receptor converge on the same glomerulus in the OB, it is also possible to use patterns of bulbar activation to investigate the receptive ranges of ORs. Mori et al. (299) screened 71 compounds against 73 OSNs of the rat and found a pattern indicating some broadly tuned and some narrowly tuned receptors. Most odorants activated more than one receptor, and most receptors responded to more than one odorant. The expression pattern of the different receptors varied across different areas of the OE, and receptors with sensitivity to similar types of compounds tended to be located in similar regions of the epithelium implying a chemotopic mapping of the epithelium. They found that the mapping of the epithelium transposed onto the bulb and so that also contains a chemotopic map.

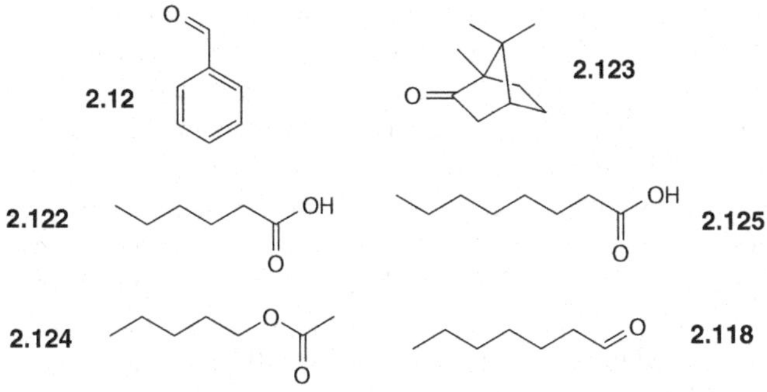

Figure 2.44   Agonists of mSR1.

2.116   **2.126**

(–)-carvone   (+)-carvone   **Figure 2.45**   Carvone enantiomers.

One example of their findings is that overlapping, but not identical, sets of glomeruli (and hence receptors) are responsible for detecting the two carvone enantiomers (**2.116**) and (**2.126**) (Figure 2.45) in rats, and thus their findings are compatible with the bulbar activation patterns for the carvone enantiomers that can be seen on the glomerular response archive of Johnson and Leon (300).

Another Japanese group also investigated the murine neurons responding to carvone (301). As with Mori et al., they found that overlapping sets of neurons responded to each of the enantiomers but in this case they also investigated the effects of varying the concentration of the agonists. They found that, at low concentrations, some neurons responded to only one enantiomer whilst others responded to both. At higher concentrations, the distinctions began to disappear and more neurons responded to both enantiomers. They also investigated the response of the carvone-responding neurones to other odorants which, to humans, smell rather like one or other carvone enantiomer. Not surprisingly, the closer to carvone an odorant was in structure, the more likely it was to activate a carvone-responding receptor.

Use of Heterologous Expression to Determine Receptive Range of Olfactory Receptors

The other main approach to identifying receptive ranges of ORs is to clone the receptors into heterologous cells such as HEK cells, human HeLa cells, Hana cells, or *Xenopus* oocytes (frog eggs) in a culture medium and then use other cloned proteins to detect changes in the cell chemistry which are indicative of the receptor having been activated. Activation of the receptor causes the second messenger train to be set in motion, as explained in more detail elsewhere, and the first step involves activation of the G-protein. This causes adenylyl cyclase to begin converting adenosine triphosphate (ATP) to cAMP, and this in turn results in an increase in the Ca^{2+} concentration in the cell. Any system that measures cAMP or Ca^{2+} can then be used as a measure of activation of the receptor and is referred to as the *reporter system*. The Ca^{2+} increase can be detected using a dye which fluoresces in the presence of calcium, and cAMP can be detected, for example, by using firefly luciferase. This enzyme also has to be introduced into the cell and normally will exist in an inactive state. It is activated by cAMP and then degrades luciferin, producing light in the process. These, and other similar mechanisms, can be used to give a quantitative

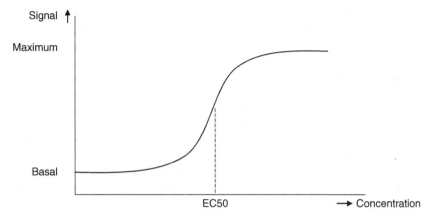

Figure 2.46 Typical dose/response curve.

measure of the degree of activation of the receptor and so, by using different concentrations of the odorant, a dose/response curve can be obtained. Normally, these are standard sigmoidal curves as shown in Figure 2.46. The concentration of the odorant is shown on the x-axis, and the response of whatever detection system is being used on the y-axis. There is always a basal level of activation, and the efficacy of the ligand is measured as the increase over that basal level. The concentration of the odorant that gives rise to half of the maximum activation is known as the *excitatory concentration* at 50% maximum response or EC_{50}. EC_{50} is the usual measure of potency of the ligand. The efficacy and potency are not related, and it is quite possible to have a very potent ligand with poor efficacy, or vice versa.

Levasseur et al. found that they obtained bell-shaped curves with heptanal (**2.118**), octanal (**2.72**) and nonanal (**2.127**) on rat I7 and with Helional (**2.128**) on OR17-40 (302) (Figure 2.47). There are many possible explanations for bell-shaped curves, and they could just be artefacts of the reporter system. The loss of signal at higher concentration could be due to inhibition of something in cell metabolism or even to poisoning of the cells. In their experiments, Levasseur et al. found that the peak of the bell was different for the three aldehydes. So they concluded that the

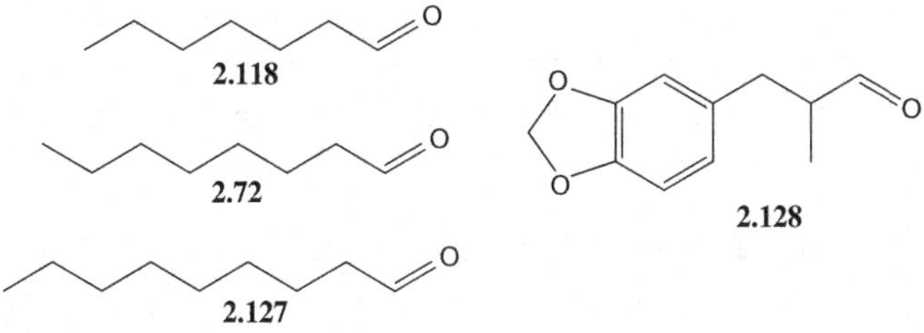

Figure 2.47 Compounds giving bell shaped curves on rORI7 or OR17-40.

phenomenon is part of the normal OR behaviour and is involved in discrimination. However, the degree of interference with the reporter system would also affect the position of the peak of the bell, and so it is still possible that the effect is merely an artefact of the reporter system.

It is also possible that small molecules such as odorants can affect the reporter system in such a way as to increase the observed signal. For example, this has been observed in the luciferase system (303). Thus, because of these possible effects of odorants elsewhere in the transduction and reporter systems, it is important to carry out sufficient blank and control experiments to ensure that the observed activity is indeed due to the odorant/OR interaction ansd nothing else.

In the OSNs of the nose, the receptor proteins are, of course, synthesised in the endoplasmic reticulum and therefore must be transported to the cell membrane and folded correctly in order to detect odorants in the olfactory mucus. Matsunami and his co-workers identified a family of accessory proteins called *receptor transporting proteins* (*RTPs*) which are involved in the process of trafficking and folding of the receptor proteins (304, 305). These proteins, RTP1 and RTP2, are only expressed in OSNs. OR proteins are notoriously difficult to functionally express in heterologous mammalian cells, and trafficking is one of the problems. It was therefore a major advance in the field when Matsunami's team showed that the RTPs improved the trafficking of ORs when co-expressed with them in cells such as HEK cells. The RTPs are also termed *chaperone proteins*. Another tecnnique which aids in functional expression of ORs in heterologous cells is the use of tags. These are fragments of other proteins that are added to the amino terminus. For example, a fragment of the rhodopsin sequence (Rho tag) often assists in expression of an OR at the cell surface, as demonstrated by Fujita et al. (306).

In a number of instances, early research on a receptor identified one ligand and so that receptor became associated with that ligand but subsequent research found many more ligands, possibly quite unrelated to the first. This is not actually surprising. If one investigates a limited number of receptors using a small library of compounds, the most likely outcome is that a broadly tuned receptor will be identified by only one or two compounds that are capable of activating it and so it might easily be considered to be narrowly tuned. In order to get a full picture of the olfactory map, it is necessary to screen thousands of odorants against hundreds of ORs. Therefore, only those equipped to do high-throughput screening at such a level will get a proper concept of the map.

An early example of the use of heterologously expressed receptors to investigate the receptive range of an OR was that of OR17-40 in both *Xenopus laevis* oocytes and HEK 293 cells by Wetzel et al. (307). They found OR17-40 to respond to Helional (**2.128**) and heliotropylacetone (**2.129**). In a later study, Jacquier et al. explored structures around the original lead and found that the tuning is relatively narrow and selective (308). Their findings are shown in Figure 2.48.

Kajiya et al. studied two selected murine ORs: mOREG and mOREV. These were chosen because they had been identified, from studies using neurons, to respond to eugenol (**2.5**) and ethylvanillin (**2.148**), respectively (309). Their results confirmed a number of important facts. First, they showed that the receptive range

Agonists

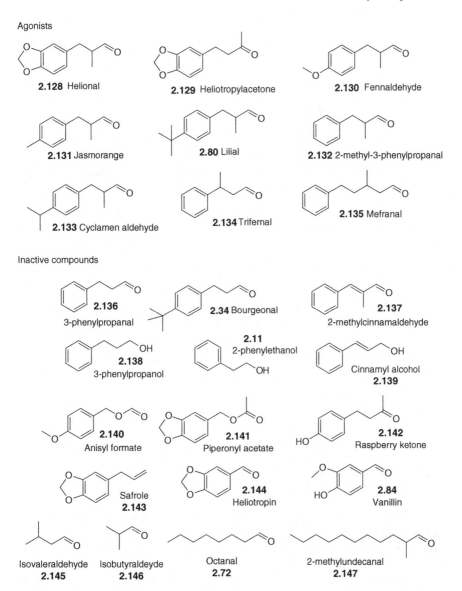

2.128 Helional

2.129 Heliotropylacetone

2.130 Fennaldehyde

2.131 Jasmorange

2.80 Lilial

2.132 2-methyl-3-phenylpropanal

2.133 Cyclamen aldehyde

2.134 Trifernal

2.135 Mefranal

Inactive compounds

2.136
3-phenylpropanal

2.34 Bourgeonal

2.137
2-methylcinnamaldehyde

2.138
3-phenylpropanol

2.11
2-phenylethanol

Cinnamyl alcohol
2.139

2.140
Anisyl formate

2.141
Piperonyl acetate

2.142
Raspberry ketone

Safrole
2.143

2.144
Heliotropin

2.84
Vanillin

Isovaleraldehyde
2.145

Isobutyraldeyde
2.146

Octanal
2.72

2-methylundecanal
2.147

Figure 2.48   Receptive range of OR17-40.

of the receptors was the same whether in OSNs or expressed heterologously. They showed that similar results were obtained using calcium imaging or detection of cAMP. These findings validate the use of the various assays in studying the receptive range. They showed that the two receptors have a breadth of tuning but are selective, and the overlapping of their ranges supports the combinatorial mechanism of olfaction. Odorants were found to be recognised by different

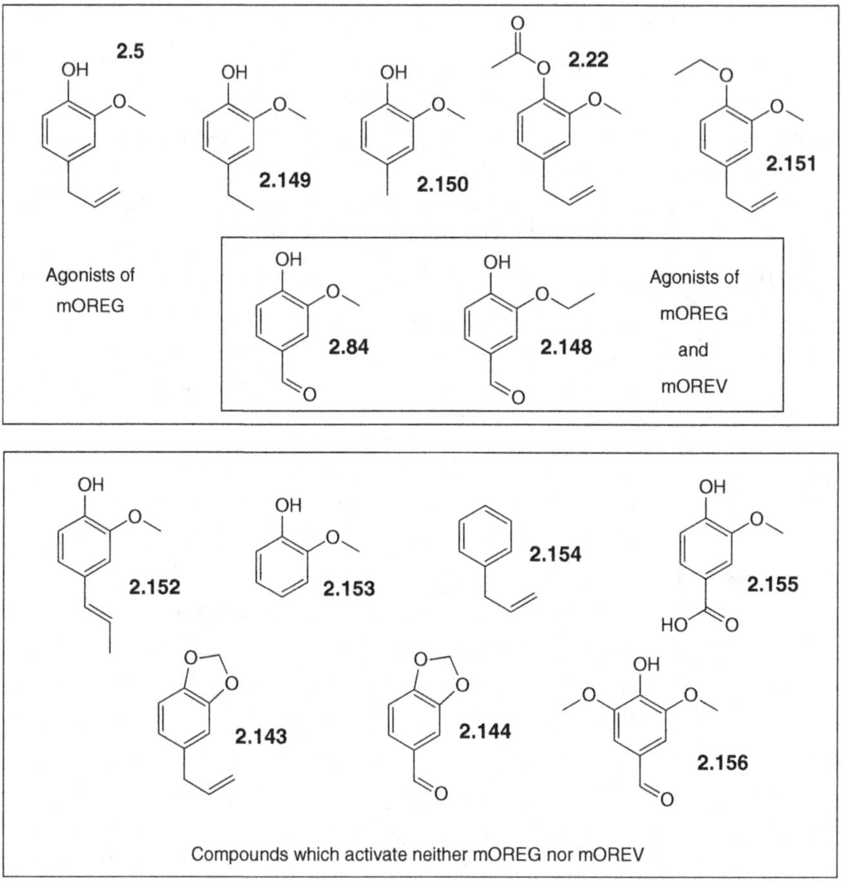

Figure 2.49   Agonists of mOREG and mOREV.

receptors at different concentrations. This could account for the fact that some odorants elicit different percepts at different concentrations. It also confirms the findings of Leinders-Zufall et al. (50) that odorant receptors, unlike the receptors of the vomeronasal system, give a concentration-dependent response. They also showed that the ligand preferences of a receptor cannot be predicted from its primary structure, even by comparison with close paralogs. Figure 2.49 shows how the receptive ranges of the two receptors compare. Eugenol (**2.5**) and vanillin (**2.84**) were found to be the most potent agonists of mOREG, and ethylvanillin (**2.148**) was somewhat more potent than vanillin (**2.84**) as an agonist of mOREV. Safrole (**2.143**), heliotropin (**2.144**) and compounds (**2.152**)–(**2.156**) did not activate either of the receptors.

In a subsequent study, the same laboratory investigated the response of mOREG to 22 shikimate derivatives related to eugenol (**2.5**) or vanillin (**2.84**) and used modelling and site-directed mutagenesis to further investigate ligand

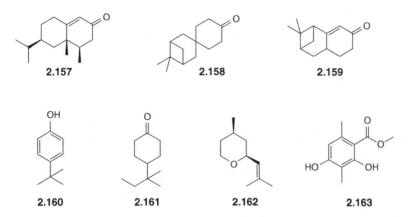

Figure 2.50 Additional agonists of mOREG.

binding (276). They found that the oxygen of the hydroxyl group of eugenol (**2.5**) (and the corresponding oxygen atom of the other agonists) is hydrogen-bonded to serine-113, with eight other amino acid residues forming the general shape of the binding pocket. They showed that recognition occurs through electrostatic (hydrogen-bonding), van der Waals and hydrophobic interactions and their results from modelling were confirmed by *in vivo* testing of the receptors.

However, the above results show how the choice of materials for screening can bias the appearance of the results. In the above two examples, Touhara's group concentrated on their lead structure, eugenol (**2.5**), and screened related structures in order to gain insight into the constraints around binding of that type of ligand to mOREG (Figure 2.50). In contrast, Vogel's team investigated a wider range of compounds and found that the receptor is much more broadly tuned than would be thought from the earlier results (310, 311). The agonists that they identified for it include nootkatone (**2.157**), Wolfwood (**2.158**), tricyclone (**2.159**), Terbutol (**2.160**), Orivone (**2.161**), rose oxide (**2.162**) and Evernyl (**2.163**). They found that different ligands could bind to (and activate) the receptor in different ways. Using site-directed mutagenesis and molecular modelling, they were able to identify the amino acids involved in forming the binding site of the receptor.

Spehr et al. found that the receptor OR1D2 (formerly known as *hOR 17-4*) was expressed in human sperm as well as in the OE, and examined its receptive range using HEK cells (120, 162). They also identified undecanal (**2.71**) as an antagonist of OR1D2. A subsequent study with a wider range of odorants continued this exploration, and the results are shown in Figure 2.51 (312). The receptor is relatively narrowly tuned and has a preference for substituted 3-phenylpropanals though it does accept some rather different structures. What is clear from the latter paper is that there is no simple correlation between activation of this particular receptor and the final odour percept that it elicits. The agonists belong to many different odour types which are not all linked to each other in any odour classification system.

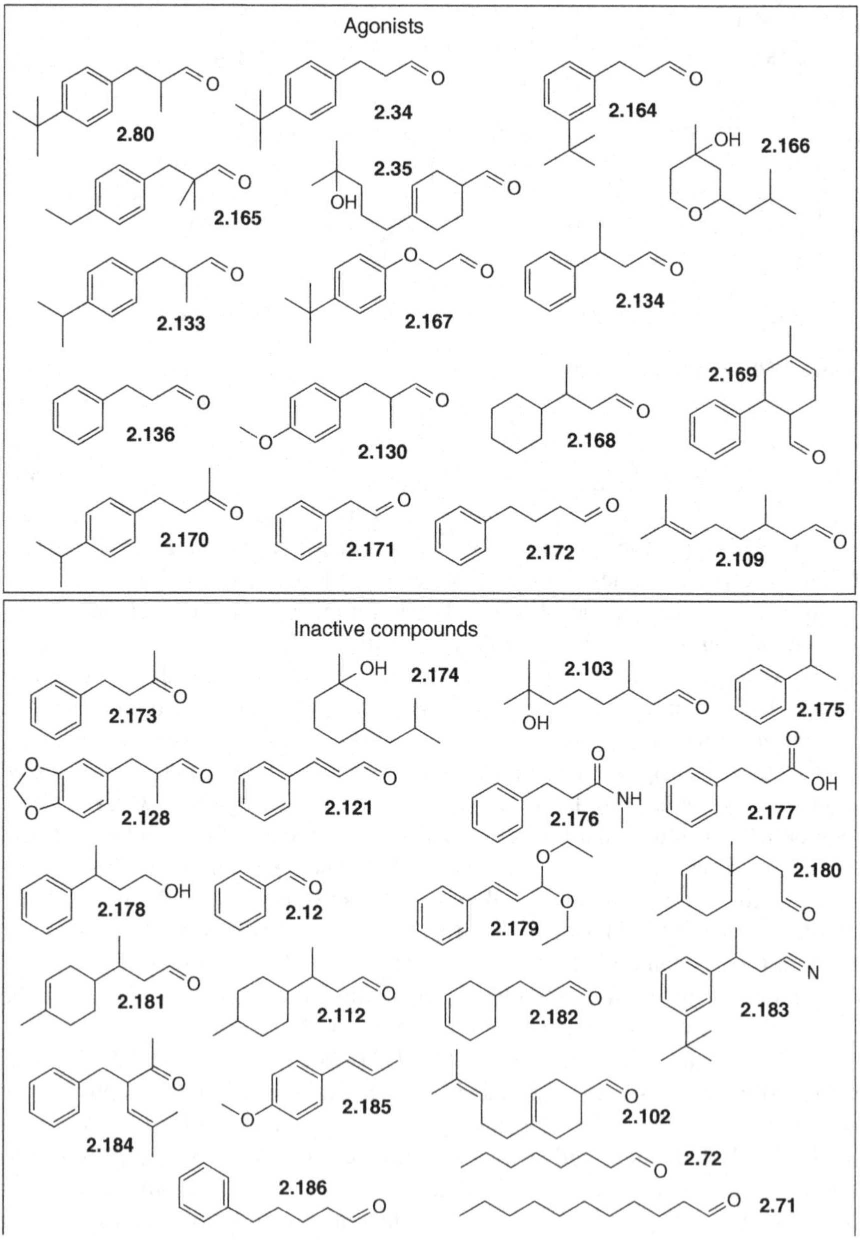

Figure 2.51 Receptive range of OR1D2.

Izewska compared the receptive ranges of OR1D2 with that of its rat ortho-
logue rOR5 (219). She screened 29 compounds and found that 4 were detected
only by OR1D2 and 7 only by rOR5; 9 activated both receptors and 9 failed to activate either of them. The results are shown in Figure 2.52. Her findings regarding
OR1D2 are in broad agreement with those of Spehr et al. (120, 162) and Triller
et al. (312).

Another two examples of de-orphanised human ORs are those of OR17-210
which Matarazzo et al. showed to respond to acetophenone (**2.119**), and
OR51E1 which Sallmann and Veithen found to respond to isovaleric acid (**2.145**)
(Figure 2.53) (313, 314).

One of the difficulties of using heterologously expressed cells is that of odorant
solubility in the aqueous media which are essential for cell culture. Sanz et al.
avoided the problem of dissolving their odorants in the culture medium by placing
them on transparent films above the culture and allowing the odorant vapour
to diffuse down into the culture and hence to the receptors. The disadvantage
of this method is that they could not know the concentration of the odorant in
the medium and therefore were unable to obtain dose/response curves (315).
Using this technique, they profiled and compared the selectivity of two human
ORs belonging to different phylogenetic classes. They found that OR52D1 has
a relatively narrow range, accepting alcohols, esters, ketones and acids with a
molecular length corresponding to a chain of eight or nine carbon atoms, whereas
OR1G1 (also known as *OR17-209*) is much more broadly tuned, responds to
a wide range of functional groups and has a preference for carbon chains of
eight or nine atoms. The strongest agonists were found to be 2-ethyl-1-hexanol
(**2.194**), 1-nonanol (**2.195**), ethyl isobutyrate (**2.196**), γ-decalactone (**2.197**) and
nonanal (**2.127**) (Figure 2.54). It also responded to such compounds as methyl
thiobutanoate (**2.198**), 2-methylpyrazine (**2.13**), benzothiazole (**2.199**), anethole
(**2.185**), piperonylacetone (**2.129**), Lyral (**2.35**) and methyl dihydrojasmonate
(**2.200**). In the series of aliphatic alcohols, aldehydes and ketones, its preference
is for compounds with 9 carbon atoms, but in the acid series its preference is for
those with 10 carbon atoms. In subsequent work, they used computer modelling,
supported by further screening, to investigate the receptive range of OR1G1 in
more detail and to propose a binding model (316). In that paper, they proposed that
the best ligands tend to be aliphatic alcohols and aldehydes and suggested a role
in perception of rosy, fatty and waxy odours (316). Their model is based on the
ligands rather than the receptor, and they propose that ligands fall into two distinct
groups, suggesting different binding modes.

However, when Matarazzo et al. investigated OR1G1, they found that it did not
respond to alcohols, aldehydes, ketones, phenols or lactones but was activated by
esters, especially isoamyl acetate (**2.88**) (313). There are several possible explanations for this discrepancy. One might lie in the way the odorants were introduced
to the culture medium. Sanz et al. used diffusion via vapour phase in a closed system, whereas Matarazzo et al. added odorant solutions directly to the medium in
an open well. The concentration of the odorant to which the receptor was exposed
could therefore be different in the two experiments due to differing solubilities and

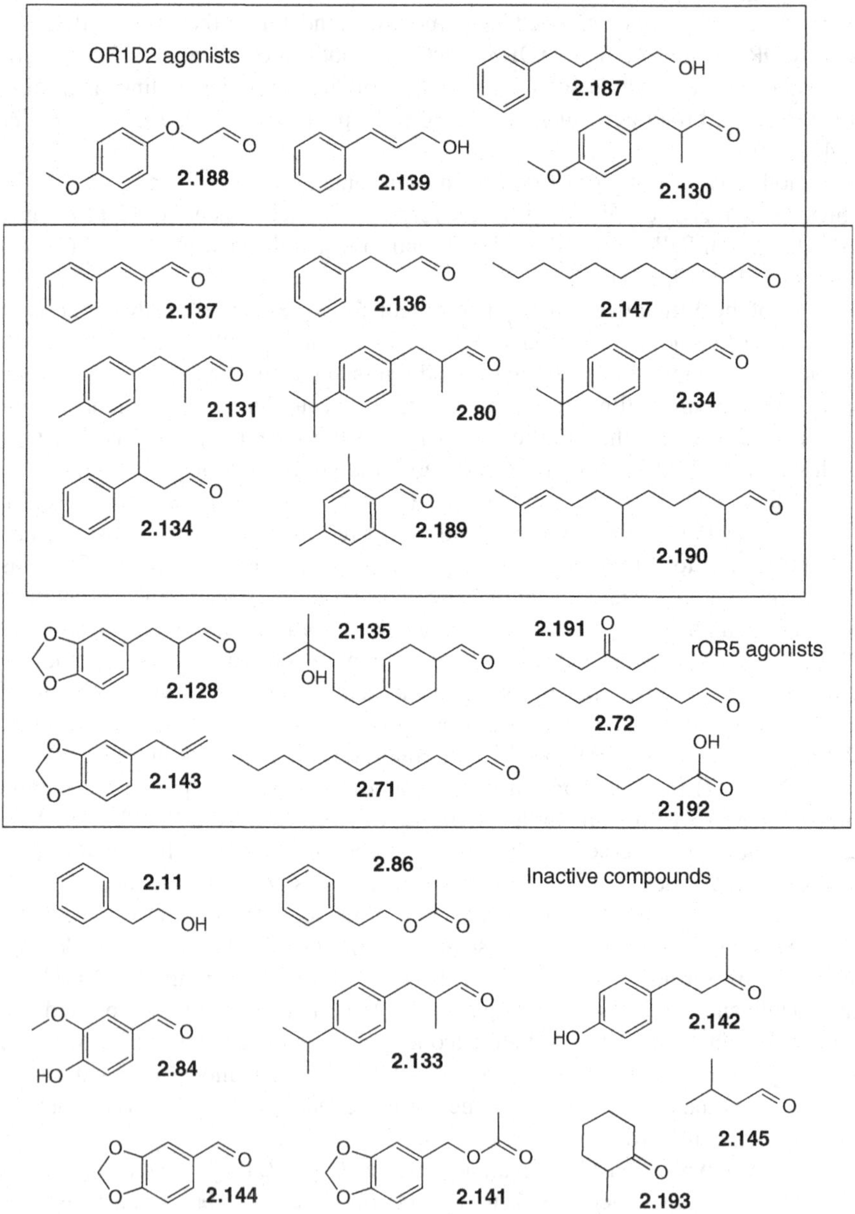

Figure 2.52 Comparison of receptive ranges of OR1D2 and rOR5.

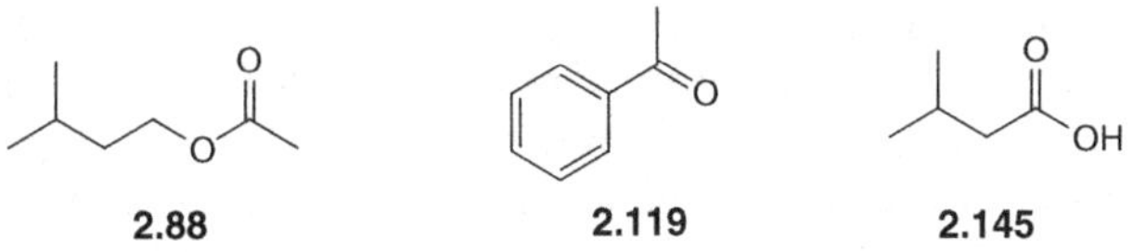

Figure 2.53   Agonists of of OR17-210 and OR51E1.

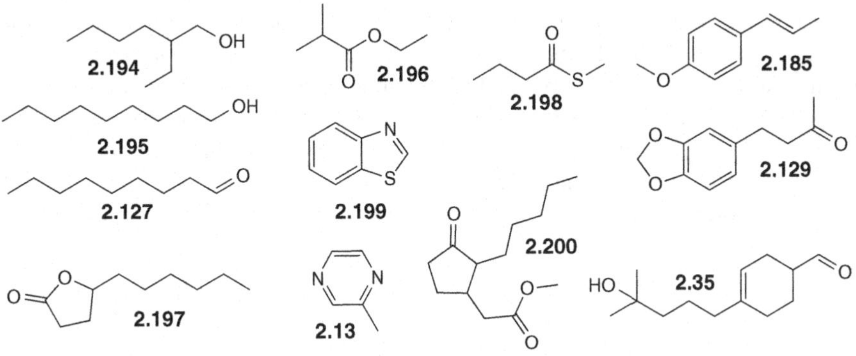

Figure 2.54   Some agonists of OR1G1 according to Sanz *et al.*

vapour pressures of the odorants. Another explanation could be that the two groups used different variants of the gene and, as described above, even one SNP can affect the ligand-binding properties of the encoded protein.

An example of a very narrowly tuned receptor is OR7D4, which was found to respond only to androstenone (**2.201**) and androstadienone (**2.202**) by Keller et al. (317) (Figure 2.55). There are two SNPs in this receptor, and Keller et al. found that there is a link between the variant found in an individual's genes and their sensitivity to androstenone (**2.201**), which is a well known malodorant found in male human sweat and the flesh of boars. The SNPs result in the amino acid substitutions R88W and T133M. The RT variant with arginine at position 88 and threonine at position 133 responds to androstenone (**2.201**) whereas the WM variant, with tryptophane at position 88 and methionine at position 133, does not. Interestingly, position 88 is at the top of TM2 and 188 is at the bottom of TM4, thus neither is in the binding pocket. In other words, the effect of the substitutions must be on a feature of the protein structure or function other than the binding

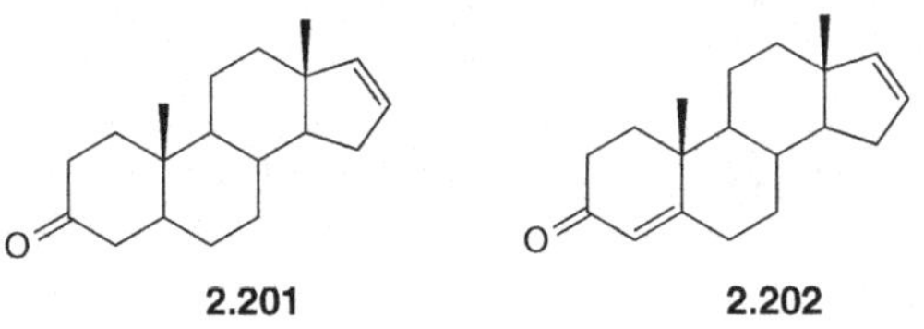

Figure 2.55   Agonists of OR7D4.

site. As would be expected in view of the binding profiles, individuals with the RT/RT genotype are more likely to be sensitive to androstenone (**2.201**) and to find it more unpleasant than those with RT/WM or WM/WM genotypes. However, other androstenone-sensitive ORs must exist since the correlation is not 100%. Of those with the RT/RT genotype, only about 20% show anosmia to androstenone (**2.201**) and, conversely, about 80% of those with other genotypes are anosmic to it.

Schmiedeberg et al. also studied a pair of receptors but they chose the closely related OR1A1 and OR1A2 (275). These were found to respond to a series of alcohols and aldehydes mostly in the C8–C10 region. In general, the potency of the ligands was very similar for both receptors. The exceptions to this are that OR1A2 is much less sensitive to nonanal (**2.127**) but more sensitive to octanol (**2.203**) than is OR1A1, The other interesting difference is that (*S*)-(−)-citronellol (**2.83**) activates only OR1A1 whereas (*R*)-(+)-citronellol (**2.204**) activates both receptors to exactly the same extent. The agonists are shown in Figure 2.56, in the order of decreasing potency towards OR1A1 (left to right and top to bottom, i.e. most potent at top left).

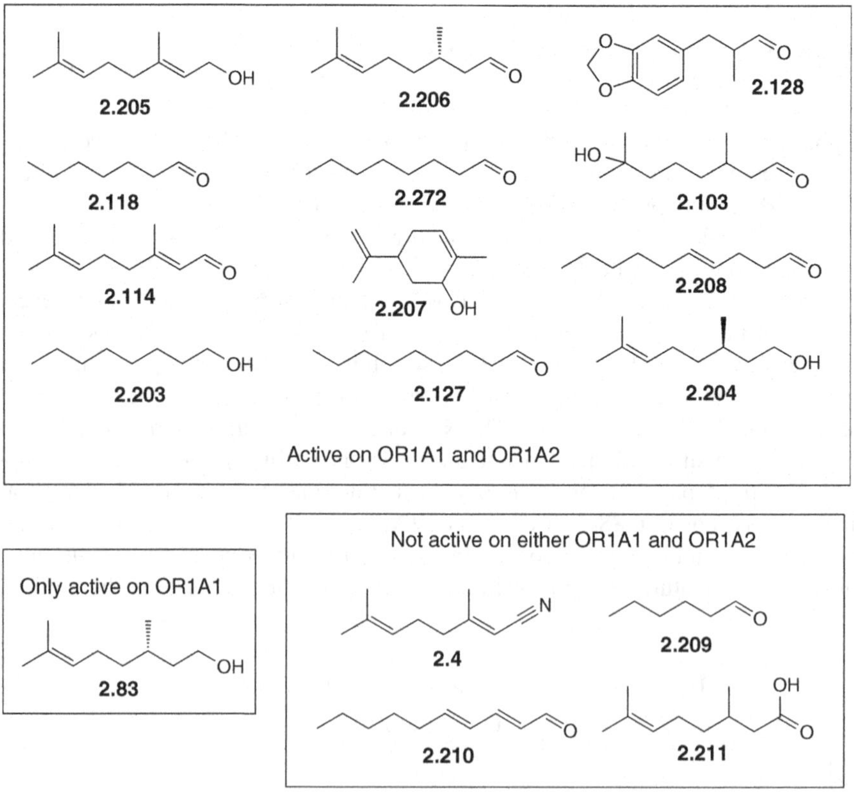

Figure 2.56   Receptive range of OR1A1 and OR1A2.

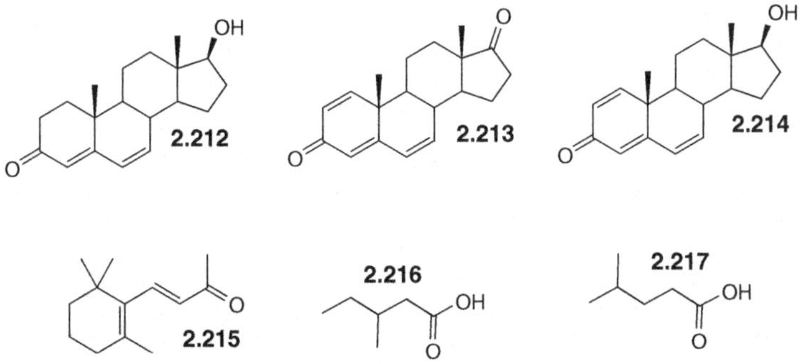

Figure 2.57   Agonists of ORs expressed in prostate cancer cells.

Neuhaus et al. investigated OR51E2, which is known to be over-expressed in prostate cancer cells (166). They found it to respond to three steroids, namely 6-dehydrotestosterone (**2.212**), 1.4.6-androstatrien-3,17-dione (**2.213**), 1.4.6-androstatrien-17 β –ol-3-one (**2.214**), as well as to β-ionone (**2.215**) but, surprisingly, not to any of the other ionones. Another receptor that is known to be expressed in prostate cancer cells as well as the OE is that known as the *Dresden GPCR (D-GPCR)*. Fujita et al. identified 3-methylvaleric acid (**2.216**) and 4-methylvaleric acid (**2.217**) as its agonists (Figure 2.57) (306).

Abaffy et al. studied the murine receptors mOR42-1 and mOR42-3 which respond to carboxylic acids in the range C8–C12. They used a combination of molecular modelling and site-directed mutagenesis to identify eight amino acids that were responsible for binding the ligands in the LBP (318).

Dahoun et al. investigated the receptive range of mOR256-17 expressed heterologously in HEK cells. The most potent ligand they found for it was *l*-carvone (**2.116**) followed by 4-*t*-butylcyclohexanone (**2.218**) and then *d*-carvone (**2.126**). Its agonists (in decreasing order of potency from left to right) and the compounds they found to be inactive are shown in Figure 2.58 (319).

Hallam and Carlson expressed *D. melanogaster* receptors into 'empty' antennal neurons and produced a map of the responses of 24 neurons versus 110 odorants (320). This showed that most of the odorants they tested were detected by several ORs and most ORs responded to a range of odorants. The breadth of tuning varied between receptors, with some of them responding to only a few odorants and others responding to a large number. The map therefore resembles those found in mammalian receptors, but the caveats mentioned earlier regarding correlating insect and mammalian results must still be borne in mind.

In an investigation of hyperosmia (that is higher than average sensitivity to odorants), Menashe et al. investigated the genotypes of individuals who showed high sensitivity to isovaleric acid (**2.145**) (321). They postulated that segregating pseudogenes might be responsible for inter-individual differences in sensitivity to odorants and found a strong link between OR11H7P and hypersensitivity to isovaleric acid (**2.145**). When the gene was cloned into *Xenopus* oocytes, it was found

Agonists

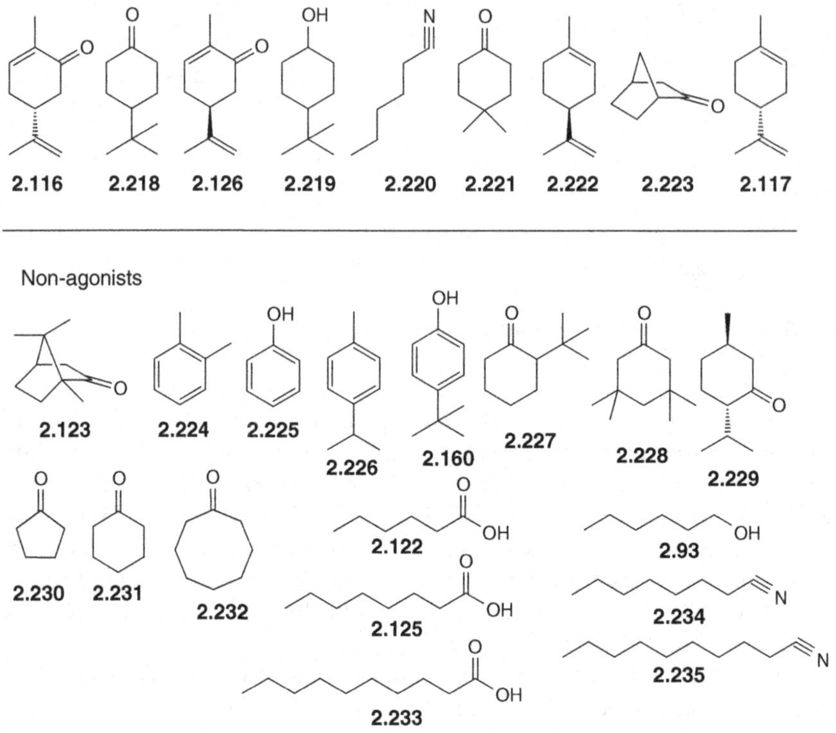
2.116
2.218
2.126
2.219
2.220
2.221
2.222
2.223
2.117
Non-agonists
2.123
2.224
2.225
2.226
2.160
2.227
2.228
2.229
2.230
2.231
2.232
2.122
2.125
2.233
2.93
2.234
2.235

Figure 2.58 Receptive range of mOR256-17.

that it did indeed respond to isovaleric acid (**2.145**). Its full receptive range was not studied but other evidence showed that hyperosmia is dependent on other factors also.

By far the largest de-orphanisation reported to date in the literature is that of Saito et al. (158). They screened 219 murine ORs and 245 human ORs against a library of 93 odorants. This corresponds to 21% of an estimated 1035 mouse ORs and 63% of an estimated 387 human ORs. Their hits produced a matrix of 63 odorants against 10 human and 52 murine ORs with an EC_{50} value for each couple. Inspection of the map shows clearly that there are more narrowly tuned than broadly tuned ORs, that ORs respond to various ranges of odorants and that odorants activate an array of receptors. The breadth of tuning of the human ORs is shown in Table 2.4.

The most broadly tuned of these receptors, OR2W1, responds to a wide variety of structural types and functional groups: terpenoids (both acyclic and cyclic), aliphatic aldehydes, aliphatic ketones, aliphatic aldehydes, carboxylic acids, coumarin (**2.16**), dihydrojasmone (**2.236**), and so on. However, despite this breadth of tuning, it is still selective: for example, it responds to hexanol (**2.93**)

Table 2.4 Breadth of Tuning of Human Olfactory Receptors De-orphanised by Saito et al

Receptor	No. of agonists	Agonists as % of agonists screened
OR2W1	37	59
OR1A1	20	32
OR5P3	6	10
OR2J2	5	8
OR2C1	2	3
OR2M7	2	3
OR51E1	2	3
OR51L1	2	3
OR10J5	1	2
OR51E2	1	2

and hexanal (**2.209**) but not hexanoic acid (**2.122**) or pentanol (**2.237**). Similarly, it responds to nonanal (**2.127**) and nonanoic acid (**2.238**) but not to decanal (**2.120**) though it does respond to decanoic acid (**2.233**). It responds to coumarin (**2.16**) but not to 2- coumaranone (**2.239**) and to benzyl acetate (**2.240**) but not to phenyl acetate (**2.241**). These selected agonists and non-agonists are shown in Figure 2.59.

At the other end of the tuning spectrum, OR2M7 responded only to the closely related structures of geraniol (**2.205**) and citronellol (**2.242**). However, the narrowly tuned receptors also have surprises, an example being OR51E1 which responds to nonanoic acid (**2.238**) but not to its homologues octanoic acid (**2.125**) and decanoic acid (**2.233**), yet it does respond to butyl butyryllactate (**2.243**) which is more structurally dissimilar. The receptive ranges of the eight more narrowly tuned human receptors are shown in Figure 2.60.

The basic pattern of this map of interactions between odorants and heterologously expressed ORs is similar in nature to the results of Nara et al. described above using OSNs to map out the range of odorant/OR couples. All the other results on the receptive range of ORs, whether determined directly or through measurements on neurons, are consistent with such a pattern. We can therefore conclude that the olfactory code is based on a combination of both broadly and narrowly tuned receptors, with the latter outnumbering the former. Of course, we have only found pairings for a tiny fraction of all the odorant/receptor couples which must exist, and the largest reported test library contains only 125 odorants. In the natural world, our sense of smell responds to an almost infinite range of molecular structures and so it is quite possible that our views on narrow tuning are there simply because we have not tested a large enough range of odorants.

When Ferrero et al. investigated the receptive ranges of a number of heterologously expressed mouse and rat TAARs, they obtained a selectivity pattern as shown in Figure 2.61 (structures of the amines tested are shown in Figure 2.62), in which the prefixes m and r indicate mouse and rat, respectively (322). They then

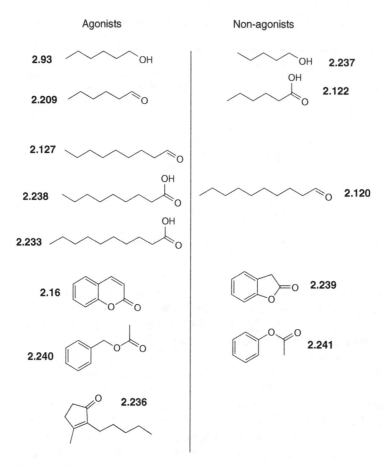

Figure 2.59 Some agonists and non-agonists of OR2W1.

compared the amino acid sequences of these receptors, looking for differences in sequence between them, and found specific sites where the presence of a particular amino acid was necessary for the receptor to respond to a specific amine. Using site-directed mutagenesis, they then exchanged these amino acids either by introducing an active one to an inactive receptor or by removing an active one from an active receptor. The effects on selectivity of the receptor were dramatic and supported the view that speficic amino acids were involved in binding of ligands. They then built computer models of the receptors based on their sequences and found that the models also supported their hypotheses. They postulated that the free carboxyl function of aspartic acid 127[3.32] is responsible for binding the basic nitrogen atom of the various amines. The position of this amino acid in the receptor is similar to that of an aspartic acid in other amine-binding receptors and to that of serine-113[3.40] which is responsible for forming a hydrogen bond to the ligand in the mouse eugenol receptor. The model shows that replacing serine-132[3.37] of

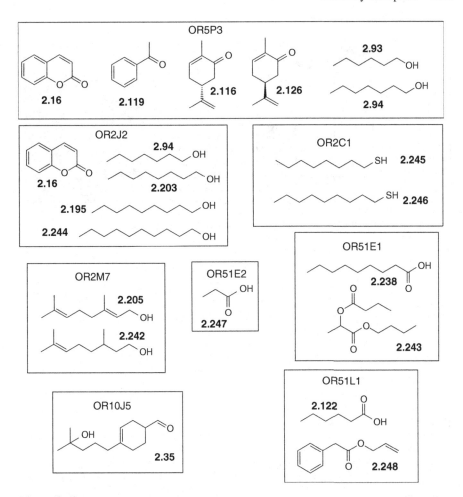

Figure 2.60 Receptive ranges of eight human olfactory receptors.

the TAARs by the larger tyrosine reduces the space available at the bottom of the binding pocket and prevents binding of *N,N*-dimethylphenylethylamine (**2.253**) to the receptor.

Broadly Tuned Receptors

The role of broadly tuned receptors is the subject of speculation. Might they simply serve to indicate the presence of odorants in general or do they play a role in discrimination? If we consider the receptive ranges of the more broadly tuned receptors described above, it is clear that, whilst broadly tuned, they are selective and respond to specific groups of compounds. So, they could very well play a role in odour discrimination. Consider the Venn diagram in Figure 2.63. This shows the

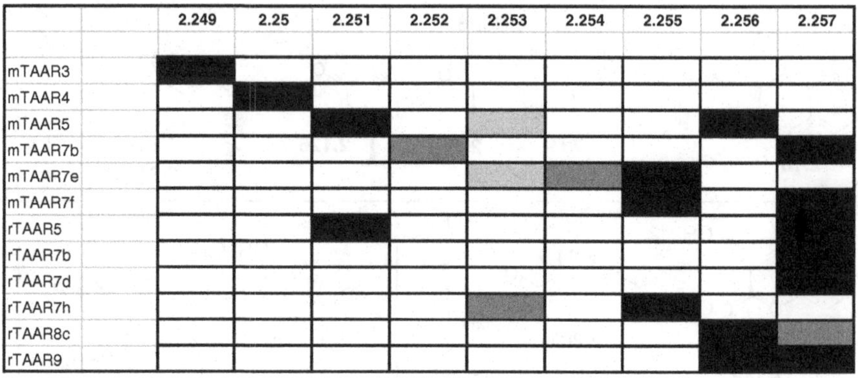

		2.249	2.25	2.251	2.252	2.253	2.254	2.255	2.256	2.257
mTAAR3		■								
mTAAR4			■							
mTAAR5				■						
mTAAR7b					▦					■
mTAAR7e						▦				■
mTAAR7f							■			
rTAAR5				■						■
rTAAR7b										■
rTAAR7d										■
rTAAR7h						▨			■	■
rTAAR8c									■	▨
rTAAR9									■	■

Figure 2.61   Response of mouse and rat TAARs to amines of Figure 2.62.

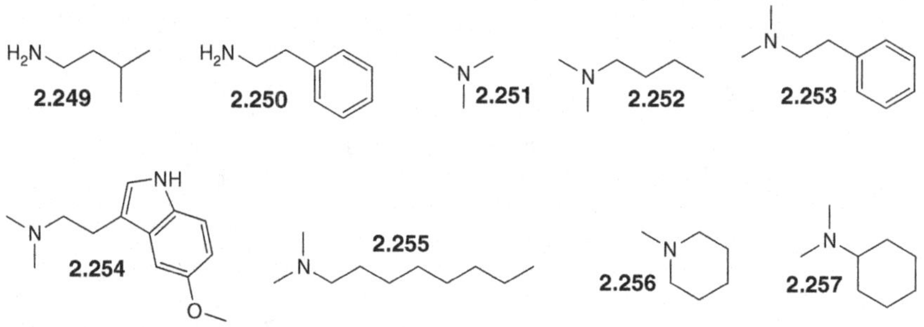

Figure 2.62   Amine ligands of mouse and rat TAARs.

receptive ranges of three hypothetical receptors A, B and C. Activation of A only could indicate the presence of any one of 30 different odorants (represented by stars in the figure). Similarly, activation of B could indicate any one of 20 different odorants, and C any one of 10. Simultaneous activation of A and B would be consistent with only 6 of the 60 compounds, and simultaneous activation of A and C would restrict the figure to only 3. Simultaneous activation of all three receptors would give an unambiguous identification of one compound. A wide range of broadly tuned receptors each with a unique selectivity profile would therefore contribute significantly to discrimination between odorants.

Receptor Amino Acid Sequence and Receptive Range

OR sequences can be identified from the human genome by searching for characteristic sequences such as MAYDRYVAIC and PMLNPFIY. They can then be compared for similarity of amino acid sequence and arranged in a cluster tree analysis known as a *cladogram* or *phylogenetic tree*. One such tree was drawn by Malnic et al., which shows all of the 339 human receptors that they had identified

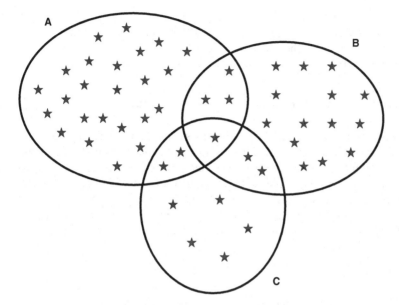

Figure 2.63 Ligand discrimination using only broadly tuned receptors.

together with 23 rodent and 23 fish ORs (323). They found that the fish receptors formed a distinct cluster at one end of the tree and that those mammalian receptors that occurred in this region responded to aliphatic carboxylic acids and aliphatic alcohols. Mammalian receptors falling into this category are usually known as *fish-like receptors*. They also found that the murine ORs responding to eugenol (**2.5**) and analogues of it were located in the same region of the tree, and also that the human ORs responding to Helional (**2.128**) and Bourgeonal (**2.34**), (hOR17-40) and (hOR17-4), respectively, lie close to each other. Both of these odorants are dihydrocinnamaldehyde derivatives and so it would seem that, not surprisingly, receptors with similar amino acid sequences have similar preferences as far as ligands are concerned. Using the older nomenclature systems for the Helional and Bourgeonal receptors also shows that the genes coding for them occur on the same chromosome, 17. In view of this, similar observations and other considerations, Malnic et al. proposed that genes encoding similar receptors, and hence similar receptive ranges, do occur in similar loci on the genome (323).

It would seem obvious that the receptive range of a receptor should be relatively stable and dependent on its structure. However, one of the many surprises in olfaction is that this is not the case. It has been found that OSNs expressing the murine receptor mOR23 respond to a wider range of odorants in neonatal mice than in mature mice. The range of the neurons narrows over the first month of the mouse pup's life, eventually becoming the very narrowly tuned receptor described earlier in this chapter. The neurons also become more sensitive to its agonists. This change in selectivity is dependent somehow on another protein, OMP, which is expressed solely in OSNs. Mice lacking the gene that encodes OMP do not show this change in selectivity and sensitivity of mOR23 and also do not show preference for their mother over other lactating females, demonstrating that the loss of olfactory maturity affects their sense of smell and consequent behaviour (324).

Interactions Between Odorants at the Receptor

The above discussion on receptive ranges describes experiments where odorants are presented one at a time to ORs. In real life, almost everything we smell is a mixture of different odorants, and so it is important to understand the interactions between odorants in mixtures. Many mixture effects result from physical-chemical effects before the receptor event and many more from interactions in neuroprocessing. It is possible that mixture effects also occur at the receptor, and two ways in which they could arise are antagonism and allosteric modulation.

Antagonism

An antagonist is something that prevents a receptor from being activated by a molecule that would normally act as an agonist. As described above, an inverse agonist is a species that binds to the active site of the receptor but stabilises the receptor in the inactive state. If an agonist and an inverse agonist are in competition for the binding site, then the inverse agonist will act as an antagonist. In experiments using intact neurons or heterologously expressed receptors in another type of cell, a molecule might be found to prevent activation of the detection system. The mechanism by which it does this could be one of many, and so various control experiments must be carried out to prove that there is a specific antagonistic effect at the receptor. An example of an antagonistic effect elsewhere in the transduction cascade is given by Ukhanov et al., who showed an effect of odorants on phosphoinositide 3-kinase (325).

Antagonism is well known for many GPCRs, and examples have been reported for ORs. One early indication that this occurs was found in 1987 when Bell et al. used deoxyglucose measurement in the OB as a measure of activation of neurons and found that one odorant could reduce the sensitivity of rat olfactory neurons to another odorant. The pairs studied correlated with sensory effects shown in humans (326). Araneda et al. found antagonistic effects in a series of receptors sensitive to aldehydes (292).

Citral (**2.114**) was found to antagonise octanal (**2.72**) at ORI7 (295), and subsequently Peterlin et al. investigated the effect of a series of aldehydes on the mouse ORI7 receptor using octanal (**2.72**) as ligand. They found that 3-cyclopentylpropanal (**2.258**), 2-cyclohexylacetaldehyde (**2.259**) and cyclo-heptanecarboxaldehyde (**2.260**) functioned as antagonists and proposed a steric relationship between the binding modes of the two classes of ligands (Figure 2.64). Their hypothesis is that the carbonyl function of each of the ligands binds to the same H-bonding centre in the binding pocket and that agonists occupy a longer

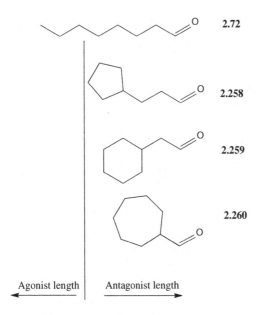

2.72

2.258

2.259

2.260

Agonist length | Antagonist length

Figure 2.64 Antagonists of mOR17.

space than occupied by antagonists and that the effect of this is to determine whether the ligand stabilises ORI7 in the active or inactive configuration (327). As with the findings of Sanz et al. (315), in this case smaller ligands rather than the agonists acted as antagonists. This is in contrast to an antagonist of mOREG, which is twice the size of the agonist as described below (328).

Spehr et al. found that undecanal (**2.71**) antagonises OR1D2 and prevents it from recognising Bourgeonal (**2.34**) (120) and showed that it translates into a sensory effect (162). Brodin et al. also found that this receptor effect does result in a sensory one (329).

Oka et al. reported antagonism of the response to eugenol (**2.5**) of the mouse receptor mOREG by isoeugenol (**2.152**), methylisoeugenol (**2.261**) and isosafrole (**2.262**) (330) (Figure 2.65). In a subsequent paper, they reported that fresh isoeugenol (**2.152**) did not have the effect and that the active antagonist was a product (**2.263**) of autoxidation and coupling, which they identified in stored samples of isoeugenol (**2.152**) (328).

Sanz et al. found that the broadly tuned receptor OR1G1 was antagonised by 1-hexanol, hexanal and cyclohexanone. Longer chain alcohols and aldehydes were amongst its agonists, as mentioned previously (315).

Isewka found that cinnamyl alcohol (**2.139**) and 2-phenylethanol (**2.11**) are antagonists of the rat receptor OR5 and that the human receptor OR17-4 is antagonised by 2-methylcyclohexanone (**2.193**), 2-phenylethyl acetate (**2.86**) and valeric acid (**2.192**) (Figure 2.66) (331). Another worker from the same laboratory, Etter, found that (+)-carvone (**2.126**) antagonises the response of mOREG to eugenol (**2.5**) (310). Whilst studying the receptive range of the murine receptor mOR42-3, Abaffy et al. found that decanedioic acid (**2.264**) did not activate it. From molecular

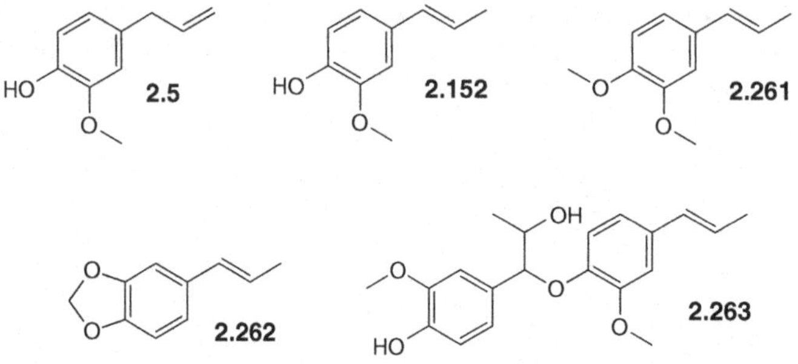

Figure 2.65   Antagonists of mOREG.

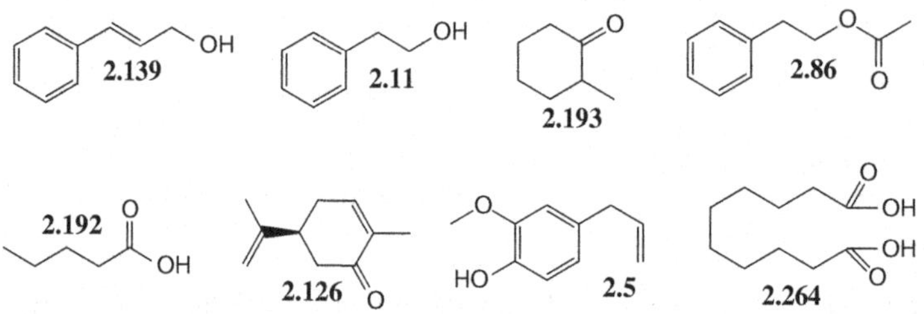

Figure 2.66   Antagonists of rOR5, hOR17-4, mOREG and mOR42-3.

modelling studies, they proposed that it might instead act as an antagonist, and this hypothesis was confirmed experimentally (318).

Since there is always a background level of firing of every receptor, a strong inverse agonist could in principle switch off the background signal and, unless inhibitory circuits in the bulb were able to quench the effect, this might constitute a change that could be interpreted positively by the brain. This seems to be an unexplored avenue of investigation.

Allosteric Modulation of GPCRs

Binding sites other than the main LBP (also known as the *orthosteric binding site*) are known as *allosteric binding sites*. When a ligand is bound to one of these sites, it can affect the binding of an agonist to the orthosteric site and this phenomenon is known as *allosteric modulation of the receptor activity*. Allosteric effects in GPCRs have been reviewed by Conn et al. (332) and by Christopoulos and Kenakin (333). Both are excellent reviews and the latter also gives an insight into the complex kinetics of allosteric modulation.

As described earlier, ORs, like all class A GPCRs, have small loops and short tails. This is in stark contrast to the sweet and umami taste receptors which belong to the class C (class 3) GPCR family and have a very large extracellular region

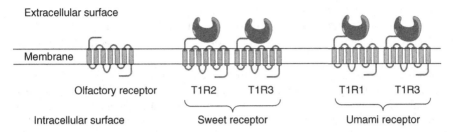

Extracellular surface

Membrane

Olfactory receptor   T1R2   T1R3   T1R1   T1R3

Intracellular surface   Sweet receptor   Umami receptor

Figure 2.67   Comparison of olfactory, sweet and umami receptors.

known as the *Venus fly trap domain* owing to its similarity in shape and mode of action to the leaves of the plant of the same name. In the case of sweet and umami, the main binding pocket is actually in the Venus fly trap domain, and allosteric modulators can act in the TM region of the receptor. The difference between these taste receptors and the ORs is even greater than this because the sweet and umami receptors function as heterodimers, as shown in Figure 2.67. The T1R1–T1R3 complex detects umami taste and T1R2–T1R3 is the sweet receptor. An excellent review by Hofmann et al. (334) describes the current state of knowledge about them.

Allosteric modulators are substances that change the response of a receptor to its cognate ligand by binding at a site (allosteric site) other than the site at which the ligand binds (orthosteric site). Examples are well known for both the sweet and umami receptors (335, 334) and for the calcium-sensing receptor (another class C GPCR). Substances eliciting a sweet perception act on the Venus fly trap domain of T1R2, and those for umami act on the Venus fly trap domain of T1R1. Allosteric modulators for these two tastes therefore can act either on the Venus fly trap domain of T1R3 or on the TM regions of either of the proteins of the heterodimer. The orthosteric site of the calcium-sensing receptor is also in the Venus fly trap domain (336) and its allosteric site is in the TM region (337). Jones et al. have identified a negative allosteric modulator for insect ORs (338). However, as stated earlier, insect receptors function as heterodimers with the receptor known as *ORCO* serving as one component, together with the ligand sensitive receptor. The allosteric modulator found by Jones et al. is known as *VU0183254* (**2.265**) and its action is due to the fact that it is an antagonist of ORCO. It is not so easy to see how allosteric modulators could act on mammalian ORs, but some possibilities have been proposed.

Examination of published structural data on GPCRs led Rosenkilde et al. (179) to postulate that a crevice near the ECS, extending from the main binding site into the space between helices 2 and 7, can also play a role in binding and therefore could also serve as a site for binding of allosteric modulators. This possible minor binding site is shown in Figure 2.16.

Another possibility for allosteric effects on class A GPCRs is that of binding of modulators to EL2, as mentioned earlier. In their study of the M2 mAChR muscarinic acetylcholine receptor using both computer modelling (based on rhodopsin homology) and site-directed mutagenesis, Christopoulos et al. investigated the role of EL2 in binding both orthosteric ligands and allosteric modulators (339–341). In M2 mAChR, the cysteine in EL2, which forms a disulfide bridge to TM3, occurs in a sequence of polar amino acids EDGECY, and there is a group of polar amino acids NTTW at the top of TM7. From the crystal structure of rhodopsin, it is known that EL2 does lie across the top of the LBP, and so these authors postulate that polar interactions between these two regions stabilise the 'gatekeeper' role of EL2 and that the allosteric modulators gallamine (**2.266**) and C7/3-phth (**2.267**) strengthen this interaction and help to hold the ligand in the LBP (Figure 2.68). They showed that the modulators do bind to EL2, and, in their presence, radiolabel experiments with tritium confirm that the modulators decrease the rate of release of the ligand from the LBP, thus supporting their postulate. Although ORs do not tend to have the EDGECY or NTTW sequences in those positions, they do have other polar amino acids at the two sites in question and so a similar mechanism could operate in ORs also. The crystal structure of the M2 muscarinic acetylcholine receptor, as described earlier, also supports the theory of positive allosteric modulation by locking the agonist in the binding pocket (214).

An alternative mechanism for gallamine's action was proposed by Dror et al., who, as mentioned earlier, proposed binding sites on the ECS to which the ligand binds initially (142). They suggest that gallamine (**2.266**) also binds to this vestibule site and blocks access to the LBP.

Another potential allosteric binding site, as described by Gloriam et al. (174), is at the edge of the TM bundle, where cholesterol (**2.52**) was found to bind by Hanson et al. (201), though that would involve penetration of the membrane by the allosteric agent.

Allosteric modulation of GPCRs has been reviewed by by Melancon et al. (342).

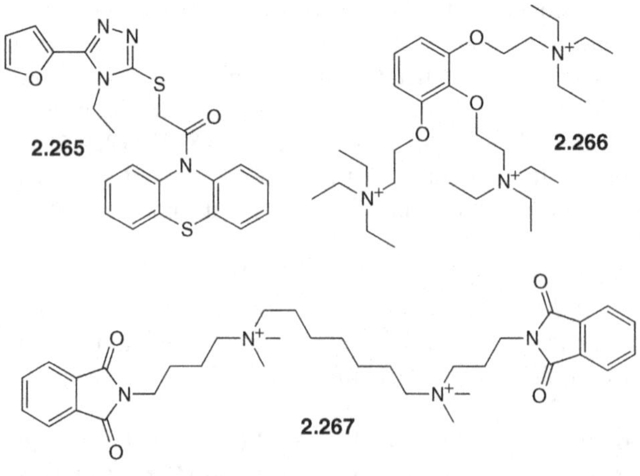

Figure 2.68   Some allosteric modulators.

ACTIVATION OF THE G-PROTEIN

As described above, the binding of an agonist to the receptor protein opens a cleft on the ICS of the latter, allowing the carboxyl terminus of Gα to bind to it. The nitrogen terminus of Gα also binds to the receptor/agonist complex, and this opens the pocket of Gα in which GDP is bound. Once GDP has dissociated from Gα, GTP can bind and the pocket closes again. This, in turn, causes dissociation of Gα from the receptor and also from Gβ and Gγ, and the free Gα/GTP complex can then activate the enzyme adenylyl cyclase ACIII. The details of the mechanism of activation of the G-protein were elucidated by Chung et al. using mass spectroscopy (H/D exchange of labile hydrogen atoms) in combination with X-ray crystallography and protein modelling (343). Their scheme is eloquently described in the original paper, and a simplified cartoon is shown in Figure 2.69 together with the structures of GDP (**2.268**) and GTP (**2.269**).

THE SECOND MESSENGER SYSTEM

Once the G-protein has been activated by the conformational changes in the receptor, a sequence of biochemical process swings into action involving what is known as a *second messenger cascade*. In OSNs, the second messenger is cAMP (**2.270**). cAMP (**2.270**) is formed from ATP (**2.271**) by an enzyme known as *adenylyl cyclase*, as shown in Figure 2.70. Adenylyl cyclase enzymes have been reviewed by Sadana and Dessauer (344). The specific form of adenylyl cyclase involved in the OSNs is the membrane-bound adenylyl cyclase III (ACIII). This enzyme is normally in an inactive state, but is converted to an active state by the dissociated G-protein. The activated ACIII converts ATP (**2.271**) into cyclic cAMP (**2.270**), which then relays the activity on to further membrane proteins: hence its being referred to as a *second messenger*.

The cascade set in motion by cAMP (**2.270**) and resulting in membrane depolarisation is shown in Figure 2.71. The role of cAMP (**2.270**) is to open the CNG channel allowing calcium to flow into the cell. The CNG channels are heterotetramers ((CNGA2)$_2$(CNGA4)(CNGB1b)) but the homotetramer (CNGA2)$_4$ is also functional. Although it belongs to a different family from the CNG channels of olfaction, the X-ray structure of the P2X4 receptor of the zebrafish shows how binding of a nucleotide to an ion channel causes opening of the central pore to allow ions to pass through (345). As a result of opening of the CNG channel, the concentration of calcium ions rises and this, in turn, opens a calcium-gated chloride channel which allows chloride ions to flow out of the cell. The calcium-activated chloride channel Anoctamin 2 (ANO2) has been found in OSNs, and it is suggested that it plays a role in amplification of the signal (346). However, later research suggests that it occurs only in the VNO and that the related channel Anoctamin2 (ANO2, also known as *TMEM16B*) is found in both the MOE and the VNO although it seems to have little effect on the response (347). In a review of the electrochemistry of transduction, Kleene describes how the calcium and chloride fluxes both

120

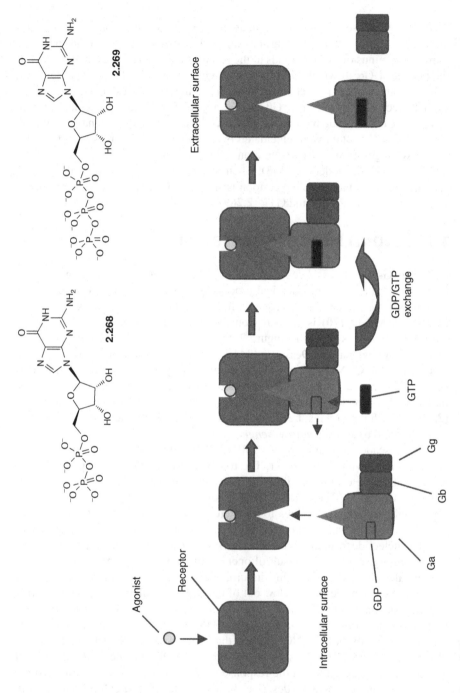

Figure 2.69 Activation of G-protein.

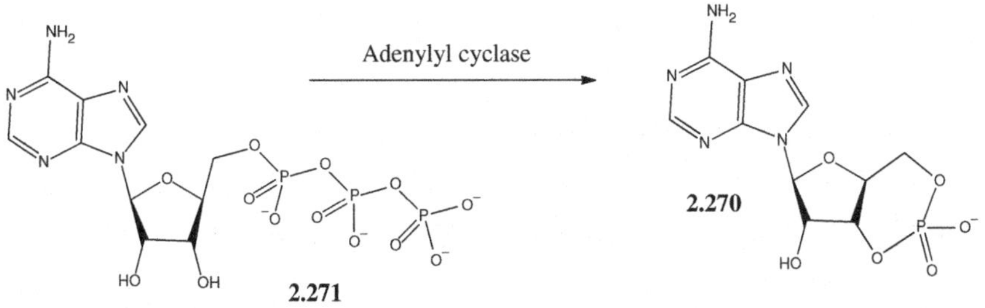

Figure 2.70 Generation of cAMP.

contribute to the signal (348). With calcium ions entering the cell and chloride ions leaving it, the resulting imbalance of charges between the cell interior (cytosol) and the surrounding mucus, results in depolarisation of the membrane. This occurs at the cilium of the cell which is located in the nasal mucus. However, the other end of the neuron is in the OB, and the depolarisation allows a signal to be generated at its γ-aminobutyric acid (GABA)ergic synapse there with the mitral and tufted cells which are the secondary neurons of the bulb.

Signal Shut-Down and Reset

As stated in the previous chapter, the sense of smell must be a temporal one and the signal must be switched off rapidly to enable the animal to continue to detect the presence of a smell and to know if it is increasing or decreasing in concentration. There are many mechanisms for doing this. There are at least 10 possible mechanisms in the sensory neuron (348), and others operate all the way through subsequent neuroprocessing.

As part of his Nobel Prize winning work, Lefkowitz showed that after break-up of GPCR complexes with the G-protein, the intracellular tail of the receptor is phosphorylated by a GPCR kinase (GRK6), which allows complexation with a protein called β-*arrestin* and this results in inactivation (349). This is a major mechanism of signal modulation. Distortion of the membrane then allows this complex to be absorbed into the cytosol as an endosome (a micelle inside the main cell membrane). The receptor is resensitised and then inserted back into the membrane.

After activation of ACIII by the G_α /GTP complex, the GTP bound to G_α is hydrolysed back to GDP and the trimeric G-protein reassembled. Ric 8B, a guanine nucleotide exchange factor, interacts with all three G-protein subunits (350) and accelerates regeneration of the GTP/GDP cycle. This makes it useful in *in vitro* work as well as in the neurons.

It was previously thought that phosphodiesterase enzymes might terminate the signal by hydrolysing cAMP (**2.270**), but knocking out the phosphodiesterase (PDE1C) found in the cilia did not affect signal termination showing that cAMP (**2.270**) degradation by PDE 1C is not rate-limiting in signal termination (351).

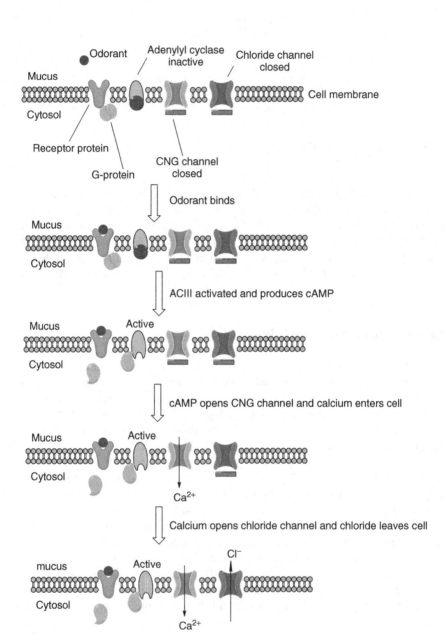

Figure 2.71   The sequence of events from odorant recognition to membrane depolarisation.

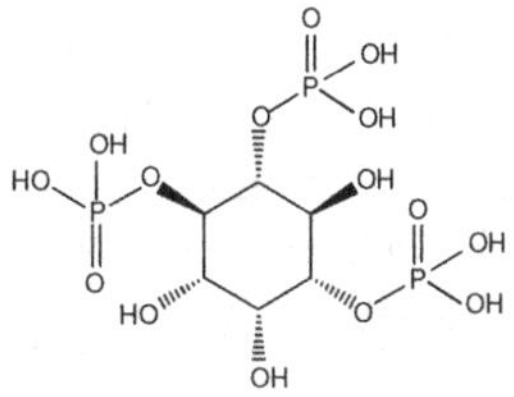

2.272

Figure 2.72 IP3.

Thanks to studies using mRNA, proteomics, knock-out mutants and so on, it is certain that most signalling in OSNs is via ACIII. However, there is occasional debate about possible involvement of an alternative pathway using inositol 1,4,5-triphosphate (IP3) (**2.272**) as second messenger (Figure 2.72). For example, Benbernou et al. found evidence that both the IP3 and cAMP pathways might operate in dogs, depending on the odorant/receptor combination (352). However, Brunert et al. found that the $G_{\beta\gamma}$ complex activates the enzyme involved, phosphatidylinositol-3 kinase (PIP3K), resulting in production of inositol triphosphate (**2.272**) which inhibits the signal initiated by the G_{α}–ACIII system (353). The role of IP3 (**2.272**) in odorant-induced inhibition of olfactory signalling has been reviewed by Ache (354).

It has been found that a key agent in signal termination is the calcium exporter NCKX4 Na^+/Ca^{2+} (355). This is an ion channel which removes calcium from the cell and admits sodium to replace it. The fall in calcium level then closes the chloride channel and ends the depolarisation. Unweaned mice that were deficient in the NCKX4 Na^+/Ca^{2+} exchanger had a poorer sense of smell than their litter mates and consequently a lower body weight because of a reduced ability to locate their mother's nipples. The work also suggests that calcium plays a role in adaptation to odour since failure to switch off the signal from the first stimulation by an odorant prevents new spikes in the electrical circuitry of the OB.

The complex between Ca^{2+} and the protein calmodulin normally serves to close CNG channels and therefore help switch off the signal, but it has been found that it plays less of a role in olfaction than in other signalling systems (356).

Modulation of Elements of the Second Messenger Cascade

Since ACIII and the various ion channels are membrane proteins, exposed to the extracellular mucus, it is quite conceivable that odorants could modulate their activity either positively or negatively. Electrochemical studies using isolated neurons (357, 358, 325) or intact animals (359, 360) clearly indicate that one odorant can affect the response of an OSN to a second odorant. In some cases, the effect is dependent on the specific identity of both odorants, suggesting that in those cases the mechanism does not involve action on a component of the signaling cascade that is common to all neurons. However, there is evidence that some odorants do interfere with these downstream elements.

ACIII Inhibitors

Chatelain et al. showed that 1-alkanols from methanol to decanol (**2.244**) suppress the activity of adenylate cyclase in the heart when present at high concentration (361). At lower concentrations, their effect could be either neutral or to positively modulate the activity. It is therefore very likely that similar modulation could occur with ACIII in the OSNs.

CNG Channel Blockers

Amiloride (**2.273**) and four related compounds were found to block ORCO in *A. gambiae* (Figure 2.73). As discussed previously, ORCO functions as an ion channel. None of the compounds would be volatile enough to reach the antenna through the gas phase, but this observation does prove the principle of blocking of ion channels (362). Breunig et al. found that the dye FM1-43 (**2.274**) also blocks olfactory CNG channels and showed that it can be used to mark OSNs that use the cAMP second messenger pathway (363). Even more significant for human olfaction, various studies have shown that odorants can suppress the activity of the CNG channels (364–366). Kurahashi et al. reported the level of blocking for a variety of odorants representing various structural classes and functional groups: for example, the terpenoid alcohols geraniol (**2.205**) and linalool (**2.275**), the terpenoid aldehyde citral (**2.114**), an aromatic alcohol (phenylethanol (**2.11**)) and an aromatic aldehyde (benzaldehyde (**2.12**)) (366). Their target odorant for masking was isovaleric acid (**2.145**), and Kurahashi et al. showed that the various CNG channel blockers did reduce the perception of isovaleric acid (**2.145**) *in vivo*. However, if they block isovaleric acid (**2.145**), they must block all odorants since all OSNs use the same CNG channel. This could be part of the explanation for the sensory effects between odorants in binary mixtures where, as will be discussed later, mutual suppression seems

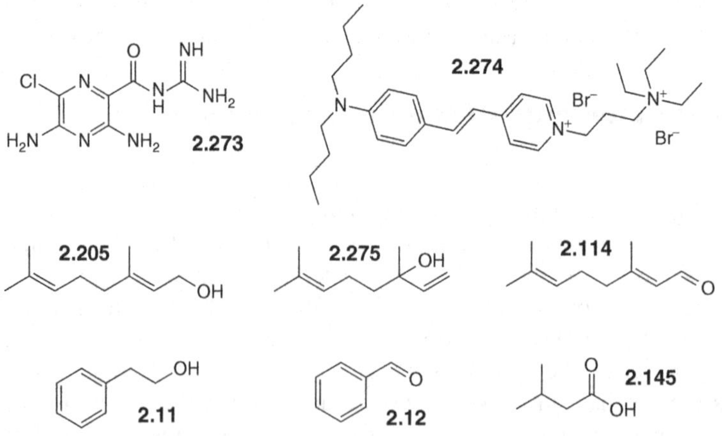

Figure 2.73 CNG channel blockers.

to be the norm. In a later paper, they showed that alcohols and esters blocked not only CNG channels but also voltage-gated sodium channels, whereas carboxylic acids did not (367). The degree of blocking correlated with the hydrophobicity of the blocking odorant. The CNG channels were actually from the newt *Cynops pyrrhogaster*. The degree of blocking increased as the length of the carbon chain of the blocker increased from four through six to eight carbon atoms. It was reported that the degree of CNG channel blocking correlated with sensory masking, but this cannot be selective since blocking the CNG channel would affect all odorants because irrespective of whichever OR is expressed in an OSN, the CNG channel will be the same.

As mentioned previously, inositol triphosphate (**2.272**) plays a role in signal transduction, and activation of PI3-kinase (which produces it) has been linked to signal reduction (368). It has been suggested that binding of the blocking odorant to an OR activates the kinase rather than triggering binding of the G-protein with subsequent activation of the cAMP cascade (358, 325).

M3 Muscarinic Acetylcholine Receptor Modulation

Although the OSNs express only one type of OR per OSN, they do also express other types of receptors and therefore their response to odorants is amenable to modulation by other agents such as hormones and neurotransmitters (4). For example, OSNs express muscarinic receptors which are activated by acetylcholine (**2.280**), and it has been shown that agonists of the muscarinic receptors do indeed modulate the response of ORs to its agonists of any OR in the same olfactory neuron (252). Thus, any odorant affecting the M3 receptor could affect the response of any of the ORs.

The Effect of Metals

As mentioned previously, it has been found that, *in vitro*, cupric copper was found to enhance the signal produced by the action of the murine semiochemical MTMT (**2.74**) on the mOR244-3 receptor (260). Concern regarding the use of nasal sprays containing zinc glucuronate (2.276) as a treatment for the common cold led Duncan-Lewis et al. to investigate the effects of zinc and other divalent metals on the ORs (369) (Figure 2.74). Their concern sprang from the known cytotoxic effects of zinc on the OE. They found that both Zn^{2+} and Cu^{2+} negatively affected the sense of smell in mice. The nasal mucus is known to contain metal ions but their role in normal olfaction is not known.

OLFACTORY NEUROPROCESSING

The number of macromolecules and small molecules involved in recognition of odorants and signal generation, and from that recognition, together with all of the interactions between different odorants in a mixture and between odorants and components of the OSNs, makes the process of signal generation a very complex one.

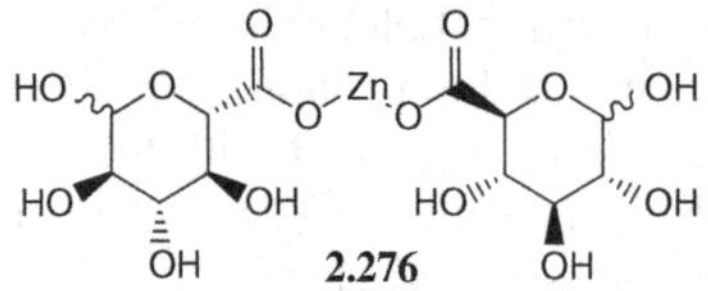

Figure 2.74   Zinc glucuronate and some neurotransmitters involved in olfaction.

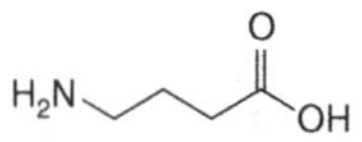

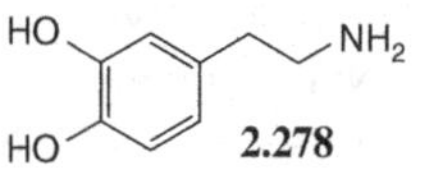

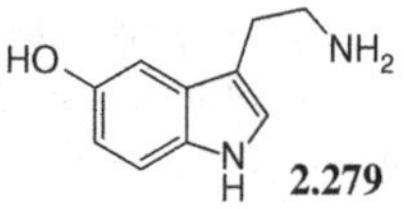

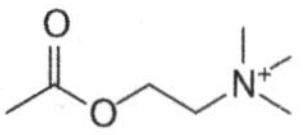

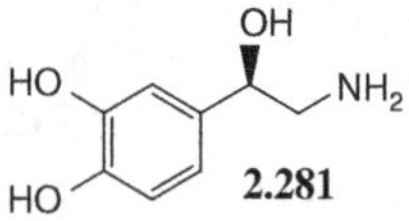

However, the processes involved in converting that initial signal into an olfactory percept (i.e. a smell) are many orders of magnitude more complex. A comprehensive description is beyond the scope of this book, and what follows serves only as an introduction. Much more detailed treatments can be found in the books by Wilson and Stevenson (3), Shepherd (5) and Hawkes and Doty (4). This section of this chapter will cover the formation of an odour percept. Higher processes and the effects of olfactory signals on our lives will be covered, also briefly, in Chapter 4. For both parts of the book, it is important to keep in mind William James' observation, 'The general law of perception, "Whilst part of what we perceive comes through our senses from the object before us, another part (and it may be the greater part) comes from inside our heads"' (370). The ORs are tuned to physical and chemical properties of molecules and not to odour, and the subsequent neuroprocessing of the signal is synthetic and affected by factors independent of the activating molecules. Thus the signal coming in from the ORs is only the start of the story, and the synthesis of the percept that we call smell is built up from many inputs, both external and internal.

Basic Flow of Signal Processing

The signals generated by activation of ORs in the OSNs result in signals in the OB. The ORs have a background level of firing. In insects, the spontaneous activity of neurons in the olfactory system stems mostly from the sensory neurons and signal discrimination is actually enhanced by combining input from many sensory neurons (371). The same is likely to be the case in mammalian olfaction.

All OSNs expressing the same OR project their axons to the same point in the OB, and this means that there is a chemotopic mapping in the bulb. The connection between the OB and the PC is known as the *olfactory tract*. In rodents and dogs, this is relatively short but in humans it is a long stretch of nerve tissue, as can be seen in Figure 2.1, with many local inhibitory circuits along its path. By the time the signal reaches the PC, the chemotopic mapping is lost and what forms in the PC is known as an *odour object*. Recognition of odours results from comparison of the new odour object with those stored in olfactory memory. Although the PC appears to be key in forming odour objects, conscious perception of odour resides in the OFC. Since the limbic system is involved in the early stages of olfactory processing, it is postulated that the direct link to the limbic system accounts for the influence of odour on memory and emotion. Processed signals from the thalamus also go to the OFC, and this also receives signals directly from the amygdala. Many of these signal pathways are two-way in nature with signals coming down from higher centres affecting the ascending signals. This accounts for the well-known effects of experience, expectancy and context in distorting odour perception.

Figure 2.75 shows some of the principal routes from the receptors to the OFC, as described by Wilson and Stevenson (3). The combination of the PC, the entorhinal cortex and the amygdala is often referred to as the *olfactory cortex*. The figure shows that many of the pathways operate in both directions, and signals coming from the receptors can be modulated by the extensive network of those descending from higher brain regions. The OB and PC are particularly affected by inputs from the cholinergic basal forebrain (372). In an excellent review of the subject, Leinwand and Chalasani describe how neuropeptides modulate the signal from the sensory neurons, how the first layer of connections from the epithelium to the bulb are stereotyped and how this is lost during forward processing and how the activity in the PC is sparse, providing much more scope for learning and experience (7).

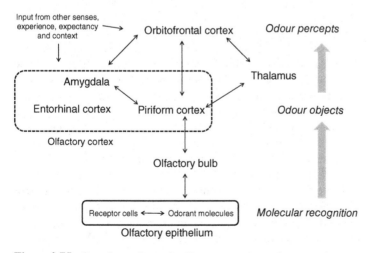

Figure 2.75   Some key pathways in olfactory neuroprocessing.

These factors account for the well-known effects of experience, expectancy and context on odour perception. Visual signals relate to wavelength of light, sound to wavelength of air waves and touch to pressure, but there are no such direct relationships with physical parameters in olfaction. Odours can only be described by reference to other odours. It is therefore clear that experience is crucial, and the more the number of odour objects stored in a person's brain, the more discriminating will their perception of smell be. For example, in perfumery, it is well known that development of an odour language is essential in training perfumers, and their powers of discrimination improve as their language improves. The effect of learning on smell perception has been demonstrated in many ways. For example, it has been shown that people trained in wine tasting perform better in identifying and describing smells than controls, although the basic sensory ability is the same in both groups (373). Expectancy also affects odour perception as the classic experiment of a red dye in a white wine proves. Addition of a tasteless red dye to a white wine caused a panel of wine experts to judge the aroma wrongly because the sight of the red colour affected the way their olfactory processing functioned (374). The dependence of odour interpretation on context is illustrated by examples such as that of isovaleric acid (**2.145**), which can be described as either that of sweaty feet or of cheese depending on the context in which it is perceived. Similarly, certain amines would be described as either fishy or leathery, depending on the context. Adaptation and habituation affect signal processing – another topic which will be explored in more detail below – and this can be a component in creating context dependency. We may expect odour perception to remain constant, but it has been shown that, in one example at least, that of androstadienone (**2.202**), both the threshold and hedonic perception change on continued exposure to the odorant (375).

Input from Other Senses and Non-Sensory Sources

The total odour percept is built up from olfactory input and also input from other senses, such as the trigeminal sense. For instance, 70% of odorants also stimulate the trigeminal system in the nasal cavity, and visual input has been shown to affect olfactory signal processing (376). Taste also affects odour perception. All cooks know that addition of salt or monosodium glutamate to a food will enhance the perception of the odorous components of the aroma. Salt and umami taste receptors and odour receptors do not interact directly but only through higher processing centres in the brain. There are also plenty of experimental results showing that the odour description given to a volatile chemical will be affected by colour cues. For instance, a simple aliphatic ester might be described as smelling of banana when a yellow colour is associated with it, or as strawberry when the colour is red. The effect also works in both directions and, for example, a towel will feel softer when paired with a lemon scent than with an animalic one (377). In the case of sight affecting smell, it has even been shown that magnetic stimulation of the neurons in the visual cortex will affect perception of an odorant presented to the nose (378).

We each build a mental model of the universe around us and, as William James so rightly observed, our internal mental processes can be as important, or even more important, than the inputs from the sensory organs. The fact that the brain synthesises our model of the world around us can easily be illustrated by a simple visual example. When we stand in a green field and look directly ahead, the field appears green ahead and also to both sides. Colours are detected by cone cells in the retina, the rod cells detecting only light as opposed to dark. There are only cone cells in the centre of the visual field, so the brain is doing a bit of extrapolation to give the illusion of seeing green at both sides. Another example comes from my own experience. After my first visit to an optician, I was told that I had developed astigmatism in my left eye. A pair of spectacles was prescribed to correct this. I remember collecting the spectacles, putting them on and finding that the counter in the optician's shop suddenly appeared curved and sloping downwards to the left instead of the straight and level counter I had seen when I entered. Going out into the street proved a challenge, as the pavement also now seemed to slope off to the left and the straight kerb appeared to curve round to my right. In the café opposite, the circular tops of cups appeared elliptical. After a day or so, my brain had adapted to the altered sensory input and everything returned to normal; straight lines were straight again, circles were circles and level surfaces level. Even more remarkably, now when I put the spectacles on or take them off, my brain adjusts faster than I am conscious of and I see correct images immediately. Such instant adjustment of the visual sense must be paralleled also in olfaction and it must be that the brain also interprets, constructs and reconstructs olfactory images with the same facility that it does visual images.

Input from the different senses is integrated in the OFC, but smell is the only sense that has a direct route into the OFC without going through the thalamus. When the input from different senses is congruent, there is an enormous amplification factor in the weight given to the sensory evidence; for example, it has been shown that when visual and olfactory signals are congruent, there is greater activity in the hippocampus and the OFC than if they are not. The direction from which a sound reaches an observer is identified more quickly if there is also an olfactory cue to help, illustrating how the senses work together (379). In the moth *Spodoptera littoralis*, the sound of a predator bat increases the moth's sensitivity to the moth's sex pheromone, giving a quite different illustration of cross-modal interaction between hearing and smell (380).

When the inputs from different senses are conflicting, the brain very rapidly constructs a compromise interpretation. One very good illustration of this is the McGurk effect, which involves conflicting visual and auditory inputs. When the observed facial movements do not correspond with the sound presented, the hearer's brain constructs a new mental image of the sound which is different from what would be experienced with the eyes closed. Readers wishing to experience this can find many examples of it on the Internet by searching for the term *McGurk Effect*.

A stimulus that is composed of input from different senses is usually easier to detect than one that relies only on one sense. There are two possible ways of processing the signals to arrive at identification of the stimulus. Probability summation (PS) involves separate analysis of the inputs, and a decision that is the sum of the probabilities of matching the sensory input to a record is stored in memory. Neural integration involves interpretation of a merging of the two different sensory inputs. Marks et al. have shown that, in the flavour system with inputs from gustation and olfaction, neural integration is more likely to account for observed results than PS. This illustrates how interaction between olfactory signals and those from other senses are combined early in the process of interpretation (381). Smell, taste and the trigeminal sense all contribute to the percept of flavour. Although all three inputs are combined to create the percept, there is no common chemical sensitivity between them and the ultimate percept is therefore dependent on the neural objects rather than the chemical properties of the substance responsible for the flavour (382).

At the neurological level, the most studied interaction between smell and other sensory inputs is probably that between the olfactory and trigeminal systems. Using fMRI, it was shown that trigeminal and olfactory signals are processed by different parts of the brain but that there is an overlap, explaining the close interaction between the two senses (383). It has been shown in rodents that SCSs in the nasal epithelium express the bitter taste receptor TRPM5 and, when activated, send their signals to the trigeminal nerve (45, 384, 385). According to Müller's Law of Specific Nerve Energies, perception is determined by the nerve carrying the signal and not by the specific means of triggering the nerve. Thus any signals from the trigeminal nerve will be interpreted as pain or irritation. Therefore, any odorant triggering the TRPM5 receptor in the nasal epithelium will carry that message, along with the odour message it transmits by way of the olfactory system. Studies using fMRI and comparing anosmic subjects with normal controls have shown how the olfactory system modulates the signals arriving via the trigeminal system (386). The temporal characteristics of trigeminal processing were shown by comparing the fMRI patterns of monomodal (olfactory only) and bimodal (olfactory and trigeminal) odorants (387). Different brain regions are involved in processing the trigeminal signals depending on the duration of sampling of the odorants. The just noticeable difference (jnd) is the change in concentration required to enable a subject to perceive a difference between two odour stimuli of different concentrations and it is lower when odorants have a stronger trigeminal component, showing that intensity changes are easier to detect for more pungent odorants (388). Presentation of a trigeminal odorant during sleep causes waking and simultaneous presentation of a pure odour stimulant increases this effect, showing how the two senses work together and probably do so at an early stage of processing (389).

As discussed previously, the input from the ORs in the sensory neurons can be modulated by the autonomous nervous system (254). Attention also alters the way in which signals are processed. Using fMRI, it was shown that cortical activity on smelling jasmine samples depended on whether subjects were asked to rate pleasantness or intensity (390). Using positron emission tomography (PET), it has also been shown that odour signals of social relevance (i.e. body odours) are not processed in the same way and not in the same organs as the other signals (391).

As will be discussed in more detail later in the section on perception of mixtures, the task in hand also affects the way signals are processed by the brain (392). The fact that non-sensory neural activity can distort sensory perception is not limited to olfaction. A study on neural activity when viewing works of art showed that the brain functions in a different way if the observer believes the picture before them to be a genuine work of a great master from the way it functions if the observer believes it to be a forgery. This difference was the same whether or not the belief in the authenticity of the painting was correct (393).

Both the amygdala and the hippocampus are closely linked to olfactory processing. The amygdala is also associated with emotion and the hippocampus with memory. Thus the well-known links of both emotion and memory to olfaction have a clear neurological basis. Pairing a pleasant odour with a bad memory can reduce the bad emotions associated with that memory, showing the involvement of both the amygdala and the OFC. However, pleasant odours stimulate activity in the medial OFC and unpleasant ones in the lateral OFC. Such separation is not seen at lower levels of processing and hence this linear quality axis is not present in the molecular structure of the odorants. Pleasantness is learnt by association, for example, by association with taste (5). Neuroticism is related to chemosensory responsivity, which predicts ratings of both intensity and threshold of odours (394). This could explain why people who perceive a particular odour as a threat tend to have lower thresholds for it than average. Viewing pictures also affects sensitivity to odours and ratings of their pleasantness (395). For example, viewing an unpleasant picture whilst smelling an odour will result in a lower pleasantness rating for the odour. Verbal cues are also well known to affect hedonic perception of odours and this has been demonstrated even in young children (396). Similarly, visual cues also affect odour perception, and both colour and shape of the visual cue are important (397, 398). In female malarial mosquitos (*A. gambiae*), the sensitivity to odorants changes after a blood meal because of up-regulation of some OR genes and down-regulation of others (399). In humans, the facial expressions displaying emotions also affect sensory perception. For example, expressing fear dilates the nostrils and thus increases olfactory acuity, whereas expressing disgust has the reverse effect (400). Thus mental state also modulates olfaction through physical effects.

In summary, because of all the inputs from other senses and the non-sensory inputs, odour is not a static, reproducible phenomenon. At the time of measurement, in addition to all of the factors affecting detection of the signal (e.g. physical chemical parameters such as humidity, genetic profile of the subject's OR repertoire, etc.), there are also many factors (e.g. previous olfactory experience, concomitant visual input, mental state, etc.) affecting the way the subject's brain processes that signal. The layman's view that odour is a simple, fixed property of the volatile molecules being sensed is far from reality.

Techniques Used to Study Olfactory Neuroprocessing

Nerve cells (or neurons) have projections called *dendrites* and *axons*. The dendrites receive signals from other neurons, whereas axons send signals to other neurons. The electrochemical connection between the axon of one neuron and the dendrite of another is known as a *synapse*. The electrical signal is carried across the synapse by neurotransmitters such as GABA (**2.277**), dopamine (**2.278**), serotonin (**2.279**) and acetylcholine (**2.280**) (FIgure 2.74). The main techniques used for studying activity in the brain are electroencaphalography (EEG), electro-olfactograms, 2-deoxy-glucose (2DG), optical imaging, PET, fMRI and optogenetics. Recently it has been suggested that IR might also be used. Each of these techniques is outlined briefly below.

Applicability of Various Neuroscience Techniques

Many of the techniques used to study neuroprocessing cannot be used on humans. The OB, PC and other organs used in early processing of olfactory signals lie at the base of the brain and are thus physically inaccessible. EEG uses electrodes placed on the exterior of the head and detects only signals that are present late in the process. fMRI can be used to detect activity in the human brain due to changes in the environment of protons (essentially those in water) but the resolution of fMRI is dependent on the strength of the applied magnetic field and there are safety limits to the field strength to which humans can be subjected. Similarly, safety constraints limit the applicability of PET to humans. Some techniques, such as optogenetics, clearly cannot be used on humans. Much of what we know about olfactory neuroprocessing is therefore obtained from work with animals and extrapolation to humans must be done with great care, as mentioned earlier for all aspects of olfaction. Differences between vertebrate and invertebrate brains and also differences between human brains and those of other mammals must always be borne in mind (7).

Electro-Olfactograms

Electrical activity in sensory neurons can be detected by electrodes placed in their vicinity, and this technique has been used to study olfaction since the 1950s. The electrodes can be used in living animals or with isolated neurons. Initially, studies were mostly carried out on rodents, fish or amphibians, but recently the technique has been extended to good effect using human volunteers. The advantage of using human subjects is, of course, that the subjects are able to give descriptions of the odours as the electrodes in their noses pick up the initial nerve signals (401). For example, using this technique, it was found that some areas of the human OE seem to be more sensitive to pleasant-smelling odorants than do others (402). Even the earliest experiments confirmed that each sensory neuron responds to a range of odorant molecules. Other distinguishing features of the olfactory system were also quickly established. The signals show a fairly narrow dose/response range, with saturation occurring within one to two orders of magnitude of the first stimulus. The latency (time before onset) reduces as the odorant concentration increases. Detection of an odorant at the receptors results in spikes of electrical activity in the neuron, and the frequency of the spikes increases with increasing odorant concentration, rising from a basal rate of 0.05 to 3 Hz, peaking at about 50 Hz and falling off as the signal reaches saturation. Hormones such as insulin and leptin affect the basal spike rate, providing a route for non-sensory factors to affect odour perception. Studies on animals suggested that the most meaningful information is transmitted in the first spikes in the activity of neurons. This view is supported by work on a model system in which patterns very close to those of the natural one were found (403).

2-Deoxy-glucose (2DG)

Although it accounts for only 3% of body weight, the human brain consumes 20% of the energy used by the body. The energy is supplied by oxidation of glucose. If 2DG is supplied instead of glucose, it is taken up in areas where energy demand is high but cannot be metabolised and therefore accumulates. Obviously, a high dose would be toxic but low doses can be tolerated. If the 2DG is radiolabelled, then its accumulation will give rise to a signal that can be quantified by use of photographic techniques similar to those used in X-ray imaging. Thus administration of 2DG can give images showing where brain activity has been the highest. When used to study the OB, these images produce what are referred to as *glomerular maps*. This technique cannot be used on humans and is normally confined to rodents.

The glomerular maps of Johnson and Leon (404) are interesting, and on looking through them, the fragrance chemist will quickly spot that often the most intense odorants, such as pyrazines, show relatively little activation whereas some weak odorants, such as saturated aliphatic hydrocarbons, produce large amounts. For example, the maps of 2,3,4-trimethyl pentane and 2,3.5-trimethylpyrazine elicit similar degrees of activation though their perceived intensities (for humans) are very different. This is rather counter-intuitive but demonstrates that it is the total pattern that matters and not individual features. Perhaps, contrast or temporal effects are more significant than total activation over an extended period.

Optical Imaging

Optical imaging of stimulated neurons is possible because of changes in light scattering during neural activity. It has been used to generate maps similar to those using 2DG. The group of Mori in Tokyo has used this technique to produce maps of mouse and rat OB activation with various sets of odorants (405, 299). In one study, they identified a region of the OB that responded to three odour types strongly associated with malodours, namely amines, acids and aldehydes. Glomeruli in neighbouring areas responded to fennel and clove odorants, and it was suggested that lateral inhibition by these might account for their ability to sensorily mask fishy off-odours (406).

Positron Emission Tomography (PET)

PET is an imaging technique that uses γ-ray emissions from a radionuclide to build a 3-D image. By tagging the radionuclide to fluorodeoxyglucose (FDG), a glucose analogue, the rate of metabolic activity can be studied. Single-photon emission computed tomography (SPECT) is a similar technique that has been used to measure the damage to the olfactory system after head trauma (407).

Functional Magnetic Resonance Imaging (fMRI)

fMRI relies on response of suitable nuclei to radio signals when there is a strong applied magnetic field. Basically, this is the technique known to chemists as NMR spectroscopy. In body imaging, it is usually the protons of water that are detected. The signals show where blood flow is highest and thus where brain activity is highest. The technique is non-invasive and so can be used on humans but with a limitation on resolution imposed by the magnetic field strengths that humans are allowed to be exposed to.

Investigation by fMRI shows that many brain regions are involved in smelling, and odour identification and the nature of the task in hand affect which regions are most active. For example, the entorhinal cortex and hippocampus are heavily involved in odour identification tasks as are the PC and OFC, but the latter two are also very active during smelling (408).

Optogenetics

Optogenetics is a relatively new technique in neuroscience and allows manipulation and observation of individual neurons in real circuits in real time (millisecond timescale). The technique involves cloning of channelrhodopsin into neurons. Channelrhodopsin is a TM protein and its active chromophore is retinal (**2.42**). Retinal (**2.42**) is bound to the protein scaffold in a very similar way and in a very similar site in the protein to those in which it is bound to opsin in rhodopsin (409). Channelrhodopsin functions as an ion channel which is activated by red light. When the light is switched on, the cell depolarises and so a nerve signal is generated. One disadvantage is that the channelrhodopsin is distributed all over the neuron and therefore the depolarisation is not localised as it is in the real situation. Halorhodopsin is a light-activated chloride pump and can be used to silence signals. Thus red light can be used to switch neurons on via channelrhodopsin and green light to switch them off, via halorhodopsin. Channelrhodopsin and halorhodopsin can be cloned selectively into OSNs by coupling them with OMP, and thus mice can be made to 'smell' light. Which neuron responds to which odorant can be detected by means of the OMP and then light can be used to imitate the effect of smelling that odorant. Which glomerulus responds to a given odorant can also be determined.

Since light can be delivered with high precision in terms both of time (submillisecond accuracy) and space (individual neurons or mitral cells), optogenetics can be used to study neural pathways between different brain structures. Blumhagen et al. used optogenetic techniques to study the effect of different activation patterns in the OB of zebrafish on activation of the dorsal telencephalon, the organ that performs a similar role to that played by the PC in humans (410). As mentioned previously (in the section on sniffing), the scale, accuracy and precision of optogenetics allowed the temporal aspects of murine responses to odour during the sniff cycle to be determined (96). Co-ordination of pressure sensors to detect sniff cycle and the administration of light to the neuron demonstrated that the mouse can detect within 10 ms accuracy, at which point in the sniff cycle the odorant was detected. This temporal patterning could therefore be part of the odour coding, for example, in determining intensity. Thus intensity can be determined using only one glomerulus since timing of mitral cell spikes correlated with odorant concentration. The OB and the hippocampus are the two brain regions where new neruons are formed, even in adults. Livneh and Mizrahi used optogenetics to show that adult born neurons in the OB do respond to odours and, contrary to former opinion, continue to form new synapses after completion of their development (411). This provides a mechanism for plasticity in olfactory neuroprocessing. Murthy and coworkers used optogenetics to confirm several theories about olfactory neuroprocessing that have been postulated some time ago but could not be tested previously. For example, they were able to show how feedback mechanisms from the PC modulate the signals coming in from the OB, for example, by suppression of firing of mitral and tufted cells (412).

Infrared Spectroscopy

Near infrared (NIR) rays can penetrate into the brain allowing NIR spectroscopy to detect changes in haemoglobin levels. Consequently, this has been shown to be a potential tool for the study of cortical responses to odour in humans (413).

Olfactory Sensory Neurons

The OSNs are situated in the OE. At the end which is in contact with the mucus layer, they have hair like projections or cilia. The OR proteins are expressed in the cilia, and regulatory mechanisms determine that only one allele is expressed in any neuron (2, 414, 415). The long axons of the OSNs pass through holes (called *foramina*) in the cribriform plate at the base of the skull and project into the OB where their dendrites form synapses with the mitral and tufted cells in a structure known as a *glomerulus*. The receptors are also expressed at the dendrite. It is thought that this is involved in the targeting of the OSN to the same glomerulus to which all other neurons expressing the same receptor also project. However, the precise mechanism by which OSNs find the correct glomerular cells is uncertain, although the ORs at the growth cone have been shown to be involved (416), and a family of proteins known as *teneurins* have been shown to be involved in matching one synaptic partner with another (417, 418). In a genetically modified mouse in which >95% of the OSNs expressed the same receptor, it was found that the mechanism broke down and the OSNs in question projected axons to almost all of the glomeruli (419).

The receptive range of the ORs is reflected in the response of the OSNs expressing them. One study showed how, in mice, an array of 59 aldehyde-responding receptors can discriminate between nine different aldehydes (292). Similarly, in *D. melanogaster*, a map of 24 neurons versus 110 odorants showed that most odorants are detected by several ORs and most ORs responding to a range of odorants. The breadth of tuning varies, with some ORs responding to only a few odorants and others responding to a large number (320). Another map plotted the response of 15 neurons of against 47 odorants (420). Grosmaitre et al. used intact OE to record the response of OSNs expressing mOR23 (164). They found a 1000-fold variation in the strength of the response. Such a large range has not been observed by other groups. Oka et al. found that the mOREG receptor responds almost equally to vanillin and eugenol but, when they looked at the pre-synaptic Ca^{2+} levels of (OSNs) expressing mOREG, the response to vanillin was much weaker than that to eugenol (91). Such effects remain unexplained at present, but they do indicate that understanding the odorant/OR interaction does not necessarily provide the complete picture.

The neurotransmitter involved in taking the signal from the OSN to the mitral and tufted cells of the OB is GABA (**2.277**). GABA may also be implicated in a presynaptic inhibition representing an inbuilt mechanism of preventing saturation of subsequent processing during repeated sniffs or persistent background odour (421). The signaling molecule dopamine (**2.278**) is probably also involved in presynaptic inhibition via dopamine D2 receptors in the OSN (422). The neurotransmitter serotonin (**2.279**) has been found to be involved in olfactory memory in *D. melanogaster* and so there is a possibility that it is also involved in mammalian, including human, olfaction (423). Human OSNs also express a sodium channel known as *Na(v)1.7*, and it has been shown that without this channel the OSNs fail to fire at the first synapse in the bulb (424). This channel is also key to pain sensation, and there exists a genetic condition in which the gene (SCN9A) coding for it is absent. Those suffering from this condition are unable to experience pain, a condition known as *channelopathy insensitivity to pain (CAIP)*, and sufferers are also anosmic.

The Olfactory Bulb

Overview

The OB sits at the base of the brain on the cribriform plate. Humans have two OBs, the right bulb receiving signals from the right OE and the left from the left. Thus smell, unlike sight, is processed on the same side of the brain that receives the signal. In the bulb, there are centres called *glomeruli*, each of which receives signals only from receptor cells (OSNs) expressing the same type of receptor protein, irrespective of where those receptors are located on the epithelium. OSNs have short lifetimes (about 2 weeks) and new receptor cells develop in such a way that each makes its connection with the correct glomerulus. Receptor types with similar substrate selectivities tend to be associated with glomeruli which are found close to each other in the OB. Thus the topographic layout of the glomerular layer on the OB results in a chemotopic map. Of course, knowing what is described above regarding receptive ranges of ORs, correlation of these maps with physical and chemical properties of odorants producing them is not straightforward.

Neural Structures and Processes

The OSNs project their axons to M/T cells in the glomerulus and the selectivity of the OSN is transferred to the M/T cell. Interneurons lying between the M/T cells have a very broad tuning, and lateral inhibition from these increases the sharpness of the signal (425). The mammalian OB contains a network of interneurons, and the level of interneurons in the OB is 50% of that of mitral cells, whereas in other brain regions the ratio is about 15%. This demonstrates the high significance of feedback in olfaction. It has been shown that even in adulthood, at least in the mouse, interneurons develop and disappear over time, with a net gain in numbers. Thus it would appear that the OB undergoes continual change (426).

Various receptors, other than ORs, are required for the sense of smell to function correctly. The role of the muscarinic acetylcholine receptor in OSNs has already been alluded to. Norepinephrine (**2.281**) affects olfactory signal processing in the OB, indicating a role for adrenergic receptors there also. The input comes from the locus coerulus and is important in forming conditioned odour preferences. A review of the functional role of norepinephrine (**2.281**) in the OB has been written by Linster et al. (427). In humans, it was found that norepinephrine is not essential for odour identification but that peripheral noradrenergic innervation is essential for that task (428). Expression of the α-7-nicotinic acetylcholine receptor in the OB is essential for the sense of smell in mice, and its absence in humans suffering from schizophrenia might account for their difficulty in odour detection and discrimination (429).

It has been found that a prion protein, PrP^c, plays a role in synaptic transmission between the granule and mitral cells of the bulb (430).

It is commonly considered that activation of a neuron results in a change in the rate of its firing. However, one study suggests that stimulation of the M/T cells in the OB results in a change of timing rather than rate of firing. Consequently, it was suggested that populations of M/T cells could encode olfactory information and allow for a trade-off between speed and accuracy of odour discrimination (431).

There is a direct synaptic interaction between the OSN and the mitral cell, but this is affected by circuits within the glomerulus (432). Evidence for lateral inhibition of mitral cells (433) led to the theory that olfaction resembled vision in having a system of centre surround inhibition leading to improved signal discrimination (434). However, more recent investigation suggests that lateral inhibition is much more sparse in olfaction than in vision and that the mitral cells of the OB perform a specific computation using inputs from a small and diverse set of glomeruli (435). When activated by a neighbouring mitral or tufted cell, periglomerular cells inhibit neighbouring cells (lateral inhibition) and thus a strongly activated glomerulus will tend to dampen activity of those around it. However, they are also capable of damping the activity of remote sites (5). Electrochemical studies have shown how inhibitory circuits in the bulb can, through temporal effects, maximize both encoding and propagation efficacy of the signals (436).

Because of the importance of smell to food signals, centres for processing olfactory signals are linked to those related to behavioural state. Thus, for example, mitral cells are less active when the subject is sated than when hungry. Similarly, mitral cell activity is affected by moods such as happy, sad and angry (437). Jadauji et al. found that stimulation of the visual cortex using a magnetic field improved performance in discrimination between odours. Sham stimulation or stimulation of the auditory cortex did not show the same effect. This suggests that the visual cortex affects the functioning of the olfactory cortex (378).

Mitral cells that respond to ethyl esters were found to each have a preferred carbon chain length for the acid component. Previous smelling experience of homologues affected the response of the cells to novel esters. Since the input is determined by the selectivity of the OSNs, this would suggest that he effect of experience stems from lateral and centrifugal OB circuits (438).

Glomeruli

Just as all of the OSNs that express the same OR converge on the same glomerulus, it is generally accepted that OSNs expressing similar ORs tend to project to neighbouring glomeruli and therefore that the ORs are involved in determining the glomerular mapping (439, 299). Optical imaging also indicates that glomeruli corresponding to receptors with similar receptive ranges are located close to each other on the OB (440) as do the maps derived from 2DG studies (441). Previous experience of an odour increases responsiveness of the bulb of young rats, not only to the odour in question but also to other odorants that activate the same region of the bulb (442). The 2DG technique has also been used to show that odorants with higher water solubility activate glomeruli at the rear of the OB (443) and that the presence of double, or even more significantly, triple, bonds also affected the region of the bulb where greatest activity was seen (444).

However, one study found that, although the distribution of glomeruli in different animals of the same species is identical to an accuracy of 1 part in 1000, the similarity between neighbouring glomeruli is no greater than that between remote ones. There are broad domains of distribution of glomeruli but no evidence of systematic organisation of odour responses or corresponding local neural circuits (445, 446). This is probably consistent with the finding that, although compounds with similar structural features do activate glomeruli in similar locations, there is no simple pattern in the chemotopic map. However, there is a hierarchical arrangement into clusters responding to compounds with related features of molecular structure (447).

The pattern of bulb activation has a temporal dimension, and the glomeruli that are activated are dependent on both the odour and its concentration. Using this extra dimension, it is possible to discriminate between odorants in milliseconds using only a subset of the available glomeruli (448). Optical imaging of neural activity in the OB of the zebrafish *Danio rero* showed that, when detecting amino acids, more receptors were involved as the length of the carbon chain increased. Similarly, the presence of polar groups resulted in activation of specific additional receptors. As would be expected, when the concentration of odorant increased, the number of receptors activated also increased. The conclusion therefore is that a number of different receptor types are necessary for discrimination of even simple odorant molecules (449).

Glomerular Maps

Johnson et al. (450) examined a range of aliphatic esters with varying structural complexity and found that, in the OB of rats, similar compounds activated the same areas of the bulb in all of the animals and that the complexity of the activation patterns increased with the complexity of molecular structures. Monocyclic hydrocarbons activated similar regions to those activated by acyclic compounds, but bicyclic compounds related to camphor activated unique regions of glomeruli.

Exposure to pure odorants was found to cause a low level of broad activation of the OB with a small number of small spots of higher activity. On examining the effects of a homologous series of odorants, related odorants activated spots close to each other in the bulb and many spots were selective to a narrow range of carbon chain length. The selectivity was also concentration dependent, with the overall activation pattern for any odorant depending on its concentration in the nasal airspace (439). Heterocyclic compounds tend to give only weak activation of a relatively small number of glomeruli in rats even though they often elicit very intense odours in human perception (450). Similarly, in mice, only one cohort of mitral cells was found to respond to the key semiochemical MTMT (**2.74**) (451). Using pairs of enantiomers, it was shown that the ease with which mice could distinguish between two odorants generating similar patterns of activation was not affected by the degree of pattern complexity in either pair (452). Taking all of these observations into consideration, it would seem that a relatively small number of glomeruli are key for each chemotopic map in the bulb.

A statistical analysis of bulbar maps from rats suggests that there are similarities with human perception and that the spatial pattern of activation in the bulb is related to perceptual parameters such as pleasantness (453). One model of bulb circuitry also supports a role of the bulb in encoding perceptual dimensions (454).

The fact that the pattern of receptor (hence glomerulus) activation is more important than the response of any individual receptor is illustrated by the effects of DEET (**2.78**) on insects. As mentioned earlier, DEET (**2.78**) confuses the insect's perception by enhancing the signals from some receptors and reducing those from others, thus distorting the pattern and preventing the insect from recognising the odour (267).

Onward Processing

In a very significant experiment, Kobayakawa et al. showed that signals from different areas of the murine OB are processed differently, a result already alluded to in Chapter 1 (143). Exposure to the predator odour TMT (**2.15**) results in activation of glomeruli in different parts of the bulb, but only those in certain regions elicit the typical 'freeze' response associated with fear. Mice lacking the fear-generating area of the bulb still smell TMT (**2.15**) but do not respond with the freeze reaction unless trained to do so.

Signals from the dorsal region of the mouse OB go to the amygdala retaining the chemotopic mapping of the bulb, whereas the signals going to the PC lose that mapping (7). This is consistent with the findings of Kobayakawa et al. (already described in Chapter 1) that mice process predator signals differently from the way they process other signals (143). In rodents, some mitral cells that respond to OSNs expressing the TRPM5 channel project directly to the amygdala (455).

Other Roles of the Olfactory Bulb

The OB also plays a role in monitoring of brain glucose and insulin levels and thus monitors metabolic state (456).

The Piriform Cortex

The output neurons from the OB lie along what is known as the *lateral olfactory tract* which is much longer in humans than in most mammals. The axons of the olfactory tract project into the PC (also known as the *olfactory cortex*), but they feed signals in at many points, each time going into a micro-circuit in the PC (5). Using optical imaging techniques, Nobel Laureate Richard Axel showed that the segregated, chemotopic mapping of signals in the OB is lost and replaced by a highly distributed pattern in the PC (457). In another elegant experiment, he generated a mouse with a 'monoclonal nose'. More than 95% of the OSNs of this mouse expressed the same receptor, M71, which is sensitive to acetophenone (**2.119**), a murine semiochemical. This signal overload resulted in loss of onward transmission to the PC and a mouse that was anosmic to acetophenone (**2.119**) although it could smell, albeit weakly, other odorants, thanks to the remaining 5% of OSNs (419). In rodents, it has been shown that signals from single glomeruli in the bulb are relayed in different ways to different higher processing centres, some retaining the chemotopic map and others reorganising it (458). Irrespective of the number of glomeruli activated in the bulb, only about 10% of neurons in the PC are activated by any odorant or mixture of odorants.

The great advantage of this change in processing technique between the OB and the PC is that it enables the brain to fill in missing parts of neural images. The phenomenon is well known for vision where partial images or fragments of images are quickly recognised and matched to a stored complete image. Thus, we can identify a face just from a photograph of the mouth or the eyes, or from a poorly resolved graphical image. The analogy between recognition of faces and recognition of odours is nicely explained by Gordon Shepherd in his book *Neurogastronomy* (459). Since the process of recognition of olfactory objects in the brain resembles that of neural objects in the visual cortex, this explains why a simple accord can give a recognisable impression of a more complex fragrance or essential oil, as will be illustrated below in the discussion of mixtures. Like the OB, the PC is affected by input from higher brain regions; for example, in humans fMRI studies suggest that expectation of an odour affects signal processing in the anterior piriform cortex (aPC) (460).

The main neurons of the PC are what are known as *pyramidal cells*. The pyramidal cells respond to changes in the input signals and exposure to the same smell for a prolonged time causes adaptation, that is, a reduction of the signal. This presented a problem with early experiments using fMRI because increased exposure to the odorant resulted in a decrease in the observed signal. The PC learns signal patterns and the more the learning, the better is the ability to distinguish between similar smells, for instance, those of members of a homologous series.

Stored images also improve the signal-to-noise ratio and enable one image (hence one odour) to be identified against a complex background.

Odour objects are formed in the PC, as elegantly demonstrated by Gottfried and his co-workers using fMRI (461). They gave samples of nine odorants: three citrus, three minty, and three woody, to subjects and monitored activity in the PC as the subjects smelt each odorant. The compounds within each group produced similar activation patterns in the PC, even though their chemical structures were dissimilar and would therefore be expected to activate different receptor proteins. PC activation is clearly more closely related to odour percept than are either chemical structure or specific OR activation. The odorants used in this landmark experiment are shown in Figure 2.76. The citrus odorants were citral (**2.114**), citronellol (**2.242**) and (+)-(*R*)-limonene (**2.117**). (It should be noted that the citrus character of commercial (+)-(*R*)-limonene (**2.117**) is mostly due to tiny traces of decanal (**2.120**) (462) but this would not materially affect the conclusions reached.) The minty odorants were (−)-(*R*)-carvone (**2.116**), *l*-menthol (**2.282**) and methyl salicylate (**2.283**). The woody odorants were not pure single entities but rather

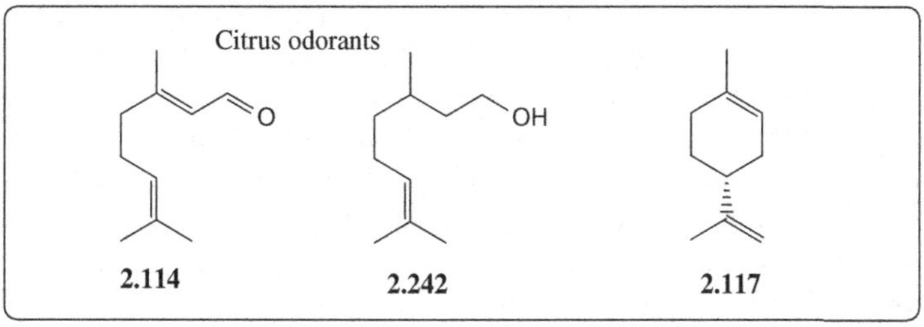

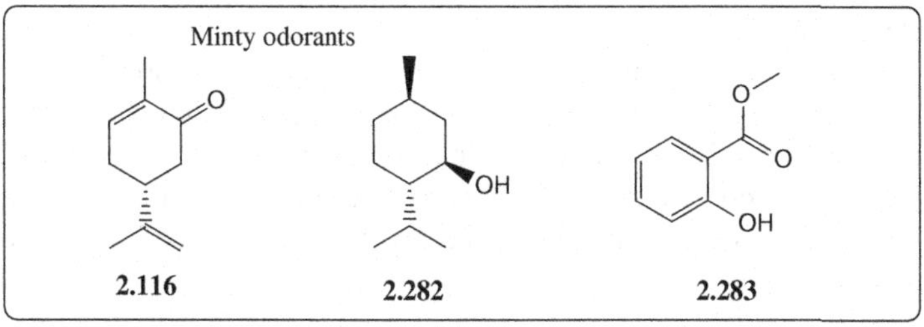

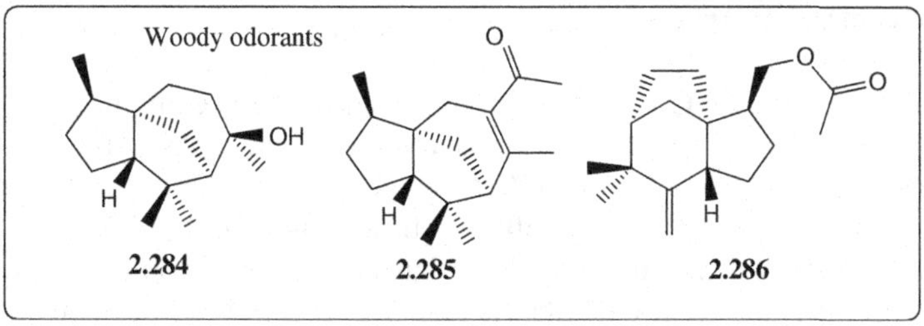

Figure 2.76 Nine odorants representing three odour categories.

were mixtures derived from cedarwood or vetiver oils, the major components being (+)-cedrol (**2.284**) (from cedarwood oil), methyl cedryl ketone (**2.285**) (acetylated cedarwood oil) and khusimol acetate (**2.286**) (acetylated vetiver oil).

Using electrophysiological recordings in rats, Wilson showed that after 10 s of exposure to a binary mixture, neither the neurons in the OB nor those in the aPC could distinguish between the mixture and its components. However, after 50 s of exposure, the aPC neurons did distinguish the mixture from its components. This shows that odour images are synthetic in nature and that learning of novel odours is rapid (463). The role of the PC in creating odour objects has been reviewed by Wilson and Sullivan (464) and by Gottfried who points out that the distributed patterns of olfactory qualities and categories are crucial for maintenance of constant percepts despite the variability of the corresponding odour input from environmental stimuli (465). Studies using a combination of fMRI and cross-adaptation to odours showed that perceptual codes for odours are degraded in the PC of sufferers of Alzheimer's disease. The results support the theory that the posterior PC is crucial for differentiating odour objects (466). Using fMRI, Gottfried et al. showed that anticipation of an odour results in formation of 'target' maps in the PC, which are then compared against incoming sensory patterns for degree of match (460).

The PC also plays a key role in forming odour memories. In mice, artificial stimulation, in the absence of odour, of random sets of PC neurons resulted in associative learning, as is the case of learnt odours. This implies that 'hard-wired' relationships are not present but that more flexible random associations are involved in odour learning and recognition (467). The ability to remember danger signals is crucial for survival, and the role of interaction between memory formation in the PC and input from higher centres following an unpleasant experience was demonstrated using fMRI. Subjects who were unable to distinguish between the odours of two enantiomers learnt to do so when exposure to one enantiomer was always paired with an electric shock (468).

The anterior and mid-dorsal insula is known as the primary taste cortex, just as the PC is the primary olfactory cortex. Magnetic resonance imaging has shown that active taste or smell tasks result in activity only in the relevant cortex; for example, actively searching for a smell activates the olfactory but not the gustatory cortex. However, one region of the far anterior insular cortex does seem to be implicated in both taste and smell (469).

The Orbitofrontal Cortex

Olfactory signals reach the OFC by various routes, and they are combined with input from other senses to form a conscious image upon which behavioural decisions can be made. By a combination of fMRI, peripheral autonomic recordings and olfactory psychophysics in a patient suffering from trauma to the right OFC, Li et al. were able to show that, although odour signals are processed in both the right and left OFCs, it is the right OFC that is responsible for conscious perception of odour (470). The coding of the signals depends on the firing rate of the OFC neurons and the codes can be read within 20 ms (471). Using fMRI, Li et al. have shown that the same input at the OE can give different activity patterns in the

OFC depending on previous experience of the odour in question (472). This clearly shows that there is not necessarily a direct relationship between odorant structure and odour percept in the conscious brain.

In non-human primates, the posterior region of the OFC is considered to be the prime target of the olfactory signals. It was therefore presumed that the same applied to humans, but more recent evidence using fMRI suggests that the rostral area is more likely to be the centre for olfactory processing in the human OFC (473). In the OFC, there is a pleasantness axis, with pleasant odours stimulating activity in the medial OFC and unpleasant ones in the lateral OFC. Such a separation is not seen at lower levels of processing and therefore this linear quality axis cannot be present in the molecules giving rise to the initial signal. The positioning of a signal along the axis is learnt in association with taste when food is concerned, and, therefore, presumably pure odour pleasantness can also be learnt by association (5). When any food is consumed to satiety, synaptic adaptation in the OFC results in a change in the hedonic rating of the food odour and hence contributes to loss of appetite for the food in question (474). The role of the OFC in determining value from the olfactory input has been reviewed by Gottfried and Delano (475).

Other Higher Brain Centres

In addition to the route through the PC, olfactory signals reach the neocortex via the mediodorsal nucleus of the thalamus (MDNT). This pathway seems to be involved in mediating olfactory attention rather than recognition, memory or discrimination (476, 477). It also would appear that a region in the anterior insular cortex is involved in attention to both smell and taste (469).

The chemotopic mapping of the bulb is lost in transmission to the PC but, in contrast, is retained to some extent in the amygdala. This leads to the proposal that the different routes allow for differences between innate and learned behaviours in response to olfactory cues (478).

One group of receptors specific to mammals is known as the *OR37 family*. These receptors are highly conserved across mammalian species and throughout evolution. They respond to long-chain aliphatic aldehydes, suggesting a possible role in identification of conspecifics, predators, and so on, through recognition of body odours resulting from oxidation of skin lipids (151). The routes from sensory neurons expressing them to higher brain centres have been traced using genetic tracing techniques, and the results suggest that signals from OR37 receptors are processed in specially designated brain areas (479, 480). Such direct connectivity is atypical for OR signal processing.

Brain Plasticity

The brain is not a static processor similar to a computer, but changes both in the way it processes signals and in its very structure. This plasticity is an important feature in evolution since it increases adaptability to changing environment and thus improves chances of survival. The amazing plasticity of the olfactory system has been reviewed by Gottfried and Wu (481). Mapping of activity in various brain regions during imagining of odours showed that the key olfactory and memory regions of the brains of experienced perfumers had undergone reorganisation relative to those of novices (482). Similar changes have been shown for other activities. For example, the volume of gray matter in the left pars opercularis (Broca's area) of the brains of male professional musicians increases in proportion with years of musical performance (483, 484). Investigations into the mechanism of such plasticity in the olfactory system suggest that sensory experience promotes the synaptic integration of new neurons into cell-type-specific olfactory circuits (485), and it has been shown that long-term odour memory (at least in *Camponotus fellah* ants) is stabilised by protein synthesis (486). It is common knowledge that blind people have greater acuity in other senses than sighted people. As far as olfaction is concerned, one part of the explanation is that the brain regions normally used for processing of visual signals have been shown to be adapted to add to the processing power for olfactory signals in people who are congenitally blind (487, 488).

Temporal Effects

In the laboratory we tend to present odours in a static situation, but in real life we are exposed to rapidly changing odour environments. The sense of smell must therefore be able to respond to these rapid changes, and it is also suggested that the dynamic temporal effects are part of the odour coding. Various groups have developed models to attempt to replicate these time-dependent aspects of olfaction (489–491). The activity of the neurons in the OB is affected by repetitive sniffing, which sets up rhythmic activity in the bulb. However, the mitral and tufted cells lock to different phases of the sniff cycle as a result of local inhibition, and this forms the basis for two distinct channels of processing (492).

The Link from Molecular Structure to Perception

Furodono et al. (493) found a statistical correlation between patterns of murine neurons activated by 12 odorants with the descriptions given to the odorants by humans. This suggests the possibility of a link from receptor to percept, in stark contrast to the results of Gottfried et al. using fMRI on human subjects. The 12 odorants used had structures that were all quite similar, despite having different odours, and so that might account for this odd result. As mentioned earlier, overlapping sets of receptors are responsible for the detection of carvone enantiomers (299). Hamana et al. isolated 2740 neurons from mice and found that >80% of carvone responding neurons responded to both the enantiomers (**2.116**) and (**2.126**). They proposed that a hierarchical code exists that allows the murine brain to distinguish between the enantiomers using the most sensitive receptors responding to single enantiomers. The selectivity was concentration dependant. As the concentration increased, those receptors accepting only one enantiomer at low concentration began to accept both enantiomers (301).

It is clear that there must be some link between the molecular structure of an odorant and the percept that it elicits in the conscious brain. Apparently conflicting observations such as those discussed above, and elsewhere in this and other chapters, tell us that the relationship is not a simple, straightforward one.

Olfactory Fatigue, Adaptation and Habituation

It is well known that exposure to an odour can result in a reduced awareness of its presence. For example, when we enter a room, its background odour is immediately apparent, but after a short time the perception is lost and we become unaware of it. Similarly, people lose awareness of the perfume they are wearing and think that it has evaporated or otherwise been lost. If we experience the same smell every day, we also tend not to notice it even when it is freshly presented. So, people do not notice the smell of their own home when entering it, even though visitors do. Similarly, someone who uses the same perfume every day loses awareness to it and often will think that the manufacturer has changed it or reduced its strength because 'It doesn't smell the way it used to'. These phenomena are easily explained by evolution. The important information for an animal concerns changes in its environment. If the animal is alive and well, then the background odour represents a safe situation with no threats and no new opportunities. The brain therefore learns to ignore it. Changes in the environmental odour could represent either a threat (e.g. approach of a predator) or an opportunity (e.g. approach of prey) and must therefore be signalled. The shorter-term loss of sensitivity serves to allow changes in concentration to be detected.

In everyday life and in the fragrance industry, the terms *olfactory fatigue*, *olfactory adaptation* and *olfactory habituation* tend to be used in different ways by different people, and their definitions are quite loose. The most usual understanding of 'olfactory fatigue' and 'olfactory adaptation' relates to the relatively short-term effect of continual exposure to an odour. For example, when presented with an odorant sample for evaluation, it is not unusual for its perceived intensity to reduce with repeated sniffing. Some odorants, such as musks, are much more likely to display this effect than others. Similarly, 'olfactory habituation' is usually taken to mean the longer term loss of awareness of, for example, the odour of one's home, place of work or favourite perfume. It would seem likely that such long-term 'habituation' occurs in higher processing centres since the individual components of the fragrance are still recognised even if the perception of the whole is dulled.

In the terminology of neuroscience, adaptation and habituation have precise meanings even though some neuroscientists use the terms more loosely or in a different sense. Versions of the stricter definitions are as follows:

Adaptation is a process in neurons, both the sensory neurons at the periphery of the olfactory system and those in higher processing circuits, that attenuates their response to a continual or repeated stimulus. This is a technique by which the cells keep their sensitivity in the optimal range for detection of changes in the signal.

Habituation is a form of learning arising from prolonged exposure to an odour. It results in reduced sensitivity to an odour, allowing static situations or non-consequential stimuli to be given lower attention. It probably occurs through adaptation of olfactory neurons, particularly those of the PC.

Adaptation

The effects of adaptation are similar to those of cellular plasticity in other use-dependent systems in the central nervous system. All of the neurons involved in olfaction—from the sensory neurons in the OE through the mitral and tufted cells and the interneurons of the OB, the pyramidal cells of the PC and right up to the neurons in the OFC—will demonstrate adaptation. Adaptation of neurons in the PC is faster and more complete than adaptation of OSNs or mitral cells (3). The role of a number of metabotropic glutamate receptors in adaptation in the PC has been identified by Young and Sun (494). Burgstaller and Tichy have shown that adaptation functions as a form of gain control in the cockroach, allowing increased sensitivity to a stimulus at low concentration and reducing sensitivity at high concentration, thus enabling the insects to detect odour signals over a wide concentration range (495).

The best and most comprehensive study of adaptation in the OSNs is that of Zufall and Leinders-Zufall (496). They identified three forms of adaptation in OSNs and the molecular mechanisms for each. They termed these forms *short-term adaptation (STA)*, *odour response desensitisation (desensitisation)* and *long-lasting adaptation (LLA)*. STA is induced by a single, short exposure to the stimulus. Onset is rapid (<1 s) and recovery is within 5 s. It is caused by calcium ions blocking the CNG channel, probably with the aid of a macromolecule but none has yet been identified. Desensitisation occurs when the stimulus is continuous, for example, an 8 s pulse. Onset occurs in 1–4 s and recovery takes 25 s. It is triggered by calcium entry to the cilium, and the mechanism involves attenuation of ACIII by CaMKII. LLA onset is in 25 s and recovery takes several minutes. It is dependent on the CO/cGMP second messenger signal pathway, but the exact mechanism is unknown. It has been suggested that adaptation in the sensory neurons is a code for concentration of the stimulating odorant (497). This could well be consistent with the observation that, in the OB of zebra fish *Danio rero*, changing concentration of odorants did not change the activation pattern but changing odorant did lead to an abrupt change in activation pattern (498).

Adaptation begins in the OSNs but, using 7 T fMRI, Schafer et al. showed that in rats a greater degree of adaptation occurred in the OB (499). The degree of adaptation was dependent on the level and duration of the stimulus.

The PC rapidly adapts to input from the lateral olfactory tract so that activity in the PC drops even when input from the OB remains constant. Best and Wilson have shown that there are two mechanisms involved, one recovers within 10–20 s, and the other is longer lasting taking about 2 min to recover (500). The latter involves both metabotropic glutamate receptors and β-adrenergic receptors.

Using both psychophysical and electrophysiological methods, Scheibe et al. showed that there is no major difference between young, healthy men and women in relation to short-term adaptation to suprathreshold chemosensory stimulation (501).

Habituation

Humans can give a verbal response indicating the intensity of the odour they perceive. This is not possible with other species, and so alternative methods must be employed. Non-invasive techniques such as fMRI can be used with both humans and animals to pinpoint brain regions involved in habituation. In animals, it is possible to use electrophysiological techniques and also behavioural methods. For example, in mice, habituation can be measured by a progressive decrease in olfactory investigation towards repeated presentation of the same odour stimulus and dishabituation as reinstatement of sniffing behaviour when a novel odour is presented (502) In flies, a Y-tube can be used with an aversive odorant in one limb of the Y. If flies show no preference for either limb of the tube, then it can be taken that they cannot smell the aversive odour (503).

The neurobiology of habituation has been well reviewed by Wilson and Linster (504). They point out that habituation is complex and spread across different brain regions. Behavioural effects of habituation are central rather than peripheral processes. Different neural mechanisms operate at different timescales and in different brain regions. The persistence of habituation memory is greater with a greater induction time, and induction time also affects the specificity in that cross-habituation to similar smells increases with increasing time of exposure (505). Hummel et al. showed that peripheral changes do not account for changes in perceived intensity by humans as they habituate to an odour (506). Electrophysiological experiments in the hamster showed that repeated or continuous exposure to an odour resulted in habituation of both the OSN and the mitral cells of the OB (507). When presentation of the stimulus was stopped, the sensory neurons recovered rapidly but the mitral cells did not. When the mitral cells were surgically isolated from their centrifugal inputs, they became hyper-responsive. They habituated more rapidly but recovered more slowly. This shows that habituation of mitral cells is not dependant on centrifugal inputs but that the latter do modulate habituation. The involvement of input from other centres is supported by a model of signal processing in the OB, in which it was shown that habituation could be modulated by both learning and attentional state (508).

In rats, repeated presentation of an odorant presented at intervals of tens of seconds leads to habituation in the olfactory cortex (PC, entorhinal cortex and amygdala), whereas behavioural habituation requires longer timescales (several minutes) and depends on effects in the OB, with the behavioural changes correlating with changes in spiking frequency of mitral cells. In both cases, the habituation is only evident in animals with functioning N-methyl-D-aspartate receptors in the OB (509).

McNamara et al. also found that repeated presentation of stimulus with spaces of 5 min between presentations led to habituation in the bulb involving N-methyl-D-aspartate receptors. This habituation is less specific than the shorter term habituation in the PC, and cross-habituation to related odorants is likely (505). They found that it is induced by a presentation of 50 s or more and lasts for 30 min and depends on MK-801, which is an N-methyl-D-aspartate receptor.

In some elegant experiments using genetic modification in *D. melanogaster*, Ramaswami and colleagues showed that olfactory habituation results from GABAergic inhibition of projection neurons by local interneurons in the fly's antennal lobe (the insect equivalent of the OB). The habituation is odorant-specific and, since the projection of the interneurons cross glomerular boundaries, it was necessary to find a mechanism to account for this. They proposed that synapse-specific potentiation by *N*-methyl-D-aspartate-type receptors mediate the inhibition, and they produced evidence to support this hypothesis (503, 510, 511). Workers from the same team had previously shown that the same habituation phenomena could be achieved using optogenetic simulation of the sensory neurons instead of activation with odorants (512). Their results are in agreement with the model proposed by Linster et al., which showed how synaptic adaptation and potentiation interact to create the observed specificity of response adaptation and the resultant habituation (513).

Habituation in the PC of mice has a 10 s onset and lasts for 2 min. It is dependent on LY341495 (a metabotropic glutamate-type receptor). It is more selective than bulbar habituation, which tends to affect signals from similar odours to that which induced it (505).

Habituation in humans occurs in the PC in 30–40 s. It results in reduced activation. It is affected by practice at the task. Even though the PC shows reduced activity, the signal is still there in the secondary cortex, suggesting that a direct route from the bulb to the OFC still operates even when the PC is habituated. It is therefore proposed that the PC looks for similarity and responds to difference, but when detecting similarity, it reduces its own activity whilst allowing a signal to go directly from bulb to OFC (514). In agreement with this, using fMRI with human subjects, Poellinger et al. showed that long-term (60 s) exposure to an odorant resulted in reduced activity in the PC but that the activity level in the OFC was sustained, indicating a degree of dissociation between the two. Their evidence also suggests that the PC, entorhinal cortex, amygdala, hippocampus and anterior insula interact closely in habituation (515).

Odorant responding neurons in the hippocampus show a response that depends on the timing of repeated stimulation, suggesting that the hippocampus keeps time between different presentations of the same odour and therefore might play a role in habituation (516).

People suffering from schizophrenia have impaired olfactory function. Their ability to identify smells is lower than in healthy controls. It is their ability to identify smells rather than their detection threshold that is affected. Schizophrenics who experience auditory or verbal hallucinations have impaired hedonic judgement, but those who suffer olfactory hallucinations do not. It would seem that impaired habituation is associated with olfactory hallucination and that those with olfactory hallucinations also tend to suffer a greater incidence of brief periods of unconsciousness (517).

In an interesting study of cognitive effects on habituation, Kobayashi et al. found that subjects who were told that the odour they were given was harmful experienced a lower degree of habituation than those who were told that the odour was beneficial. The effect was found only when the odour was presented in short bursts (60 bursts of 0.2 s with 9 min intervals between bursts) and did not occur when the presentation of the odour was continuous (20 min) (518). Habituation to an irritant odour reduces the level of self-reported symptoms of exposure such as the sense of well-being. Subjects who were exposed to ammonia vapour in their workplace were less likely to report adverse effects of ammonia than those who were unaccustomed to it (519).

Male guinea pigs (*Cavia porcellus*) were found to habituate to the smell of the urine from a female even if only exposed to the odour for two periods of 2 min, separated by 7 days. Dishabituation occurred rapidly when exposed to the odour from the urine of a different female. This indicates a long-lasting effect of habituation and also a modulatory effect of input from higher brain centres (520).

Pattern Recognition

The ability to recognise patterns is one of the great strengths of the human brain. We can recognise patterns even if they are degraded or corrupted, and this is clearly a very important survival technique. Animals that do not recognise a pattern of sensory information that is indicative of the presence of a predator will be less likely to leave descendants than those that do see the imminent danger. There are many well-known examples, and some will be found in the books by Wilson and Stevenson (3), and by Shepherd (5). Two simple examples are shown in Figure 2.77. The first sentence contains words in which some letters have been replaced by numbers, and the second contains words in which the initial and final letters are in the correct place but other letters have been jumbled. In both examples, the brain has little difficulty in reading the sentences because it recognises the patterns as they should be and not as they are. This ability is important in recognition of the smell of mixtures, especially those in which there is a degree of variation in composition. The elegant experiment of Chapuis and Wilson described below illustrates pattern recognition in action in olfaction (392). Other examples in the following section include the construction of simple perfume accords which immediately recall the odour of a much more complex mixture.

MIXTURES

Much of the research described in this book relates to smelling of single chemical entities. This is what would be expected in research since it simplifies a complex process as much as possible and therefore improves the chances of understanding it. However, outside the laboratory, in real life, virtually everything we smell is a mixture, and we know that even a small trace of one odorant can dramatically alter the odour of a mixture to which it is added. For example, tiny traces of rose

7H15 53N73NC3 C4N B3 R34D R3L471V3LY 3451LY 3V3N 7HOUGH M4NY
OF 7H3 L3773R5 H4V3 B33N R3PL4C3D BY NUMB3R5.

In tihs scnetnee, mnay of the ltetres in wdros hvae been jbmueld. Hevewor, it is sltil rtaleivley esay to raed bausece the barin rogicsenes penttras of wrdos.

Figure 2.77 Two examples of pattern recognition.

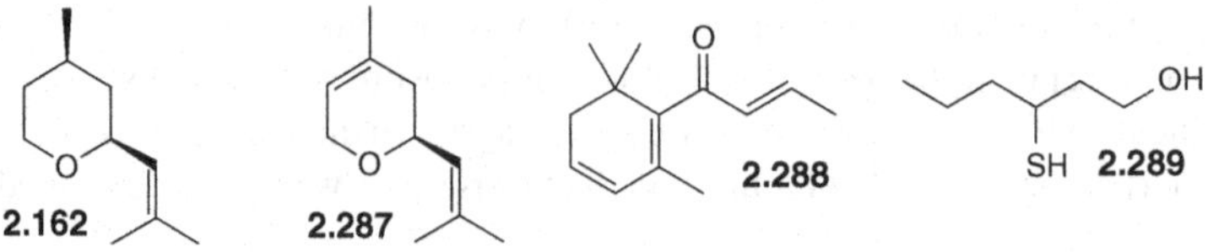

Figure 2.78   Some character impact compounds of wines.

oxide (**2.162**), nerol oxide (**2.287**) and damascenone (**2.288**) provide the characteristic rosy note of Gewürztraminer wines and 3-sulfanylhexan-1-ol (**2.289**) gives the citrus character to Sauvignon Blanc (521) (Figure 2.78). Like all animals, we need to be able to recognise that two different mixtures are similar yet at the same time be able to identify small differences between two complex odour mixtures. For example, such abilities are essential for recognition of self, kin or non-kin. Equally, an animal that cannot recognise the odour of a predator against a complex odour background is unlikely to pass its genes on to another generation. The predator odour will be due to a complex chemical mixture as explained in Chapter 1, and the background odour of the natural environment will be an even more complex mixture of chemicals coming from plants and other animals. The sense of smell must have evolved to deal with all these issues and to do so very rapidly. It will be clear from some of the preceding and following discussion that odours (whether individual molecules or mixtures) that have been learnt to be associated with danger tend to have a greater impact than those that have not. Everyone who has worked in the fragrance industry is only too aware of the difficulty of hiding bad odours. Our brains are clearly conditioned to identifying them and responding to them despite all the obstacles that can be put in the way.

It must also be remembered that volatile molecules, whether or not they activate ORs, can act as inverse agonists, stabilising a receptor in an inactive state. Thus, inhalation of a mixture of an agonist and an inverse agonist of the same receptor type will result in a lower level of firing of that receptor than would be observed if only the agonist were inhaled. Since receptors have a spontaneous background rate of firing, an inverse agonist will reduce the background activity and generate a negative signal. As described above, volatile odorants or odourless materials can also interfere with the transduction cascade, for example, by blocking CNG channels, or serve as allosteric modulators. All of these interactions between the volatile components of a mixture will be dependent on molar ratios of the components, binding efficacies of each component to its binding site and relative on- and off-rates. So, even at the receptor level, interaction between components of a mixture is very complex.

Simple odorants and complex mixtures both give rise to a similar degree of activation in the PC. Thus it is difficult to see how mixtures can be deconvoluted since each odour image is a distinct picture in the PC and all odour images are of similar sizes (7). Stettler and Axel found that about 10% of pyramidal cells in the PC are activated by an odour, irrespective of the chemical complexity of the odour (457). Similarly, Leinwand and Chalasani show how in other senses components of the sensory image are represented separately and therefore can be deconvoluted but the sense of smell is different and deconvolution is not possible. Olfaction is much more related to memory, and odour recognition is the result of matching a new image to stored images in memory. Odour signals in the PC are sparse, probably as a result of inhibitory circuits from interneurons which reduce the sensitivity of the pyramidal cells to odour. About 50% of the pyramidal cells in the PC that respond to the odour of a pure odorant fail to respond to it when it is present in a mixture.

Laing and collaborators have carried out various experiments suggesting that it is impossible to distinguish individual components of mixtures if more than three components are present (522, 523). These sensory experiments therefore tend to support what has been observed using neurological techniques. However, anyone who has worked in the perfumery industry knows that perfumers are capable of identifying ingredients in a fragrance and can even deconstruct a fragrance using smell alone. Of course, perfumers and trained fragrance chemists use various 'tricks of the trade' to analyse perfume formulae by nose. These include smelling the perfume as it dries down (ages) on a blotter. The first notes to disappear are the more volatile ones, and the dry-down after 24 h will give a better indication of the less volatile components. Clearly, there is also a need to build up knowledge of ingredients, mixtures containing them and the effects of one odorant on a mixture. Confirmation of the value of the latter has been demonstrated in psychophysical experiments. For instance, the ability to evaluate the odours of mixtures was found to be improved both by learning the odours of the components and by the odour types into which the odours of the mixtures fell (524).

A simple mixture of α-hexylcinnamic aldehyde (**2.290**), benzyl acetate (**2.240**) and indole (**2.291**) in the ratio of 40 : 20 : 1 elicits a percept that is instantly recognisable as jasmine (Figure 2.79). The percepts elicited by the pure ingredients are fatty, fruity and fecal, respectively. There are two possible mechanisms by which a perfumer might smell jasmine and recognise these three odour components, each of which mechanisms depends on a different *modus operandi* of the brain. In the first model, the brain would analyse the final percept by looking for small differences between different stored images. The other model relies on the brain's ability to compare stored patterns and look for similarities between them. Thus in this case, using the first approach, the brain would look for differences between the new percept and those stored for similar odours and then concentrate on the differences in order to detect the individual patterns α-hexylcinnamic aldehyde (**2.290**), benzyl acetate (**2.240**) and indole (**2.291**). An oversimplified description of the logic of such a process would be, for example, having spotted a characteristic of a pattern that tends to indicate the presence of indole (**2.291**) in a mixture; that part of the pattern could then be ignored in order to identify another component. In the second model, the brain would compare the pattern of the new percept with those of similar odours and look for previously experienced patterns that could be linked back to other mixtures. The oversimplified logic here would be to identify similar

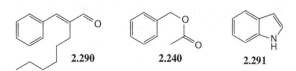

Figure 2.79   Components of a simple jasmine accord.

patterns elicited by mixtures of known composition and look for commonalities between these. So, in this instance the three components would be recognised as occurring in the majority of those images recalled from the memory bank. The trained perfumer would therefore know from experience that the jasmine percept is a combination of fatty, fruity and fecal percepts and would find it much easier to identify the components responsible for each. For example, he would know that the pattern was more likely to result from the presence of benzyl acetate (**2.240**) than another fruity odorant such as ethyl isobutyrate (**2.196**). To put it another way, in the first model, the perfumer would be able to work backwards from the final odour image to reconstruct the individual patterns that combined to produce it, whereas, in the second model, experience of previous mixtures would tell him that the final image is likely to be the result of certain inputs. So, when a wine connoisseur picks out the fruitiness of damascenone (**2.288**) in a Gewurztraminer wine, is it because his brain has succeeded in disentangling the receptor inputs to the complex odour percept or is it that he knows from experience that certain aspects of an odour image are due to the presence of damascenone (**2.288**) or similar odorants? The second model offers a possible explanation for the ability of someone who is anosmic to a given odorant to detect its presence in accords.

In order to investigate these possibilities, Chapuis and Wilson investigated the ability of rats to distinguish between two similar mixtures (392). Two different pairs of odours were studied. In each case, one odour was a mixture of 10 components. In the first pair, one of the 10 components was omitted to form a second mixture containing only 9 of the original components. In the second pair, one of the original 10 components was replaced by a different odorant to give two similar but not identical 10-component mixtures. They showed that the animals found it easier to distinguish the new 10-component mixture from the original than they did the original from the mixture with one missing component. The neural processes involved pattern completion in the case of the 9-component mixture and pattern separation in the case of the two different 10-component mixtures. The fact that the discrimination of the 9-component from the 10-component mixtures was the easier task shows the facility with which the brain can recognise and complete a degraded pattern. In this elegant series of experiments, they also showed that the ability of rats to perform the tasks was dependent on experience and also on the context of the task in terms of reward. If a reward was given for discrimination between the 9- and 10-component mixtures, the rats learnt to distinguish between these two, which they had previously treated as the same. However, if the reward was given for being able to associate the two 10-component mixtures and treat them as the same against a fourth, different, odour, then the rats learnt to group the two 10-component mixtures and select the fourth odour as the different one. After this training, the rats showed reduced ability to distinguish between the two 10-component mixtures. Using electrophysiological methods, Chapuis and Wilson were then able to determine where in the brain this pattern recognition and plasticity occurred. The activation patterns in the OB for each odorant mixture remained constant throughout, and each of the patterns for each of the three mixtures (original 10 components, original minus one component and original with one component replaced) remained clearly distinct from those of the other two. Discrimination and plasticity therefore did not occur in the OB. However, activity patterns in the aPC did show differences. Originally, the response of the aPC to the original and the nine-component mixtures was the same, showing that pattern completion had already occurred in the aPC. When rats were trained to distinguish between the original and the nine-component mixtures, the activity patterns elicited by each in the aPC diverged. Conversely, training to associate the two 10-component accords as similar resulted in previously distinct patterns of aPC activity to resemble each other more closely. This clearly demonstrates that the animals' brains are able to look for either similarities or differences in odours depending on need in a given situation. It is also evident that the process of converting output from the OB into images for comparison with those in memory begins in the aPC. This is in contrast to the work of Gottfried discussed previously, in which the identity of odour objects in humans would appear to be coded in the posterior PC. The differences between these two sets of results could be due to either the differences between the two species or the different experimental techniques employed. Subsequently, Lovitz et al. demonstrated that the ability of rats to discriminate between a 10-component mixture and one in which one of the components had been replaced by a different substance was dependent on the relative amount of the new component in the mixture (525).

Binary Mixtures

Mixtures of two components are obviously the simplest mixtures to study. Chemists tend to think in terms of physical concentration when studying mixtures. However, if physical concentration is used as a reference, then it is clear that an intense odorant will dominate in any mixture with one that is significantly less intense. Therefore, sensory scientists prefer to use isointense concentrations. Thus the concentration used for each component of the mixture is adjusted so that the 'single odorants' (actually a solution of the odorant in an odourless diluent) are given equal ratings of intensity. Use of actual concentrations would make more sense at the receptor level but isointense ratios give more indication of what is happening in neuroprocessing. Both approaches are therefore limited since they each concentrate on one part of a multi-stage process.

Patte and Laffort developed a system for studying binary mixtures at isointensity using parameters which they designated as σ and τ (526). In their system for a mixture of two components A and B, σ is the ratio of the perceived intensity of the mixture (I_{AB}) to that which would be expected by simple addition of the intensities of the two components (I_A and I_B). This is shown as Equation 1 in Figure 2.80. The τ value of a component is the ratio of the perceived intensity of that component in the mix compared to the sum of the intensities of both components. This is illustrated by Equation 2 in Figure 2.80.

Using these parameters, it is possible to draw a graph of perceived intensities for mixtures containing different ratios of A and B. The framework of such a graph is shown in Figure 2.81. The ratio of components is plotted on the horizontal axis

$$\sigma = \frac{I_{AB}}{I_A + I_B}$$ Equation 1

$$\tau_A = \frac{I_A}{I_A + I_B}$$ Equation 2

Figure 2.80   Patte and Laffort's σ and τ factors.

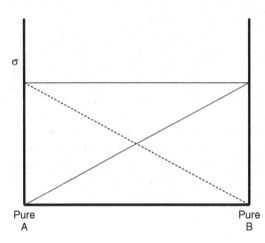

Pure A — Pure B

Figure 2.81   Patte and Laffort's σ and τ diagram.

with 100% A on the left and 100% B on the right. The measured value of σ is then placed on the vertical axis. Perfect addition between ingredients is defined as instances where the intensity of the mixture equals the sum of the intensities of its components, thus giving a σ value of 1. If perfect addition occurs for all combinations of A and B, then the graph will be the solid horizontal line shown in Figure 2.81. The dashed line in Figure 2.81 shows where the value of τ_A would fall if component B were absent. Similarly, the dotted line in Figure 2.81 shows where the value of τ_B would fall if component A were absent.

In an excellent meta-analysis of binary mixtures, Ferreira demonstrated that, in the majority of cases, the σ values of binary mixtures follow the line shown in Figure 2.82 (527). In other words, for most cases, the intensity of the mixture seems to be determined mainly by that of the major component (remember that this is based on isointensity and not on concentration).

Synergy between components is defined as instances where the intensity of the binary mixture is higher than would be expected based on the sum of its components. Ferreira gives some examples, and a further one comes from musk odorants. It has been shown that any 1 : 1 mixture of two musks selected from cyclopentadecanone (**2.292**), cyclopentadecanolide (**2.293**) and cyclohexadecanolide (**2.294**) has a higher perceived intensity than either of its individual components (Figure 2.83) (528).

Addition of sub-threshold odour and taste stimuli can give rise to conscious perception. Furthermore, such congruency can be either innate or learnt (529).

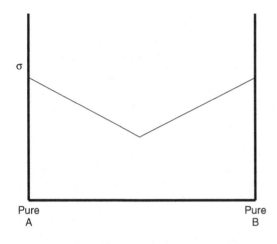

Figure 2.82 Typical behaiour of binary mixtures in Patte and Laffort's σ and τ diagram.

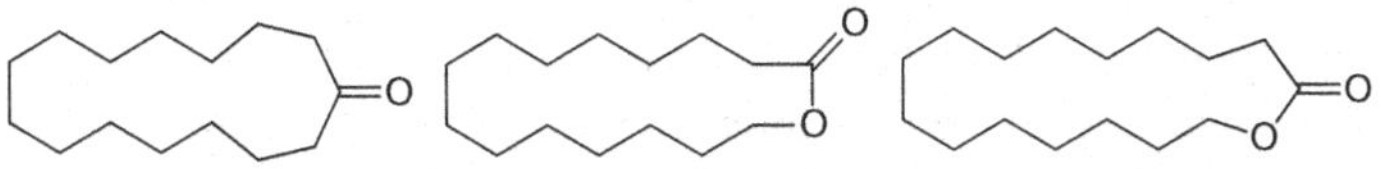

2.292   2.293   2.294

Figure 2.83 Musks giving synergistic mixtures.

Similarly, a substance to which a subject is anosmic can also affect perception of a mixture containing it (530). These and other such counter-intuitive observations imply a central mechanism and show that we still have much to learn about the sense of smell.

Suppression, in which one component reduces the perceived intensity of another, is much more common than either perfect addition or synergy, and in binary mixtures it is normally the less pleasant of the two that exerts the greater suppression effect on the other component (527). This also would tend to suggest a central effect rather than a peripheral one since there is a hedonic component involved that would not be present at the receptor level. When all of the components of a mixture are malodorants, suppression usually still occurs. For example, the intensity of mixtures of three or four unpleasant odorants has been found to be approximately equal to the intensity of the dominant component of the mix (531). Examples of suppression include the effect of small amounts of 1-propanol (**2.295**) in strongly reducing the intensity of amyl butyrate (**2.85**) (532), and the suppression of the odour of (*E*)-2-hexenal (**2.297**) by (*E*)-2-decenal (**2.296**) even though the former has a more striking odour character, green grass as opposed to tallow (533) (Figure 2.84). Suppression of perceived intensity caused by mixtures of odorants is similar to saturation at high concentration of single odorants, and the resultant compression of the signal is consistent with the findings of fMRI that the percentage of neurons activated in the PC is relatively independent of the number of components in a mixture.

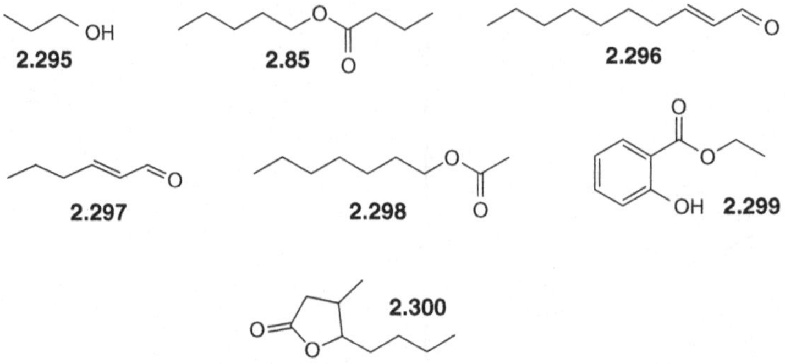

2.295

2.85

2.296

2.297

2.298

OH 2.299

2.300

Figure 2.84 Substances providing examples of suppression.

The behaviour of heptyl acetate (**2.298**) and ethyl salicylate (**2.299**) is remarkable in that some mixtures of the two show synergy and others suppression. Small amounts of ethyl salicylate (**2.299**) enhance heptyl acetate (**2.298**), but at higher amounts of ethyl salicylate (**2.299**) the total intensity is suppressed to a level below that of the ethyl salicylate (**2.299**) alone (534, 535).

In contrast to his findings for suprathreshold mixtures, Ferreira found that additivity is the norm at the peri-threshold level (527). A useful parameter at the peri-threshold level is the odour activity value (OAV). This is defined as the actual concentration of the odorant divided by its threshold concentration. In mixtures of odorants at peri-threshold level, it has been found that the OAV of the mixture is the simple sum of those of its components. For example, addition of two odorants, each at a concentration below its threshold value, can result in a detectable odour. A possible explanation for this dichotomy would be that, at peripheral level, there is essentially no competition between odorants at low concentration and so more receptors are activated by the mixture than by a single component, whereas at higher concentrations odorants will compete for receptors and antagonism could result in a lower overall level of activation than expected. On the other hand, the explanation could be that lateral inhibition in neural circuits increases as the overall signal intensity from the receptor sheet increases. Another interesting question raised by the observation of additivity at the peri-threshold level is that of distinguishing signal from noise. The receptors fire spontaneously, and so there is always a degree of background noise. If two odorants activate different sets of receptors and each is present at a concentration which, by itself, would be insufficient to enable the signal to be detected above background noise, then how is it that mixing the two would give rise to a signal that is above the noise level since every receptor would still be activated only below the level of the noise? Could there be some mechanism in the bulb that detects total level of activity rather than the activity of one set of glomeruli?

It has been shown that the ability of rabbit kits to recognise binary mixtures of odorants depends on the proportion of the odorants in the mixture relative to the original odours (either other blends of the same two odorants or the pure odorants) to which they were trained (536). This implies that the more closely the receptor response pattern of the mixture is to that of one of the components, the easier it is for the animal to relate it to the individual component. Barkat et al. found that, in human olfaction, some binary and tertiary mixtures formed blended percepts more readily than others and it was therefore more difficult for subjects to recognise their components (537). Presumably, this could be due either to interactions between specific odorants at the receptor level or to differences in interaction between the neural circuits involved in the different categories of mixtures. They also found that panellists who were experienced in smelling were more able to detect individual components in binary and tertiary mixtures. Of course, this could reflect the ability to memorise and recognise odour mixtures as being mixtures of specific components rather than any ability to deconstruct a mixture into its component parts.

Fletcher found that the glomerular patterns of binary mixtures were simple additions of those of the individual components. He thus concluded that the bulb contains a chemotopic map and therefore retains an analytical nature as opposed to beginning the process of synthesis of an odour image through lateral inhibition by interneurons (538). However, Johnson et al. found that, in some instances, the bulbar activation pattern resulting from smelling of a mixture was less complex than would be expected based on the activation patterns of the individual components (539). Kuebler et al. used various odorant mixtures containing 2–7 components and found that the resultant activity in the antennal lobes of moths represented distinct patterns from which the components of the mixture could not be identified (540). Since the antennal lobes of insects correspond to the OBs of mammals, this implies that processing with lateral inhibition and percept synthesis begins in the sensory neurons of insects. Similar findings have also been reported by Su et al. (541). Chaput et al. studied responses to isoamyl acetate (**2.88**) and whiskey lactone (**2.300**) in three different systems, namely, isolated receptors expressed in HEK cells, electrophysiological recordings in rat OSNs and psychophysical measurements in humans (542). They found that interactions at the level of both receptors and neurons could be either mutually enhancing or suppressing, and that the combined response was not the simple sum of the individual inputs. Thus the odour mixture information is established after the peripheral stage. Thus it would seem that effects such as antagonism of one odorant by another could provide an explanation of some of the conflicting observations above. Consequently, selection of components of mixtures might be significant in determining whether such interactions are observed. Indeed, Rospars et al. investigated the response of rat OSNs to single chemicals and to mixtures and found that both competitive and non-competitive effects existed where the neuron responded to different odorants (360). Similarly, as described previously, DEET (**2.78**) was found to interfere with both the inhibitory and activation processes at the receptor level, and thus distort the activation pattern and prevent insects from smelling their food source, fruit in the case of *D. melanogaster* and humans in the case of *Anopheles gambae* or *Aedes egyptii*. Such interactions at the receptor must play a role in the coding of odour mixtures and will reduce the ability to determine components of a mixture.

THE SUBJECTIVITY OF ODOUR

"If it were not for the great variability among individuals, medicine might as well be a science and not an art."

Sir William Osler, 1892.

Sir William Osler's comment is even more applicable to perfumery than to medicine. It is well known that odour is subjective in that different individuals might describe the odour of the same molecule in different ways. The same individual might even describe it differently on different days or even at different times on the same day. This variability has been, and often still is, a serious obstacle in many aspects of fragrance research. Variations in one individual's descriptions of the same odour can be due to environmental factors such as relative humidity or psychological factors such as experience and expectancy, as are described in Chapter 3. Variations between individuals are often attributed to semantics, training and suchlike. Poor test protocol is also often seen as a factor in subjectivity, and many papers report the use of protocols that have been devised to reduce or eliminate subjectivity. It has even been claimed that subjectivity is used by lazy fragrance chemists as an excuse for not doing the necessary work to use their structure/odour correlations to provide insight into biological mechanisms. Poor protocol and inattention to detail might be responsible for some reported cases of subjectivity, but subjectivity is clearly apparent in even the best experiments where a high level of intellectual rigour has been used to devise the most stringent test conditions. Some examples are described in Chapter 3.

We cannot escape from the conclusion that odour is subjective. Each of us lives in our own unique sensory universe and each of us will never know how our universe compares with that of another person, even if the two do happen to be quite similar. This is a profound phenomenon and some people seem to have a serious emotional barrier to the acceptance of it.

In vision, we all know that the colour of a ripe strawberry is red whereas the sky is blue. We will never know if our experience of red is the same as that of another person, but we know that each of us has learnt to call the colour of light of one certain wavelength red and that of another, blue. So when any of us looks at the colour of the top light in traffic lights, we call it red and no one would say that it is blue. Proof of the fact that our sensory universes are unique is more likely to be found in olfaction. Given (Z)-α-santalol (**2.95**) as a sandalwood reference odour and cyclopentadecanolide (**2.294**) as a musk standard, some subjects, when presented with a 'blind' (i.e. with no hint of identity or molecular structure) sample of Radjanol (**2.98**), will describe its odour as sandalwood whilst others will find it to be musk (Figure 2.85) (543). This is a genuine perceptual phenomenon and not just a question of semantics since the subjects have the two reference standards for comparison. As we learn more detail of the mechanism of olfaction, reasons for such differences are becoming apparent. As described previously, differences in nasal metabolism can account for such subjectivity. Variation in OR profile between individuals is another source of such subjectivity.

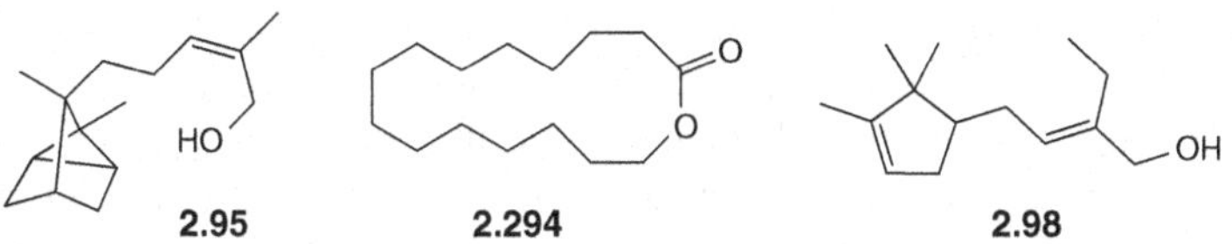

Figure 2.85 Sandalwood or musk?

It is well known that SNPs affect OR genes. Menashe et al. investigated those that resulted in the gene becoming a ψ-gene and showed that the degree of variation between individuals is very high (544). Similarly, Hasin-Brumshtein et al. have shown how copy number variation (CNV) also produces differences in individual genotypes (545). The link between genotype and phenotype has been established in a number of cases. For example, in a study of twins and their siblings, Knaapila et al. showed that intensity ratings of androstenone (**2.201**) and Galaxolide (**2.301**) had a genetic component, whereas ratings of amyl acetate (**2.124**), eugenol (**2.5**), mercaptans and rose did not (546). Similarly, two SNPs in the receptor OR2J3 have been found to affect sensitivity to the green odorant *cis*-3-hexenol (**2.302**) (Figure 2.86). The mutations T113A and R226Q each lowered sensitivity to (**2.302**) in people with either of those variants, and those with both variants were essentially anosmic to it (547).

In a group of 391 subjects, 13 different SNPs were observed in OR7D4 which responds to androstenone (**2.201**) (317). The most frequent of these are R88W and T133M. The protein with arginine in position 88 and threonine at position 133 responds much more strongly to androstenone than the variant with tryptophane at position 88 and methionine at position 133. People with the RT/RT genotype are more likely to be able to smell androstenone than those with either RT/WM or WM/WM genotypes. Another SNP elsewhere in the structure, S84N, even resulted in an increase in sensitivity to androstenone over the more common variant. Horde is a database of known human OR genes (548), and it shows that such variation in genes across the population is the norm. For example, for OR1D2 it lists 13 SNPs and 2 CNVs. Similarly, one of five haplotypes of OR2J3 identified in a group of 52 subjects results in reduced perception of *cis*-3-hexen-1-ol (**2.302**) (547). At the 16th International Symposium on Olfaction and Taste, Mainland et al. reported that 86% of the receptor genes they found in a group of subjects had polymorphisms which affected the functionality of the receptors and that, on average, the functionality of

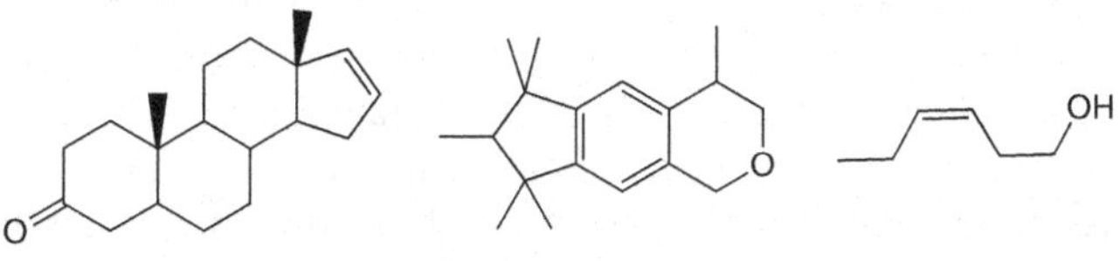

Figure 2.86 Genetic variation in perception.

OR alleles of any two of the individuals differed by 42% (549). On the basis of such facts, the following fictitious example is certainly not an exaggeration.

Let us take a simple, purely hypothetical, case to illustrate how such factors might affect perception. In this fictitious example, we will study two men, Fred and Jim, and their perceptions of three pure aroma chemicals, molone, moltwo and molthree. Screening of all ORs reveals that there are 10 receptors (A–J) that are sensitive to these odorants. Naturally, the sensitivity of each receptor will be different for each of the odorants. However, SNPs mean that there are four variants for each receptor and these modify the sensitivity of that receptor to each of the odorants. The existence of four variants of each OR is not an unreasonable hypothesis in view of the results on OR7D4 and OR2J3 described above. We genotyped Fred first and so labelled his variants as variant one for each of the 10 ORs. Jim has the same variant as Fred for five of the ORs, a different variant for four of them and, thanks to one SNP resulting in Jim's version of ORC being a ψ-gene, Jim is lacking an active form of that receptor. Our two subjects' genotypic profiles are shown in Figure 2.87.

When we measure the sensitivities of the four variants of each of the 10 ORs to the three odorants using a 5-point scale, we see the patterns shown in Figure 2.88.

Now consider the receptor activation patterns that are elicited when Fred and Jim are exposed to the three odorants. These are shown in Figure 2.89.

From Figure 2.89, we can see that the pattern activated in Fred's OE when he smells molone is not the same as that which is found in Jim's OE under the same conditions. However, each will learn that the pattern he perceives is the odour described as molone. Neither will ever know that the other experiences a different activation pattern and, presumably different experience, but each will be able to recognise the odour of the molecule molone and give it the same agreed name as the other. Exactly the same will happen with moltwo and molthree. When comparing the odours, Fred's brain will see that the patterns activated by all three odorants are similar in that there is stronger activation at the left and only weak activation at the right. He will therefore judge all three odours to be similar. On the other hand, Jim's brain will see a similarity between the odours of molone and moltwo but that of molthree will seem quite different to him and so he will disagree with Fred. Neither is right and neither is wrong. Each is describing his own perception of the similarities and differences between the molecules, and each lives in his own olfactory universe.

As far as correlating molecular structure with odour is concerned, the implications of this phenomenon are indicated in Figure 2.90. The figure shows how the odorant molecule elicits different percepts in the brains of the two men, for the reasons explained above. However, when they apply semantic processing in order to associate a name and a description to the odour, they have both been trained to give it the same name. So, for example, when anyone smells geraniol and says, 'This is geraniol, it has a rosy odour', he or she is giving the name and description they have been trained to associate with whatever percept has formed in their brain. It is not the same percept as that of the next person, but the fact that everyone agrees on the descriptor deludes us into thinking that the odour is an intrinsic property of the molecule.

Fred has the following genetic profile

Receptor	A	B	C	D	E	F	G	H	I	J
Variant 1	#	#	#	#	#	#	#	#	#	#
Variant 2										
Variant 3										
Variant 4										

Jim has the following genetic profile

Receptor	A	B	C	D	E	F	G	H	I	J
Variant 1		#		#	#	#			#	
Variant 2			absent				#			
Variant 3								#		#
Variant 4	#									

Figure 2.87   Genetic profiles of two subjects.

162

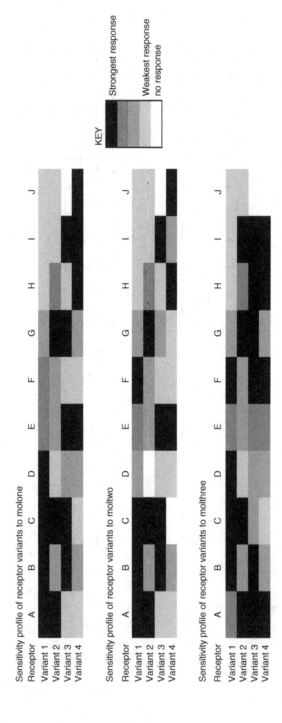

Figure 2.88 Sensitivity of OR variants to odorants.

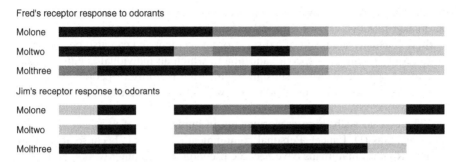

Fred's receptor response to odorants

Molone
Moltwo
Molthree

Jim's receptor response to odorants

Molone
Moltwo
Molthree

Figure 2.89   Receptor response patterns activated when subjects smell odorants.

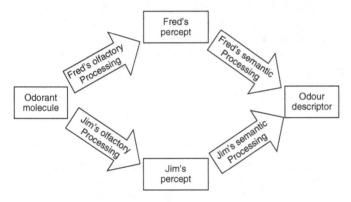

Figure 2.90   Same stimulus, different percepts but same descriptor.

In subsequent neuroprocessing, more differences will be found that will introduce further discrepancies between Fred's perception and Jim's. For example, when he was a child, Fred's mother used to make delicious apple pies and season them with cloves. Jim's mother used cinnamon instead. Jim has a lot of trouble with his teeth and had some traumatic experiences at the dentist. Jim's dentist used oil of cloves in his work but Fred's did not. So if molone is eugenol, then Jim is likely to show quite an aversive reaction to it, whereas Fred will find it to have a pleasant and reassuring odour.

Subjectivity of odour is a real phenomenon and not a question of poor communication or laziness of researchers. It will continue to present issues for fragrance chemists, perfumers and marketers alike.

ANOSMIA

Introduction

Anosmia is the inability to smell. However, it not necessarily a simple on/off phenomenon but often takes the form of hyposmia, a reduced ability to smell, as evidenced, for example, by a higher than average detection threshold. Anosmia can be divided into two classes. General anosmia is the inability to smell any odours at all. It is usually the result of a disease or an accident and can also be genetic in origin. More common is specific anosmia in which an individual is either unable to detect a specific chemical substance which the majority of people can detect, or displays a threshold of detection for it which is significantly above the normal range.

There is no such thing as normosmia in humans; all of us suffer from specific anosmias (D. Lancet, Personal communication). This is because about 60% of the OR genes in humans have become pseudogenes. In other words, each of us only uses about 350–400 of the 1000 potential receptor protein types. Each of us (with the possible exception of homozygous twins) uses a unique combination of receptors, and this is one basis for the observation that we each perceive odours differently. The specific individual profiles are genetically determined, and there are patterns across ethnic groups. The differences are mostly in thresholds for individual notes. Of course, nothing in the body is determined purely by genetics, and environmental factors also play a part.

General Anosmia

Innate general anosmia means an inborn inability to smell anything. Since there is a multiplicity of OR types, general anosmia must be related to a genetic defect in production of proteins other than the ORs. Suspects included the G-proteins, adenylylcyclase and the ion channels. Work in Lancet's team in the Weizman Institute has shown that the c-AMP channel and the proteins coded in Kalmann's region of the X chromosome are not implicated in general anosmia (D. Lancet, Personal communication). A study of the Hutterite community in the United States showed that general hyposmia was linked to a region on chromosome 4q in which there are no genes for OR proteins (550). A more specific result came from the team of Zufall, who found that loss of function of the gene SCN9A resulted in general anosmia. Loss of this gene is known to cause inability to sense pain in humans because it codes for the voltage-gated sodium channel Na(v)1.7 which is crucial for signalling of the pain-sensing neurons. Using mice, they were able to show that the same channel is also involved in synaptic signalling of OSNs (424).

The yellow colour of the OE is due in part to the presence of vitamin A, and its level does appear to correlate with sensitivity to odours (551). Vitamin A deficiency was an early suggestion for the cause of general anosmia (552), but its role in olfaction remains unknown.

Anosmia as a result of head injury is well known, and the author has met a number of people who have suffered this. In one of these cases, the subject recovered the sense over time, but in all the others the loss was permanent. Techniques for evaluating the extent of such damage include SPECT (407).

Anosmia can also result from brain damage due to neurodegenerative diseases, and this will be discussed in more detail in Chapter 4. There is a correlation between loss of gray matter volume in brain regions, such as the medial prefrontal cortex and nucleus accumbens, associated with olfaction in patients suffering anosmia as a result of neurodegenerative diseases (553). In addition to brain scanning techniques, procedures using sensory tests are also used for identifying such anosmias. For instance, the Lyon Clinical Olfactory Test (LCOT) was used to measure the rate of hyposmia in a local community. It was found that the incidence of hyposmia was 6% amongst healthy subjects and 16% in an elderly control group. The test was also able to distinguish between anosmia resulting from neurodegenerative diseases and other causes (554).

Specific Anosmia

Specific anosmias have received much attention from those interested in the sense of smell. Definitions and understanding of these definitions vary, and so semantic issues have added to the confusion around the topic. The term *specific anosmia* has been used to describe both a total inability to smell a given substance at any concentration or a matter of higher than average threshold to it. The latter is more properly referred to as *specific hyposmia*. About two-thirds of odorants also stimulate the trigeminal nerves in the nasal cavity. Generally, trigeminal threshold concentrations are higher than those of olfactory thresholds. This means that sensory tests in which the concentration of the odorant is above the trigeminal threshold will give misleading results if the odorant under study is a trigeminal stimulant. This has been proved for androstenone (**2.201**) anosmia (555). The exact experimental protocol used to identify specific anosmia is therefore of crucial importance. For example, if a test odorant is presented to subjects alongside other odorants and the subjects are asked to indicate which substances they can smell, one value of incidence of specific anosmia will be obtained. If the same test odorant is presented to the same subjects but this time in a forced choice triangle test in which the test material is hidden amongst odourless blanks and the subjects are asked to pick the odd sample out, then a lower value will almost certainly be obtained. In a review of reported values for the incidence of anosmia to androstenone (**2.201**), Triller et al. found that, depending on experimental method, estimates ranged from 1.8% to 75% of the population (530). The best method for detecting specific anosmia is to use an olfactometer to plot dose/response curves for a group of subjects. Any one with a specific hyposmia or anosmia will then be clearly identifiable by having a threshold value above the range of the normal Gaussian distribution of the population. However, such experiments require specialised equipment and are time consuming and costly.

The following is a list of substances for which specific anosmias have been reported in the literature between 1895 and 2008. Since these have been determined using different experimental protocols, care must be taken when drawing conclusions from the results. That said, we can try to glean some high level of information from the list. Fragrance chemists will instinctively associate specific anosmia with larger and more rigid molecules. The explanation for this would probably lie in the importance of musk and amber ingredients in fragrance. In the list below, the substances are arranged in increasing order of molecular weight and these vary from 29 to 394 in a relatively even distribution pattern. There are slight peaks around those molecular weights which are typical of biogenetic pathways, such as 154 which is the commonest weight of mono-oxygenated monoterpenoids. Thus natural synthetic pathways have built in a bias to that aspect of the data set. Examination of the structures does suggest a higher proportion of more rigid molecules, but this could also be an intrinsic bias because of the materials selected for study, for example, a disproportionately high proportion of musk and amber molecules. The conclusion must be that specific anosmia does not depend on either molecular weight or rigidity of molecular structure.

Hydrogen cyanide, formaldehyde, formic acid, trimethylamine, acetic acid, dimethylsulphide, 1-pyrroline, 2,4,5-triazole, isobutyraldehyde, *t*-butylcarbinol, benzene, pyridine, methyl cyclopropyl ketone, thiophene, diacetyl, isoamyl alcohol, putrescine, isobutyric acid, propenylsulfenic acid, dimethyldisulfide, ethylene dichloride, allyl isothiocyanate, *N*-methylpyrrolidone, glutaric dialdehyde, hexylamine, isovaleric acid, phenylisocyanide, methional, caproic acid, benzylamine, benzyl alcohol, cyclotene, 2-heptanone, indole, phenylethanol, salicylaldehyde, 3-methyl-2-hexenoic acid, naphthalene, amyl acetate, isoamyl acetate, scatole, cinnamadehyde, trichloroethylene, phenyl isothiocyanate, phenylacetic acid, adamantane, anisaldehyde, anisole, 2-nonenal, Melonal, *p*-dichlorobenene, geranyl nitrile, L-carvone, thymol, geranial (citral), vanillin, geraniol, 1,8-cineole, isoborneol, menthone, menthol, allicin, eugenol, phenethyl methyl ethyl carbinol, geosmin, γ-undecalactone, β-damascenone, β-ionone, γ-dodecalactone, Lyral, farnesol, cyclopentadecanone, Timberol (Norlimbanol), benzyl salicylate, Ambercore, Ambrox, cedryl methyl ether, muscone, cyclopentadecanolide, hexadcecanolide, Musk R1, Galaxolide, Tonalid, Traseolide, Versalide, cedryl acetate, Ysamber K, Karanal, ethylene brassylate, androstadienone, androstenone, androstenol, Ambrocenide, Jeger's ketal, Helvetolide, Okoumal, Musk Ketone, Musk Xylene, iodoform. Specific anosmias to various unspecified mercaptans and alkylpyrazines have also been reported.

In a study of specific anosmia to musk and amber ingredients, Triller et al. found that there is no evidence for general anosmia to either odour type but rather anosmia to individual compounds (530). Cluster tree analysis showed that the relationship between structure and anosmia is not a simple one and that anosmia of an individual to one representative molecule of a given odour class did not predict their sensitivity to other compounds eliciting the same odour for the general population. Plots of number of subjects versus number of substances of a given odour class to which they were anosmic showed an approximately exponential decline for amber but a strange grouping for musks. Thus, the largest number of subjects reporting an anosmia to amber ingredients reported anosmia to only one and only a tiny percentage reported anosmia to all amber odorants tested. With musk, there seemed to be groups of subjects who were anosmic to a number of musks. The study also confirmed that anosmia to an ingredient does not necessarily affect an individual's ability to distinguish its presence in a perfume accord. This was not an effect of physical chemistry since the concentration of the other ingredients in the headspace was the same whether using the test substance, Galaxolide (**2.301**), or an

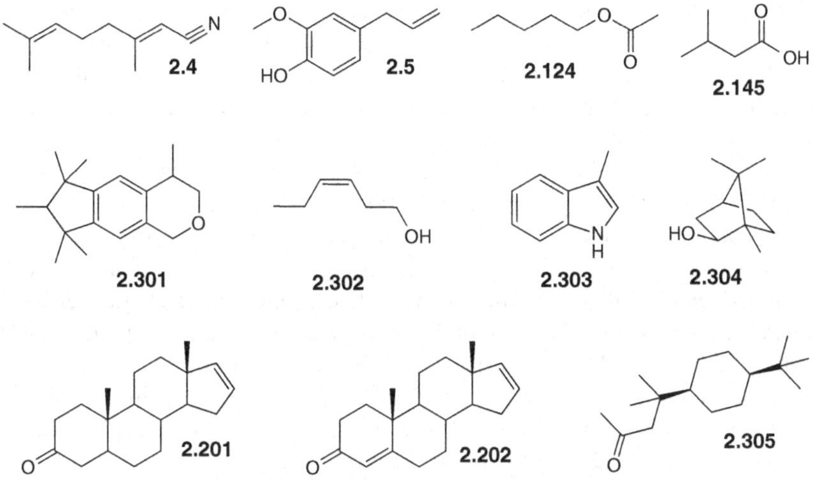

Figure 2.91   Some molecules related to anosmias.

odourless diluent, dipropylene glycol, to replace it in the formula. The results also showed no differences in rates of anosmia between the sexes or between experienced and naïve panellists for musks. For amber materials, males were more likely to present anosmia but smelling experience had no effect. Gilbert and Kemp found that when tested for sensitivity to 11 different musks, subjects fell into four categories, one of which were anosmic to all musks (556). This grouping is reminiscent of that of Triller et al. (530).

The concept of primary odours was proposed by analogy with the primary colours of vision (557). This idea was taken up by John Amoore who postulated that specific anosmias could be used to identify the primary odours on the assumptions that a specific anosmia indicated the absence of a basic receptor type and that these basic receptor types would correspond to the primary odours (558). His search was joined by Paolo Pelosi, and amongst the primary odours that they proposed were malty (559), urinous (560), musky, minty (561) and camphor (562). Of course, the unequivocal establishment of the combinatorial mechanism of olfaction has rendered the primary odour concept obsolete.

One likely cause of specific anosmia lies in variation of the receptor array used by individuals. If someone lacks a key receptor for a molecule or group of molecules, then it is reasonable to expect that that individual might present a specific anosmia for that specific molecule or group of molecules. The receptor array is determined by the individual's genotype, and therefore specific anosmia would be expected to show a genetic effect. This is indeed the case, and genetic effects have been found for specific anosmias to, for example, androstenone (**2.201**) (563–565, 546), Galaxolide (**2.301**) (564, 546) and isovaleric acid (**2.145**) (566) (Figure 2.91). There is a link between reduced sensitivity to cis-3-hexen-1-ol (**2.302**) (the odour impact compound of freshly cut grass) and variations in the receptor OR2J3 (547).

It has been shown that subjects having the T113A and R226Q SNPs in their genetic code for OR2J3 were less likely to be able to perceive it and, of the 5 haplotypes identified in a group of 52 individuals, this one accounted for 26.4% of the variation in sensitivity to *cis*-3-hexen-1-ol (**2.302**). The ability to smell the change in odour of urine after consumption of asparagus is hereditary and linked to a single SNP close to OR2M7 on chromosome 1q44, though the exact location is not known (567). The production of the, as yet unidentified, odorant responsible is also hereditary and the two hereditary effects are independent of each other.

The most studied specific anosmia is probably that to androstenone (**2.201**). As discussed earlier in the section on receptive ranges of ORs, the team of Matsunami showed that OR7D4 responds very selectively to androstenone (**2.201**) and androstadienone (**2.202**) (317). The two SNPs result in three allelic variations, and there is a correlation between genotype and phenotype, though only to the extent of about 80%, indicating that other receptors are also likely to play a role in androstenone (**2.201**) perception. In their study of androstenone (**2.201**) anosmia, Knaapila et al. also found that other receptors must be involved in addition to OR7D4 (568). In another study, Knaapila et al. investigated 126 pairs of monozygotic twins, 264 pairs of dizygotic twins and 137 twin individuals (338 males and 579 females with ages ranging from 10 to 83 years and from countries as far apart as Finland and Australia) for their ratings of intensity and pleasantness of androstenone (**2.201**) (569). They found that genetic effects contributed 28% to variation in perceived intensity and 21% to pleasantness of the odour. There was a genetic correlation between pleasantness and intensity but not between environmental factors and perception. However, in a study of the link between OR7D4 genotype and the ability to detect boar taint[†] in pork, it was found that the correlation between those with RT/RT genotype and sensitivity to androstenone (**2.201**) was much higher (571). Their evidence suggests that the ability to overcome specific anosmia is also dependent on having the correct allelic make-up. In a study of twins and their siblings, Knaapila et al. showed that intensity ratings of androstenone (**2.201**) and Galaxolide (**2.301**) had a genetic component, whereas ratings of amyl acetate (**2.124**), eugenol (**2.5**), mercaptans and rose did not, indicating that there could be different underlying mechanisms for anosmias to the two classes of ingredients (546).

One study of specific anosmia to isovaleric acid (**2.145**) in mice showed a recessive inheritance, and postulated that the protein involved was not an OR (572), whilst another found that three loci (hence three proteins) were involved, again not ORs but possibly transcription factors or regulatory proteins (573). However, in humans, anosmia to isovaleric acid does have a genetic component related to an OR, the pseudogene OR11H7P, although the relationship is not a simple one and involves other factors also (321).

[†]Boar taint is a combination of androstenone (**2.201**) and scatole (**2.303**). The level of androstenone (**2.201**) is related to the general level of male sex hormones, and boars metabolise tryptophane (Figure 2.12) differently from sows, resulting in higher levels of scatole (**2.303**) in their flesh (570).

Overcoming Specific Anosmia

Interestingly, it has been demonstrated that exposure to a substance can affect anosmia to it, and individuals can begin to smell materials to which they were previously anosmic. This effect has been demonstrated for androstenone (**2.201**), amyl acetate (**2.124**), geranyl nitrile (**2.4**) and isoborneol (**2.304**) (574–578). Intriguingly, Stevens and O'Connell found that sensitivity to androstenone (**2.201**) could be increased in both anosmic and normosmic subjects by exposure to Pemenone (**2.305**) (579). It has also been shown that mice who are anosmic to isovaleric acid (**2.145**) can be taught to smell it (580).

In an attempt to elucidate the mechanism for this effect, Sobel and co-workers exposed one nostril to an odorant and found that the resultant sensitivity increase applied to samples administered to either nostril. Therefore, a central mechanism is indicated. The effect was consistent across all subjects in the anosmic group studied (74). One possible explanation is that of learning of a pattern that had not been recognised previously. Another might be a direct or hormonal effect of the central olfactory cortex on the epithelium in the nose. Another group found a correlation between the perceived intensity of androstenone (**2.201**) and both the evoked olfactory potential (EOP) in the epithelium and electroencalography (EEG) measurements on the skin of the head (581). In the case of anosmics learning to smell androstenone (**2.201**), both EOP and EEG signals increased in line with the decrease in threshold. This tends to imply a peripheral mechanism.

In summary, the issue of specific anosmia gives us further insight into the complexity of olfaction. Whilst it is tempting to assume that specific anosmia is the result of missing receptors, this might be a part of the explanation but there is plenty of evidence to show that it is certainly not the whole story.

REFERENCES

1. E. H. Polak, *J. Theor. Biol.* **1973**, *40 (3)*, 469–484.
2. B. Malnic, J. Hirono, T. Sato and L. B. Buck, *Cell*, **1999**, *96*, 713–723.
3. D. A. Wilson and R. J. Stevenson, *Learning to Smell*, The Johns Hopkins University Press, Baltimore, 2006 ISBN 0-8018-8368-7.
4. C. H. Hawkes and R. L. Doty, *The Neurology of Olfaction*, Cambridge University Press, Cambridge, 2009 ISBN 978-0-521-68216-9.
5. G. M. Shepherd, *Neurogastronomy: How the Brain Creates Flavour and Why it Matters*, Columbia University Press, New York, 2012, ISBN 978-0-231-15910-4.
6. C. Delano and N. Sobel, *Neuron*, **2005**, *48(3)*, 431–454.
7. S. G. Leinwand and S. H. Chalasani, *Curr. Opin. Genet. Dev.*, **2011**, *21*, 806–811.
8. P. C. Brunjes, R. B. Kay and J. P. Arrivillaga, *J. Comp. Neurol.*, **2011**, *519*, 2870–2886.
9. K. W. Ashwell, *Brain Behav. Evol.*, **2012**, *79 (1)*, 45–56 doi: 10.1159/000332804.
10. U. B. Kaupp, *Nat. Rev. Neurosci.*, **2010**, *11*, 188–200 doi: 10.1038/nrn2789.
11. M. Nei, Y. Nimura and M. Nozawa, *Nat. Rev. Genet.*, **2008**, *9*, 951–963.
12. M. Ai, S. Min, Y. Grosjean, C. Leblanc, Bell, R. R. Benton and G. S. B. Suh, *Nature*, **2010**, *468*, 691–695 doi: 10.1038/nature09537.
13. J. Chandrashekar, D. Yarmolinsky, L. von Buchholtz, M. Goulding, W. Sly, N. J. P. Ryba, and C. S. Zuker, *Science*, **2009**, *326*, 443–445 doi : 10.1126/science.1174601.

14. J. Fan, F. Francis, Y. Liu, J. L. Chen and D. F. Cheng, *Genet. Mol. Res.*, **2011**, *10*, 3056–3069.
15. E. Grosse-Wilde, A. Svatos and J. Krieger, *Chem. Senses*, **2006**, *31*, 547–555 doi: 10.1093/chemse/bjj059.
16. B. Pophof, *Chem. Senses*, **2004**, *29*, 117–125 doi: 10.1093/chemse/bjh012.
17. S. Swarup, T. I. Williams and R. R. Anholt, *Genes Brain Behav.*, **2011**, *10(6)*, 648–657, doi: 10.1111/j.1601-183x.2011.00704.x.
18. M.-S. Kim and D. P. Smith, *Chem. Senses*, **2001**, *26*, 195–199.
19. P. Xu, R. Atkinson, D. N. M. Jones and D. P. Smith, *Neuron*, **2005**, *45(2)*, 193–200.
20. J. D. Laughlin, T. S. Ha, D. N. Jones and D. P. Smith, *Cell*, **2008**, *133*, 1255–1265.
21. S. Spinelli, A. Lagarde, I. Iovinella, P. Legrand, M. Tegoni, P. Pelosi, and C. Cambillau, *Insect Biochem. Mol. Biol.*, **2012**, *42*, 41–50.
22. S. Forêt and R. Maleszka, *Genome Res.*, **2006**, *16*, 1404–1413 doi: 10.1101/gr.5075706.
23. A. Lagarde, S. Spinelli, M. Tegoni, X. He, L. Field,J. J. Zhou and C. Cambillau, *J. Mol. Biol.*, **2011**, *414*, 401–412.
24. S. Vandermoten, F. Francis, E. Haubruge and W. S. Leal, *PLoS One*, **2011**, *6*, e23608.
25. T. Zhong, J. Yin, S. Deng, K. Li and Y. Cao, *J. Insect Physiol.*, **2012**, *58 (6)*, 771–781 doi: 10.1016/j.jinsphys.2012.01.011.
26. A. Lagarde, S. Spinelli, H. Qiao, M. Tegoni, P. Pelosi and C. Cambillau, *Biochem. J.*, **2011**, *437*, 423–430.
27. C. Lautenschlager, W. S. Leal and J. Clardy, *Structure*, **2007**, *15 (9)*, 1148–1154 doi: 10.1016/j.str.2007.07.013.
28. S. A. Hoffman, L. Aravind and S. Velmurugan, *Parasit. Vectors*, **2012**, *5*, 27.
29. J. Pelletier and W. S. Leal, *J. Insect Physiol.*, **2011**, 57, 915–929.
30. M. Wistrand, L. Kall and E.L. Sonnhammer, *Protein Sci.*, **2006**, *15*, 509–521.
31. U. B. Kaupp, *Nat. Neurosci.*, **2010**, *11*, 188–200.
32. K. Sato, M. Pellegrino, T. Nakagawa, L. B. Vosshall and K. Touhara, *Nature*, **2008**, *452*, 1002–1007 doi: 10.1038/nature06850.
33. D. Wicher, R. Schäfer, R. Bauernfeind, M. C. Stensmyr, R. Heller, S. H. Heinemann and B. S. Hanson, *Nature*, **2008**, *452*, 1007–1012 doi: 10.1038/nature06861.
34. A.S. Nichols, S. Chen and C. W. Luetje, *Chem. Senses*, **2011**, *36(9)*, 781–790 doi: 10.1093/chemse/bjr053.
35. F. Martin, J. Riveron and E. Alcorta, *J. Insect Physiol.*, **2011**, *57(12)*, 1631–1642 doi: 10.1016/j.jinsphys.2011.08.016.
36. Y. Oka, L. R. Saraiva and S. I. Korsching, *Chem. Senses*, **2012**, *37(3)*, 219–227 doi: 10.1093/chemse/bjr095.
37. G. Daghfous, M. Smargiassi, P.-A. Libourel, R. Wattiez and V. Bels, *Chem. Senses*, **2012**, *37 (9)*, 883–896 doi: 10.1093/chemse/bjs072.
38. M. Ma, *Crit. Rev. Biochem. Mol. Biol.*, **2007**, *42*, 463–480 doi: 10.1080/10409230701693359.
39. M. Spehr and S. D. Munger, *J. Neurochem.*, **2009**, *109*, 1570–1583 doi: 10.1111/j.1471-4159.2009.06085.x.
40. S. D. Munger, T. Leinders-Zufall and F. Zufall, *Ann. Rev. Physiol.*, **2009**, *71*, 115–140 doi: 10.1146/annurev.physiol.70.113006.100608.
41. K. Touhara and L. B. Vosshall, *Annu. Rev. Plant. Physiol. Plant. Mol. Biol.*, **2009**, *71*, 307–332 doi: 10.1146/annurev.physiol.010908.163209.
42. S. D. Liberles and L. B. Buck, *Nature*, **2006**, *442*, 645–650 doi:10.1038/nature05066.
43. W. Lin, R. Margolskee, G. Donnert, S. W. Hell, W. Stefan and D. Restreppo, *Proc. Natl. Acad. Sci. U. S. A.*, **2007**, *104 (7)*, 2471–2476.
44. W. Lin, E. A. D. Ezekwe Jr., Z. Zhao, E. R. Liman and D. Restreppo, *BMC Neurosci.*, **2008**, *9*, 114 doi:10.1186/1471-2202-9-114.
45. W. Lin, T. Ogura, R. F. Margolskee, T. E. Finger and D. Restreppo, *J. Neurophysiol.*, **2008**, *99*, 1451–1460 doi: 10.1152/jn.01195.2007.
46. S. D. Liberles, L. F. Horowitz, D. Kuang, J. J. Contos, K. L. Wilson, J. Siltberg-Liberles, D. A. Liberles and L. B. Buck, *Proc. Natl. Acad. Sci. U. S. A.*, **2009**, *106(24)*, 9842–9847 doi: 10.1073/pnas.0904464106.

47. S. Zibman,G. Shpak, and S. Wagner, *Neuroscience*, **2011**, *189*, 51–67.
48. P. Chamero, V. Katsoulidou, P. Hendrix, B. Bufe, R. Roberts, H. Matsunami, J. Abramowitz, L. Birnbaumer, F. Zufall and T. Leinders-Zufall, *Proc. Natl. Acad. Sci. U. S. A.*, **2011**, *108*, 12898–12903.
49. K. Hagino-Yamagishi and H. Nakazawa, *J. Comp. Neurol.*, **2011**, *519*, 3189–3201.
50. T. Leinders-Zufall, A. P. Lane, A. C. Puche, W. Ma, M. V. Novotny, M. T. Shipley and F. Zufall, *Nature*, **2000**, *405*, 792–796 doi: 10.1038/35015572.
51. Y. Isogai, S. Si, L. Pont-Lezica, T. Tan, V. Kapoor, V. N. Murthy and C. Dulac, *Nature*, **2011**, *478*, 241–245 doi: 10.1038/nature10437.
52. S. Rivière, L. Chalet, D. Fluegge, M. Spehr and I. Rodriguez, *Nature*, **2009**, *459*, 574–577 doi: 10.1038/nature08029.
53. E. Shirokova, J. D. Raguse, W. Meyerhof and D. Krautwurst, *FASEB J.*, **2008**, *22*, 1–6 doi:10.1096/fj.07-9233com.
54. D. Trotier, *Eur Ann. Otorhinolaryngol. Head Neck Dis.*, **2011**, *128*, 184–190.
55. T. Rodolfo-Masera, *Arch. Ital. Anat. Embryol.*, **1943**, *48*, 157–212.
56. J. Fleischer, K. Schwarzenbacher and H. Breer, *Chem. Senses*, **2007**, *32*, 623–631 doi: 10.1093/chemse/bjm032.
57. K. Mamasuew, N. Hofmann, H. Breer and J. Fleischer, *Chem. Senses*, **2011**, *36*, 271–282 doi: 10.1093/chemse/bjq124.
58. K. Mamasuew, N. Hofmann, V. Kretzschmann, M. Biel, R.B. Yang, H. Breer, and J. Fleischer, *Neurosignals*, **2011**, *19*, 198–209.
59. M. J. Storan and B. Key, *J. Comp. Neurol.*, **2005**, *494(5)*, 834–844.
60. G. M. Shepherd, *PLOS Biol.*, **2004**, *2(5)*, e146 doi: 10.1371/journal.pbio0020146.
61. G. M. Shepherd, *Chem. Senses*, **2005**, *30 (Suppl 1)*, i3–i5 doi: 10.1093/chemse/bjh085.
62. M. Laska, D. Genzel and A. Wieser, *Chem. Senses*, **2005**, *30*, 171–175 doi: 10.1093/chemse/bji013.
63. C. Blakemore and S. Jennett, *The Oxford Companion to the Body*, Oxford University Press, Oxford, 2001, ISBN 0 19 852403 X.
64. D. T. Moran, J. C. Rowley III, B. W. Jafek and M. A. Lovell, *J. Neurocytology*, **1982**, *11*, 721–746.
65. J. N. Lundström, J. A. Boyle and M. Jones-Gotman, *Chem. Senses*, **2006**, *31*, 249–252 doi: 10.1093/chemse/bjj025.
66. J. N. Lundström, J. A. Boyle and M. Jones-Gotman, *Chem. Senses*, **2008**, *33*, 23–33 doi: 10.1093/chemse/.
67. D. M. Lipnicki, *Chem. Senses*, **2008**, *33*, 223–224 doi: 10.1093/chemse/bjm076.
68. D. A. Werntz, R. G. Bickford and D. Shannahof-Khalsa, *Human Neurobiol.*, **1983**, *2(1)*, 39–43.
69. D. A. Werntz, R. G. Bickford and D. Shannahof-Khalsa, *Human Neurobiol.*, **1987**, *6(3)*, 165–171.
70. A.N. Gilbert and A. M. Rosenwasser, *Acta Otolaryngol.*, **1987**, *104*, 180–186.
71. M. M. Mozell, P. F. Kent and S. J. Murphy, *Chem. Senses*, **1991**, *16*, 721.
72. N. Sobel, R. M. Khan, A. Saltman, E. V. Sullivan and J. D. S. Gabrieli, *Nature*, **1999**, *402*, 35.
73. R. E. Frye and D. L. Doty, in *Chemical Signals in Vertebrates 6*, Eds. R. L. Doty and D. Müller-Schwarze, Plenum Press, New York, 1992, pp. 595–596.
74. J. D. Mainland, E. A. Bremner, N. Young, B. N. Johnson, R. M. Khan, M. Bensafi, and N. Sobel, *Nature*, **2002**, *419*, 802.
75. A.N. Gilbert, M. S. Greenberg and G. K. Beauchamp, *Neuropsychologia*, **1989**, *27(4)*, 505–511.
76. M. D. Kass, J. Pottackal, D. J. Turkel and J. P. McGann, *Chem. Senses*, **2013**, *38 (1)*, 77–89 doi: 10.1093/chemse/bjs081.
77. J. A. Gottfried, *Curr. Biol.*, **2009**, *19(18)*, R862–R864.
78. S. C. Kinnamon and S. D. Reynolds, *Science*, **2009**, *325*, 1081–1082.
79. A.S. Shah, Y. Ben-Shahar, T. O. Moninger, J. N. Kline and M. J. Welsh, *Science*, **2009**, *325*, 1131–1134.
80. A. Montell, Sci STKE, 2005:re3, I. S. Ramsey, M. Delling and D. E. Clapham, *Ann. Rev. Physiol.*, **2006**, *68*, 619–647.
81. T. Braun, B. Mack and M. F. Kramer, *Rhinology*, **2011**, *49*, 507–512.

82. B. D. Gulbransen, T. R. Clapp, T. E. Finger and S. C. Kinnamon, *J. Neurophysiol.*, **2008**, *99(6)*, 2929–2937 doi: 10.1152/jn.00066.2008.

83. R. L. Doty and J. E. Cometto-Muñiz, Trigeminal chemosensation, in *Handbook of Olfaction and Gustation*, Ed. R. L. Doty, Marcel Dekker, New York, 2003, pp. 981–1000.

84. D. N. Willis and J. B. Morris, *Chem. Senses*, **2013**, *38 (1)*, 91–100 doi: 10.1093/chemse/bjs085.

85. M. H. Abraham, R. Sánchez-Moreno, J. E. Cometto-Muñiz and W. S. Cain, *Chem. Senses*, **2007**, *32*, 711–719 doi: 10.11093/chemse/bjm038.

86. M. Laska, H. Distel and R. Hudson, *Chem. Senses*, **1997**, *22* 447–456.

87. J. Frasnelli, G. Charbonneau, O. Collignon and F. Lepore, *Chem. Senses*, **2009**, *34*, 139–144 doi: 10.1093/chemse/bjn068.

88. N. Sobel, V. Prabhakaran, J. E. Desmond, G. H. Glover, R. L. Goode, E. V. Sullivan and J. D. E. Gabrieli, *Nature*, **1998**, *392*, 282–286.

89. M. J. Lawson, B. A. Craven, E. G. Paterson and G. S. Settles, *Chem. Senses*, **2012**, *37*, 553–566 doi: 10.1093/chemse/bjs039.

90. Rodríguez, M. A. M. Teixeira and A. Rodrigues, *Flav. Frag. J.*, **2011**, *26(6)*, 421–428 doi: 10.1002/ffj.2076.

91. Y. Oka, S. Katada, M. Omura, M. Suwa, Y. Yoshihara and K. Touhara, *Neuron*, **2006**, *52*, 857–869.

92. Y. Oka, Y. Takai and K. Touhara, *J. Neurosci.*, **2009**, *29*, 12070–12078.

93. M. Damm, H. E. Eckel, M. Jungehulsing and T. Hummel, *Ann. Otol. Rhinol. Laryngol.*, **2003**, *112*, 91–97.

94. D. W. Wesson, T. N. Donahou, M. O. Johnson and M. Wachowiak, *Chem. Senses*, **2008**, *33*, 581–596 doi: 10.1093/chemse/bjn029.

95. A.S. Ghatpande, and J. Reisert, *J. Physiol.*, **2011**, *589*, 2261–2273.

96. M. Smear, R. Shusterman, R. O'Connor, T. Bozza and D. Rinberg, *Nature*, **2011**, *479*, 397–400 doi: 10.1038/nature10521.

97. E. Courtiol, C. Hegoburu, P. Litaudon, S. Garcia, N. Fourcaud-Trocme and N. Buonviso, *J. Neurophysiol.*, **2011**, *106*, 2813–2824.

98. E. Courtiol, C. Amat, M. Thevenet, S. Messaoudi, S. Garcia, and N. Buonviso, *PLoS One 6*, **2011**, e16445.

99. M. A. Rosero, and M. L. Aylwin, *Eur. J. Neurosci.*, **2011**, *34*, 787–799

100. R. M. Carey, and M. Wachowiak, *J. Neurosci.*, **2011**, *31*, 10615–10626.

101. N. Sobel, M. E. Thomason, I. Stappen, C. M. Tanner, J. W. Tetrud, J. M. Bower, E. V. Sullivan and J. D. Gabrieli, *Proc. Natl. Acad. Sci. U. S. A.*, **2001**, *98*, 4154–4159.

102. R.L. Doty, D. A. Deems, R. E. Frye, R. Pelberg and A. Shapiro, *Arch. Otolaryngol. Head Neck Surg.*, **1988**, *114*, 1422–1427.

103. R. Eccles, M. S. Jawad and S. Morris, *Acta Otolaryngol.*, **1989**, *108*, 268–273.

104. R. Teghtsoonian, M. Teghtsoonian B. Berglund and U. Berglund, *J. Exp. Psychol. Hum. Percept. Perform.*, **1978**, *4*, 144–152.

105. D. E. Hornung, C. Chin, D. B. Kurtz, P. F. Kent and M. M. Mozell, *Chem. Senses*, **1997**, *22*, 177–180.

106. A.M. Kleeman, R. Kopietz, J. Albrecht, V. Schöpf, O. Pollatos, T. Schreder, J. May, J. Linn, H. Brückmann and M. Wiesmann, *Chem. Senses*, **2009**, *34*, 1–9 doi: 10.1093/chemse/bjn042.

107. A. Tourbier and R. L. Doty, *Chem. Senses*, **2007**, *32*, 515–523 doi: 10.1093/chemse/bjm020.

108. O.Gladysheva, D. Kukushkina and G. Martynova, *Acta Histochem.*, **1986**, *78*, 141–146.

109. A. Tromelin and E. Guichard, *Flavour Fragr. J.*, **2006**, *21*, 13–24 doi: 10.1002/ffj.1696.

110. H. Debat, O. C. Eloit, F. Blon, B. Sarazin, C. Henry, J.-C. Huet, D. Trotier, J.-C. Pernollet, *J. Proteome Res.*, **2007**, *6(5)*, 1985–1996.

111. A.P. M. Menco and E. E. Morrison, in *Handbook of Olfaction and Gustation*, Ed. R. L. Doty, Marcel Dekker, New York, 2003, pp. 17–49.

112. T. T. Solbu and T. Holen, *Chem. Senses*, **2012**, *37(1)*, 35–46 doi: 10.1093/chemse/bjr063.

113. T. G. Păunescu, A. C. Jones, R. Tyszkowski and D. Brown, *Am. J. Physiol. Cell Physiol.*, **2008**, *295*, C293–C930 doi: 10.1152/ajpcell.00237.2008.

114. N. E. Baldaccini, A. Gagliardo, P. Pelosi, and A. Topazzini, *Comp. Biochem. Physiol. B*, **1986**, *84(3)*, 249–253.

115. (a) E. Napolitano, and P. Pelosi, *Bioorg. Med. Chem. Lett.*, 1992, *2(12)*, 1603–1606. (b) P. Pelosi, *Crit. Rev. Biochem. Mol. Biol.*, 1994, *29(3)*, 199–228.

116. F. Vincent, S. Spinelli, R. Ramoni, S. Grolli, P. Pelosi, C. Cambillau and M. Tegoni, *J. Mol. Biol.*, 2000, *300(1)*, 127–39.

117. Tegoni, M., Pelosi, P., Vincent, F., Spinelli, S., Campanacci, V., Grolli, S., Ramoni, R., Cambillau, C. *Biochim. Biophy. Acta Protein Struct. Mol. Enzymol.*, 2000; *1482(1–2)*, 229–240.

118. M. A. Bianchet, G. Bains, P. Pelosi, J. Pevsner, S. H. Snyder, H. L. Monaco and L. M. Amzel, *Nat. Struct. Biol.*, 1996, *3 (11)*, 934–939.

119. L. Briand, C. Eloit, C. Nespoulous, V. Bézirard, J.-C. Huet, C. Hemry, F. Blon, D. Trotier and J.-C. Pernollet, *Biochemistry*, 2002, *41*, 7241–7252.

120. M. Spehr, G. Gisselmann, A. Poplawski, J. A. Riffell, C. H. Wetzel, R. K. Zimmer, and H. Hatt, *Science*, 2003, *299*, 2054–2058.

121. M. Yabuki, D. J. Scott, L. Briand and A. J. Taylor, *Chem. Senses*, 2011, *36*, 659–671, doi:10.1093/chemse/bjr037.

122. K. Tsuchihara, K. Fujikawa, M. Ishiguro, T. Yamada, C. Tada, K. Ozaki and M. Ozaki, *Chem. Senses*, 2005, *30*, 559–564 doi: 10.1093/chemse/bji049.

123. L. Tcatchoff, C. Nespoulous, J.-C. Pernollet and L. Briand, *FEBS Lett.*, 2006, *580*, 2102–2108 doi: 10.1016/j.febslet.2006.03.017.

124. A.Nespoulous, L. Briand, M.-M. Delage, V. Tran and J.-C. Pernollet, *Chem. Senses*, 2004, *29*, 189–198, doi: 10.1093/chemse/bjh017.

125. H. Débat, C. Eloit, F. Blon, B. Sarazin, C. Henry, J. C. Huet, D. Trotier and J. C. Pernollet, *J. Proteome Res.*, 2007, *6*, 1985–1996.

126. U. Mager, A. Küller, P. C. Daiber, J. Neudorf, U. Warnken, M. Schnölzer, S. Frings and F. Möhrlen, *Proteomics*, 2009, *9*, 322–334.

127. X. Ding and A. R. Dahl in *Handbook of Olfaction and Gustation*, 2nd Edn. Ed. R. L. Doty, Marcel Dekker, New York, 2003, pp. 51–58.

128. X. Ding and L. S. Kaminsky, *Annu. Rev. Pharmacol. Toxicol.*, 2003, *43*, 149–173.

129. X. Zhang, Q. Y. Zhang, D. Liu, T. Su, Y. Weng, G. Ling, Y. Chen, J. Gu, B. Schilling and X. Ding, *Drug Metab. Disp.*, 2005, *33*, 1423–1428.

130. A.E. Hornung and M. M. Mozell, Preliminary data suggesting alteration of odorant molecules by interaction with receptors in *Olfaction and Taste*. Information Retrieval, London, Eds. J. Le Magnen and P. MacLeod, 1977, p. 63.

131. A. Lazard, N. Tal, M. Rubenstein, M. Khen, D. Lancet K. Zupko, *Biochem.*, 1990, *29(32)*, 7433–7440 doi: 10.1021/bi00484a012.

132. M. Maïbèche-Coisne, A. A. Nikonov, Y. Ishida, E. Jacquin-Joly and W. S. Leal, *Proc. Natl. Acad. Sci. U. S. A.*, 2004, *101(31)*, 11459–11464 doi: 10.1073/pnas.0403537101.

133. A. Nagashima and K. Touhara, *J. Neurosci.*, 2010, *30 (48)*, 16391–16398.

134. B. Schilling, R. Kaiser, A. Natsch and M. Gautschi, *Chemoecology*, 2010, *20*, 135–147 doi: 101007/s00049-009-0035-5 and references cited therein.

135. B. Schilling, CH 2005/000412, 2005 to Givaudan.

136. B. Schilling, WO 2006/007751, 2006 to Givaudan.

137. B. Schilling, WO 2006/007752, 2006 to Givaudan.

138. B. Schilling, T. Granier, G. Fráter, A. Hanhart, WO 2008 116338, 2008 to Givaudan.

139. J. E. Cometto-Muñiz and M. H Abraham, *Exp. Brain Res.*, 2010, *207*, 75–84 doi: 10.1007/s00221-010-2430-0.

140. J. E. Cometto-Muñiz and M. H. Abraham, *Chem. Senses*, 2010, *35*, 289–299 doi: 10.1093/chemse/bjq018.

141. J. E. Cometto-Muñiz, W. S. Cain, M. H. Abraham and J. Gil-Lostes, *Physiol. Behav.*, 2008, *95*, 658–667 doi: 10.1016/j.physbeh.200809.

142. R. O. Dror, A. C. Pan, D. H. Arlow, D. W. Borhani, P. Maragakis, Y. Shan, H. Xu and D. E. Shaw, *Proc. Natl. Acad. Sci. U. S. A.*, 2011, *1008 (32)*, 13118–13123 doi: 10.1073/pnas.1104614108.

143. K. Kobayakawa, R. Kobayakawa, H. Matsumoto, Y. Oka, T. Imai, M. Ikawa, M. Okabe, T. Ikeda, S. Itohara, T. Kikusui, K. Mori and H. Sakano, *Nature*, 2007, *450*, 503–508.

144. P. Nef, I. Hermans-Borgmeyer, H. Artières-Pin, L. Beasley, V. E. Dionne and S. F. Heinemann, *Proc. Natl. Acad. Sci. U. S. A.*, **1992**, *89*, 8948–8952.

145. J. Strotman, I. Wanner, T. Helfrich, A. Beck and H. Breer, *Cell Tissue Res.*, **1994**, *278*, 11–20.

146. R. Vassar, J. Ngai and R. Axel, *Cell*, **1993**, *74*, 309–331.

147. K. J. Ressler, S. L. Sullivan and L. B. Buck, *Cell*, **1993**, *73*, 597–609.

148. T. Abaffy and A. R. Defazio, *BMC Res. Notes*, **2011**, *4*, 137.

149. R. Hoppe, H. Breer and J. Strotman, *Genomics*, **2003**, *82*, 355–364.

150. J. Strotman, S. Conzelman, A. Beck, P. Feinstein, H. Breer and P. Mombaerts, *J. Neurosci.*, **2000**, *20*, 6927–6938.

151. V. Bautze, R. Bär, B. Fissler, M. M. Trapp, D. Schmidt, U. Beifuss, B. Bufe, F. Zufall, H. Breer and J. Strotman, *Chem. Senses*, **2012**, *37*, 479–493, doi: 10.1093/chemse/bjr130.

152. L. B. Buck and R. A. Axel, *Cell*, **1991**, *65*, 175–187.

153. A. Feldmesser, T. Olender, M. Khen, I. Yanai, R. Ophir and D. Lancet, *BMC Genomics*, **2006**, *7*, 121–129 doi: 10.1186/1471-2164-7-121.

154. Y. Gilad, O. Man, S. Pääbo and D. Lancet, *Proc. Natl. Acad. Sci. U. S. A.*, **2003**, *100(6)*, 3324–3327.

155. H. Breer, *Biochem. Soc. Trans.*, **2003**, *31(1)*, 113–116.

156. G. Glusman, I. Yanai, I. Rubin and D. Lancet, *Genome Res.*, **2001**, *11*, 685–702 doi: 10.1101/gr.171001.

157. I. Menashe and D. Lancet, *Cell. Mol. Life Sci.*, **2006**, *63*, 1485–1493, doi: 10.1007/s00018-006-6661-x.

158. H. Saito, Q. Chi, H. Zhuang, H. Matsunami and J. D. Mainland, *Sci. Signal.*, **2009**, *2 (60)*, ra9 doi: 10.1126/scisignal.2000016.

159. Y. Gilad, O. Man and G.A. Glusman, *Genome Res.*, **2005**, *15*, 224–230.

160. http://bioinfo.weizmann.ac.il/HORDE.

161. X. Zhang, O. De la Cruz, J. M. Pinto, D. Nicolae, S. Firestein and Y. Gilad, *Genome Biol.*, **2007**, *8*, R86 doi: 10.1186/gb-2007-8-5-r86.

162. M. Spehr, K. Schwane, S. Heilmann, G. Gisselmann, T. Hummel and H. Hatt, *Curr. Biol.*, **2004**, *14 (19)*, 832–833.

163. N. Fukuda, K. Yomogida, M. Okabe and K. Touhara, *J. Cell Sci.*, **2004**, *117*, 5835–5845.

164. X. Grosmaitre, A. Vassalli, P. Momberts, G. M. Shepherd and M. Ma, *Proc. Natl. Acad. Sci. U. S. A.*, **2006**, *103*, 1970–1975.

165. A. Touhara, S. Sengoku, K. Inaki, A. Tsuboi, J. Hirono, T. Sato, H. Sakano and T. Haga, *Proc. Natl. Acad. Sci. U. S. A.*, **1999**, *96*, 4040–4045.

166. E. M. Neuhaus, W. Zhang, L. Gelis, Y. Deng, J. Noldus and H. Hatt, *J. Biol. Chem.*, **2009**, *284 (24)*, 16218–16225 doi: 10.1074/jbc.M109.012096/DC1.

167. A.L. Pluznik, D.-J. Zou, X. Zhang, Q. Yan, D. J. Rodriguez-Gil, C. Eisner, E. Wells, C. A. Greer, T. Wang, S. Firestein, J. Schnermann and M. J. Caplan, *Proc. Natl. Acad. Sci. U. S. A.*, **2009**, *106(6)*, 2059–2064 doi: 10.1073/pnas0812859106.

168. J. L. Pluznik, R. J. Protzko, H. Gevorgyan, Z. Peterlin, A. Sipos, J. Han, I. Brunet, L.-X. Wan, F. Rey, T. Wang, S. J. Firestein, M. Yanagisawa, J. I. Gordon, A. Eichmann, J. Peti-Peterdi and M. J. Caplan, *Proc. Natl. Acad. Sci. U. S. A.*, **2012**, *110(11)*, 4410–4415 doi: 10.1073/pnas.1215927110.

169. G. K. Pavlav, *Cell Adh. Migr.*, **2010**, *4(4)*, 502–506 doi: 10.4161/cam4.4.12291.

170. D. M. Rosenbaum, S. G. F. Rasmussen and B. K. Kobilka, *Nature*, **2009**, *459*, 356–363.

171. S. Zozulya, F. Echeverri and T. Nguyen, *Genome Biol.*, **2001**, *2*, RESEARCH0018.

172. J. C. Venter, M. D. Adams, E. W. Myers, P. W. Li, R. J. Mural, G. G. Sutton, H. O. Smith, M. Yandell et al., *Science*, **2001**, *291*, 1304–1351.

173. F. Fredriksson, M. C. Lagerstrom, L. G. Lundin and H. B. Schioth, *Mol. Pharmacol.*, **2003**, *63*, 1256–1272.

174. D. E. Gloriam, S. M. Foord, F. E. Blaney and S. L. Garland, *J. Med. Chem.*, **2009**, *52 (14)*, 4429–4442 doi: 10.1021/jm900319e.

175. R. Fredriksson and H. B. Schiöth, G-protein-coupled receptors in the human genome in *Ligand Design for G-protein Coupled Receptors*, Ed. D. Rognan, 2006, Wiley-VCH, Weinheim, ISBN-10:3-527-31284-6, ISBN-13:978-3-527-31284-6.

176. J. A. Ballesteros and H. Weinstein, *Methods Neurosci.*, **1995**, *25*, 366–428.
177. O. Man, Y. Gilad and D. Lancet, *Protein Sci.*, **2004**, *13*, 240–254.
178. Y. Pilpel and D. Lancet, *Protein Sci.*, **1999**, *8*, 969–977.
179. M. M. Rosenkilde, T. Benned-Jensen, T. M. Frimurer and T. W. Schwartz, *Trends Pharmacol. Sci.*, **2010**, *31(12)*, 567–574, doi 10.1016/j.tips.2010.08.006.
180. H. Matsui, R. J. Lefkowitz, M. G. Caron and J. W. Regan, *Biochemistry*, **1989**, *28*, 4125–4130.
181. J. W. Regan, T. S. Kobilka, T. L. Yang-Feng, M. G. Caron, R. J. Lefkowitz and B. K. Kobilka *Science*, **1987**, *238*, 650–656.
182. H. G. Dohlman, M. G. Caron, C. D. Strader, N. Amlaiky and R. J. Lefkowitz, *Biochemistry*, **1988**, *27*, 1813–1817.
183. G. Tikhonova, C. S. Sum, S. Neumann, C. J. Thomas, B. M. Raaka, S. Costanzi and M. C. Gershengorn, *J. Med. Chem.*, **2007**, *50*, 2981–2989.
184. D. M. Garrido, D. F. Corbett, K. A. Dwornik, A. S. Goetz, T. R. Littleton, S. C. McKeown, W. Y. Mills, T. L. Smalley Jr., C. P. Briscoe and A. J. Peat, *Bioorg. Med. Chem. Lett.*, **2006**, *16*, 1840–1845.
185. K. Corin, P. Baaske, D. B. Ravel, J. Song, E. Brown, X. Wang, C. J. Wienken, M. Jerabek-Willemsen, S. Duhr, Y. Luo, D. Braun and S. Zhang, *PLoS One*, (**2011**, *6*, e25067.
186. H. Kandori, *Chem. Ind.*, **1995**, *18*, 735–739.
187. R. Henderson, J. M. Baldwin, T. A. Laska, F. Zemlin, E. Beckmann and K. H. Downing, *J. Mol. Biol.*, **1990**, *213*, 899–929.
188. K. Palczewski, T. Kumasaka, T. Hori, C. A. Behnke, H. Motoshima, B. A. Fox, I. Le Trong, and M. Miyano, *Science*, **2000**, *289*, 739–745.
189. J. Li, P. C. Edwards, M. Burghammer, C. Villa and G. F. X. Schertler, *J. Mol. Biol.*, **2004**, *343(5)*, 1409–1438.
190. T. Okada, M. Sugihara, A.-N. Bondar, M. Elstner, P. Entel, and V. Buss, *J. Mol. Biol.*, **2004**, *342*, 571–583.
191. J. H. Park, P. Scheerer, K. P. Hofmann, H.-W. Choe and O. P. Ernst, *Nature*, **2008**, *454*, 183–188, doi: 10.1038/nature07063.
192. P. Scheerer, J. H. Park, P. W. Hildebrand, Y. J. Kim, N. Krauss, H.-W. Choe, K. P. Hofmann and O. P. Ernst, *Nature*, **2008**, *455*, 497–503, doi: 10.1038/nature07330
193. H.-W. Choe, Y. J. Kim, J. H. Park, T. Morizumi, E. F. Pai, N. Krauss, K. P. Hofmann, P. Scheerer and O. P. Ernst, *Nature*, **2011**, *471*, 651–655. doi: 10.1038/nature09789.
194. J. Standfuss, P. C. Edwards, A. D'Antona, M. Fransen, G. Xie, D. D. Oprian and G. F. X. Schertler, *Nature*, **2011**, *471*, 656–660 doi: 10.1038/nature09795.
195. T. Warne, M. J. Serrano-Vega, J. G. Baker, R. Moukhametzianov, P. C. Edwards, R. Henderson, A. G. W. Leslie, C. G. Tate and G. F. X. Schertler, *Nature*, **2008**, *454*, 486–492, doi: 10.1038/nature 07101.
196. T. Warne, R. Moukhametzianov, J. G. Baker, R. Nehme, P. C. Edwards, A. G. W. Leslie, G. F. X. Schertler and C. G. Tate, *Nature*, **2011**, *469*, 241–244, doi 10.1038/nature 09746.
197. R. Moukhametzianov, T. Warne, P. C. Edwards, M. J. Serrano-Vega, A. G. W. Leslie, C. G. Tate and G. F. X. Schertler, *Proc. Natl. Acad. Sci. U. S. A.*, **2011**, *108 (20)*, 8228–8232.
198. S. G. F. Rasmussen, H.-J. Choi, D. M. Rosenbaum, T. S. Kobilka, F. S. Thian, P. C. Edwards, M. Burghammer, V. R. P. Ratnala, R. Sanishvili, R. F. Fischetti, G. F. X. Schertler, W. I. Weis and B. K. Kobilka, *Nature*, **2007**, *450*, 383–388. doi: 10.1038/nature06325.
199. V. Cherezov, D. M. Rosenbaum, M. A. Hanson, S. G. F. Rasmussen, S. F. Thian, T. S. Kobilka, H.-J. Choi, P. Kuhn, W. I. Weis, B. K. Kobilka and R. C. Stevens, *Science*, **2007**, *318*, 1258–1265 doi: 10.1126/science.1150577.
200. D. M. Rosenbaum, V. Cherezov, M. A. Hanson, S. G. F. Rasmussen, F. S. Thian, T. S. Kobilka, H.-J. Choi, X. J. Yao, W. I. Weis, R. C. Stevens and B. K. Kobilka, *Science*, **2007**, *318*, 1266–1273 doi: 10.1126/science.1150609.
201. A. A. Hanson, V. Cherezov, M. T. Griffith, C. B. Roth, V.-P. Jaakola, E. Y. T. Chien, J. Velasquez and R. C. Stevens, *Structure*, **2008**, *16 (6)*, 897–905.
202. S. M. Pontier, Y. Percherancier, S. Galandrin, A. Breit, C. Galés and M. Bouvier, *J. Biol. Chem.*, **2008**, *283 (36)*, 24659–24672.

203. D. Wacker, G. Fenalti, M. A. Brown, V. Katritch, R. Abagyan, V. Cherezov and R. C. Stevens, *J. Am. Chem. Soc.*, **2010**, *132*, 11443–11445.
204. S. Devanathan, Z. Yao, Z. Salamon, B. Kobilka and G. Tollin, *Biochemistry*, **2004**, *43*, 3280–3288.
205. S. G. F. Rasmussen, H.-J. Choi, J. J. Fung, E. Pardon, P. Casarosa, P. S. Chae, B. T. DeVree, D. M. Rosenbaum, F. S. Thian, T. S. Kobilka, A. Schnapp, I. Konetzki, R. K. Sunahara, S. H. Gellman, A. Pautsch, J. Steyaert, W. I. Weis and B. K. Kobilka, *Nature*, **2011**, *469*, 175–181 doi: 10.1038/nature09648.
206. D. M. Rosenbaum, C. Zhang, J. A. Lyons, R. Holl, D. Aragao, D. H. Arlow, S. G. F. Rasmussen, H.-J. Choi, B. T. DeVree, R. K. Sunahara, P. S. Chac, S. H. Gellman, R. O. Dror, D. E. Shaw, W. I. Weis, M. Caffrey, P. Gmeiner and B. K. Kobilka, *Nature*, **2011**, *469*, 236–242 doi: 10.1038/nature09665.
207. V.-P. Jaakola, M. T. Griffith, M. A. Hanson, V. Cherezov, E. Y. T. Chien, J. R. Lane, A. P. Ijzerman and R. C. Stevens, *Science*, **2008**, *322*, 1211, doi: 10.1126/science.1164772.
208. F. Xu, H. Wu, V. Katritch, G. W. Han, K. A. Jacobson, Z.-G. Gao, V. Cherezov and R. C. Stevens, *Science*, **2011**, *332*, *(6027)*, 322–327 doi: 10.1126/science.1202793.
209. S. J. Mantell, P. T. Stephenson, S. M. Monaghan, G. N. Maw, M. A. Trevethick, M. Yeadon, D. K. Walker and F. Macintyre, *Bioorg. Med. Chem. Lett.* **2009**, *19 (15)*, 4471–4475.
210. G. Lebon, T. Warne, P. C. Edwards, K. Bennett, C. J. Langmead, A. G. W. Leslie and C. G. Tate, *Nature*, **2011**, *474*, 521–526, doi: 10.1038/nature10136.
211. T. Shimamura, M. Shiroishi, S. Weyand, H. Tsujimoto, G. Winter, V. Katritch, R. Abagyan, V. Cherezov, W. Liui, G. W. Han, T. Kobayashi, R. C. Stevens and S. Iwata, *Nature*, **2011**, *475*, 65–70 doi: 10.1038/nature10236.
212. E. Y. T. Chien, W. Liu, Q. Zhao, V. Katritch, G. W. Han, M. A. Hanson, L. Shi, A. H. Newman, J. A. Javitch, V. Cherezov and R. C. Stevens, *Science*, **2010**, *330*, 1091–1095.
213. B. Wu, E. Y. T. Chien, C. D. Mol, G. Fenalti, W. Liu, V. Katritch, R. Abagyan, A. Brooun, P. Wells, F. C. Bi, D. J. Hamel, P. Kuhn, T. M. Handel, V. Cherezov and R. C. Stevens, *Science*, **2010**, *330*, 1066–1071.
214. K. Haga, A. C. Kruse, H. Asada, T. Yurugi-Kobayashi, M. Shiroishi, C. Zhang, W. I. Weis, T. Okada, B. K. Kobilka, T. Haga and T. Kobayashi, *Nature*, **2012**, *482*, 547–551, doi: 10.1038/nature10753.
215. A. C. Kruse, J. Hu, A. C. Pan, D. H. Arlow, D. M. Rosenbaum, E. Rosemond, H. F. Green, T. Liu, P. S. Chae, R. O. Dror, D. E. Shaw, W. I. Weis, J. Wess and B. K. Kobilka, *Nature*, **2012**, *482*, 552–559 doi: 10.1038/nature10867.
216. H. Wu, D. Wacker, M. Mileni, V. Katritch, G. W. Han, E. Vardy, W. Liu, A. A. Thompson, X.-P. Huang, F. I. Carroll, S. W. Mascarella, R. B. Westkaempfer, P. D. Mosier, B. L. Roth, V. Cherezov and R. C. Stevens, *Nature*, **2012**, *485*, 327–334 doi: 10.1038/nature10939.
217. A. Manglik, A. C. Kruse, T. S. Kobilka, F. S. Thian, J. M. Mathiesen, R. K. Sunahara, L. Pardo, W. I. Weiss, B. K. Kobilka and S. Granier, *Nature*, **2012**, *485*, 321–327 doi:10.1038/nature10954.
218. S. Granier, A. Manglik, A. C. Kruse, T. S. Kobilka, F. S. Thian, W. I. Weis and B. K. Kobilka, *Nature*, **2012**, *485*, 400–404 doi: 10.1038/nature11111
219. A.A. Thompson, W. Liu, E. Chun, V. Katritch, H. Wu, E. Vardy, X.-P. Huang, C. Trapella, R. Guerrini, G. Calo, B. L. Roth, V. Cherezov and R. C. Stevens, *Nature*, **2012**, *485*, 395–400 doi: 10.1-38/nature11085.
220. J. F. White, N. Noinaj, Y. Shibaka, J. Love, B. Kloss, F. Xu, J. Gvozdenovic-Jeremic, P. Shah, J. Shiloach, C. G. Tate and R. Grisshammer, *Nature*, **2012**, *490*, 508–513, doi:10.1038/nature11558.
221. S. Ahuja, E. Crocker, M. Eilers, V. Hornak, A. Hirshfeld, M. Ziliox, N. Syrett, P. J. Reeves, H. G. Khorana, M. Sheves and S. O. Smith, *J. Biol. Chem.*, **2009**, *284 (15)*, 10190–10201 doi: 10.1074/jbc.M805725200.
222. M. P. Bokoch, Y. Zou, S. G. F. Rasmussen, C. W. Liu, R. Nygaard, D. M. Rosenbaum, J. J. Fung, H.-J. Choi, F. S. Thian, T. S. Kobilka, J. D. Puglisi, W. I. Weis, L. Pardo, R. S. Prosser, L. Müller and B. K. Kobilka, *Nature*, **2010**, *463*, 108–114, doi: 10.1038/nature08650.
223. L. Shi, I. Kawamura, K.-H. Jung, L. S. Brown and V. Ladizhansky, *Angew. Chem. Int. Ed.*, **2011**, *50*, 1302–1305 doi: 10.1002/anie.201004422.

224. J. J. Liu, R. Horst, V. Katritch, R. C. Stevens and K. Wüthrich, *Science*, **2012**, *335 (6072)*, 1106–1110 doi: 10.1126/science.1215802.
225. V. A. Higman, K. Varga, L. Aslimovska, P. J. Judge, L. J. Sperling, C. M. Rienstra and A. Watts, *Angew. Chem. Int. Ed.*, **2011**, *50*, 8432–8435 doi:10.1002/anie.201100730.
226. A. Gautier, J. P. Kirkpatrick and D. Nietlisprach, *Angew. Chem. Int. Ed.*, **2008**, *47*, 7297–7300.
227. A. Gautier, H. R. Mott, M. J. Bostock, J. P. Kirkpatrick and D. Nietlisprach, *Nat. Struct. Mol. Biol.*, **2010**, *17*, 768–774.
228. S. Reckel, D. Gottstein, J. Stehle, F. Löhr, M.-K. Verhoefen, M. Takeda, R. Silvers, M. Kainosho, C. Glaubitz, J. Wachtveitl, F. Bernhard, H. Schwalbe, P. Güntert and V. Dötsch, *Angew. Chem. Int. Ed.*, **2011**, *50*, 11942–11946.
229. S. H. Park, B. B. Das, F. Casagrande, Y. Tian, H. J. Nothnagel, M. Chu, H. Kiefer, K. Maier, A. A. De Angelis, F. M. Marassi and S. J. Opella, *Nature*, **2012**, *491*, 779–783 doi:10.1038/nature11580.
230. S. H. Park, F. Casagrande, L. Cho, L. Albrecht and S. J. Opella, *J. Mol. Biol.*, **2011**, *414*, 194–203 doi: 10.1016/j.jmb.2011.08.025.
231. S. Ye, E. Zaitseva, G. Caltabiano, G. F. X. Schertler, T. P. Sakmar, X. Deupi and R. Vogel, *Nature*, **2010**, *464*, 1386–1390 doi: 10.1038/nature.08948.
232. K. Katayama, Y. Furutani, H. Imai and H. Kandori, *Angew. Chem. Int. Ed.*, **2010**, *49*, 891–894 doi: 10.1002/anie.200903837.
233. C. Altenbach, A. K. Kusnetzow, O. P. Ernst, K.P. Hofmann and W. L. Hubbell, *Proc. Natl. Acad. Sci. U. S. A.*, **2008**, *105*, 7439–7444.
234. F. Wade, A. Espagne, M. A. Persuy, J. Vidic, R. Monnerie, F. Merola, E. Pajot-Augy and G. Sanz, *J. Biol. Chem.*, **2011**, *286*, 15252–15259.
235. A. G. Beck-Sickinger and N. Budisa, *Angew. Chem. Int. Ed.*, **2012**, *51*, 310–312 doi: 10.1002/anie201107211.
236. S. R. Hawtin, J. Simms, M. Conner, Z. Lawson, R. A. Parslow, J. Trim, A. Sheppard and M. Wheatley, *J. Biol. Chem.*, **2006**, *281 (50)*, 38478–38488.
237. V. A. Avlani, K. J. Gregory, C. J. Morton, M. W. Parker, P. M. Sexton and A. Christopoulos, *J. Biol. Chem.*, **2007**, *282 (35)*, 25677–25686.
238. H. G. Dohlman, M. G. Caron, A. DeBlasi, T. Frielle and R. J. Lefkowitz, *Biochemistry*, **1990**, *29*, 2335–2342.
239. A. E. Elling, T. M. Frimurer, L.-O. Gerlach, R. Jorgenson, B. Holst and T. W. Schwartz, *J. Biol. Chem.*, **2006**, *281*, 17337–17346 doi: 10.1074/jbc.M512510200.
240. T. W. Schwartz, T. M. Frimurer, B. Holst, M. M. Rosenkilde and C. E. Elling, *Annu. Rev. Pharmacol. Toxicol.*, **2006**, *46*, 481–519 doi: 10.1146/annurev.pharmtox.46.120604.141218.
241. T. E. Angel, M. R. Chance and K. Palczewski, *Proc. Natl. Acad. Sci. U. S. A.*, **2009**, *106 (21)*, 8555–8560 doi: 10.1073/pnas.0903545106.
242. M. D. Kurland, M. B. Newcomer, Z. Peterlin, K. Ryan, S. Firestein and V. S. Batista, *Biochemistry*, **2010**, *49*, 6302–6304 doi: 10.1021/bi100976w.
243. A. Kato, S. Katada, and K. Touhara, *J. Neurosci.*, **2008**, *107*, 1261–1270 doi: 10.1111/j.1471-4159.2008.05693.x.
244. C. D. Strader, T. Gaffney, E. E. Sugg, M. R. Candelore, R. Keys, A. A. Pratchett and R. A. Dixon, *J. Biol. Chem.*, **1991**, *266*, 5–8.
245. H.-J. Böhm and G. Klebe, *Angew. Chem. Int. Ed.*, **1996**, *35*, 2588–2614.
246. J. A. Ballosteros, A. D. Jensen, G. Liapakis, S. G. F. Rasmussen, L. Shi, U. Gether and J. A. Javitch, *J. Biol. Chem.*, **2001**, *276*, 29171–29177.
247. S. Madathil and K. Fahmy, *J. Biol. Chem.*, **2009**, *284 (42)*, 28801–28809.
248. S. G. F. Rasmussen, B. T. DeVree, Y. Zou, A. C. Kruse, K. Y. Chung, T. S. Kobilka, F. S. Thian, P. S. Chae, E. Pardon, D. Calinski, J. M. Mathiesen, S. T. A. Shah, J. A. Lyons, M. Caffrey, S. H. Gellman, J. Steyaert, G. Skiniotis, W. I. Weiss, R. K. Sunahara and B. K. Kobilka, *Nature*, **2011**, *477*, 549–557 doi: 10.1038/nature10361.
249. T. Hino, T. Arakawa, H. Iwanari, T. Yurugi-Kobayashi, C. Ikeda-Suno, Y. Nakada-Nakura, O. Kusano-Arai, S. Weyand, T. Shimamura, N. Nomura, A. D. Cameron, T. Kobayashi, T. Hamakubo, S. Iwata and T. Murata, *Nature*, **2012**, *482*, 237–240 doi: 10.1038/nature10750.

250. A. Shirokova, K. Schmiedeberg, P. Bedner, H. Niessen, K. Willecke, J.-D. Rageuse, W. Meyerhof and D. Krautwurst, *J. Biol. Chem.*, **2005**, *280 (12)*, 11807–11815.
251. S. L. Ritter and R. A. Hall, *Nat. Revi. Mol. Cell Biol.*, **2009**, *10*, 819–830.
252. Y. R. Li and H. Matsunami, *Sci. Signal.*, **2011**, *4 (155)*, 1–11, doi: 10.1126/scisignal.2001230.
253. T. Ogura, S. A. Szebenyi, K. Krosnowski, A. Sathyanesan, J. Jackson and W. Lin, *J. Neurophysiol.*, **2011**, *106(3)*, 1274–1287.
254. A. Hall, *Sci. Signal.*, **2011**, *4*, e1.
255. W. Liu, E. Chun, A. A. Thompson, P. Chubukov, F. Xu, V. Katritch, G. W. Han, C. B. Roth, L. H. Heitman, A.P. Ijzerman, V. Cherezov and R. C. Stevens, *Science*, **2012**, *337*, 232–236, doi: 10.1126/science.1219218.
256. A. Bockenhauer, A. Fürstenberg, X. J. Yao, B. K. Kobilka and W. E. Moerner, *J. Phys. Chem.*, **2011**, *115*, 13328–13338.
257. P. Kolb and G. Klebe, *Angew. Chem. Int. Ed.*, **2011**, *50*, 11573–11575 doi: 10.1002/anie.201105869.
258. M. Audet and M. Bouvier, *Cell*, **2012**, *151*, 14–23 doi: 10.1016/j.cell.2012.09.003.
259. S.-Z. Xu, P. Sukumar, F. Zeng, J. Li, A. Jairaman, A. English, J. Naylor, C. Ciurtin, Y. Majeed, C. J. Milligan, Y. M. Bahmasi, E. Al-Shawaf, K. E. Porter, L.-H. Jiang, P. Emery, A. Sivaprasadarao and D. J. Beech, *Nature*, **2008**, *451*, 69–73 doi: 10.1038/nature06414.
260. X. Duan, E. Block, Z. Li, T. Connelly, J. Zhang, Z. Huang, X. Su, Y. Pan, L. Wu, Q. Chi, S. Thomas, S. Zhang, M. Ma, H. Matsunami, G.-Q. Chen and H. Zhuang, *Proc. Natl. Acad. Sci. U. S. A.*, **2012**, *109 (9)*, 3492–3497 doi: 10.1073/pnas.1111297109.
261. A. Natsch and H. Gfeller, *Toxicol. Sci.*, **2008**, *106(2)*, 464–478 doi: 10.1093/toxsci/kfn194.
262. H. Dods, J. A. Mosely and J. M. Sanderson, *Org. Biomol. Chem.*, **2012**, *10*, 5271–5378.
263. W. Kahsai, K. Xiao, S. Rajagopal, S. Ahn, A. K. Shukla, J. Sun, T. G. Oas and R. J. Lefkowitz, *Nat. Chem. Biol.*, **2011**, *7*, 692–700 doi: 10.1038/nchembio.634.
264. M. Ditzen, M. Pellegrino and L. B. Vosshall, *Science*, **2008**, *319*, 1838–1842.
265. J. A. Pickett, M. A. Birkett and J. G. Logan, *Proc. Natl. Acad. Sci. U. S. A.*, **2008**, *105(36)*, 13195–13196.
266. Z. Syed and W. S. Leal, *Proc. Natl. Acad. Sci. U. S. A.*, **2008**, *105(36)*, 13598–13603.
267. M. Pellegrino, N. Steinbach, M. O. Stensmyr, B. S. Hansson and L. B. Vosshall, *Nature*, **2011**, *478*, 511–514 doi:10.1038/nature10438.
268. S. Katada, T. Hirokawa and K. Touhara, *Curr. Comput. Aided Drug Des.*, **2008**, *4*, 123–131.
269. N. Vaidehi, *Drug Discov. Today*, **2010**, *15 (21–22)*, 951–957 doi: 10.1016/j.drudis.2010.08.018.
270. M. Michino, E. Abola, GPCR Assessment Participants, C.L. Brooks, J.S. Dixon, J. Moult, R.C. Stevens *Nature Rev. Drug Disc.*, **2009**, *8*, 455–463.
271. M. S. Singer and G. M. Shepherd, *Neuroreport*, **1994**, *5*, 1297–1300.
272. M. Afshar, R. E. Hubbard and J. Demaille, *Biochimie*, **1998**, *80*,129–135.
273. J. A. Bajgrowicz and C. Broger, Molecular Modelling in Design of New Odorants: Scope and Limitations. In Flavours, Fragrances and Essential Oils. Proceedings of the 13th International Congress of Flavours, Fragrances and Essential Oils, Istanbul, Turkey, AREP, Istanbul. 1995. pp. 1–5.
274. L. Doszczak, P. Kraft, H.-P. Weber, R. Bertermann, A. Triller, H. Hatt and R. Tacke, *Angew. Chem. Int. Ed.*, **2007**, *46*, 3367–3371 doi: 10.1002/anie.200605002.
275. K. Schmiedeberg, E. Shirokova, H.-P. Weber, B. Schilling, W. Meyerhof and D. Krautwurst, *J. Struct. Biol.*, **2007**, *159 (3)*, 400–412 doi: 10.1016/jsb.2007.04.013.
276. S. Katada, T. Hirokawa, Y. Oka, M. Suwa, and K. Touhara, *J. Neurosci.*, **2005**, *25 (7)*, 1806–1815 doi: 10.15223/JNEUROSCI.4723-04.2005.
277. P. C. Lai, M. S. Singer and C. J. Crasto, *Chem. Senses*, **2005**; *30(9)*, 781–792 doi: 10.1093/chemse/bji070.
278. L. Gelis, S. Wolf, H. Hatt, E. M. Neuhaus, and K. Gerwert, *Angew. Chem Int. Ed.*, **2011**, *51*, 1274–1278 doi: 10.1002/anie.201103980.
279. M. H. Abraham, R. Sanchez-Mooreno, J. E. Cometto-Muniz and W. S. Cain, *Chem. Senses*, **2012**, *37*, 207–218 doi: 10.1093/chemse/bjr094.

280. N. Vaidehi, W. B. Floriano, R. Trabanino, S. E. Hall, P. Freddolino, E. J. Choi, G. Zamanakos and W. B. Goddard III, *Proc. Natl. Acad. Sci. U. S. A.*, **2002**, *99(20)*, 12622–12627.
281. W. B. Floriano, N. Vaidehi, W. B. Goddard III, M. S. Singer and G. M. Shepherd, *PNAS*, **2000**, *97(20)*, 10712–10716.
282. W. B. Floriano, N. Vaidehi and W. B. Goddard III, *Chem. Senses* **2004**, *29(4)*, 269–290. doi: 10.1093/chemse/bjh030.
283. S. E. Hall, W. B. Floriano, N. Vaidehi and W. B. Goddard III, *Chem. Senses*, **2004**, *29 (7)*, 595–616 doi: 10.1093/chemse/bjh063.
284. P. Hummel, N. Vaidehi, W. B. Floriano, S. E. Hall and W. B. Goddard III, *Prot. Sci.*, **2005**,*14*, 703–710.
285. D. A. Goldfeld, K. Zhu, T. Beuming, and R. A. Friesner, *Proc. Natl. Acad. Sci. U. S. A.*, **2011**, *108 (20)*, 8275–8280.
286. S. Bhattacharya and N. Vaidehi, *J. Am. Chem. Soc.*, **2010**, *132*, 5205–5214.
287. A. Klabunde, C. Giegerich, A. Evers, *J. Med. Chem.*, **2009**, *52(9)*, 2923–2932.
288. S. S. Phatak, E. A. Gatica and C. N. Cavasotto, *J. Chem. Inf. Model.*, **2010**, *50(12)*, 2119–2128 doi 10.1021/ci100285f.
289. M. P. A. Sanders, S. Verhoeven, C. de Graaf, L. Roumen, B. Vroling, S. B. Nabuurs, J. de Vlieg and J. P. G. Klomp, *J. Chem. Inf. Model.*, **2011**, *51 (9)*, 2277–1192 doi: 10.1021/ci200088d.
290. A. Veithen, F. Wilkin, M. Philippeau and P. Chatelain, *Perfumer & Flavorist*, 2009, *34 (6)*, 36–44.
291. J. Reisert and D. Restrepo, *Chem. Senses*, **2009**, *234*, 535–545 doi: 10.1093/chemse/bjp028.
292. R. C. Araneda, Z. Peterlin, X. Zhang, A. Chesler and S. Firestein, *J. Physiol*, **2004**, *555–3*, 743–756.
293. S. Bieri, K. Monastyrskaia and B. Schilling, *Chem. Senses*, **2004**, *29*, 483–487 doi: 10.1093/chemse/bjh050.
294. K. Nara, L. R. Saraiva, X. Ye and L. B. Buck, *J. Neurosci.*, **2011**, *31 (25)*, 9179–9191 doi: 10.1523/jneurosci.1282-11.2011.
295. R. C. Araneda, A. D. Kini and S. Firestein, *Nat. Neurosci.*, **2000**, *3(12)*, 1248–1255.
296. D. Krautwurst, K. W. Yau and R. R. Reed, *Cell*, **1998**, *95(7)*, 917–926.
297. T. Bozza, P. Feinstein, C. Zheng and P. Mombaerts, *J. Neurosci.*, **2002**, *22*, 3033–3043.
298. X. Grosmaitre, S. H. Fuss, A. C. Lee, K. A. Adipietro, H. Matsunami, P. Mombaerts and M. Ma, *J. Neurosci.*, **2009**, *29*, 14545–14552.
299. K. Mori, Y. K. Takahashi, K. M. Igarashi and M. Yamaguchi, *Physiol. Rev.*, **2006**, *86*, 409–433 doi: 10.1152/physrev.00021.2005.
300. http://gara.bio.uci.edu/.
301. A. Hamana, J. Hirono, M. Kizumi and T. Sato, *Chem. Senses*, **2003**, *28*, 87–104.
302. A. Levasseur, M.-A. Persuy, D. Grebert, J.-J. Remy, R. Salesse and E. Pajot-Augy, *Eur. J. Biochem.*, **2003**, *270*, 2905–2912 doi: 10.1046/j.1432-1033.2003.03672.x.
303. D. S. Auld, N. Thorne, D.-T. Nguyen and J. Inglese, *ACS Chem. Biol.*, **2008**, *3(8)*, 463–470 doi: 10.1021/cb8000793.
304. A. Saito, M. Kubota, R. W. Roberts, Q. Chi and H. Matsunami, *Cell*, **2004**, *119*, 679–691.
305. H. Matsunami, J. D. Mainland and S. Dey, *Ann. N. Y. Acad. Sci.*, **2009**, *1170 (1)*, 153–156 doi:10.1111/j.1749-6632.2009.03888.x.
306. Y. Fujita, T. Takahashi, A. Suzuki, K. Kawashima, F. Nara and R. Koishi, *J. Recept. Signal Transduct. Res.*, **2007**, *27 (4)*, 323–334.
307. H. Wetzel, M. Oles, C. Wellerdieck, M. Kuczkowiak,G. Gisselmann and H. Hatt, *J. Neurosci.*, **1999**, *19 (17)*, 7426–7433.
308. V. Jacquier, H. Pick and H. Vogel, *J. Neurochem.*, **2006**, *97*, 537–544 doi: 10.1111/j.1471-4159.2006.03771.
309. K. Kajiya, K. Inaki, M. Tanaka, T. Haga, H. Kataoka and K. Touhara, *J. Neurosci.*, **2001**, *21 (16)*, 6018–6025.
310. S. Etter, Ph.D. thesis, 3810 École Polytechnique Fédérale de Lausanne, 2007.
311. Baud, O., S. Etter, M. Spreafico, L. Bordoli, T. Schwede, H. Vogel and H.M. Pick, *Biochemistry*, **2011**, *50*, 843–853 doi: 10.1021/bil017396.

312. A. Triller, E. A. Boulden, A. Churchill, H. Hatt, J. Englund, M. Spehr and C. S. Sell, *Chem. Biodivers.*, **2008**, *5*, 862–886.

313. V. Matarazzo, O. Clot-Faybesse, B. Marcet, G. Guiraudie-Capraz, B. Atanasova, G. Devauchelle, M. Cerutti, P. Etiévant and C. Ronin, *Chem. Senses*, **2005**, *30*, 195–207 doi: 10.1093/chemse/bji015.

314. F. Sallmann and A. Veithen, WO2006094704 (A2), 2006, assigned to ChemCom S. A.

315. G. Sanz, C. Schlegel, J.-C. Pernollet and L. Briand, *Chem. Senses*, **2005**; *30*, 69–80. doi:10.1093/chemse/bji002.

316. G. Sanz, C. Schlegel, J.-C. Pernollet and L. Briand, *Chem. Senses*, **2008**, *33*, 639–653.

317. A. Keller, H. Zhuang, Q. Chi, L. B. Vosshall and H. Matsunami, *Nature*, **2007**, *449*, 468–472 doi: 10.1038/nature06162.

318. T. Abaffy, A. Malhotra and C. W. Luetje, *J. Biol. Chem.*, **2007**, *282*, 1216–1224.

319. T. Dahoun, L. Grasso, H. Vogel and H. Pick, *Biochemistry*, **2011**, *50*, 7228–7235 doi: 10.1021/bi2008596.

320. E. A. Hallem and J. R. Carlson, *Cell*, **2006**, *125*, 143–160 doi: 10.1016/j.cell.2006.01.050.

321. Menashe, I. T. Abaffy, Y. Hasin, S. Goshen, V. Yaholom, C. W. Luetje and D. Lancet, *PLoS Biol.*, **2007**, *5 (11) e284*, 2462–2468 doi: 10.1371/journal.pbio.0050284.

322. M. Ferrero, D. Wacker, M. A. Roque, M. W. Baldwin, R. C. Stevens and S. D. Liberles, *Chem. Biol.*, **2012**, *7*, 1184–1189 doi: 10.1021/cb3000111e.

323. B. Malnic, P. A. Godfrey and L. B. Buck, *PNAS*, **2004**, *101 (8)*, doi: 10.1073/pnas.0307882100.

324. A. C. Lee, J. He, and M. Ma, *J. Neurosci.*, **2011**, *31*, 2974–2982.

325. K. Ukhanov, D. Brunert, E. A. Corey, and B. W. Ache, *J. Neurosci.*, **2011**, *31*, 273–280.

326. G. A. Bell, D. G. Laing and H. Panhuber, *Brain Res.*, **1987**, *426*, 8–18.

327. Z. Peterlin, Y. Li, G. Sun, R. Shah, S. Firestein and K. Ryan, *Cell Chem. Biol.*, **2009**, 1317–1327.

328. Y. Oka, A. Nakamura, H. Watanabe and K. Touhara, *Chem. Senses*, **2004**, *29*, 815–822 doi: 10.1093/chemse/bjh247.

329. A. Brodin, M. Laska and M. J. Olsson, *Chem. Senses*, **2009**, *34 (7)*, 625–630 doi: 10.1093/chemse/bjp044.

330. Y. Oka, M. Omura, H. Kataoka and K. Touhara, *EMBO J.*, **2004**, *23 (1)*, 120–126.

331. P. Izewska, Investigating the molecular basis of ligand binding to and activation of receptors in living cells. Ph.D. thesis, 3668 École Polytechnique Fédérale de Lausanne, 2006.

332. A. J. Conn, A. Christopoulos and C. W. Lindsley, *Nat. Rev. Drug Discov.*, **2009**, *8*, 41–54 doi: 10.1038/nrd2760.

333. A. Christopoulos and T. Kenakin, *Pharmacol. Rev.*, **2002**, *54*, 323.

334. M. Behrens, W. Meyerhof, C. Hellfritsch and T. Hofmann, *Angew. Chem. Int. Ed.*, **2011**, *50*, 2220–2242 doi: 10.1002/anie.201002094.

335. G. Servant, C. Tachdjian, X. Q. Tang, S. Wemer, F. Zhang, X. Li, P. Kamdar, G. Petrovic, T. Ditschun, A. Java, P. Brust, N. Brune, G. E. DuBois, M. Zoller and D. S. Karanewsky, *Proc. Natl. Acad. Sci. U. S. A.*, **2010** *107*, 4746–4751.

336. M. Bai, S. Quinn, S. Trivedi, O. Kifor, S. H. Pearce, M. R. Pollak, K. Krapcho, S. C. Herbert and E. M. Brown, *J. Biol. Chem.*, **1996**, *271 (32)*, 19537–19545.

337. S. U. Miedlich, L. Gama, K. Seuwen, R. M. Wolf and G. E. Breitwieser, *J. Biol. Chem.*, **2004**, *279 (8)*, 7254–7263.

338. P.L. Jones, G. M. Pask, I. M. Romaine, R. W. Taylor, P. R. Reid, A. G. Waterson, G. A. Sulikowski and L. J. Zwiebel, *PLoS One 7*, **2012**, e30304.

339. V. A. Avlani, K. J. Gregory, C. J. Morton, M. W. Parker, P. M. Sexton and A. Christopoulos, *J. Biol. Chem.*, **2011**, *282 (35)*, 25677–25686.

340. A. Lanzafame, A. Christopoulos and F. Mitchelson, *J. Pharmacol. Exp. Ther.*, **1997**, *282*, 278–285.

341. A. Christopoulos, A. Lanzafame, and F. Mitchelson, *Clin. Exp. Pharmacol. Physiol.*, **1998**, *25*, 184–194.

342. B. J. Melancon, C. R. Hopkins, M. R. Wood, K. A. Emmitte, C. M. Niswender, A. Christopoulos, P. J. Conn and C. W. Lindsley, *J. Med. Chem.*, **2012**, *55*, 1445–1464 doi: 10.1021/jm201139r.

343. K. Y. Chung, S. G. F. Rasmussen, T. Liu, S. Li, B. T. DeVree, P. S. Chae, D. Calinski, B. K. Kobilka, V. L. Woods Jr. and R. K. Sunahara, *Nature*, **2011**, *477*, 611–617.
344. R. Sadana and C. W. Dessauer, *Neurosignals*, **2009**, *17*, 5–22 doi: 10.1159/000166277.
345. M. Hattori and E. Gouaux, *Nature*, **2012**, *484*, 207–212 doi: 10.1038/nature11010.
346. A. B. Stephan, E. Y. Shum, S. Hirsh, K. D. Cygnar, J. Reisert and H. Zhao, *Proc. Natl. Acad. Sci. U. S. A.*, **2009**, *106(28)*, 11776–11781.
347. G. M. Billig, B. Pal, P. Fidzinski and T. J. Jentsch, *Nat. Neurosci.*, **2011**, *14*, 763–769.
348. S. J. Kleene, *Chem. Senses*, **2008**, *33*, 839–859 doi:10.1093/chemse/bjn048.
349. J. D. Violin, L. M. DiPilato, N. Yildirim, T. C. Elston, J. Zhang and R. J. Lefkowitz, *J. Biol. Chem.*, **2008**, *283(5)*, 2949–2961.
350. S. Kerr, L. E. C. Von Dannecker, M. Davalos, J. S. Michaloski and B. Malnic, *Mol. Cell. Neurosci.*, **2008**, *38*, 341–348.
351. K. D. Cygnar and H. Zhao, *Nat. Neurosci.*, **2009**, *12*, 454–462 doi: 10.1038/nn.2289.
352. N. Benbernou, S. Robin, S. Tacher, M. Rimbault, M. Rakotomanga, and F. Galibert, *J. Hered.*, **2011**, *102 Suppl. 1*, S47–S61.
353. D. Brunert, K. Klasen, E. A. Corey and B. W. Ache, *Chem. Senses*, **2010**, *35*, 301–308.
354. B. W. Ache, *Chem. Senses*, **2010**, *35*, 533–539 doi: 10.1093/chemse/bjq045.
355. A. B. Stephan, S. Tobochnik, M. Dibattista, C. M. Wall, J. Reisert and H. Zhao, *Nat. Neurosci.*, **2012**, *15 (1)*, 131–139 doi: 10.1038/nn.2943.
356. Y. Sone, K. D. Cygnar, B. Sagdullaev, M. Valley, S. Hirsh, A. Stephan, J. Reisert and H. Zhao, *Neuron*, **2008**, *58(3)*, 374–386.
357. T. Kurahashi, G. Lowe, and G. H. Gold, *Science*, **1994**, *265*, 118–120.
358. K. Ukhanov, E. A. Corey, D. Brunert, K. Klasen and B. W. Ache, *J. Neurophysiol.*, **2010**, *103*, 1114–1122.
359. P. Duchamp-Viret, A. Duchamp and M. A. Chaput, *Eur. J. Neurosci.*, **2003**, *18*, 2690–2696.
360. J. P. Rospars, P. Lansky, M. Chaput and P. Duchamp-Viret, *J. Neurosci.*, **2008**, *28*, 2659–2666.
361. P. Chatelain, P. Robberecht, M. Waelbroek, J. C. Camus and J. Christophe, *J. Membrane Biol.*, **1986**, *93*, 23–32.
362. G. M. Pask, Y. U. Bobkov, E. A. Corey, B. W. Ache and L. J. Zwiebel, *Chem. Senses*, **2013**, *38*, 221–229 doi: 10.1093/chemse/bjs100.
363. Breunig, E. Kludt, E. Czesnik, D. and D. Schild,*J. Biol. Chem.*, **2011**, *286*, 28041–28048.
364. H. Yamada and K. Nakatani, *Chem. Senses*, **2001** *26*, 25–34.
365. T. Y. Chen, H. Takeuchi and T. Kurahashi, *J. Gen. Physiol.*, **2006**, *128*, 365–371.
366. H. Takeuchi, H. Ishida, S. Hikichi and T. Kurahashi, *J. Gen. Physiol.*, **2009**, *133 (6)*, 583–601.
367. Y. Kishino, H. Kato, T. Kurahashi and H. Takeuchi, *J. Physiol. Sci.*, **2011**, *61*, 231–245 doi: 10.1007/s12576-011-0142-2.
368. M. Spehr, C. H. Wetzel, H. Hatt and B. W. Ache, *Neuron*, **2002**, *33*, 731–739.
369. A. Duncan-Lewis, R. L. Lukman and R. K. Banks, *Comp. Med.*, **2011**, *61*, 361–365.
370. W. James, *Principles of Psychology*, Macmillan, London, 1891.
371. J. Joseph, F. A. Dunn, and M. Stopfer, *J. Neurosci*, **2012**, *32*, 2900–2910.
372. M. L. Fletcher and W. R. Chen, *Learn. Mem.*, **2010**, *17(11)*, 561–570.
373. A. S. Marino-Sanchez, I. Alobid, S. Cantellas, C. Alberca, J. M. Guilemany, J. M. Canals, H. J. De and J. Mullol, *Rhinology*, **2010**, *48*, 273–276.
374. A. Morrot, F. Brochet, D. Dubourdieu, *Brain Lang.*, **2001**, *79*, 309–320.
375. N. Boulkroune, L. Wang, A. March, N. Walker and T. J. C. Jacob, *Neuropsychopharmacology*, **2007**, *32*, 1822–1829 doi: 10.1038/sj.npp.1301303.
376. J. A. Gottfried and R. J. Dolan, *Neuron*, **2003**, *39 (2)*, 375–386.
377. M. Luisa Demattè, D. Sanabria and C. Spence, *Acta Psychol. (Amst)*, **2007**, *124(3)*, 332–343.
378. J. B. Jadauji, J. Djordjevic, J. N. Lundström and C. C. Pack, *J. Neurosci.*, **2012**, *32(9)*, 3095–3100 doi: 10.1523/jneurosci.6022-11.2012.
379. V. La Buissonniere-Ariza, J. Frasnelli, O. Collignon and F. Lepore, *Neurosci. Lett.*, **2011**, *506(2)*, 188–192 doi: 10.1016/j.neulet.2011.11.002.
380. S. Anton, K. Evengaard, R. B. Barrozo, P. Anderson, and N. Skals, *Proc. Natl. Acad. Sci. U. S. A.*, **2011**, *108*, 3401–3405.

381. L. E. Marks, M. G. Veldhuizen, T. G. Shepard and A. Y. Shavit, *Chem. Senses*, **2012**, *37(3)*, 263–277 doi: 10.1093/chemse/bjr103.

382. J. N. Lundstrom, A. R. Gordon, P. Wise and J. Frasnelli, *Chem. Senses* **2012**, *37*, 371.

383. J. A. Boyle, M. Heinke, J. Gerber, J. Frasnelli and T. Hummel, *Chem. Senses*, **2007**, *32*, 343–353 doi: 10.1093/chemse/bjm004.

384. T. E. Finger, B. Böttger, A. Hansen, K. T. Anderson, H. Alimohammadi and W. L. Silver, *Proc. Natl. Acad. Sci. U. S. A.*, **2008**, *100(15)*, 8981–8986 doi: 10.1073/pnas.1531172100.

385. M. Tizzano, B. D. Gulbransen, A. Vandenbeuch, T. R. Clapp, J. P. Herman, H. M. Sibhatu, M. E. A. Churchill, W. L. Silver, S. C. Kinnamon and T. E. Finger, *Proc. Natl. Acad. Sci. U. S. A.*, **2010**, *107(7)*, 3210–3215 doi: 10.1073/pnas.0911934107.

386. E. Iannilli, T. Bitter, H. Gudziol,H. P. Burmeister, H. J. Mentzel, A. P. Chopra and T. Hummel, *Rhinology*, **2011**, *49*, 458–463.

387. E. Billot, A. Comte, E. Galliot, P. Andrieu, V. Bonnans, L. Tatu, T. Gharbi, T. Moulin and J. L. Millot, *Neuroscience*, **2011**, *189*, 370–376 doi: 10.1016/j.neuroscience.2011.05.035.

388. L. Jacquot, J. Hidalgo and G. Brand, *Rhinology*, **2010**, *48*, 281–284.

389. B. A. Stuck, J. Baja, F. Lenz, R. M. Herr and C. Heiser, *Neuroscience*, **2011**, *176*, 442–446.

390. A.T. Rolls, F. Grabenhorst, C. Margot, M. A. A. P. Da Silva and M. I. Velazco, *J. Cogn. Neurosci.*, **2008**, *20*, 1815–1826.

391. J. N. Lundström, J. A. Boyle, R. J. Zatorre and M. Jones-Gotman, *Cereb. Cortex*, **2008**, *18*, 466–1474 doi: 10.1093/cercor/bhm178.

392. J. Chapuis and D. A. Wilson, *Nat Neurosci.*, **2012**, *15(1)*, 155–163 doi: 10.1038/nn.2966.

393. M. Huang, H. Bridge, M. J. Kemp and A. J. Parker, *Front. Hum. Neurosci.*, **2011**, *5*, 134 doi: 10.3389/fnhum.2011.00134.

394. C. Kärnekull, F. U. Jönsson, M. Larsson and J. K. Olofsson, *Chem. Senses*, **2011**, *36*, 641–648 doi: 10.1093/chemse/bjr028.

395. O. Pollatos, R. Kopietz, J. Linn, J. Albrecht, V. Sakar, A. Anzinger, R. Schandry and M. Wiesmann, *Chem. Senses*, **2007**, *32*, 583–589 doi: 10.1093/chemse/bjm027.

396. M. Bensafi, F. Rinck, B. Schaal and C. Rouby, *Chem. Senses*, **2007**, *32*, 855–862 doi: 10.1093/chemse/bjm055.

397. A. Levitan, M. Zampini, R. Li and C. Spence, *Chem. Senses*, **2008**, *33*, 415–423 doi: 10.1093/chemse/bjn008.

398. M. L. Demattè, *Chem. Senses*, **2009**, *34*, 103–109 doi: 10.1093/chemse/bjn055.

399. Y. T. Qiu, J. J. A. van Loon, W. Takken, J. Meijerink and H. M. Smid, *Chem. Senses*, **2006**, *31*, 845–863 doi: 10.1093/chemse/bjl027.

400. J. M. Susskind, D. H. Lee, A. Cusi, R. Feiman, W. Grabski and A. K. Anderson, *Nat. Neurosci.*, **2008**, *11*, 843–850 doi: 10.1038/nn.2138.

401. H. Lapid and T. Hummel, *Chem. Senses*, **2013**, *38(1)*, 3–17 doi: 10.1093/chemse/bjs073.

402. H. Lapid, S. Shushan, A. Plotkin, H. Voet, Y. Roth, T. Hummel, E. Schneidman and N. Sobel, *Nat. Neurosci.*, **2011**,*14*, 1455–1461 doi: 10.1038/nn2926.

403. E. Martinelli, D. Polese, F. Dini, R. Paolesse, D. Filippini, I. Lundström and C. Di Natale, *Front Neuroeng.*, **2011** *4*, 16 doi: 10.3389/fneng.2011.00016.

404. http://gara.bio.uci/.edu/.

405. Y. K. Takahashi, M. Kurosaki, S. Hirono and K. Mori, *J. Neurophysiol.*, **2004**, *92*, 2413–2427 doi: 10.1152/jn.00236.2004.

406. Y. K. Takahashi, S. Nagayama and K. Mori, *J. Neurosci.*, **2004**, *24(40)*, 8690–8694 doi: 10.1523/jneurosci.2510.-04.2004.

407. Gerami, H. Nemati, S., Abbaspour, F. and R. Banan, *Acta Med. Iran.*, **2011**, *49*, 13–17.

408. Kjelvik, G. Evensmoen, H. R., V. Brezova and A. K. Håberg, *J. Neurophysiol.* **2012**, *108*, 645–657.

409. H. E. Kato, F. Zhang, O. Yizhar, C. Ramakrishnan, T. Nishizawa, J. Ito, Y. Aita, T. Tsukazaki, S. Hayashi, P. Hegemann, A. D. Maturana, R. Ishitani, K. Deisseroth and O. Nureki, *Nature*, **2012**, *482*, 369–374 doi: 10.1038/nature10870.

410. F. Blumhagen, P. Zhu, J. Shum, Y.-P. Zhang Schärer, E. Yaksi, K. Deisseroth and R. W. Friedrich, *Nature*, **2011**, *479*, 493–498 doi:10.1038/nature10633.

411. Y. Livneh and A. Mizrahi, *Nat. Neurosci.*, **2011**, *15(1)*, 26–28.
412. A. Markopoulos, D. Rokni, D. H. Gire and V. N. Murthy, *Neuron*, **2012**, *76(6)*, 1175–1188 doi: 10.1016/j.bneuron2012.10.028.
413. N. Kokan, N. Sakai, K. Doi, H. Fujio, S. Hasegawa, H. Tanimoto and K. Nibu, *Am. J. Rhinol. Allergy*, **2011**, *25*, 163–165.
414. S. Serizawa, K. Miyamichi, H. Nakatani, M. Suzuki, M. Saito, Y. Yoshihara and H. Sakano, *Science*, **2003**, *302*, 2088–2094.
415. W. Lewcock and R. R. Reed, *Proc. Natl. Acad. Sci. U. S. A.*, **2004**, *101*, 1069–1074.
416. M. Maritan, G. Monaco, I. Zamparo, M. Zaccolo, T. Pozzan and C. Lodovichi, *Proc. Natl. Acad. Sci. U. S. A.*, **2009**, *109(9)*, 3537–3542 doi: 10.1073/pnas.0813224106.
417. W. Hong, T. J. Mosca and L. Luo, *Nature*, **2012**, *484*, 201–207 doi: 10.1038/nature10926.
418. T. J. Mosca, W. Hong, V. S. Dani, V. Favaloro and L. Luo, *Nature*, **2012**, *484*, 237–241 doi:10.1038/nature10923.
419. A. Fleischmann, B. M. Shykind, D. L. Sosulski, K. M. Franks, M. E. Glinka, D. F. Mei, Y. Sun, J. Kirkland, M. Mendelsohn, M. W. Albers and R. Axel, *Neuron*, **2008**, *60*, 1068-1-81 doi: 10.1016/j.neuron.2008.10.046.
420. M. de Bruijne, K. Foster and J. R. Carlson, *Neuron*, **2001**, *30*, 537–552.
421. M. Wachowiak, J. P. McGann, P. M. Heyward, Z. Shao, A. C. Puche and M. T. Shipley, *J. Neurophysiol.*, **2005**, *94*, 2700–2712.
422. M. Ennis, F. M. Zhou, K. J. Ciombor, V. Aroniadou-Anderjaska, A. Hayar, E. Borrelli, L. A. Zimmer, F. Margolis, and M. T. Shipley, *J. Neurophysiol.*, **2001** *86*, 2986.
423. Johnson, O., J. Becnel and C. D. Nichols, *Neuroscience*, **2011**, *192*, 372–381.
424. J. Weiss, M. Pyrski, E. Jacobi, B. Bufe, V. Willnecker, B. Schick, P. Zizzari, S. J. Gossage, C. A. Greer, T. Leinders-Zufall, C. G. Woods, J. N. Wood, and F. Zufall, *Nature*, **2011**, *472*, 186–190 doi: 10.1038/nature09975.
425. J. Tan, A. Savigner, M. Ma and M. Luo, *Neuron*, **2010**, *65*, 912–926 doi: 10.1016/j.neuron.2010.02.11.
426. Y. Adam and A. Mizrahi, *J. Neurosci.*, **2011**, *31*, 7967–7973.
427. C. Linster, Q. Nai and M. Ennis, *J. Neurophysiol.*, **2011**, *105*, 1432–1443 doi: 10.1152/jn.00960.2010.
428. E. M. Garland, S. R. Raj, A. C. Peltier, D. Robertson and I. Biaggioni, *Neurology*, **2011**, *76*, 456–460.
429. J. L. Hellier, N. L. Arevalo, M. J. Blatner, A. K. Dang, A. C. Clevenger, C. E. Adams and D. Restrepo, *Brain Res.*, **2010**, *1358*, 140–150 doi: 10.1016/j.brainres.2010.08.027.
430. C. E. LePichon, M. T. Valley, M. Polymenidou, A. T. Chesler, B. T. Sagdullaev, A. Aguzzi and S. Firestein, *Nat. Neurosci.*, **2009**, *12(1)*, 60–69 doi: 10.1038/nn.2238.
431. O. Gschwend, J. Beroud and A. Carleton, *PLoS One*, **2012**, *7*, e30155 doi: 10.1371/journal.pone.0030155.
432. M. Najac, J. D. De Saint, L. Reguero, P. Grandes and S. Charpak, *J. Neurosci.*, **2011**, *31*, 8722–8729.
433. D. A. Wilson and M. Leon, *Brain Res.*, **1987**, *417*, 175–180.
434. M. Luo and L. C. Katz, *Neuron*, **2001**, *32*, 1165–1179.
435. A.L. Fantana, E. R. Soucy and M. Meister, *Neuron*, **2008**, *59*, 802–814 doi: 10.1016/j.neuron.2008.07.039.
436. S. Giridhar, B. Doiron and N. N. Urban, *Proc. Natl. Acad. Sci. U. S. A.*, **2011**, *108(14)*, 5843–5848 doi: 10.1073/pnas.1015165108.
437. M. Shepherd, *Neurogastronomy: How the Brain Creates Flavour and Why it Matters*, Columbia University Press, New York, 2012, p. 98, ISBN 978-0-231-15910-4.
438. L. Fletcher and D. A. Wison, *J. Neurosci.*, **2003**, *23(17)*, 6946–6955.
439. T. Bozza, A. Vassalli, S. Fuss, J.-J. Zhang, B. Weiland, R. Pacifico, P. Feinstein and P. Mombaerts, *Neuron*, **2009**, *61*, 220–233 doi: 10.1016/j.neuron.2008.11.101.
440. M. Meister and T. Bonhoeffer, *J. Neurosci.*, **2001**, *21(4)*, 1351–1360.
441. B. A. Johnson and M. Leon, *J. Comp. Neurol.*, **2007**, *503(1)*, 1–34 doi: 10.1002/cne.21396.

442. C. C. Woo, E. E. Hingco, B. A. Johnson and M. Leon, *Chem. Senses*, **2006**, *32*, 51–55 doi: 10.1093/chemse/bjl035.

443. B. A. Johnson, S. Arguello and M. Leon, *J. Comp. Neurol.*, **2007**, *502(3)*, 468–482.

444. B. A. Johnson, J. Ong, K. Lee, S. L. Ho, S. Arguello and M. Leon, *J. Comp. Neurol.*, **2007**, *500(4)*, 720–733.

445. E. R. Soucy, D. F. Albeanu, A. L. Fantana, V. N. Murthy and M. Meister, *Nat Neurosci.*, **2009**, *12(2)*, 210–220 doi: 10.1038/nn.2262.

446. V.N. Murthy, *Annu. Rev. Neurosci.*, **2011**, *34*, 233–258.

447. L. Ma, Q. Qiu, S. Gradwohl, A. Scott, E. Q. Yu, R. Alexander, W. Wiegraebe and C. R. Yu, *Proc. Natl. Acad. Sci. U. S. A.*, **2012**, *109(14)*, 5481–5486.

448. B. Bathellier, D. L. Buhl, R. Accolla and A. Carleton, *Neuron*, **2008**, *57*, 586–5989 doi: 10.1016/j.neuron.2008.02.011.

449. S. H. Fuss and S. I. Korsching, *J. Neurosci.*, **2001**, *21(21)*, 8396–8407.

450. B. A. Johnson, C. C. Woo and M. Leon, *J. Comp. Neurol.*, **1998**, *393*, 457–471.

451. D. Y. Lin, S.-Z. Zhang, E. Block and L. C. Katz, *Nature* , **2005**, *434*, 470–477.

452. B. Slotnick, *Chem. Senses*, **2007**, *32*, 721–725 doi: 10.1093/chemse/bjm039.

453. B. Auffarth, A. Gutierrez-Galvez and S. Marco, *Front. Syst. Neurosci.*, **2011**, *5*, 82.

454. B. Auffarth, B. Kaplan and A. Lansner, *Front Syst. Neurosci.*, **2011**, *5*, 84.

455. J. A. Thompson, E. Salcedo, D. Restrepo and T. E. Finger, *J. Comp. Neurol.*, **2012**, *520(8)*, 1819–1830.

456. K. Tucker, M. A. Cavallin, P. Jean-Baptiste, K. C. Biju, J. M. Overton, P. Pedarzani and D. A. Fadool, *Results Probl. Cell Differ.*, **2010**, *52*, 147–157.

457. D. D. Stettler and R. Axel, *Neuron*, **2009**, *63*, 854–864 doi: 10.1016/j.neuron.2009.09.005.

458. S. Ghosh, S. D. Larson, H. Hefzi, Z. Marnoy, T. Cutforth, K. Dokka and K.K. Baldwin, *Nature*, **2011**, *472*, 217–220 doi: 10.1038/nature09945.

459. G. M. Shepherd, *Neurogastronomy: How the Brain Creates Flavour and Why it Matters*, Columbia University Press, New York, 2012, p. 83, ISBN 978-0-231-15910-4.

460. C. Zelano, A. Mohanty and J. A. Gottfried, *Neuron*, **2011**, *72*, 178–187.

461. J. D. Howard, J. Plailly, M. Grueschow, J.-D. Haynes and J. A. Gottfried, *Nat. Neurosci.*, **2009**, *12*, 932–938.

462. C. S. Sell, *Chem. Biodivers.* **2004**, *1*, 1899–1920.

463. D. A. Wilson, *J. Neurophysiol.*, **2003**, *90(1)*, 65–72.

464. D. A. Wilson and R. M. Sullivan, *Neuron*, **2011**, *72*, 506–519.

465. a(a) J.A. Gottfried, *Curr. Opin. Neurobiol.*, **2009**, *19(4)*, 422–429.b(b) J. A. Gottfried, *Nat. Rev. Neurosci.*, **2010**, *11(9)*, 628–641.

466. W. Li, J. D. Howard and J. A. Gottfried, *Brain*, **2010**, *133(9)*, 2714–2726.

467. L.M. Kay, *Curr. Biol.*, **2011**, *21*, R928–R929.

468. W. Li, J. D. Howard, T. B. Parrish and J. A. Gottfried, *Science*, **2008**, *319*, 1842–1845.

469. M. G. Veldhuizen and D. M. Small, *Chem. Senses*, **2011**, *36(8)*, 747–760. doi: 10.1093/chemse/bjr043.

470. W. Li, L. Lopez, J. Osher, J. D. Howard, T. B. Parrish, and J. A. Gottfried, *Psychol. Sci.*, **2010**, *21 (10)*, 1454–1463.

471. E. T. Rolls and A. Treves, *Prog. Neurobiol.*, **2011**, *95(3)*, 448–490 doi: 10.1016//j.pneurobio.22011.08.002.

472. W. Li, E. Luxenberg, T. Parrish and J. A. Gottfried, *Neuron*, **2006**, *52*, 1097–1108 doi: 10.1016/j.neuron.2006.10.026.

473. A. Gottfried and D. H. Zald, *Brain Res. Rev.*, **2005**, *50*, 287–304 doi: 10.1016/j.brainresrev.2005.8.004.

474. E. T. Rolls, *Front. Syst. Neurosci.*, **2011**, *5*, 78 doi: 10.3389/fnsys.2011.00078.

475. J. A. Gottfried and C. Delano, *Annals N. Y. Acad. Sci.*, **2011**, *1239(1)*, 138–148.

476. J. Plailly, J. D. Howard, D. R. Gitelman and J. A. Gottfried, *J. Neurosci.*, **2008**, *28(20)*, 5257–5267.

477. W. W. Tham, R. J. Stevenson and L. A. Miller, *Neurocase*, **2011**, *17(2)*, 148–159 doi: 10.1080/13554794.2010.504728.

478. D. L. Sosulski, M. L. Bloom, T. Cutforth, R. Axel and S. R. Datta, *Nature*, **2011**, *472*, 213–218 doi: 10.1038/nature09868.
479. A. Bader, B. Klein, H. Breer and J. Strotman, *Front. Neural Circuits*, **2012**, *6*, 84 doi: 10.3389/fncir.2012.00084.
480. A. Bader, H. Breer and J. Strotman, *Histochem. Cell Biol.*, **2012**, *137*, 615–628.
481. J. A. Gottfried and K. N. Wu, *Ann. N. Y. Acad. Sci.*, **2009**, *1170*, 324–332.
482. J. Plailly, C. Delon-Martin and J.P. Royet, *Hum. Brain Mapp.*, **2012**, *33*, 224.
483. I. Abdul-Kareem, A. Stancak, L. Parkes and V. Sluming, *J. Magn. Reson. Imaging*, **2011**, *33(1)*, 24–32.
484. V. Sluming, T.R. Barrick, M.A. Howard, A.R. Mayes and N. Roberts, *Neuroimage*, **2002**, *17*, 1613–1622.
485. B.R. Arenkiel, H. Hasegawa, J. J. Yi, R.S. Larsen, M.L. Wallace, B.D. Philpot, F. Wang and M.D. Ehlers, *PLoS One 6*, **2011**, e29423.
486. F.J. Guerrieri, P. d'Ettorre, J. M. Devaud and M. Giurfa, *J. Exp. Biol.*, **2011**, *214*, 3300.
487. M. Beaulieu-Lefebvre, F.C. Schneider, R. Kupers and M. Ptito, *Brain Res. Bull.*, **2011**, *84*, 206–209.
488. R. Kupers, M. Beaulieu-Lefebvre, F.C. Schneider, T. Kassuba, O.B. Paulson, H.R. Siebner and M. Ptito, *Neuropsychologia*, **2011**, *49*, 2037–2044.
489. M. Leon and B. A. Johnson, *Cell Mol. Life Sci.*, **2009**, *66*, 2135–2150 doi: 10.1007/s00018-009-0011-9.
490. M. N. Geffen, B. M. Broome, G. Laurent and M. Meister, *Neuron*, **2009**, *61*, 570–586 doi: 10.1016/j.neuron. 2009.01.021.
491. M. E. Frank, H. F. Goyert and T. P. Hettinger, *Chem. Senses*, **2010**, *35*, 777–787 doi: 10.1093/chemse/bjq078.
492. I. Fukunaga, M. Berning, M. Kollo, A. Schmaltz and A. T. Schaefer, *Neuron*, **2012**, *75(2)*, 320–329 doi: 10.1016/j.neuron.2012.05.017.
493. Y. Furodono, Y. Sone, K. Takizawa, J. Hirono and T. Sato, *Chem. Senses*, **2009**, *34*, 151–158 doi: 10.1093/chemse.bjn071.
494. A. Young and Q.-Q. Sun, *Chem. Senses*, **2007**, *32*, 783–794 doi: 10.1093/chemse/bjm046.
495. M. Burgstaller and H. Tichy, *Eur. J. Neurosci.*, **2012**, *35*, 519–526.
496. F. Zufall and T. Leinders-Zufall, *Chem. Senses*, **2000**, *25*, 473–487.
497. J. Lecoq, P. Tiret and S. Charpak, *J. Neurosci.*, **2009**, *29(10)*, 3067–3072 doi: 10.1523/jneurosci.6187-08-2009.
498. J. Niessing, R. W. Friedrich, *Nature*, **2010**, *465*, 47–54 doi: 10.1038/nature08961.
499. J. R. Schafer, I. Kida, D. L. Rothman, Hyder, F. and F. Xu, *Magne. Reson. Medi.*, **2005**, *54(2)*, 443–448.
500. A. R. Best and D. A. Wilson, *J. Neurosci.*, **2004**, *24(3)*, 652–660.
501. M. Scheibe, O. Opatz and T. Hummel, *Eur. Arch. Otorhinolaryngol.*, **2009**, *266(8)*, 1323–1326.
502. M. Yang and J. M. Crawley, *Curr. Protoc. Neurosci.*, **2009**, 8.24 doi: 10.1002/0471147301.ns0824s48.
503. S. Das, M. K. Sadanandappa, A. Dervan, A. Larkin, J. A. Lee, I. P. Sudhakaran, R. Priya, R. Heidari, E. E. Holohan, A. Pimentel, A. Gandhi, K. Ito, S. Sanyal, J. W. Wang, V. Rodrigues and M. Ramaswami, *Proc. Natl. Acad. Sci. U. S. A.*, **2011**, *108*, E646–E654, doi: 10.1073/pnas.1106411108.
504. D. A. Wilson and C. Linster, *J. Neurophysiol.*, **2008**, *100(1)*, 2–7 doi: 10.1152/jn.90479.2008.
505. A. M. McNamara, P. D. Magidson, C. Linster, D. A. Wilson and T. A. Cleland, *Learn. Mem.*, **2008**, *15*, 117–125 doi:10.1101/lm.785608.
506. T. Hummel, M. Knecht and G. Kobal, *Brain Res.*, **1996**, *717*, 160–164.
507. H. Potter and S. L. Chorover, *Brain Res.*, **1976**, *116(3)*, 417–429.
508. B. Ma, S. Wang, Y. Li, C. Feng and A. Guo, *Sci. China Ser. C: Life Sci.*, **2003**, *46(4)*, 358–369.
509. D. Chaudhury, L. Manella, A. Arellanos, O. Escanilla, T. A. Cleland and C. Linster, *Behav. Neurosci.*, **2010**, *124(4)*, 490–499.
510. C. McCann, E. E. Holohan, S. Das, A. Dervan, A. Larkin, J. A. Lee, V. Rodrigues, R. Parker and M. Ramaswami, *Proc. Natl. Acad. Sci. U. S. A.*, **2011**, *108*, E655–E662, doi: 10.1073/pnas.1107198108.

511. D. L. Glanzman, *Proc. Natl. Acad. Sci. U. S. A.*, **2011**, *108*, 14711–14712, doi: 10.1073/pnas.1111230108.
512. A. Larkin, S. Karak, R. Priya, A. Das, C. Ayyub, K. Ito, V. Rodrigues and M. Ramaswami, *Learn. Mem.*, **2010**, *17(12)*, 645–653.
513. C. Linster, A. V. Menon, C. Y. Singh and D. A. Wilson, *Learn. Mem.*, **2009**, *16(7)*, 452–459.
514. N. Sobel, V. Prabhakaran, Z. Zhao, J. E. Desmond, G. H. Glover, E. V. Sullivan and J. D. E. Gabrieli, *J. Neurophysiol.*, **2000**, *83*, 537–551.
515. A. Poellinger, R. Thomas, P. Lio, A. Lee, N. Makris, B. R. Rosen and K. K. Kwong, *Neuroimage* **2001**, *13(4)*, 547–60.
516. S. S. Deshmukh and U. S. Bhalla, *J. Neurosci.*, **2003**, *23(5)*, 1903–1915.
517. R. J. Stevenson and R. Langdon, *Cogn. Neuropsychiatry*, **2012**, *17 (4)*, 315–333, doi: 10.1080/13546805.2011.633748.
518. T. Kobayashi, N. Sakai, T. Kobayakawa, S. Akiyama, H. Toda and S. Saito, *Chem. Senses*, **2008**, *33(2)*, 163–171 doi: 10.1093/chemse/bjm075.
519. A. Ihrig, J. Hoffman and G. Triebig, *Int. Arch. Occup. Environ. Health*, **2006**, *79(4)*, 332–338.
520. G. K. Beauchamp and J. L. Wellington, *Physiol. Behav.*, **1984**, *32(3)*, 511–514.
521. R. Kaiser, *Meaningful Scents Around the World*, Wiley-VCH/VCHA, Weinheim/Zürich, 2006, ISBN 3-906390-37-3.
522. D. G. Laing and Francis, G. W. *Physiol. Behav.*, **1989**, *46*, 809–814.
523. D. G. Laing and B. A. Livermore, Perceptual analysis of complex chemical signals by humans, in *Chemical Signals in Vertebrates VI*, Eds. R. L. Doty and D. Müller-Schwartz, Plenum Press, New York, 1992, pp. 587–593.
524. E. Le Berre, E. Jarmuzek, N. Beno, P. Etievant, J. Prescott, T. Thomas-Danguin, *Chemosens. Percept.*, **2010**, *3*, 156–166.
525. A. M. Lovitz, A. M. Sloan, R. L. Rennaker and D. A. Wilson, *Chem. Senses*, **2012**, *37 (6)*, 533–540. doi: 10.1093/chemse/bjs006.
526. F. Patte and P. Laffort, *Chem. Sens. Flav.*, **1979**, *4(4)*, 267–274.
527. V. Ferreira, *Flavour Fragr. J.*, **2012**, *27*, 124–140.
528. A. J. A. van der Weerdt and I. M. Payne, EP 1 007 610 B1, assigned to Quest International BV, 1998.
529. G. M. Shepherd, *Neurogastronomy: How the Brain Creates Flavour and Why it Matters*, Columbia University Press, New York, 2012, p. 122, ISBN 978-0-231-15910-4.
530. A. Triller, E. A. Boulden, A. Churchill, H. Hatt, J. Englund, M. Spehr and C. S. Sell, *Chem. Biodivers.*, **2008**, *5 (6)*, 862–887.
531. D. G. Laing, A. Eddy and D. J. Best, *Physiol. Behav.*, **1994**, *56*, 81–93.
532. W. S. Cain, *Chem. Sens. Flav.*, **1975**, *1*, 339–352.
533. D. G. Laing and M. E. Willcox, *Chem. Senses.*, **1983**, *7*, 249–264.
534. T. Thomas-Danguin and M. Chastrette, *C.R. Biol.*, **2002**, *325(7)*, 767–772.
535. H. Moskowitz, C. N. Dubose and M. J. Reuben in *Flavor Quality: Objective Measurement*, Ed. R. A. Scanlan, ACS Symposium Series 51, American Chemical Society, Washington DC, 1977, pp. 29–44.
536. G. Coureaud, D. Gibaud, E. Le Berre, B. Schaal and T. Thomas-Danguin, *Chem. Senses*, **2011**, *36(8)*, 693–700. doi: 10.1093/chemse/bjr049.
537. S. Barkat, E. Le Berre, G. Couread, G. Sicard and T. Thomas-Danguin, *Chem. Senses*, **2011**, *37*, 159–166 doi: 10.1093/chemse/bjr086.
538. M. L. Fletcher, *PLoS One*, **2011**, *6*, e29360.
539. B. A. Johnson, J. Ong and M. Leon, *J. Comp. Neurol.*, **2010**, *518(9)*, 1542–1555 doi: 10.1002/cne.22289.
540. L. S. Kuebler, S. B. Olsson, R. Weniger, and B. S. Hansson, *Front Neural Circuits*, **2011**, *5*, 7.
541. C. Y. Su, C. Martelli, T. Emonet, and J. R. Carlson, *PNAS*, **2011**, *108*, 5075–5080.
542. M. A. Chaput, M. F. El, B. Atanasova, T. Thomas-Danguin, A. M. Le Bon, A. Perrut, B. Ferry and P. Duchamp-Viret, *Eur. J. Neurosci.*, **2012**, *35*, 584–597.
543. C. S. Sell, *Perfumer & Flavorist*, 2000, *25 (1)*, 67–73.
544. I. Menashe, O. Man, D. Lancet and Y. Gilad, *Nat. Genet.*, **2003**, *34 (2)*, 143–144.

545. Y. Hasin-Brumshtein, D. Lancet and T. Olender, *Trends Genet.*, **2009**, *25(4)*, 178–184, doi: 10.1016/j.tig.2009.02.002.
546. A. Knaapila, G. Zhu, S. E. Medland, C. J. Wysocki, G. W. Montgomery, N. G. Martin, M. J. Wright and D. R. Reed, *Chem. Senses*, **2012**, *37 (6)*, 541–552 doi: 10.1093/chemse/bjs008.
547. J. F. McRae, J. D. Mainland, S. R. Jaeger, K. D. Adipietro, H. Matsunami and R. D. Newcomb, *Chem. Senses*, **2012**, *37(7)*, 585–593 doi: 10.1093/chemse/bjs049.
548. http://genome.weizman.ac.il/horde/.
549. J. D. Mainland, J. R. Willer, A. Lindstrand, A. Keller, L. B. Vosshall, N. Katsanis and H. Matsunami, 2012, Proceedings XVI International Symposium on Olfaction and Taste.
550. J. M. Pinto, S. Thanaviratananich, M. G. Hayes, R. M. Naclerio and C. Ober, *Chem. Senses*, **2008**, *33(4)*, 319–329 doi: 10.1093/chemse/bjm092.
551. M. H. Briggs and R. B. Duncan, *Nature* , **1961**, *191*, 1310–1311.
552. W. Ogle, *Med. Chir. Trans.*, **1870**, *53*, 263.
553. T. Bitter, H. Gudziol, H. P. Burmeister, H.-J. Mentzel, O. Guntinas-Lichius and C. Gaser, *Chem. Senses*, **2010**, *35(5)*, 407–415 doi: 10.1093/chemse/bjq028.
554. C. Rouby, T. Thomas-Danguin, M. Vigouroux, G. Ciuperca, T. Jiang, J. Alexanian, M. Barges, I. Gallice, J.-L. Degraix and G. Sicaard, *Int. J. Otolaryngol.*, **2011**, Article ID 203805 doi: 10.1155/2011/203805.
555. J. A. Boyle, J. N. Lundström, M. Knecht, M. Jones-Gotman, B. Schaal and T. Hummel, *J. Neurobiol.*, **2006**, *66*, 1501–1510.
556. A. N. Gilbert and S. E. Kemp, *Chem. Senses*, **1996**, *21*, 411–416.
557. M. Guillot, *Compt. Rend. Acad. Sci.*, **1948**, *26*, 143.
558. J. E. Amoore, *Nature*, **1967**, *214*, 1095–1098.
559. J. E. Amoore, L. J. Forrester, and P. Pelosi, *Chem. Senses & Flavour*, **1976**, *2(1)*, 17–25.
560. J. E. Amoore, P. Pelosi and L. J. Forrester, *Chem. Senses & Flavour*, **1977**, *2(4)*, 401–425.
561. P. Pelosi and R. Viti, *Chem. Senses & Flavour*, **1978**, *3(3)*, 331–337.
562. P. Pelosi, and A. M. Pisanelli, *Chem. Senses*, **1981**, *6(2)*, 87–93.
563. C. J. Wysocki and G. K. Beauchamp, *Proc. Natl. Acad. Sci. U. S. A.*, **1984**, *81*, 4899–4902.
564. A. Baydar, M. Petrzilka and M.-P. Schott, *Chem. Senses*, **1993**, *18(6)*, 661–668.
565. E. A. Bremner, J. D. Mainland, R. M. Khan and N. Sobel, *Chem. Senses*, **2003**, *28*, 423–432.
566. I. Menashe, T. Abaffy, Y. Hasin, S. Goshen, V. Yahalom, C. W. Luetje and D. Lancet, *PLoS Biol.*, **2007**, 5e284 doi: 10.1371/journal.pbio.0050284.
567. M. L. Pelchat, C. Bykowski, F. F. Duke and D. R. Reed, *Chem. Senses*, **2010**, *36(1)*, 9–17 doi: 10.1093/chemse/bjq081.
568. A. Knaapila, G. Zhu, S. E. Medland, C. J. Wysocki, G. W. Montgomery, N. G. Martin, M. J. Wright and D. R. Reed, *Chem. Senses*, **2012** *37*, 541–552 doi:10.1093/chemse/.
569. Knaapila, A. H. Tuorila, K. Silventoinen, M. J. Wright, K. O. Kyvik, L. F. Cherkas, K. Keskitaolo, J. Hansen, N. G. Martin, T. D. Spector, J. Kaprio and M. Perola, *Chem. Percept.*, **2008**, *1*, 34–42 doi: 10.1007/s12078-007-9005-x.
570. G. V. Kochetov, Y. G. Kostenko and A. N. Ivankii, *Tehnologija Mesa*, **2006**, *46 (5–6)*, 279–282.
571. K. Lunde, B. Egelandsdal, E. Skuterud, J. D. Mainland, T. Lea, M. Hersletha and H. Matsunami, *PLoS One*, **2012**, *7(5)*, e35259.
572. S. N. Novikov, O. S. Gladysheva and G. A. Churakov, *Doklady Biol. Sci.*, **2002**, *387*, 505–507.
573. I.C. Griff and R. R. Reed, *Cell*, **1995**, *83(3)*, 407–414.
574. C. J. Wysocki, K. M. Dorries and G. K. Beauchamp, *Proc. Natl. Acad. Sci. U. S. A.*, **1989**, *86*, 7976–7978.
575. P. Dalton, and C. J. Wysocki, *Percept. Psychophys.*, **1996**, *58*, 781–792.
576. B. M. Pause, K. P. Rogalski, B. Sojka and R. Ferstl, *Physiol. Behav.*, **1999**, *68*, 129–137.
577. K. K. Yee and C. J. Wysocki, *Physiol. Behav.*, **2001**, *72*, 705–711.
578. L. Wang, L. Chen and T. Jacob, *J. Physiol.*, **2004**, *554*, 236–244.
579. D. A. Stevens and R. J. O'Connell, *Chem. Senses*, **1995**, *20*, 413–419.
580. L. Wang, C. J. Wysocki and G. H. Gold, *Science*, **1993**, *260*, 998–1000 doi: 10.1126/science.8493539.
581. L. Wang, L. Chen and T. Jacob, *J. Physiol.*, **2003**, *554 (1)*, 236–244.

Chapter 3

Analysis and Characterisation of Odour

CHEMICAL ANALYSIS OF ODOUR

Since this book is intended for readers who are well-versed in chemistry, it is unnecessary to give a detailed explanation of the various methods of chemical analysis referred to. Any reader who needs more detail can find it in Chapter 6 of the book by Sell (1). Similarly, any reader who wishes to see more detail on experimental protocols in both chemical and sensory analysis can find them in the book edited by Goodner and Rouseff (2). Simple physical methods of analysis such as refractive index, density and colour are inexpensive techniques that can be used in quality control (QC) of essential oils by comparison of the properties of the analyte against those of reference samples. Optical rotation is particularly useful in QC of essential oils since most contain optically active components. Flashpoint is a key measure where transport regulations are concerned. Titrimetric analysis is also used in QC to check such values as acid content and water content (Karl–Fischer method), and back titration after treatment with appropriate reagents such as a hydroxide or a primary amine can be used to measure the ester content or the aldehyde content, respectively. Chemical oxygen demand (COD) using back titration after digestion with chromic acid is used for waste streams to assess the biological oxygen demand (BOD) in water treatment plants, and redox titration for peroxide content is important as a safety measure. Metal content is usually determined by techniques such as atomic absorption.

Essentially, all odorants are relatively small molecules with a molecular weight below 300 Da and most are liquids. The most efficient technique for the structural determination of an unknown compound is therefore nuclear magnetic resonance (NMR) spectroscopy. Infrared and ultraviolet spectroscopies are also used but are very much less important. Not surprisingly, proton and ^{13}C spectroscopy are the commonest NMR methods used. For the determination of natural status, deuterium

NMR would also be used, together with ^{14}C analysis. This is much more important for flavours than for fragrances because of food safety regulation and labelling. The high price differential between natural and nature-identical makes this an important activity for any flavour company. It amuses me that, if a flavour ingredient is not sufficiently radioactive (i.e. the natural compound has been diluted by a petrochemically derived nature-identical material), it is not judged by some of the public to be as safe to eat.

Since all odorants must be volatile enough to reach the olfactory receptors in the nose, it is clear that gas chromatography (GC) is the ideal chromatographic method of separation of their components in a complex mixture such as an essential oil or food flavour. When GC is coupled with mass spectrometry (MS), the result is a very powerful tool namely gas chromatography/mass spectrometry (GC/MS) for the analysis of odorant mixtures. Generally, MS does not give sufficient information to unambiguously determine the structure of an unknown compound, but those working regularly in odorant analysis acquire extensive libraries of reference MS spectra, which allows them to easily compare the MS of any unknown sample against the library of MS spectra. Fragrance companies and others have invested a great deal in the development of GC/MS tools, and a modern GC/MS laboratory will produce an analysis of even a complex mixture such as an essential oil in less than an hour, complete with a list of all compounds identified (with a numerical score for level of confidence in the identification) and its relative concentration in the sample. Multiple MS scans across GC peaks are used to detect overlapping peaks, and algorithms for the determination of the individual MS spectra are then employed. Of course, such automated techniques can lead to misidentification because of the similarity of the mass spectra of two different compounds, especially if the analyst's library of spectra is not large enough (as is the case with many academic spectral libraries). Those using GC/MS analyses to determine the components of a new essential oil must therefore apply intelligence and think about the likely biosynthetic pathways of the plant they are studying in order to spot mistakes introduced by machines and computers. The implications of such errors are also commented upon in Chapter 5.

However, such an analysis of a perfume is only the beginning of its analysis. For example, if the GC/MS shows the presence of limonene (**3.1**) in the fragrance, the next question is 'where did the limonene (**3.1**) come from?' Limonene (**3.1**) is the major component of all citrus oils and, so, it is almost certain that the presence of limonene in a perfume indicates the use of a citrus oil in the formulation. The perfume analyst would therefore look for specific marker compounds for the various citrus oils. For example, citral (**3.2**) would suggest lemon and decanal (**3.3**), orange. By continuing in this way and looking for a variety of markers, the perfume analyst would be able to identify which citrus oils were used as well as their relative proportions. Interpretation of the GC/MS result is therefore a more time-consuming activity than obtaining the spectra and requires a good deal of skill and experience (Figure 3.1).

3.1   3.2   3.3   **Figure 3.1**   Citrus oil components.

One recent extension of GC/MS is headspace analysis. The technique involves drawing air from an odorous source through a filter which traps the volatiles in the air. The air sample can be an environmental sample, for example from a forest or a room, or it can be from a single flower. In the latter case, the live flower is usually enclosed in a suitable vessel and the air entering the vessel will be made free of any environmental volatiles by first passing it through another filter. Various polymeric filter materials are now available commercially. The filter containing the trapped volatiles of interest is then taken to the laboratory, and the volatiles are thermally desorbed directly into a GC/MS instrument. Nature has always served as an inspiration for the chemist. Modern technology enables us to go to previously inaccessible places such as the top of the canopy in a tropical rainforest and to sample the headspace around flowers without the need to pick them. Thus, the odours of even endangered species can be analysed and reconstituted in the laboratory. The results of such research can be used to provide alternatives to otherwise inaccessible oils and also to identify opportunities for replacement of one natural by another more sustainable one. Similarly, botany and chemistry can work together to support ethical sourcing of naturals.

Another development of great use in the study of odorant mixtures is that of GC-olfactometry. In this technique, the odorant mixture is passed through a GC instrument, and the effluent from the column is split with one part going to a detector such as a flame ionization detector or, better, a mass spectrometer, and the other part passing into a smelling port. Thus an evaluator can describe the smell of the compound being eluted and annotate the GC or GC/MS output accordingly. There are traps which the analyst must be aware of regarding this technique. If the effluent gas stream is dry, the odorant often cannot be detected. Olfactory fatigue and cross-adaptation are other causes of the inability to detect the odour of an odorant leaving the instrument. For example, if two musks are eluted in quick succession, the evaluator might fail to detect the odour of the second eluent because of cross-adaptation to the odour of the first. If such an effect is suspected, the evaluator will perform two analyses and, in the second one, will not expose his/her nose to the first eluted peak. Interaction between odorants at the epithelium (e.g. competitive antagonism at the receptor) or in neurotransmission will not take place if the components of a mixture are presented individually, and so compounds that are not detected by smell in the mixture might be detected in GC-olfactometry. Thus, as with perfume analysis, skill and experience in application of the technique are important.

Chemical Purity and Odour Purity

Chemical purity and organoleptic purity are not synonymous. A small amount of a compound with a low odour threshold and/or high perceived intensity can distort the odour of the major compound and will have more effect on odour than an analytically significant level of a compound with a similar odour, one with a low odour intensity or one with no odour. For example, smelling a sample of (S)-3-thiomethylbutanal (**3.4**) with 99.9% ee from a GC olfactometer fitted with a column containing an achiral stationary phase would give the impression of a potato odour even though (S)-3-thiomethylbutanal (**3.4**) itself is odourless, because of the 0.02 ng of the enantiomeric (R)-3-thiomethylbutanal (**3.5**) contained in the 20 ng sample injected onto the column (3). Another illustration of the effect of highly odorous impurities is in the reduction of the ester (**3.6**) by lithium aluminium hydride (4,5).

The major product was the expected alcohol (**3.7**), but GC analysis showed that some isomerisation of the double bond into the ring had occurred and so there was a significant amount of the isomeric alcohol (**3.8**) in the product mixture. However, whilst the product mixture elicited a strong muguet odour, GC-olfactometry revealed that both of these components were odourless and that the odour was entirely due to a third component that was almost undetectable by GC. Investigation of the third component revealed it to be the aldehyde (**3.9**), presumably formed by the isomerisation of the double bond in the intermediate lithium complex to give the enolate of the aldehyde, which would give the aldehyde on hydrolysis (Figure 3.2).

Incorrect odour descriptions due to trace impurities can then be perpetuated in the literature by subsequent authors citing one error. For instance, it is often reported that (R)-(+)-limonene (**3.10**) elicits an orange odour, whereas its enantiomer (S)-(−)-limonene (**3.11**) elicits one of lemon. Where citation trails exist, they usually lead back to the paper of Friedman and Miller (6). Whilst this chapter describes sterling research on the carvone enantiomers, it also includes a comment that (S)-(−)-limonene (**3.11**) elicits a lemon odour percept. Inspection of the experimental detail reveals that the sample of '(S)-(−)-limonene' was extracted from lemons. However, lemons, like all citrus fruit, produce (R)-(+)-limonene (**3.10**) and so one is forced to conclude that the lemon odour elicited by Friedman and Miller's sample was due to contamination by traces of citral (**3.2**).

MEASUREMENT AND CHARACTERISATION OF ODOUR

Most people have a poor language of odour and have difficulty in describing a new odour, and this is a major issue in measurement and characterisation of odour. Moreover, it will become clear from Chapter 4 that deterioration in semantic ability is a significant factor in age-related smelling loss. It is not surprising that one of the first things a perfumer or fragrance evaluator has to do in their training is to develop a good odour vocabulary. When it comes to attempts to measure odour

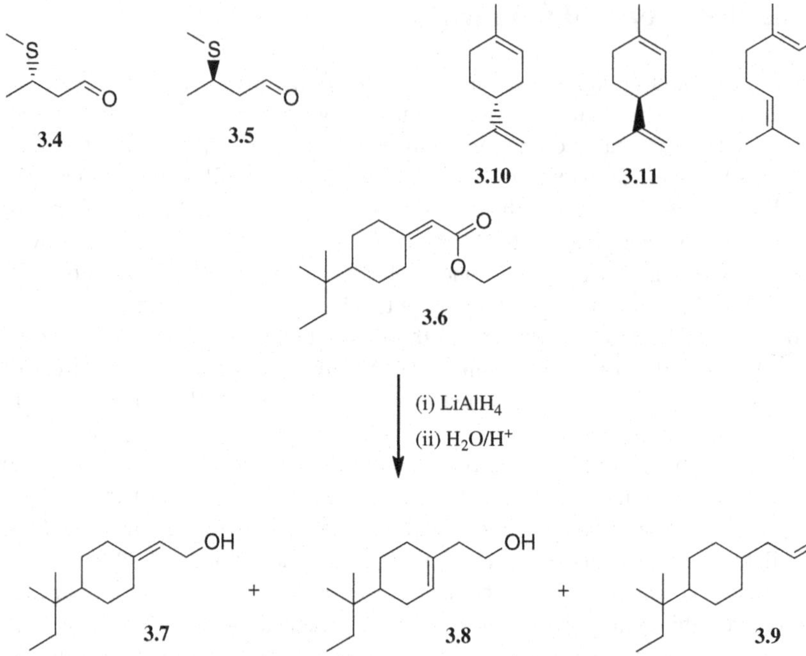

Figure 3.2   Chemical purity and odour purity.

on a more scientific basis, the problems become even more acute. The following sections of this chapter describe some of the issues and the techniques used to try to overcome them.

Subjectivity of Odour

As explained in Chapter 2, odour is subjective even at the most basic level and it is unlikely that any two humans (except identical twins) use the same set of receptors in their epithelial array (7). Differences between individuals in subsequent neuroprocessing, owing to physiological and experiential factors, further increase inter-individual differences in odor perception. Therefore, published odour data tend to be the average from a group of subjects. However, the average figure does not necessarily represent that of any of the individuals in a group. For example, to take a simple analogy, the average family in a country might have 1.8 children but no family in that country has 1.8 children. Additionally, as with all sensory magnitude estimation, odour intensity measurement must take into account the fact that humans unconsciously adjust mental scales to suit the task in hand. For example, the statement that 'a large person was seen entering a small house' causes no problems at all in our minds since the scales for people and houses are different. As a result of all of these considerations, care must always be taken when using average figures for any odour measurement.

A good example of subjectivity in odor character measurement is provided by Ohloff et al. (8). When 27 panellists were asked to allocate the odour of the cyclic ether (**3.12**) to one of various odour categories, 14 described it as minty/camphoraceous, 6 as fruity, 3 as balsamic and 4 as musky/woody. Therefore, classification as minty (based on the largest subject group) would only be correct for 50% of the panel. Similarly, it is easy to demonstrate that Radjanol (**3.13**) is perceived by some subjects as sandalwood in character but by others as musk (5). Odour intensity is also subjective. For example, the average odour threshold for (−)-geosmin (**3.14**) was found to be 1/10th of that for the (+)-enantiomer (**3.13**). However, some individuals were 40 times more sensitive to one enantiomer than the other, some experienced similar thresholds for both enantiomers and some were more sensitive to the (+)-enantiomer (9). Czerny et al. measured the odour detection thresholds and odour character descriptions for a series of simple alkyl phenols using a panel of 13 people. In most cases, agreement on descriptor terms was less than 50% (Figure 3.3) (10).

The variability of odour perception depends on many biological parameters in addition to the innate genetic factors described in Chapter 2. For example, Ketterer et al. found that hyperinsulinemia (e.g. the increase in blood insulin level after a meal) causes an increase in odour detection thresholds (11). This could be a mechanism to reduce attractiveness of food after satiation, but it is also a factor that should be borne in mind by anyone trying to measure odour data.

The method used to determine odour properties can also affect the result, as illustrated in the following section for the incidence of androstenone (**3.18**) anosmia. Different experimenters also report results using different units. For example odour thresholds are sometimes reported in nanograms per litre (ng/l) air and sometimes as detectability in the headspace of an aqueous solution of given concentration and sometimes using other units. This makes comparison between different sets of results difficult, especially when one involves partitioning from an aqueous

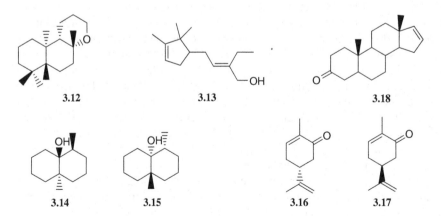

Figure 3.3   Subjectivity and measurement of odour.

medium and the other does not. However, we should see consistency of comparisons within each set of results, but even this is not always the case. For example, the ratio of detection thresholds of carvone enantiomers should be the same irrespective of the measurement technique. All reports indicate a lower threshold for (*R*)-(−)-carvone (**3.16**). Polak et al. found that the threshold for (*R*)-(−)-carvone (**3.16**) varied from 0.7 to 2000 ppb in water and that for (*S*)-(+)-carvone (**3.17**) from 1 to 4000 ppb. The ratio of the difference in threshold between the two enantiomers in each individual subject varied from 28 to 230 (12). In contrast, Leitereg et al. (13,14) found the ratio to be vary from 42 to 65; Padrayuttawat et al. found the ratio as 25 (15); Christoph as 3–4 (16) and Pelosi andViti found as 14 (17). In the case of the carvone enantiomers, there is, at least, more consistency in qualitative descriptors, and both Russell and Hills (18) and Friedman and Miller (6) described (*R*)-(−)-carvone (**3.16**) as eliciting a spearmint odour and (*S*)-(+)-carvone (**3.17**) one of caraway. However, Leitereg et al. found that only about 70% of the population can discriminate between the two (13,14).

Techniques of Odour Measurement

In order to understand olfaction at a chemical level, it is necessary to have good data linking the chemical structure to the odour properties. This is much more difficult than it would appear at first sight, for instance, to a chemist sniffing a sample that he has just synthesised in the laboratory and applying an odour descriptor to the molecular structure of his synthetic target. These difficulties stem from both chemical and sensory issues. Techniques for odour measurement and the difficulties involved have been reviewed by Neune-Jehle and Etzweiler (19), and, as already mentioned, more details on experimental protocols in both chemical and sensory analysis can be found in the book edited by Goodner and Rouseff (2).

As mentioned in the preceding section, the example of anosmia to androstenone (**3.18**) provides a good example of how different measurement techniques can give very different results (20). Table 3.1 gives an indication of the degree of variation seen in this case, and it is remarkable that the reported incidence of anosmia to androstenone (**3.18**) ranges from 75% down to 1.8% of the population depending on how it was measured. In my own experience, if a subject is given a sample and asked, 'Can you smell this?', the reported level of anosmia will be higher than if the subject is given three samples to smell, one with the compound in question and the other two with odourless blanks and (s)he is asked to select the odd sample out (20). The dependence on experimental procedure makes comparison between different sets of results difficult, and care must be taken when drawing conclusions from such comparisons. The expertise of the panellists is also important; for example, it has been demonstrated that short-term odour memory improves with experience in odour assessment (21).

When describing odour quality, the most important thing is language, and any sensory panel will be trained in a specific set of odour descriptors using reference standards. Of course, each research group will have its own set of standards and so comparison between the output from different laboratories remains an issue.

Table 3.1   Reported Levels of Anosmia to Androstenone (**3.18**)

Percentage of subjects showing anosmia			References
Females	Males	Both sexes	
		75	(22)
		46	(23)
7.6	44.3		(24)
24	40		(25)
24	33		(26) (USA)
15.8	26.8	21.5	(27)
14	22		(26) (Africa)
22			(28)
20	9		(29)
9	13		(30)
		11	(31)
10.6			(32)
1.8–5.96			(33)
	5		(34)
		1.8	(35)

One simple technique that is useful in both qualitative and quantitative measurement is the triangle test. In this test, subjects are presented with three visually identical samples and asked to determine the odd one out. The samples could be liquids in jars, perfumers' blotters or other supported fragrance samples. If determining threshold, one sample will contain the odorant at a set concentration, and the other two odourless controls. If the subject can distinguish the odorous sample, then the concentration to which (s)he was exposed is above her/his threshold. Obviously, the test could use two odour samples and one odourless control instead of the other way around. To determine the odour purity, one test substance would be a pure standard and the other the sample to be evaluated. To make the test more rigorous, five samples could be used, two of one substance and three of the other. This significantly reduces the probability of finding the correct 'odd one out' by chance.

More sophisticated test methods use olfactometers in which a predetermined concentration of the odorant in clean air is sampled by the subject. Such equipment is expensive and often tailor-made by the laboratory doing the work.

Whatever, the technique used for sample delivery, it is important to use as large a group of subjects as possible, because of the subjectivity of odour. Testing is best carried out under controlled environmental conditions. For example, the evaluation room should be uncluttered, free of environmental odours, decorated in bland colours and so on. This helps to reduce interference from input from other senses. The time of day is also important: for instance, evaluation should not be performed shortly after a meal and certainly not after a heavily spiced meal. The raw results should then be handled using proper statistical methods. Indeed, the order of sample presentation should also be determined by a sound statistical randomisation. Blind assessment is crucial since expectation exerts a strong effect on olfactory perception. In other words, the subjects should not be aware of the identity of the samples. A double-blind protocol, in which the person presenting the samples to the panellist is also unaware of the sample identity, is even more preferable, since that prevents inadvertent giving of non-verbal cues. Laboratories, such as those in fragrance companies, that carry out olfactory testing regularly, have custom-built panel suites and standardised experimental protocols. Others wishing to have good sensory evaluation of odours but who cannot afford to set up such facilities can use the services of specialist olfactory testing companies.

Examples of good practice in sensory measurement include the papers, discussed above, by Polak et al. (12) and Leitereg et al. (13,14) on the carvones and of Polak and Provasi (9) on geosmin. Anyone wishing to learn how to carry out odour measurement would be well advised to study these and similar papers.

Odour Character

Odour character, also referred to as *odour quality*, is the property through which we identify an odour and relate it to a source, such as fruits and flowers. Most of the published work on odour/structure correlation is on character, and simple assumptions about the nature of this property have led to issues in research as will be explained in Chapters 7 and 8. The brain centres involved in processing odour signals are close to those involved in memory and emotion. This means that memories and emotions are often triggered by an odour before the person is consciously aware of the odour percept. The description of odour character can therefore be modulated by the emotions or memories that it has already triggered. Some of the issues and effects of odour characterisation are described below.

No Fixed Reference Points

All our other senses have reference points associated with measurable physical properties. In vision, the frequency/wavelength of electromagnetic radiation gives a physical reference just as frequency/wavelength of sound waves in air does for hearing. Touch can be measured by pressure. Taste is less easy to measure physically but the sour taste can be correlated to pH and salt by sodium ion concentration. Sweet can be compared on a scale related to concentration of a standard sweet tastant such as glucose (**3.19**), umami to the concentration of monosodium glutamate (**3.20**) and bitter to that of quinine (**3.21**) (Figure 3.4). However, in olfaction there are no fixed reference points or standard odorants and one odour can only be compared with another rather than to an independent reference. The fact that odour descriptors are invariably associative, that is, dependent on recalled memories of similar odours, led Stevenson and Boakes to propose a mnemonic theory of odour perception and they established consistency of their model with neurophysiological and neuropsychological data (36).

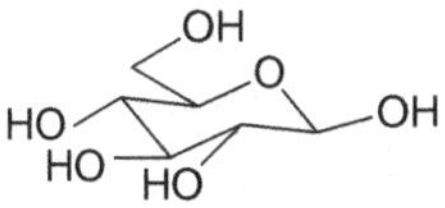

Figure 3.4   Standard tastants.

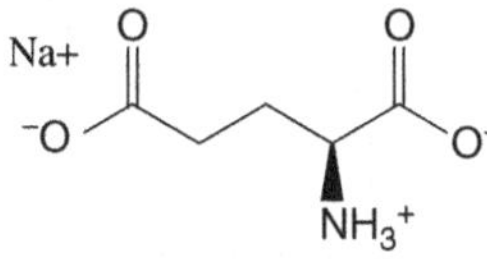

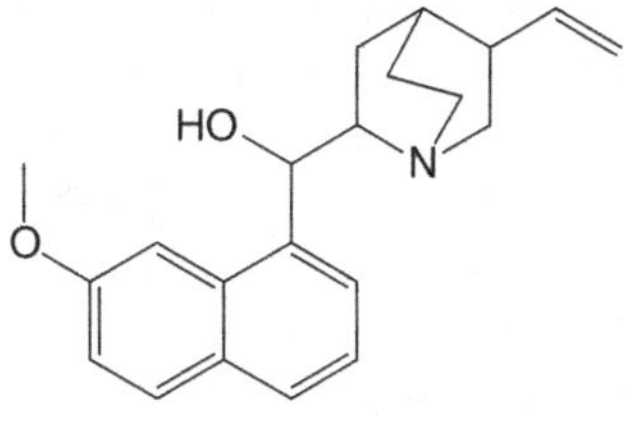

Concentration Dependence of Odour Character

When giving odour character descriptions, very few authors state the concentration at which the character was determined. Indeed, many are probably unaware of the fact that odour character does change according to concentration in many cases. As will be discussed also in Chapter 8, most chemists are aware that sulfurous compounds, especially thiols, tend to smell foul at high concentration but develop more pleasant odours at low concentration. Examples are the character impact compounds of grapefruit (**3.22**) and passionfruit (**3.23**) both of which elicit unpleasant odours reminiscent of burnt rubber at high concentration. However, few chemists seem to be aware that the odour character elicited by many other odorants are also concentration dependent. The series of differently functionalised compounds with a C_7 chain (**3.24**)–(**3.28**) have all been shown to elicit different odours at different concentrations (37) (Figure 3.5). Calone (**3.29**) elicits an intense marine, sea breeze odour at low concentrations but also develops a vanilla character when concentrated. Since there is some structural resemblance to vanillin (**3.30**), a possible explanation for this would be that, as the concentration in the nasal mucus rises, more receptor types respond. Receptors that are selective for vanillin (**3.30**) might fail to respond to Calone (**3.29**) at low concentration but become activated by it at higher concentration. So if this is the correct explanation, at low concentration Calone only activates one set of receptors but at higher concentrations those associated with vanillin also begin to respond to it and a different pattern of activation results in the olfactory bulb.

It has been shown that physiochemical factors and odorant delivery methods can seriously affect the actual concentration of odorant delivered to an insect antenna in a laboratory situation (38). In view of the concentration dependence of odour, this is a significant factor that should be taken into consideration in studies with humans, other animals and also with electronic 'noses'.

Head–Heart–Base

Those in the fragrance industry talk about perfume notes as head, heart and base notes. The less romantic might use the terms top, middle and bottom (or end) instead. These terms are based largely on volatility. For example, top notes would

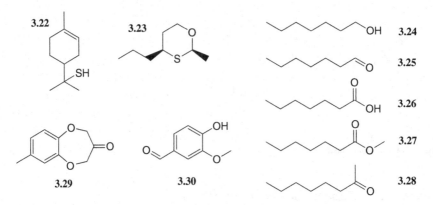

Figure 3.5   Concentration dependence of odour character.

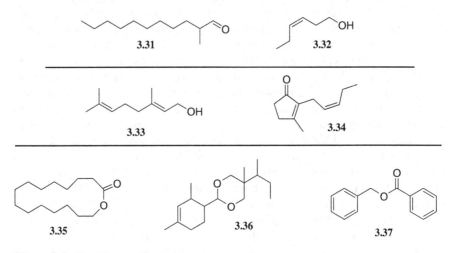

Figure 3.6   Head, heart and base notes.

include the aldehydic odour family typified by 2-methylundecanal (**3.31**), and green notes such as *cis*-3-hexenol (**3.32**). Typical heart notes include the floral categories such as rose, with geraniol (**3.33**) as a typical example, and jasmine, typified by jasmone (**3.34**). Base notes would include musks such as Thibetolide (**3.35**), ambers such as Karanal (**3.36**) and balsams such as benzyl benzoate (**3.37**) (Figure 3.6).

  cis-3-Hexenol (**3.32**) is more volatile than benzyl benzoate (**3.37**), and the first impression of a mixture of the two is that of *cis*-3-hexenol (**3.32**). Because it is more volatile, more molecules of it will reach the nose in the first sniff on opening a bottle. Similarly, if the mixture is left on a strip, the *cis*-3-hexenol (**3.32**) will evaporate more quickly, and after a day, only the benzyl benzoate (**3.37**) will remain. However, if a base note has a very low odour threshold, it might well be perceived in the top note of the perfume. An example for me is Karanal (**3.36**). Normally when it is present in a fragrance, it is one of the first things I smell. A perfumer colleague once pointed out to me that, similarly, pyrazines are head notes in the olfactory sense but base notes on volatility. The head–heart–base classification must therefore be treated with caution.

Classification of Odour Character

'Efforts to advance knowledge that is based on ill-conceived classifications can prove futile. At best, they might result in wasted time spent arguing over terminology. More seriously, they can misdirect research efforts and funding. J. Parsons and Y. Wand' (39).

Each fragrance company has its own system of odour classification and these are useful tools for perfumery. However, these systems should not be treated as tools for scientific research. Odour character descriptions are all associative and, consequently, odour classifications are based largely on these associations. We talk about woody odours because each of them is reminiscent of a specific wood. However, other than that association, what relationship is there between a cedarwood chemical such as cedrol (**3.38**) and a sandalwood such as (Z)-α-santalol (**3.39**), or even between cedrol (**3.38**) that elicits an odour of Texan cedarwood and α-atlantone (**3.40**) with a clearly Atlas cedarwood character? In chemical terms, these natural odorants are all sesquiterpenoids but they all come from quite different structural classes: cedrol (**3.38**) from cedranes, (Z)-α-santalol (**3.39**) from santalanes and α-atlantone (**3.40**) from bisabolanes. In terms of chemical functionality, cedrol (**3.38**) is a hindered tertiary alcohol, whereas the alcohol function of (Z)-α-santalol (**3.39**) is allylic and relatively unhindered and α-atlantone (**3.40**) is an α,β-α',β'-unsaturated ketone (Figure 3.7).

Similarly, it might make sense to see apples and pears as subclasses of fruit in botanical terms, but in terms of their odours putting apple and pear under the general heading of fruity odours leads to difficulties in structure/odour correlation. In the case of ester odorants, the structural requirements for pear are for a relatively unhindered ester function, whereas an ester needs to be fairly hindered sterically to elicit and apple odour (40). The brain sees each new odour as a new percept rather than as a combination of existing percepts (41) and therefore odour classification such as that shown in Figure 3.8, though a useful tool in perfumery, is essentially meaningless scientifically. For example, mixing together in suitable proportions,

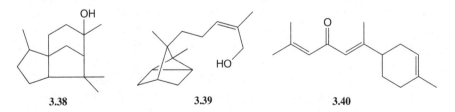

3.38   **3.39**   **3.40**

Figure 3.7   Woody odorants.

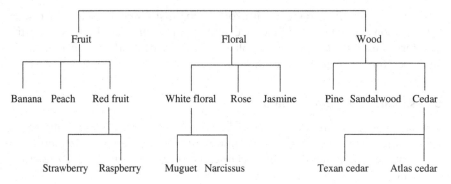

Figure 3.8   Fragment of a typical odour classification scheme.

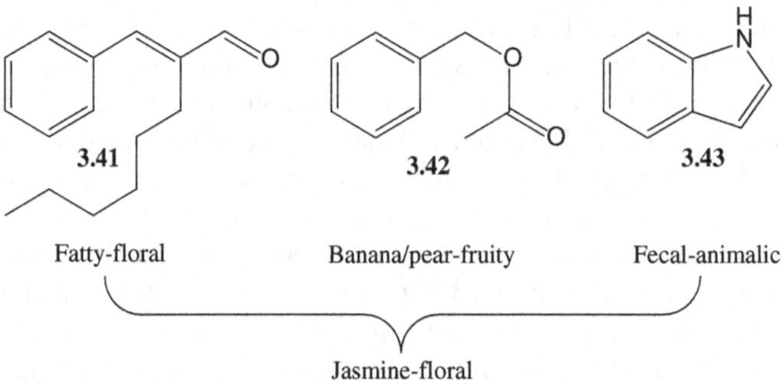

Figure 3.9   Jasmine components and odour classification.

hexylcinnamic aldehyde (**3.41**) (classified as fatty, floral), benzyl acetate (**3.42**) (classified as banana/pear, fruity) and indole (**3.43**) (classified as fecal, animalic) will give a perfume that is recognisable as jasmine in character and therefore classified as floral, thus illustrating that, though these classifications individually have significance in perfumery, they do not have a scientific significance in terms of molecular recognition (Figure 3.9).

Odour Space

It is often stated that the human nose can distinguish 10,000 different odours. In his book *What the Nose Knows*, Avery Gilbert traces the origin of this figure back through a citation trail to some estimates made by two chemists, Ernest Crocker and Lloyd Henderson, who worked for the management consultancy firm of Arthur D. Little (42). The figure is a fairly crude estimate but is usually cited as a serious one. I think most olfaction researchers nowadays see odours as spread across an infinite continuum in multi-dimensional semantic space. Maps such as that shown in Figure 3.10 are useful tools for perfumers and fragrance chemists, but the reality is much more complex. The linkages between odour types in this map reflect

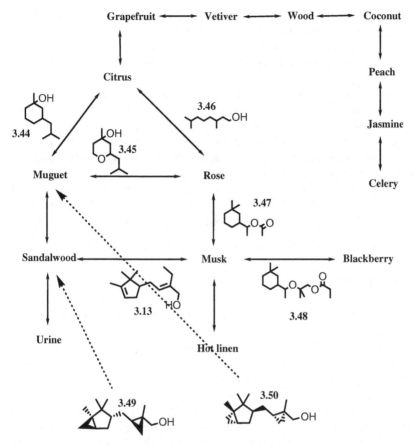

Figure 3.10   A typical chemist's map of odour space.

chemical structural relationships. These exist in three levels. For example, Rossitol (**3.44**) elicits mainly a muguet percept but there are hints of grapefruit in its total profile. Similarly, Florosa (**3.45**) has both muguet and rose elements in its profile, and cirtronellol (**3.46**) has an odour that lies between rose and citrus. Similarly, Rosamusk (**3.47**) combines rose and musk notes, and Helvetolide (**3.48**) has musky and blackberry notes. In a second mode of action, Radjanol (**3.13**) elicits a sandalwood perception for some people and musk for others. In the third linkage mechanism, a small structural change will flip an odorant from one percept to another. For example, Javanol (**3.49**) is a potent sandalwood odorant, yet its enantiomer (**3.50**) has a muguet odour (43). Fragrance discovery chemists are well aware that, if a ketone elicits a urinous odour, the corresponding alcohol is likely to elicit a sandalwood one, and, similarly, a compound eliciting a 'hot linen' odour is usually close in structure to one eliciting a musk percept. Such links are not random or isolated. There are usually numerous examples of each.

One attempt to produce a mathematical map of odour space suggested that the odour descriptors representing the human sense of smell lie on a curved surface within the mathematical space defined by the parameters used (44). A mathematical mapping of the semantic space related to the emotional response to odours suggested that the emotional effects of odours are primarily involved with well-being, social interaction, avoidance of danger, arousal, relaxation and emotional memories (45). Several authors have suggested an innate link between odour and pleasantness, but this goes against common knowledge that pleasantness is experience dependent. For example, as described in Chapter 2, those who associate clove with dentistry have a rather different hedonic response to it from those who associate it with food. Similarly, it is well known that, because of the cultural background and dietary experience, people who grew up in South East Asia have a greater tolerance for sulfurous components in food than Europeans or North Americans. Is it possible that the apparent link between odour and hedonic preference exists because of the need for semantic processing?

ODOUR THRESHOLD

The odour detection threshold is the lowest concentration at which a compound can be detected, and the recognition threshold, which is higher, is that at which the odorant can be recognised and named, and there are techniques for measuring the size of the gap between them (46). The odour detection and odour recognition thresholds of ingredients are often treated as if they are molecular properties but, as explained above, the thresholds are subjective and can only be reported either as pertaining to one individual or as an average value across a group. Usually, if a large number of individual thresholds are plotted on a graph showing number of people with a given threshold, the result is a Gaussian distribution. In some cases, such a graph will show two or even three Gaussian distributions, as the population divides into two or three groups in those instances. The threshold values are usually obtained by presenting different concentrations of the test compound in equal steps from an easily detectable concentration down to one where the subject can detect it. One study showed that halving (from 16 to 8) of the dilution steps used to determine thresholds reduced the time required without significantly affecting the accuracy of the result (47). One fascinating finding is that detection thresholds are lower when the left nostril is more occluded (48).

Odour detection thresholds are often given as single concentrations. An alternative measure is the probability of detection (49). Thus, subjects are presented with various randomised concentrations of an odorant and blanks with no odorant present, and asked to indicate whether they can smell it or not. Each concentration is presented many times, and the probability of the subject giving the correct answer can be plotted against concentration. Interestingly, the slopes of the resultant sigmoidal graphs vary considerably from one compound to another. For instance, Cometto-Muniz et al. measured detectability of various odorants (50–52). They plotted results on a graph of concentration against probability of detection and found that different odorants could have very different slopes for the sigmoidal curves that they produced. For example, the concentration range over which butyric acid detectability goes from a predictability of 0 to 1 (i.e. correct every time) is one order of magnitude, whereas that for ethyl heptanoate is about five orders of magnitude.

Many examples for calculation of thresholds exist (53). Interestingly, they use only physicochemical properties of molecules. The most important of these properties relate to volatility and hydrophobicity. These are clearly of importance to the transport of the odorant from nasal air to the receptor, and the hydrophobicity is also a significant factor in the potency of binding of a ligand to a receptor. This partly explains why odour character and odour potency are independent parameters.

ODOUR INTENSITY

Odour intensity is often confused with detectability, and sometimes intensity comparisons are made using threshold values. Care should be taken to avoid this because threshold and intensity are distinct and different phenomena. As with threshold, intensity varies from one person to another. For one individual, we can measure threshold fairly easily by presenting increasing concentrations until the subject is able to accurately say whether he or she can smell something. However, to measure intensity at concentrations above the threshold (super-threshold intensity) is not so easy. With light, we can physically measure the amount of light emitted or absorbed by a source, but for smell we have to rely on someone's judgement of whether sample A smells stronger than sample B or not and if it is, what is the ratio. Intensity is measured by panellists trained in sensory techniques such as magnitude estimation, and the perceived intensity of a compound is measured relative to that of a reference standard.

The relationship between concentration and perceived intensity does vary from one fragrance material to another. For one material a 10-fold drop in concentration might result in a 100-fold reduction in intensity, whereas for another a 10-fold drop in concentration might have hardly any affect whatsoever on perceived intensity. Methyl dihydrojasmonate (**3.51**) is an example of the latter. The perceived intensity at saturation concentration is close to that around the recognition threshold. Materials like that tend to be very effective fragrance ingredients. This interplay is known as the *psychophysical function* and is illustrated in Figure 3.11. Compound A has a lower threshold and so will be recognised at lower concentrations than compound B. At concentration X, A will have a more intense odour than B, but at the higher concentration Y, B will appear much more intense than A. Kamadia et al. (54) plotted the perceived intensity against concentration for five odorants and found that all obeyed Stevens's power law (55). The results show crossing of the lines as in Figure 3.11. For example, at 312.5 ppm, dimethyl trisulfide (**3.52**) and methional (**3.53**) were found to be isointense, whereas at 0.5 ppm the intensity of methional (**3.53**) was significantly higher than that of dimethyl trisulfide (**3.52**) (Figure 3.12).

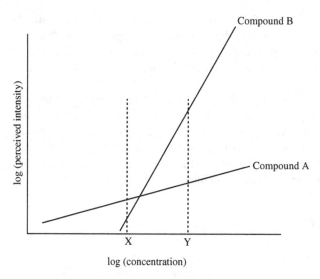

Figure 3.11   Stevens's Law and perceived intensity.

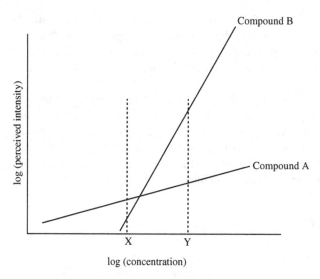

Figure 3.12   Odour intensity.

ODOUR TENACITY

Odour tenacity, also known as *persistence*, is the length of time that the odour of a compound remains perceptible after application to a surface. It is dependent on the vapour pressure of the compound. Obviously, compounds with a high vapour pressure will evaporate more rapidly than those with a low vapour pressure and therefore be lost from the surface more quickly. The tenacity will also be affected by the affinity of the compound for the surface to which it is applied, and so, for example, a hydrocarbon would probably have a lower tenacity than a highly functionalised molecule with similar vapour pressure. When evaluating potential new ingredients, the compound will be evaluated from perfumers' blotters that were impregnated at various time intervals (e.g. 0 min, 1 h, 5 h, 1 day) before the time of evaluation. This not only gives information about the tenacity of the compound but is also an easy test of the olfactive purity since trace highly odorous impurities are likely to have different tenacities from the test compound and so changes in odour quality over the dry out could be an indication that the olfactive purity is not satisfactory.

ODOUR RADIANCE, BLOOM AND TRAIL

Radiance, bloom and trail are increasingly popular, but not very well defined, terms relating to the perception of an odour. All are essentially about the perceptibility of the odour at a distance from the source. Radiance usually refers to the ability of a compound to fill the space around a sample, for example, a perfumers' blotter or air freshener placed in a room. Bloom is usually used to describe the ability of a perfumed object, such as a bar of soap, to perfume the space where it is present, for instance, the ability of a perfume in a bar of soap to fragrance the bathroom. It is also sometimes used to describe the release of a burst of odour when the object is used: for example, the increase in perceptible odour when the soap bar is wetted and used. Trail (silage in French) refers to the trail of perceptible perfume left in the air by a wearer of the perfume passing through a space. All three properties depend heavily on the threshold of the compound or perfume and also on its super-threshold intensity. A compound with a very low threshold and low slope on the Stevens's law plot will give a similar perception of intensity whether the sample is concentrated or dilute, and so a small amount in the air will give the same impression as if a neat sample is being smelt. A compound with high vapour pressure will dissipate more quickly from the source and be lost by diffusion. So, compounds with low threshold, low Stevens's law slope and low vapour pressure will tend to be more radiant than one with a high threshold, high Stevens's law slope and high vapour pressure since the latter will be lost quickly by evaporation, the falling concentration will result in rapid decrease of perceived intensity and the concentration will also quickly fall below the threshold value. Bloom also depends on the affinity of the compound for the substrate in which it is incorporated.

ELECTRONIC NOSES

There is considerable current interest in developing odorant detectors to replace the human or canine nose or cumbersome equipment such as a GC/MS instrument. Applications include environmental monitoring, medical diagnostics, land mine detection, searching damaged buildings for survivors or bodies and so on. Early versions of electronic noses (e-noses) were based on a series of semiconducting polymers, the resistance of which changes in the presence of adsorbed volatile molecules. Using an array of sensors with different sensitivities, an activation pattern can be produced and compared with reference standards. The principle is based on the combinatorial mechanism of olfaction but the number and diversity of the sensors achievable in practice are much smaller than those of the receptors in the nose. Other forms of sensor, such as microbalances, have also been tried. One detector system uses an array of brightly coloured metalloporphyrins which change colour on exposure to volatile compounds and therefore give an odorant-dependent colour change pattern that can be detected by a digital camera or even by the naked eye (56–58). In order to improve e-noses and bring them closer to human olfaction, odour-binding proteins have been used as sensors (59), and attention has also been directed towards sensors that use human olfactory receptors for odorant detection. We now know the sequence of all of the human olfactory receptors and are beginning to build a map of their receptive ranges (60–62), and advances in cloning and reproducing quantities of human olfactory receptors (63,64) and of the human trace amine receptors (TAARs) have been reported (65,66). Other laboratories have developed interfaces allowing an electrical signal to be generated by activation of a G-protein coupled receptor (GPCR) (67,68), thus opening up the possibility of incorporating them into computer chips (69–72). Together, these advances might lead to the dream of using human olfactory receptors in e-noses. However, although such instruments would represent a great step forward in odour analysis, they will still fall short of the human sense of smell because of the fact that the ultimate odour percept in the brain involves much more than the receptor activation pattern, as explained in Chapter 2.

REFERENCES

1. C. S. Sell, *Understanding Fragrance Chemistry*, Allured, Carol Stream, 2008, ISBN: 978-1-932633-38-2.
2. K. Goodner and R. Rousseff, *Practical Analysis of Flavor and Fragrance Materials*, Wiley, New York, 2011, ISBN: 978-1-4051-3916-8.
3. B. Weber and A. Mosandl, *Z. Lebensm.-Unters.-Forsch., A*, **1997**, *204(3)*, 194–197.
4. D. Munro, Eur. Pat. 1029845 A1, 1999, to Quest Int.
5. C. S. Sell, *Perfumer & Flavorist*, 2000, *25(1)*, 67–73.
6. L. Friedman and J. G. Miller, *Science*, **1971**, *172*, 1044–1046.
7. Y. Gilad, O. Man, S. Pääbo, and D. Lancet, *PNAS*, **2003**, *100(6)*, 3324–3327.
8. G. Ohloff, C. Vial, H. R. Wolf, K. Job, E. Jegou, J. Polonsky, and E. Lederer, *Helv. Chim. Acta*, **1980**, *63*, 1932–1946.
9. E. Polak and J. Provasi, *Chem. Senses*, **1992**, *17*, 23–26.
10. M. Czerny, R. Brueckner, E. Kirchoff, R. Schmitt, and A. Buettner, *Chem. Senses*, **2011**, *36*, 539–553.
11. C. Ketterer, M. Heni, C. Thamer, S. A. Herzberg-Schafer, H. U. Haring, and A. Fritsche, *Int. J. Obes.*, **2010**, *35(8)*, 1135–1138 doi: 10.1038/ijo.2010.251.
12. E. H. Polak, A. M. Fombon, C. Tilquin, and P. H. Punter, *Behav. Brain Res.*, **1989**, *31*, 199–206 doi: 10.1016/0166-4328(89)90002-8.
13. T. J. Leitereg, D. G. Guadagni, J. Harris, T. R. Mon, and R. Teranishi, *J. Ag. Food Chem.*, **1971**, *19(4)*, 785–787.
14. T. J. Leitereg, D. G. Guadagni, J. Harris, T. R. Mon, and R. Teranishi, *Nature*, **1971**, *230*, 455–456 doi: 10.1038/230455a0.
15. A Padrayuttawat, T Yoshizawa, H Tamura, and T Tokunaga, *Food Sci. Technol. Int. Tokyo*, **1997**, *3(4)*, 402–408.
16. N. Christoph, Ph.D. thesis, München, 1983, p. 57.
17. P. Pelosi and R. Viti, *Chem. Senses Flavour*, **1978**, *3*, 331–337 doi: 10.1093/chemse/3.3.
18. G. F. Russell and J. I. Hills, *Science*, **1971**, *172*, 1043–1044.
19. N. Neuner-Jehle and F. Etzweiler, *The Measuring of Odor*, pp. 153–212 in *Perfumes: Art Science and Technology*, Eds P. M. Müller and D. Lamparsky, Elsevier, London, 1991.
20. A. Triller, E. A. Boulden, A. Churchill, H. Hatt, J. Englund, M. Spehr, and C. S. Sell, *Chem. Biodiv.*, **2008**, *5(6)*, 862–866.

21. D. Valentin, C. Dacremont, and I. Cayeux, *Flav. Frag. J.*, **2011**, *26(6)*, 408–415 doi: 10.1002/ffj.2069.
22. D. A. Stevens and R. J. O'Connell, *Chem. Senses*, **1995**, *20(4)*, 413–419 doi: 101093/chemse/20.4.413.
23. J. E. Amoore and L. J. Forrester, *J. Chem. Ecol.*, **1976**, *2*, 49–56.
24. N. M. Griffiths, R. L. S. Patterson, *J. Sci. Food Agric.*, **1970**, *21(1)*, 4–6 doi: 10.1002/jsfa.2740210102.
25. K. M. Dorries, H. J. Schmidt, G. K. Beauchamp, and C. J. Wysocki, *Dev. Psychobiol.*, **1989**, *2(5)*, 423–435 doi: 10.1002/dev.420220502.
26. C. J. Wysocki, J. D. Pierce, and A. N. Gilbert, in *Smell and Taste in Health and Disease*, Eds.T. V. Getchell, L. M. Bartoshuk, R. L. Doty and J. B. Snow, Raven Press, New York, 1991, p. 297.
27. A. Baydar, M. Petrzilka, and M. P. Schott, *Chem. Senses*, **1993**, *18(6)*, 661–668 doi 10.1093/chemse/18.6.661.
28. M. Morofushi, K. Shinohara, T. Funabashi, and F. Kimura, *Chem. Senses*, **2000**, *25*, 407–411 doi: 10.1093/chemse/25.4.407.
29. D. B. Gower, S. Bird, P. Sharma, and F. R. House, *Experientia*, **1985**, *41*, 1134–1136.
30. E. E. Filsinger, J. J. Braun, W. C. Monte, and D. E. Linder, *J. Comp. Psych.*, **1984**, *98*, 219–222.
31. M. G. J. Beets and E. T. Theimer, in *Taste and Smell in Vertebrates*, Eds. G. E. W. Wostenholme and J. Knights, J&A Churchill, London, 1970, p. 313.
32. B. M. Pause, K. P. Rogalski, B. Sojka, and R. Ferstl, *Physiol. Behav.*, **1999**, *68*, 129-137 doi: 10.1016/S0031-9384(99)00158-4.
33. E. A. Bremner, J. D. Mainland, R. M. Khan, and N. Sobel, *Chem. Senses*, **2003**, *28*, 423–432 doi: 10.1093/chemse/28.5.423.
34. P. Sirota, B. Davidson, T. Mosheva, R. Benhatov, J. Zohar, and R. Gross-Isseroff, *Psychiatry Res.*, **1999**, *86(2)*, 143–153 doi: 10.1016/S0165-1781(99)00025-6.
35. J. N. Labows and C. J. Wysocki, *Perfum. Flavor*, **1984**, *9(2)*, 21–26.
36. R. J. Stevenson and R. A. Boakes, *Psychol. Rev.*, **2003**, *110(2)*, 340–364 doi: 10.1037/0033-295X.110.2.340.
37. D. G. Laing, P. K. Legha, A. L. Jinks, and I. Hutchinson, *Chem. Senses*, **2003**, *28*, 57–69.
38. M. N. Andersson, F. Schlyter, S. R. Hill, and T. Dekker, *Chem. Senses*, **2012**, *37*, 403–420 doi: 10.1093/chemse/bjs009.
39. J. Parsons and Y. Wand, *Nature*, **2008**, *455(23)*, 1040–1041.
40. C. S. Sell, *Seifen, Öle, Fette, Wachse*, **1986**, *112(8)*, 267–270.
41. D. A. Wilson and R. J. Stevenson, *Learning to Smell*. 2006. The Johns Hopkins University Press, Baltimore, ISBN 0-8018-8368-7.
42. Avery Gilbert, *What the Nose Knows*, pp. 1–4, Crown Publishers, New York, 2008, ISBN 978-1-4000-8234-6.
43. J. A. Bajgrowicz, I. Frank, and G. Fráter, *Helv. Chim. Acta*, **1998**, *81*, 1349–1358.
44. A. A. Koulakov, B. E. Kolterman, A. G. Enikolopov, and D. Rinberg, *Front. Systems Neurosci.*, **2011**, *5*, 65 doi: 10.3389/fnsys.2011.00065.
45. C. Chrea, D. Grandjean, S. Delplanque, I. Cayeux, B. Le Calvé, A. Aymard, M. I. Valazco, D. Sander, and K. L. Scherer, *Chem. Senses*, **2009**, *34*, 49–62 doi: 10.1093/chemse.bjn052.
46. L. Delahaye, M. S. Le Gac, C. Martins-Carvalho, L. Vazel, G. Potard, and R. Marianowski, *Eur. Ann. Otorhinolaryngol. Head Neck Dis.*, **2010**, *127*, 130–136.
47. I. Croy, K. Lange, F. Krone, S. Negoias, H.-S. Seo, and T. Hummel, *Chem. Senses*, **2009**, *34*, 523–527 doi: 10.1093/chemse/bjp029.
48. R. E. Frye and D. L. Doty, pp. 595–596 in *Chemical Signals in Vertebrates 6*, Eds. R. L. Doty and D. Müller-Schwarze1992, Plenum Press, New York.
49. V. Ferreira, *Flavour Fragr. J.*, **2012**, *27*, 124–140.
50. J. E. Cometto-Muniz, W. S. Cain and M. H. Abraham, *Behav. Brain Res.*, **2005**, *156*, 115–123 doi: 10.1016/j.bbr2004.05.014.
51. J. E. Cometto-Muniz and M. H. Abraham, *Chem. Senses*, **2010**, *35(4)*, 289–299 doi: 10.1093/chemse/bjq018.

52. J. E. Cometto-Muniz and M. H. Abraham, *Exp. Brain Res.*, **2010**, *207*, 75–84 doi: 10.1007/s00221-010-2430-0.

53. M. H. Abraham, R. Sanchez-Mooreno, J. E. Cometto-Muniz, and W. S. Cain, *Chem. Senses*, **2012**, *37*, 207–218 doi: 10.1093/chemse/bjr094.

54. V. V. Kamadia, Y. Yoon, M. W. Schilling, and D. L. Marshall, *J. Food Sci.*, **2006**, *71(3)*, S193–S197.

55. S. S. Stevens, *Psychol. Rev.*, **1971**, *778(5)*, 426–450.

56. K. S. Suslick and N. A. Rakow, *Nature*, **2000**, *406*, 710–713 doi: 10.1038/35021028.

57. K. S. Suslick and N. A. Rakow, U.S. Patent 6, 368, 558; April 9, 2002.

58. K. S. Suslick, N. A. Rakow, and A. Sen, U.S. Patent 6, 495, 102 B1; Dec. 17, 2002 .

59. S. Sankaran, S. Panigrahi, and S. Mallik, *Biosens. Bioelectron.*, **2011**, *26*, 3103–3109.

60. H. Saito, Q. Chi, H. Zhuang, H. Matsunami, and J. D. Mainland, *Sci. Signalling*, **2009**, *2(60)*, ra9 doi: 10,1126/scisignal.2000016.

61. A. Veithen, F. Wilkin, M. Philippeau, and P. Chatelain, *Perfumer & Flavorist*, 2009, *34(6)*, 36–44.

62. A. Veithen, F. Wilkin, M. Philippeau, C. Van Osselaer, and P. Chatelain, *Perfumer & Flavorist*, 2010, *35(1)*, 38.

63. L. Kalser, J. Graveland-Bikker, D. Steuerwald, M. Vanberghem, K. Herlihy, and S. Zhang, *PNAS*, **2008**, *105(41)*,15726–15731.

64. B. L. Cook, D. Steuerwald, L. Kalser, J. Graveland-Bikker, M. Vanberghem, A. P. Berke, K. Herlihy, H. Pick, H. Vogel, S. Zhang, *PNAS* Early Edn., doi: 10.1073/pnas.0811089106 .

65. X. Wang, K. Corin, C. Rich, and S. Zhang, *Sci. Rep.*, **2011**, *1*, 102.

66. K. Corin, P. Baaske, D. B. Ravel, J. Song, E. Brown, X. Wang, C. J. Wienken, M. Jerabek-Willemsen, S. Duhr, Y. Luo, D. Braun, and S. Zhang, *PLoS One*, **2011**, *6*, e25067.

67. H. Yoon, S. H. Lee, O. S. Kwon, H. S. Song, E. H. Oh, T. H. Park, and J. Jang, *Angew. Chem. Int. Edn.*, **2009**, *48*, 2755–2758.

68. B. R. Goldsmith, J. J. Mitala, J. Josue, A. Castro, M. B. Lerner, T. H. Bayburt, S. M. Khamis, R. A. Jones, J. G. Brand, S. G. Sligar, C. W. Luetje, A. Gelperin, P. A. Rhodes, B. M. Discher, and A. T. Johnson, *ACS Nano*, **2011**, *5*, 5408.

69. T. H. Kim, S. H. Lee, J. Lee, H. S. Song, E. H. Oh, T. H. Park, and S. Hong, *Adv. Mater.*, **2009**, *21*, 91–94.

70. S. H. Lee, H. J. Jin, H.S. Song, S. Hong, and T. H. Park, *J. Biotechnol.*, **2012**, *157(4)*, 467–472.

71. S. H. Lee, O. S. Kwon, H. S. Song, S. J. Park, J. H. Sung, J. Jang, and T. H. Park, *Biomaterials*, **2012**, *33*, 1722–1729.

72. C. Wu L. Du, D. Wang, L. Zhao, and P. Wang, *Biosens. Bioelectron*, **2012**, *31*, 44–48.

Chapter 4

The Sense of Smell in Our Lives

Freud described humans as microsmatic, that is, having only a poor sense of smell. We are visually dominant unlike almost all other mammals. We have fewer types of receptor than most other mammals but we also have much greater processing power in our brains which could well compensate for the poorer input from the periphery. In his book *Neurogastronomy: How the Brain Creates Flavour and Why It Matters*, Gordon Shepherd argues that our sense of smell is not nearly so poor as Freud believed and that, on the contrary, it plays a vital role in our daily life and has also helped to shape human evolution (1).

We know that perfume has played a part in human life for millennia. Evidence includes the existence of a 4000-year-old perfume factory on Crete, numerous Egyptian tomb paintings and written records in ancient documents. For example, the Bible contains references of perfume use in religious rites (Leviticus 2,1; Exodus 30, 1–10) and in various secular applications. The Song of Songs is littered with references to perfume in the relationship between the lover and his beloved and, in addition to personal use (Esther 2,12), there are references to use in laundry (Song of Songs 4,11; Psalm 45,8) and in household (Proverbs 7,17) applications. The modern perfumery industry continues to enrich our lives with pleasant smells, and unpleasant smells continue to remind us of dangers such as spoiled food and leaking gas. Nowadays, there is a growing interest in the role of smell in health and illness, and olfaction is a topic of increasing interest to medical research. As the average age of the world's population rises to levels never encountered before, there is also a growing interest in preserving our sense of smell and also in the effects of ageing on olfactory acuity. Loss or distortion of the sense of smell leads to a reduced quality of life and is an important factor leading to malnutrition in the elderly and in those undergoing cancer chemotherapy. This chapter will concentrate on some of the more recent developments in these areas; the role of perfume in consumer goods is already well covered in books such as those of Calkin and

Chemistry and the Sense of Smell, First Edition. Charles S. Sell.
© 2014 John Wiley & Sons, Inc. Published 2014 by John Wiley & Sons, Inc.

Jellinek (2), Curtis and Williams (3), Gilbert (4), Jellinek and Jellinek (5), Kaiser (6), Müller and Lamparsky (7), Sell (8) and van Toller and Dodd (9).

PERFUME AND SMELL IN OUR DAILY LIVES

One of the roles of smell in our day-to-day life is in determining food preferences. For example, foods were rated as more pleasant and stronger when descriptive labels were present on the containers than if a purely numerical code was used. The difference was greatest when the label description and the food flavour were congruous (10). Bitter taste is usually blamed for disliking for some vegetables but is has been shown that the odour of the vegetables also plays a part (11). Strangely, it has been found that, while hunger increases olfactory sensitivity, it only does so for neutral odours and not those associated with food. The increased sensitivity was greater in those with a low body mass index (BMI) (12).

One of the traditional uses of perfume is to hide unpleasant odours. Fine fragrance is worn to hide body odour, air fresheners are used to hide cooking odours or bathroom malodours, and so on. The perfume industry was established in Grasse because the town was a centre of glove manufacture and perfume was needed to hide the unpleasant odours left from the leather tanning process. The section on mixtures in Chapter 2 shows how the usual result of adding two odorants together is a reduced perception of both components. For example, it has been shown that various citrus components can reduce the odour of dimethyl sulfide (**4.1**) for mice (13). Malodours are usually warning signals, and our brains are necessarily adept at perceiving them even in the presence of other competing odours. The traditional means of covering a malodour was to add such intensity of perfume that the malodour was swamped. However, even then, the malodour might still be perceived to a degree and then the whole perfume/malodour combination becomes a malodour. Malodour counteractancy is, therefore, a major challenge, and effective fragrances are the result of skilful exploitation of understanding of the complex mechanisms involved both in malodour formation and its perception. The most effective malodour-counteracting fragrances are likely to use a combination of techniques including removal of malodorants at source, removal of malodorants in the air, blocking of malodorant detection at the olfactory epithelium and blocking of malodour perception during neuroprocessing. Major fragrance companies usually possess an armoury of proprietary techniques for such purposes.

The finding that the smell of methional (**4.2**) is suppressed by octanal (**4.3**) but not by other aldehydes might open a route for identification of malodour counteraction mechanism in that case (14). Identification of the receptors involved in perception of methional (**4.2**) could help to determine whether blocking by octanal (**4.3**) occurs at the receptor level or in the subsequent neuroprocessing. Answering such questions could lead to ever-more effective agents for counteraction of malodours (Figure 4.1).

While it is clear that smell plays an important part in our daily lives, it must be remembered, as illustrated in Chapter 2, that smell is a complex mental phenomenon and is influenced by other senses and non-sensory input. In the famous

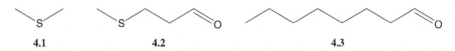

4.1   4.2   4.3

Figure 4.1 Malodour suppression.

Slossen experiment, a sample of water was described as an industrial solvent and poured onto a surface at the start of a lecture. Within minutes, people at the front of the lecture theatre could 'smell' it and the lecture eventually had to be abandoned because some listeners were becoming physically ill as a result of the 'smell' (15). More recently, an experiment was set up in a shopping mall, which involved giving passers by two samples of jam and tea and asking them which they preferred. They were then given their "preferred" sample again (it had actually been swapped for the other) and asked to describe why they preferred it. Only about one-third of the subjects noticed that the second sample was not the one they had selected initially (16). Both of these experiments show how the cognitive mind can override sensory input through a top-down mechanism. The top-down effect on odour decisions has also been demonstrated by Nienborg and Cumming (17).

Conversely, odour can, unconsciously, affect cognitive judgements. For example, odour affects the judgement of facial attractiveness. When female subjects were asked to rate the facial attractiveness of photographs of males, the ratings were significantly lower if an unpleasant odour was present compared to a pleasant odour or a control with no odour. Interestingly, the pleasant odour had no effect, and the observations in both cases were independent of whether or not the odour had any body relevance, for example, sweat malodour versus rubber malodour (18). Introduction of a complex odour was found to reduce the accuracy in a task involving lexical decisions (19), and telling subjects that an odorant was hazardous resulted in them perceiving it as more intense (20).

Our senses do not work independently, and there are many examples of inter-action between them. For example, towels feel softer if paired with a lemon scent than with an animalic one (21). There is also a cross-modal interaction between smell and music, and specific odour qualities link to type of musical instrument. For instance, vanilla seems to relate to piano and woodwind rather than to strings or brass. The complexity and pleasantness of an odour, but not its intensity, link to pitch of a musical note. Musk and chocolate odours link to pitches in the low octaves, whereas lemon and apple link to the highest pitches (22). There is also a correlation between the flavour of a flavoured milk sample and the pitch of musical notes and the instrument producing them. Interestingly, there was no correlation between notes and the fat content of the milk (23).

Similarly, words related to tastes had an effect on improvisation of music, with bitter leading to low pitched legato playing, salty to staccato, sour to high pitched notes and dissonant chords and sweet to soft and slow concordant impro-visations. In the opposite direction, untrained listeners were able to correctly guess the taste stimulus that had inspired the playing, with a relatively high degree of accuracy (24).

A study of people from four different countries (Mexico, Korea, The Czech Republic and Germany) showed that attitudes and affective responses to odour did differ from one cultural background to another, the greatest difference being between the Mexicans and the other nationalities (25). There has been much discussion in the literature and popular press about the role of sex and hormones on olfaction. Women are generally considered to have a more acute sense of smell than men, though this could be due to factors such as attention rather than any direct hormonal action. For example, one study indicates that women have lower detection thresholds in the follicular phase of the menstrual cycle than in the luteal phase, and oral contraceptives also affect their sense of smell (26). Another study suggested that women tend to rate odours as more intense than do men and are better at identifying them. However, it also indicated that there is no difference between the sexes in terms of affective response to odours, and that smell experience affects pleasantness rating of pleasant smells more than it does unpleasant ones (27). A review of the subject by Doty and Cameron suggests that the link between sex and olfactory acuity is, like everything else in olfaction, complex and that simple correlations do not exist (28).

It is well known that odour can affect emotions (short-term reaction to a stimulus), mood (a longer duration mental state) and sense of well-being (the way in which someone perceives him- or herself and their relationship to the external world). Odour also can affect both the parasympathetic nervous system (which restores calm and homeostasis) and the sympathetic nervous system (which initiates either the fight or flight response to a stimulus). For example, the effects of lavender oil on the parasympathetic system resulted in a reduced heart rate versus control on subjects undergoing facial injections, even though no effect was found on the level of pain experienced (29).

It is not surprising that odour memories are encoded more effectively if two training sessions are used instead of one. The reinforcement of the second session seems to be slightly greater for less pleasant or less familiar odours (30).

A study on autobiographical odour memory was carried out at Stockholm University and is described in the doctoral thesis of Johan Willander (31). It was found that memories that are evoked by odour stem from earlier in life than those evoked by words or pictures and have a stronger link to emotions. Memories recalled by those in their 70s or 80s tend to have been formed in their teens, for those memories triggered by words or pictures. But, when the memory is triggered by a smell, it is more likely to stem from the years up to teenage. Another study found that recall of autobiographical memories was dependent on both age and sex (32). Odour-induced autobiographical memories have been shown to result in slower, deeper breathing and the triggering of emotions (33). One study has shown that odour-evoked memories of aversive events are more detailed, unpleasant and arousing than memories evoked by auditory triggers. Visual triggers gave similar results to olfactory ones (34).

Another study showed that odour-induced autobiographical nostalgic memories resulted in increased positive mood states, decreased negative mood states and a decrease in heart rate, but an increase in skin conductance and interleukin-2 levels. Thus the memory resulted in changes in the autonomic nervous and immune activities (35).

The emotions most likely to be recalled by odour triggers are happiness, disgust and anxiety (36) and odour-induced mood induction seems to be different for the two sexes (37). In an investigation using functional magnetic resonance imaging (fMRI), Gottfried et al. found that the brain areas responsible for mediating emotional responses were affected differently by pleasant or unpleasant odours, thus demonstrating physical evidence of the link between odour and emotion (38). fMRI also showed that body odours are processed in brain regions that are not strongly involved in other odours and are more closely linked to emotional processing regions (39). Exposure to human sweat produced by donors experiencing fear or disgust elicited appropriate behaviour in subjects perceiving the smell, thus suggesting that humans do communicate emotion through chemosensory signals (40).

Odours affect our emotions even when asleep. The emotional content of dreams is affected by delivery of odorants to the nose of sleepers during rapid eye movement (REM) sleep. A pleasant odour (rose) influences the dream to a more positive emotional content, whereas an unpleasant one (rotten eggs) has the opposite effect although the odours themselves do not become part of the dream (41). Odour associated with fear was found to affect the sleep pattern of rats, giving an indication of how odour interacts with memory consolidation during sleep (42).

It is often claimed that blind people have a better sense of smell than their sighted counterparts. One report suggests that, although they have lower detection thresholds, their discrimination and identification ability is similar to that of sighted people. Blind people do tend to have increased attention to smell, presumably because they lack visual information about the environment (43). A study of those who have lost their sight early in life showed that they had larger olfactory bulbs than average and a keener sense of smell (44). Similarly, it was shown by fMRI that congenitally blind people use brain regions for olfaction that are not used for it in sighted subjects (45).

OLFACTION AND HEALTH

The interaction between olfactory and other brain regions makes it clear that smell could relate to mental health and that physical illness is likely to produce changes in biochemistry that would have an effect on volatile metabolite production. It is, therefore, not surprising that olfaction is a subject of increasing significance in medical research.

The fact that some scents relieve stress has been known for some time, and physical evidence for the effect is now being reported. For example, smelling of rose oils was found to reduce stress-induced trans-epidermal water loss and salivary cortisol in humans (46). The effect of smell on emotions was mentioned above, and this, together with stress reduction, shows that olfaction can improve well-being. Well-being is associated with increased activity in the left frontal cortex, a decrease in serum cortisol and an increase in immunoglobulin-A, showing that there is a link between well-being and the immune system. Those with a higher level of well-being are less likely to suffer from the common cold, stroke and AIDS mortality, demonstrating genuine physiological effects of the mental state.

The following paragraphs describe olfactory impairment as an illness in itself and also some of the medical conditions where olfaction is involved. In many instances, there are conflicting reports in the literature about how different conditions affect the sense of smell. These differences might be due to the test methodology since measurement of odour is inherently difficult as discussed in Chapter 3.

Smell Disorders

Men and older people are more likely to display olfactory impairments. Common causes of olfactory dysfunction include nasal polyps, septal deviation, lesions in the nasal airways, upper respiratory tract infection, sinonasal disease, head trauma and low ankle-brachial index (ratio of blood pressure in legs to that in arms–an indicator of peripheral vascular disease).

One study found that, in a particular population, smoking led to a higher level of impairment only for women (47). Inhalation of irritants and toxins such as ammonia, gasoline and zinc can also lead to olfactory impairment, and it has been shown that air pollution in large cities can adversely affect both olfaction and the trigeminal senses (48). The uremic state caused by kidney malfunction causes impaired olfactory function. Dialysis does not completely restore the sense of smell to normal, and renal transplantation has been found to be the only really effective treatment (49).

Drugs such as opioids and cannabinoids can interfere with the second messenger system in olfactory sensory neurons and drugs affecting the central nervous system (CNS) can affect neuroprocessing of olfactory signals. However, it would seem that delta-9-tetrahydrocannabinol (THC) (**4.4**) can alleviate the distortion of olfactory perception in cancer patients (50). Cancer and cancer chemotherapeutic agents also modulate the sense of smell and lead to altered food preferences and potential malnutrition in patients (51). Alcoholism is another cause of olfactory impairment (52). The effects of drugs on the sense of smell have been reviewed (53).

A group of genetic disorders known as ciliopathies affect the formation and function of cilia and, therefore, can affect the sense of smell by damaging olfactory cilia (54). There is some hope for sufferers in that it has been found that in mice the condition can be cured by gene therapy (55). Rheumatoid arthritis also seems to result in olfactory dysfunction (56).

Loss or distortion of the sense of smell is traumatic and can lead to complications such as malnutrition. If the condition cannot be improved by medical treatment, the sufferer may require counselling or anti-depressant medication. Nasal obstructions can be remedied by surgery, which might also be necessary in the case of polyposis if medication proves ineffective. Topical and systemic steroids are used for the treatment of polyposis and inflammatory conditions. Current treatments for olfactory dysfunction have been reviewed (57).

Olfaction and Other Medical Conditions

Larsson et al. found that diseases of the CNS affected olfaction. Interestingly, they also found that the personality type affects smell perception. People with high neuroticism scores had better than average olfactory capability, but impulsivity and lack of assertiveness reduced scores in identification tests (58). Sufferers from the Tourette syndrome have been found to have higher than normal sensitivity to odours (59), and anosmia is also a predictor for hypogonadotropic hypogonadism in patients with CHARGE syndrome (60). Odour identification was found to be poorer in those with obsessive-compulsive disorder (61).

The nature of olfaction makes it difficult to determine whether or not olfactory illusions occur. Stevenson has reviewed the evidence for and against, and concluded that the balance is in favour of their existence (62). Patients with Type 1 myotonic dystrophy (DM1) have reduced odour detection thresholds and some also have impairment of recognition. They also tend to give lower than normal pleasantness scores for pleasant odours (63).

Dimethyl trisulfide (**4.5**) has been found to be responsible for the odour of fungating cancer wounds (64). There are many reports that dogs can detect melanoma as well as prostate, breast, ovary and lung cancers. This could be the result of altered patterns of odorants, or perhaps there are specific marker compounds such as dimethyl trisulfide (**4.5**) (65,66). The over-expression of the olfactory receptor OR1D2 in prostate cancer cells was used to improve fluorescence and magnetic resonance imaging (MRI) of tumour cells by adding Bourgeonal (**4.6**), a known agonist of the receptor (67) or undecanal (**4.7**), an antagonist (68). One interesting recent application of odour is in triage of samples in point-of-care assays. In this triage, the presence of an enzyme indicator of a disease state results in the release of an easily detectable odorant from a test medium. If the triage does not produce an odour, then it is unnecessary to carry out expensive confirmatory analysis and quantification, resulting in significant savings and increased overall speed of screening (69).

Smell can give a warning signal, and it has been shown that the scent of a ferret produces stress in rats. However, on continued exposure, the rats habituate to the odour and the endocrinological response becomes lower (70). Similarly, rats with no olfactory bulb give a different response, as measured by heart rate, to a stressor than do normal rats (71). Similar effects of smell on stress exist in humans. For instance, smelling of rose oils was found to reduce stress-induced trans-epidermal water loss and salivary cortisol in humans (46). Investigation of possible mechanisms indicated one involving higher brain action rather than a direct pharmacological activity, but the overall result does give support to the claims of aromatherapists regarding the ability of fragrance to reduce stress. In contrast, incensole acetate (**4.8**) is a component of frankincense and has been shown to be anxiolytic through its action on TRPV3 channels in neurons in the brain. The feeling of calm that it induces might help to account for the historical use of frankincense in religious ceremonies (72) (Figure 4.2).

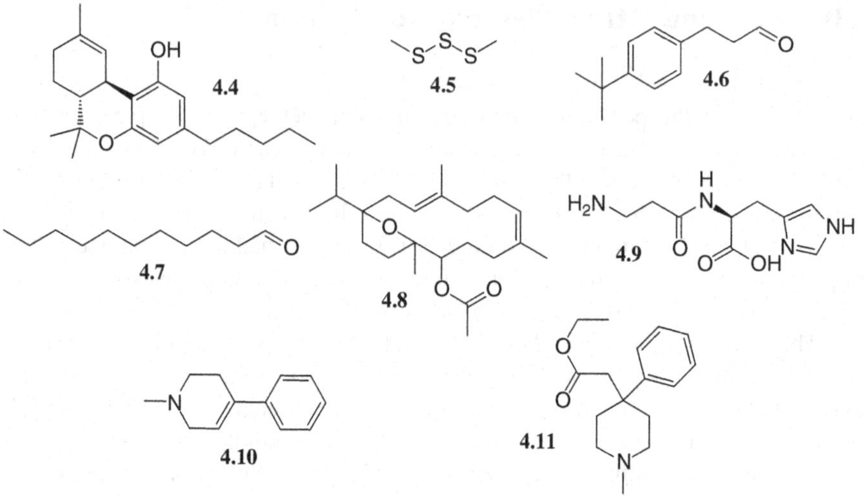

Figure 4.2   Olfaction and other medical conditions.

Anorexia

Early theories that anorexia nervosa is related to poor olfactory acuity have been dispelled by several studies showing that anorexics have similar sense of smell to healthy controls (73–75).

Autism

One study found that Asperger's syndrome and high-functioning autism resulted in higher odour detection thresholds in children but odour identification was not affected (76). A study of adults with autism spectrum conditions found no difference from normal controls (77).

Bipolar Disorder

Different studies using control groups matched for age, sex and demographic origin have found that olfaction is compromised in subjects with bipolar disorder. The effects vary depending on the phase of the condition, that is, manic phase or depressive phase. Odour detection thresholds were higher in the manic phase and during social avoidance, whereas sensitivity increased in the depressive phase (78). Subjects with bipolar disorder tended to rate odours as more pleasant than normal controls (79).

Dementia

In view of the degree of neuroprocessing involved in olfaction, it is not surprising that many forms of dementia affect the sense of smell and the literature on the subject is growing very significantly at present. Loss of the sense of smell can be used as an indicator of various forms of dementia such as Alzheimer's and Parkinsons's diseases (80). The effects of these conditions in ageing are discussed further in a later section of this chapter. The extent and specific nature of olfactory impairment varies from one type of dementia to another and, therefore, allows olfactory tests to be used as a diagnostic tool: for example, in distinguishing between those with dopamine beta-hydroxylase deficiency (no olfactory impairment) and those with pure autonomic failure but not in multiple system atrophy (impaired olfaction) (81) or between those with Alzheimer's disease and those with amnestic mild cognitive impairment (82). At present, there is no cure for either Alzheimer's or Parkinson's diseases and so the use of olfactory tests as early predictors does raise some difficult ethical questions for medical practitioners.

Alzheimer's Disease

The commonest form of dementia is Alzheimer's disease, and the e4 allelle of ApoE gene is a genetic risk for it. This condition is commonest in the elderly and starts in the area of the hippocampus and amygdala. It primarily affects cognition in the early stages and, because of the location of the initial damage, one of the very earliest symptoms is loss of the sense of smell. The early loss of smell in Alzheimer's disease has been confirmed by fMRI (83,84) and positron emission tomography (85) studies. Studies in different countries show that olfactory tests might prove useful in early diagnosis of the disease (86,87). Olfactory impairments resulting from physical damage tend to produce deficits in both detection and identification. Patients with Alzheimer's disease tend to be close to normal on detection of odorants but give a poorer result when asked to identify the odour (88). In a mouse model of Alzheimer's disease, Wesson et al. showed that build-up of amyloid plaques in the olfactory regions of the brain accounts for early olfactory dysfunction related to learning of odours. Their results suggest that altered learning of and adaptation to odours could be an early indicator of the disease and could be used to monitor the efficacy of potential therapies (89).

Carnosine (β-alanyl- L-histidine) (**4.9**) is present in the olfactory regions of the brain at higher levels than average and is known to be able to counteract many of the biochemical processes involved in the neurodegeneration resulting from Alzheimer's disease. It has, therefore, been suggested as a possible therapeutic agent (90), and studies in rats suggest that smelling the volatile oil of *Acorus gramineus* can help to ameliorate olfactory loss in the disease (91).

Parkinson's Disease

Parkinson's disease (PD) is a neurodegenerative disease characterised by tremor, stiffness and bradykinesia (slow movement). There is a genetic component, for example R1441G and G2019S mutations in LRRK2 (92), and it is more prevalent in men, unlike Alzheimer's disease which is commoner in women. PD was first defined by Parkinson in 1870 and, about 100 years later, it was realised that

Table 4.1   Hawkes' Timescale of PD Progress

Braak stage	Symptom	Time in years
0	Loss of smell	0
1–2	Sleep disturbance, obesity, constipation	10
3–4	Clinical symptoms	20
5	Loss of taste	35
6	Death	40

olfactory impairment is an early symptom of the condition and it is now clear that there is a significant association between the olfactory deficits of PD and cognitive dysfunction (93).

Progress of PD is measured by neuropathological states known as *Braak stages*. The motor symptoms appear only between Braak stages 3 and 4 which occur about 20 years after the disease starts since more than 50% (and possibly more than 80%) of the locus coerulus must be destroyed before the motor symptoms appear. Pre-clinical symptoms include olfactory loss, constipation and obesity (Midlife obesity indicates a three-fold risk of PD.) The locus coerulus is also involved in control of diet and the waking/sleeping pattern. Hawkes proposes that PD follows the course shown in Table 4.1 (94).

The disease is caused by accumulation of a rogue protein, α-synuclear, in the brain of sufferers. It affects many parts of the brain and these include the olfactory cortex. The protein forms clumps, known as *Lewy bodies*, and the spread of these as the disease progresses follows the axonal pathways of olfaction. The long, thin neurons of the piriform cortex might explain the sensitivity of this region to the build-up of Lewy bodies. Baba et al. found that cognitive deficit in olfactory perception is an important aspect of hyposmia in PD and that this deficit is caused by altered brain metabolism in the amygdala and piriform cortex (95–98), and there is evidence that the olfactory bulb reduces in volume in sufferers from PD, also contributing to the loss of olfactory acuity (99,100).

Dysphagic gastric function (bloating and nausea) and constipation are also potential early warnings of PD, which implies the involvement of the vagus nerve. The genetic predisposition could indicate either an autoimmune disorder or a genetic weakness towards an infection. The dual hit theory of Hawkes is based on the fact that the disease seems to start almost simultaneously in the nose and pharynx and that there is a link between allergic rhinitis and PD (101). The postulate is that a pathogen enters via the nose and the nasal secretions, and it travels to the stomach where the distance between the outside of the body (i.e. the stomach contents) and the nearest nerve ending (that of the vagus nerve) is about the same as the distance from nasal mucosa to the olfactory nerve. Thus the infection of the olfactory nerve and the vagus nerve will occur very close in time. The pathogen could be a virus, a prion or a simple chemical entity. In support of this, it is known that one form of PD can be caused by exposure to drugs. For instance, exposure to 1-methyl-4-phenyl-1,2,3-6-tetrahydropyridine (**4.10**), which is a contaminant of illicit pethidine (**4.11**), will invariably lead to PD (102).

There are nine different forms of Parkinson's disease and three other conditions that have similar symptoms. The nature and extent of olfactory loss might, therefore, have potential use not only in diagnosis of PD but also in distinguishing between different forms of the disease, for example, in distinguishing between idiopathic PD and essential tremor. Loss of taste also occurs in PD but it does not correlate with loss of smell. If both taste and smell are normal, then the chance that the patient has PD is only about 16% (J. Deeb, personal communication). Smell tests can be used as a biomarker for diagnostic purposes, a predictor of clinical outcomes and a potential measure of disease progression (103–105). Deeb et al. have found that smell tests are as sensitive as a dopamine transporter scan in detecting PD (106). The University of Pennsylvania smell identification test (UPSIT) has been used to distinguish between different types of PD (107), and other smell tests have been used to distinguish between PD and multiple system atrophy and progressive supra nuclear palsy (108) and to predict appearance of other non-motor symptoms (109). It has laso been suggested that smell tests could be combined with MRI in the diagnosis of PD (110).

Depression

Impairment of olfactory memory, identification and recognition are symptoms of major depressive disorder (111), and impaired olfaction has been associated with depression and reduced quality of life (112). In a rat model of depression, the volume of the olfactory bulb was found to shrink because of reduction in the formation of new neurons, and there was also pre-synaptic dysfunction (113). The anxiolytic activity of incensole acetate has already been mentioned, and this suggests a potential role for it in treatment of depression (72).

Diabetes

The insulin receptor IR kinase is expressed at high levels in the olfactory bulb, and delivery of insulin to the noses of mice has been shown to improve both short- and long-term odour object memory, identification and discrimination. Olfactory thresholds are much less affected. Mice made pre-diabetic did not show the same effects. These results suggest that the effects of diabetes on olfactory abilities are worth further studies (114). Conversely, physiological state affects olfaction, and odour perception is modulated by the endocrine system and hormones and other messengers such as ghrelin, orexins, neuropeptide Y, insulin, leptin and cholecystokinin. Normal eating patterns and also disorders of the endocrine system cause fluctuations in the levels of hormones such as insulin. These fluctuations affect the sense of smell, and disruption of the olfactory system can affect energy homeostasis. In view of the increasing incidence of diabetes, these interactions are significant; the interplay between olfaction and the endocrine system has been reviewed (115).

Epilepsy

There is some evidence that dogs can be trained to recognise onset of a seizure in humans and alert them to it. However, the evidence is not conclusive as yet, and it is not known whether the dog is responding to an odour or to a behavioural trigger (116). In human sufferers, epilepsy negatively affects the odour identification ability (58). There is a fascinating report by Efron concerning the prevention of seizures using odour (117). A jazz singer suffering from epilepsy was treated by conditioning so that when the singer felt a seizure to be imminent, smelling of a jasmine oil prevented the seizure. Eventually, just thinking about the smell could avert the seizure. Therefore, the effect is not a pharmaceutical one but results initially from the relaxing effect of jasmine then through cognitively induced relaxation.

Migraine

During a migraine attack, the sufferer is likely to have heightened awareness of odours and to experience osmophobia. Even when not suffering from a headache, some odours are more offensive to migraine sufferers than to normal controls (118). A study using fMRI found evidence of a relationship between the olfactory and the trigemino-nociceptive pathway in the pathophysiology of migraine disease (119). Olfactory hallucinations (phantosmias) occur in a very low percentage (0.66%) of migraine sufferers. Phantosmias, usually of unpleasant smells, are commoner in women than in men and usually last for 5–60 min and occur before onset of headache (120).

Multiple Sclerosis

Olfactory threshold detection levels are higher in those suffering from multiple sclerosis (MS), and the olfactory loss occurs early in the disease (121,122). Odour identification is associated with the extent of lesions in the olfactory brain of MS sufferers (123), and smell tests could offer a means of distinguishing between secondary progressive and relapsing-remitting forms of MS (124). Impaired neurogenesis in MS sufferers may contribute to olfactory dysfunction (125) and some sufferers were found to have decreased olfactory bulb volume and impaired olfaction (126). An MRI study confirmed that the correlation between MS lesions in the olfactory brain with a decreased OB and olfactory brain volume could help explain olfactory dysfunction in MS sufferers (127).

Schizophrenia

People suffering from schizophrenia have impaired olfactory function, and it is their ability to identify smells rather than their detection threshold that is most affected. One study found that odour identification was more affected in male schizophrenics and odour sensitivity more in females (128). A study comparing young and older schizophrenics found that those at clinical risk of schizophrenia were impaired in both identification and discrimination of odours, whereas those at genetic risk were only impaired on identification. Thus it could be that impairment in odour identification might only represent a genetic marker of vulnerability for schizophrenia, while impairment in odour discrimination might be a biomarker associated with the development of psychosis (129). One study found that the rating of pleasant odours was found to be reduced in schizophrenics relative to those of normal controls, but their rating of unpleasant odours was as for normal subjects (130). However, another study found that schizophrenics rated odours to be more pleasant than did healthy controls, although the identification accuracy was lower in schizophrenics (79). A third found that schizophrenics were less accurate in the identification of pleasant and neutral odours than healthy controls but were equal in the ratings of unpleasant odours (131).

Schizophrenics who experience auditory or verbal hallucinations have impaired hedonic judgement, but those who suffer olfactory hallucinations do not. It would seem that impaired habituation is associated with olfactory hallucination and that those with olfactory hallucinations also tend to suffer a greater incidence of brief periods of unconsciousness (132). Olfactory hallucinations in schizophrenia are dependent on social and cultural factors and also on hallucinations in the other senses (133).

MRI showed that the olfactory bulb volume and sulcus depth were reduced in a group of schizophrenics and they also scored lower than normal in a smell test (134). α-7-Nicotinic acetylcholine receptors in the hippocampus and other brain regions contribute strongly to olfactory discrimination and detection in mice, and, therefore, the fact that expression of these receptors is reduced in schizophrenia might be one of the mechanisms producing olfactory dysfunction in schizophrenics (135).

ADVERSE REACTION TO ODOUR

Identification of unpleasant odours, especially those associated with food that has gone bad, is a necessity for survival and so it is not surprising that unpleasant odours are identified faster and more accurately than pleasant or neutral odours (136). Following on from that, it is not surprising to find that odours tend to provoke the alarm signal at subconscious level and so will tend to bias towards warning rather than indicator of opportunity. This might help explain why some, such as MCS (multiple chemical sensitivity) or IEI (idiopathic environmental intolerance) sufferers, react adversely to smells when there is no real pharmacological risk (137). Indeed, Kärnekull et al. found that response to environmental odours correlated with neuroticism and also to response to environmental noise. Thus the higher an individual's level of neuroticism, the more negatively they would respond to both noise and odour in the environment (138). However, habituation to an irritant odour reduces the level of self-reported symptoms of exposure such as sense of well-being. Subjects who were exposed to ammonia vapour in their workplace were less likely to report adverse effects of ammonia than those who were unaccustomed to it (139).

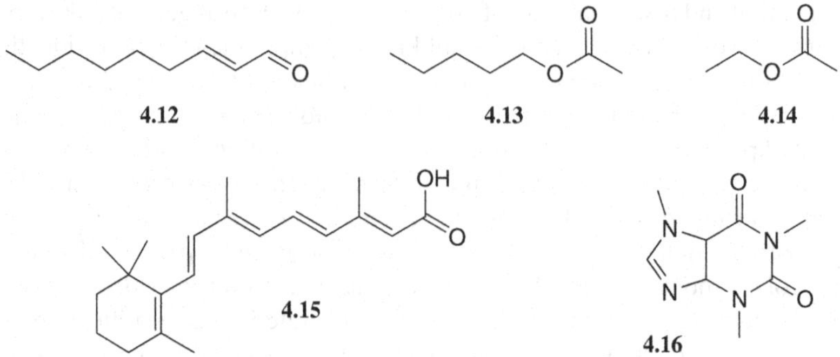

Figure 4.3   Olfaction and ageing.

OLFACTION AND AGEING

There is a common belief that elderly people have a different body odour from their younger counterparts. 2-Nonenal (4.12) is a normal skin odorant, but Yamazaki et al. found that the level of it in skin headspace increased with the age of the donor. They, therefore, speculate that it is part of the "smell of old age" (140). However, the vast majority of published literature on olfaction and ageing relates to the loss of the sense of smell rather than changes in body odour (Figure 4.3).

In view of the extreme complexity of the olfactory system, its direct exposure to the external environment, its high rate of protein synthesis, the requirement for a considerable amount of neuroprocessing, etc., there are many points at which things can go wrong. Therefore, it would not be surprising to find that things do go wrong and that the rate of failure increases with age as damage accumulates and the body's ability to repair and renew cells deteriorates.

Does Smell Decline with Age?

There are a number of key questions concerning age-related loss of the sense of smell. Does the sense deteriorate with age? Is the deterioration due to disease or to the ageing process? Is the loss homogeneous or heterogeneous? Does a subject's sex affect the loss? Is the loss due to central or peripheral damage/deterioration? In looking for answers to these questions, we are confronted, in the most part, by conflicting evidence.

One of the most quoted studies on age-related smell loss is that of Doty et al., carried out in 1984 using the UPSIT. These workers found that 50% of subjects in the 65–80 age bracket showed a measurable degree of smell loss and that this figure increased to 75% for people over 80 (141). Similarly, Murphy et al. showed, in a study of 2491 inhabitants of the town of Beaver Dam, Wisconsin, that while 24.5% of the whole group had some olfactory impairment, 62.5% of those in the 80–97 year bracket had some olfactory impairment (142). These researchers also found that self-reported olfactory impairment was much lower (9.5%) than measured impairment (24.5%). Stevens et al. found that older subjects adapt more quickly to odours and resensitise more slowly, thus demonstrating a degree of impairment (143). In a study of the next generation of Beaver Dam residents, it was found that there was an overall level of 3.8% olfactory impairment. Across the age groups, the average rose from 0.8% of those under 35 years of age to 13.9% of those over 65. Factors such as smoking and nasal polyps were associated with increased impairment, and high household income with a lower level. General health, depression and diet were not found to affect the level of impairment. The overall result is, therefore, encouraging because many of the factors related to olfactory impairment can be controlled (47).

On the other hand, Almkvist et al. compared a group of 16 people aged 77–87 with a similar group of 20–25-year-olds and found that both groups performed equally well in smell tests (144). Similarly, Elsner found that 21 healthy centenarians had test scores in the same range as young subjects (145). In a study of the effect of genetics on odour perception among 297 members of the Hutterite community of South Dakota, Pinto et al. found that there was no correlation of hyposmia with age (146). Interestingly, they also found that the incidence of hyposmia (8.8%) in the Hutterite community was much lower than that (24.5%) reported by Murphy et al. (142) for a broader community, thus suggesting that the Hutterite lifestyle (e.g. no smoking) might have a less detrimental effect on the sense of smell during the ageing process. They did find that there are genetic factors but that these are not straightforward and interact with environmental factors.

Is Age-Related Olfactory Impairment Due to Disease or Normal Ageing Processes?

Many diseases are known to affect the sense of smell, as will be discussed later, and most researchers agree that both disease and ageing are involved in loss of the sense of smell in the elderly. There is also good agreement that there is a significant difference between healthy elderly people and those with disease, an example being the study by Schiffman (147). She also includes malnutrition as a disease and, since loss of the sense of smell leads to poorer dietary intake, this represents a vicious circle of cause and effect (147). Several studies conclude that disease is a more significant cause of olfactory loss than normal ageing (148–150).

Is Age-Related Olfactory Impairment Central or Peripheral?

Adam et al. (151) compared 20–40-year-olds with 55–75-year-olds in their response to 30 odorants. Average thresholds for the older group were 85% of those for the younger and, similarly, intensity ratings were 94%. When it came to description of the odorants, the average descriptors chosen by either group differed for 13 of the 30 odorants. Since semantic descriptors were affected more than threshold and intensity ratings, this would tend to suggest that the difference between the two age groups is more due to central (i.e. cognitive) than peripheral (receptor-based) effects. Taking all of the literature into account, it would seem that both central and peripheral effects are involved, but the balance between them is not certain.

Is Age-Related Olfactory Impairment Homogeneous or Heterogeneous?

Once again, there is conflicting evidence in the literature on this question. The results of Stevens et al. suggest that all elderly people lose their sense of smell and lose it uniformly across odour types (152). However, Hawkes et al. found that the loss is more acute for pleasant smells (153).

Larsson et al. (58) showed that olfactory deficits are different for different odours. They also found that older people were better at smelling amyl acetate (**4.13**), a finding in keeping with that of Adam et al. who found that the older group in their study were better at smelling ethyl acetate (**4.14**) (151). The work of Wysocki and Gilbert (154) and that of Pelchat (155) also lead to the conclusion that the loss is heterogeneous.

Is Age-Related Odour Loss Dependent on the Subject's Sex?

There are many reports that women have a superior sense of smell to men, yet Buschhütter et al. (156) have found that men have larger olfactory bulbs than women and olfactory performance correlates with bulb size. The olfactory bulbs shrink on ageing and the percentage loss in volume is greater in women than in men. This is in contrast to the results such as those of Doty et al. (141) who found that women outperform men at all ages and especially as they grow older. Their figures suggest that the UPSIT score for women falls from 37 at age 20 to 22 at age 90, whereas for men of the same ages the scores fall from 35 to 16. However, Yousem et al. (157) found that, although the olfactory bulb volume does decline from the fourth decade, there was no correlation with UPSIT score. The UPSIT score, of course, requires cognitive processing and so the answer to this apparent contradiction might lie in semantics, attention or other higher processes.

The community-based study of Murphy et al. (158) found that age-related olfactory impairment was twice as likely in men as in women. However the genetic study of Pinto et al. (149) found that hyposmia correlated with female sex and that this had a much greater effect than age, for which they found no correlation. Larsson et al. (58) found that the general decline with age was equal between the sexes.

As far as sex- and disease-related olfactory loss is concerned, men are more likely to get PD, whereas Alzheimer's Disease is more common among women.

Diseases/Conditions as a Cause of Olfactory Dysfunction

The following is a list of some of the conditions that are known to be capable of adversely affecting the sense of smell:

- physical damage
- stroke
- rhinosinusitis
- polyposis
- epilepsy
- upper respiratory tract infections
- smoking
- malnutrition
- dentures
- high content of body fat
- liver disease
- chronic renal disease
- non-otolaryngological cancer
- Kreuzfeld–Jakob's disease
- Alzheimer's disease
- Parkinson's disease.

It is obvious that diseases such as polyposis or viral infections in the upper respiratory tract could affect the sense of smell. Less obvious is the fact that renal disease (159), liver disease and non-otolaryngological cancers can also adversely affect olfaction (160).

Chemoreception at the olfactory epithelium is only the start of the process of olfaction. Considerable signal processing in the brain is required in order to form an odour percept. Therefore, it is not surprising that any condition that damages the brain might affect the sense of smell. Thus olfactory dysfunction can be associated with epilepsy and stroke (158,161,58). Larsson et al. showed that CNS diseases generally can affect odour identification rather than detection (58).

The two commonest neurodegenerative diseases are Alzheimer's and Parkinson's and both of these are known to be associated with olfactory impairment; both are reviewed in the book by Hawkes and Doty (94). Alzheimer's disease affects many parts of the limbic system that receive signals from the olfactory system and, therefore, can affect olfaction. In fact, Nordin et al. (162) suggest odour tests as a measure of the degree and rate of progression of Alzheimer's and also as a way of distinguishing between sub-types of the disease. Olfactory loss can be a very early symptom of PD, occurring as much as 10 years before clinical symptoms appear (94). Doty et al. (163) found that 90% of PD patients had demonstrable smell loss but only 28% were aware of this, and Hawkes et al. (164) found that weight loss in idiopathic PD was related to hyposmia.

The relationship between hyposmia and malnutrition is important in ageing, disease and during treatment with certain drugs (e.g. chemotherapeutic agents) and there is a danger of a vicious circle setting in with smell loss causing malnutrition leading to further smell loss leading to further malnutrition, and so on (147). Other lifestyle factors are also associated with olfactory dysfunction. These include smoking (158), general poor health, wearing of dentures and a high content of body fat (150). The damage to the olfactory system has a cumulative effect and it takes years after giving up before the sense returns to near normal (94). As stated previously, the Hutterite lifestyle does appear to offer better prospects of maintaining a good sense of smell into old age than does a more conventional American lifestyle (149).

Ageing Processes as a Cause of Olfactory Dysfunction

Starting right at the beginning of the process, we can see causes for age-related decline in olfactory acuity. Elderly people secrete less epithelial mucus than their younger counterparts (165) and it is more viscous (94). There is less protein synthesis in the epithelium, of both receptor proteins and enzymes such as P450 cytochromes (94). The vascularity of the epithelium is altered, there is decreased mitotic activity and decreased intra-mucosal blood flow. Two independent groups of researchers have shown that epithelial damage from environmental xenobiotics accumulates, causing patches of the olfactory epithelium to resemble more closely the respiratory epithelium (166,167). Olfactory neurons survive only for 30 days and the rate of replacement slows with age (147).

The number of receptors decreases with age in mice (168) and rats (169,170). Both the pattern and degree of expression of olfactory receptors were found to reduce in aged mice (171). In mice, there is less resistance to epithelial damage and poorer recovery in older animals (172,173), and related effects have been seen in humans (174). The receptive ranges of their olfactory sensory neurons appear to broaden as a person ages, and this might account for some loss of specificity in olfaction (175).

Olfactory neurons run from the epithelium to the olfactory bulb through holes called *foramina* in the cribriform plate of the ethmoid bone. With age, oppositional bone growth in this plate closes the holes, causing pinching of the axons of the olfactory receptor neurons (176).

Young adults have about 8000 glomeruli and 40,000 mitral cells in the olfactory bulbs. The numbers decline by about 10% per decade (177,178) and this results in measurable reduction in the size of the bulbs (156,157). Such a decline has also been observed in rats (179). Unlike the olfactory bulb, the temporal lobes do not appear to reduce in volume over time (157) but neurodegenerative conditions in the CNS structures can result in olfactory impairment (94). Nigrostriatal dopaminergic denervation is a key patholobiological feature of PD but it can also occur with normal ageing and can lead to olfactory loss (180).

Treatment

Some of the diseases affecting olfaction can be treated. For example, steroids or surgery can be used to restore normal olfaction in cases of polyposis. However, others cannot, and this is particularly the case, for the present at least, with Alzheimer's and Parkinson's diseases. Similarly, there is little that can be offered at present for many of the symptoms of ageing. The risk of suffering olfactory impairment increases with age. Risk factors include nasal polyps, deviated septum and a history of heavy alcohol use. Decreased risk is associated with the use of lipid-lowering agents, exercising at least once a week and oral steroid use. The fact that some of the risk factors can be modified suggests that some impairment could be amenable to treatment or prevention (181).

Retinoic acid (**4.15**) is known to be involved in the development of the olfactory system, and its repair after injury (182–184) and has been shown to improve odour memory and learning and to alleviate age-related deficits in mice (185). To date, there have been no studies to determine whether it has a real effect in humans, so no therapy based on it is forthcoming. Caffeine (**4.16**) improves odour discrimination in ageing rats, but there are many possible mechanisms and it is not known which is responsible (186). The fact that caffeine (**4.16**) and adenosine receptor antagonists help restore olfactory memory after hypoxia in mice suggests a role of the adenosine receptors (187). Caffeine (**4.16**) has also been found to counteract the memory impairment induced in rats by ethanol consumption (188). As with retinoic acid (**4.15**), there are no studies of the effect in humans. Selective activation of 5-HT(4) receptors by a partial 5-HT(4) receptor agonist helped in improving olfactive memory performance in aged rats (189).

One study has shown that odour exposure improves the life of olfactory neurons (190), and, although there has been no work to determine whether or not this produces a real benefit in humans, exercising the sense of smell would seem consistent with the view that exercise generally has a beneficial effect on our bodies and brains. The Sense of Smell Institute provides the following suggestions for keeping the olfactory function working well. Train your mind (smell is a mental thing, so keep it active by analysing everyday smells); smell often, but not a lot in order to avoid fatigue; create odour associations; vary fragrance types since variety creates interest and avoids long-term fatigue. Constant creative use of the sense of smell and increased awareness of the scents around us prolong olfactory performance (191).

Benefits of Maintaining a Healthy Sense of Smell

The sense of smell evolved in order to provide warning of dangers and to be alert to opportunities. Therefore, it is obvious that maintaining a good sense of smell will help us to avoid dangers such as spoiled food or leaking gas and to maintain a healthy diet by enjoying a variety of foods. Also, olfactory impairment in the elderly has been shown to be independently associated with depressive symptoms and poorer quality of life (112), functional disability and reduced independence among older adults (192). Elderly people with olfactory impairment are more likely to die within 5 years than those whose sense of smell is closer to that of the population as a whole. Of course, the explanation could lie in diseases, such as Alzheimer's, that cause olfactory impairment (193).

Fragrance can also improve the quality of life in other ways, such as its ability to affect mood (194). For example, fragrance is known to be capable of inducing relaxation, improving feelings of happiness, sensuality and general well-being. It can help reduce stress, improve alertness, enforce positive moods and counteract irritation, depression and apathy. Different odours have been found to exert different effects. For instance, muguet is reported to reduce depression, apathy and irritation and enhance relaxation and happiness; Douglas Fir to improve relaxation; tuberose to promote relaxation and happy and sensuous moods; osmanthus to stimulate and promote happiness while reducing apathy and depression; and hyacinth to enhance happy, sensuous and relaxed moods while reducing all negative moods. One experiment demonstrated how fragrance's ability to improve happiness also made people more willing to help a stranger (195). Mood problems often arise in mid-life for women during menopause and, on average, the most stressful time for women is between the ages of 40 and 49. The most stressful period for men is between 50 and 59 when they often suffer mood problems because of decline in potency, career frustration and so on (196). Fragrance can therefore contribute to improving their quality of life by promoting positive moods and reducing negative ones. In addition to effects on mood, fragrance has also been shown to be capable of promoting sleep and, since some people develop sleep disorders on ageing, fragranced products could also prove beneficial for them (196). As people age, they tend to suffer more health problems and hence require more medical treatment. Fragrance has been shown to reduce stress during medical procedures. Patients demonstrated 63% less anxiety during MRI scanning when smelling heliotropin than a control group who had no odour present at the time (197). Fragrances could, therefore, be used in medicine to support treatment.

What Could the Fragrance and Cosmetics Industries do to Help Older People?

The modern fragrance industry is heavily oriented towards young people. At the 2008 Annual Fragrance Symposium of the American Society of Perfumers, Anne Gottlieb commented 'If we're looking to grow our business, we have to find a way to talk to the over-35 consumer'. She pointed out that this group has money, loves fragrance, but has been abandoned by the mainstream industry. Marian Bendeth added that 'Many fragrances dubbed as "universal" are in fact aimed straight at the youth market' (198).

The 'Baby Boom' generation is defined as those born between 1946 and 1964. They are now moving into their 50s and 60s and thus this generation represents a growing market. Older consumers tend to be more loyal to a brand than are their younger counterparts (199). By formulating products for this market, the industry could not only benefit commercially but also improve the quality of life for these consumers and, by stimulating their interest in smell, possibly help them to keep it for longer.

In order to formulate for the older consumer, it is important to understand how their sense of smell has changed. As discussed earlier, the weight of evidence suggests that age-related smell loss is heterogeneous. More study is necessary to define exactly how odour perception changes and what formulation adjustments are necessary in order to compensate. Odour preferences also change; for example, it has been shown that preference for strawberry and vanilla increases with age (200). Therefore, targeted market research is vital. Some of the diseases affecting olfaction show ethnic patterns of incidence (201,202) and so regional market research is also important.

Autobiographical memories, particularly those of the first decade of life, are triggered by odour, as discussed previously (31). So, should fragrances formulated for specific age groups be based on odours that would relate to happy events from the first decade of that group's life? For example, fragrances for 50-year-olds might recall the good things of the 1960s, whereas those for sexagenarians should relate to the 1950s. One report tells us that certain fragrances 'are being celebrated as true classics rather than just old fragrances and are intended to be worn as they were 30 or 40 years ago' (203). Perhaps this is the Willander effect already at work on baby boomers.

When we think of fragranced products for the elderly, we tend to think of such things as incontinence pads, products for colostomy patients and so on. These are important but there is also a growing market for more pleasant and exciting applications. As discussed earlier, medical applications of fragrance could be extended to reduce stress during procedures. Mood-enhancing fragrances could also be used to improve alertness in the morning, maintain a happy and positive mood during the day and promote sleep at night. Using a variety of different fragrances, consumers would be encouraged to smell actively and increase their awareness of odour. This could play a role in keeping their sense of smell healthier for longer and consequently improving their quality of life.

REFERENCES

1. G. M. Shepherd, *Neurogastronomy: How the Brain Creates Flavour and Why It Matters*, Columbia University Press, New York, 2012, ISBN: 978-0-231-15010-4, 98-0-231-53031-6 (e-book).
2. R. R. Calkin and J. S. Jellinek, *Perfumery, Practice and Principles*, John Wiley & Sons, Inc., New York, 1994, ISBN 0-471-58934-9.
3. T. Curtis and D. G. Williams, *Introduction to Perfumery*, Ellis Horwood, Chichester, 2nd Edn., 2001, ISBN 1-870228-24-3.

4. A. Gilbert, *What the Nose Knows*, Crown Publishers, New York, 2008, ISBN: 978-1-4000-8234-6.
5. P. Jellinek, Ed., J. S. Jellinek, *The Psychological Basis of Perfumery*, 4th Edn., Blackie Academic and Professional, London, 1997, ISBN 0-7514-0368.
6. R. Kaiser, *Meaningful Scents Around the World*, Wiley-VCH, Weinheim, 2006, ISBN: 3-906390-37-3.
7. P. M. Müller and D. Lamparsky, *Perfumes: Art, Science and Technology*, Elsevier, London, 1991.
8. C. S. Sell. *The Chemistry of Fragrances From Perfumer to Consumer* 2nd Edn. Royal Society of Chemistry, Cambridge, 2006 ISBN-10: 0-85404-824-3 ISBN-13: 978-0-85404-824-3.
9. S. van Toller and G. H. Dodd, *Perfumery, The Psychology and Biology of Fragrance*, Chapman and Hall, London, 1988 ISBN 0-412-30010-9.
10. M. Okamoto, Y. Wada, Y. Yamaguchi, A. Kimura, H. Dan, T. Masuda, A. K. Singh, L. Clowney, and I. Dan, *Chem. Senses*, **2008**, *34*, 187–194 doi:10:1093/chemse/bjn075.
11. J. Lim and A. Padmanabhan, *Chem. Senses*, **2013**, *38(1)*, 45–55 doi: 10.1093/chemse/bjs080.
12. L. D. Stafford and K. Welbeck, *Chem. Senses*, **2011**, *36*, 189–198 doi: 10.1093/chemse/bjq114.
13. K. Osada, M. Hanawa, K. Tsunoda, and H. Izumi, *Chem. Senses*, **2013**, *38(1)*, 57–65 doi: 10.1093/chemse/bjs077.
14. K. Burseg and C. De Jong, *J. Agric. Food Chem.*, **2009**, *57*, 9086–9090 doi: 10.1021/jf9016866
15. E. E. Slossen, *Psychol. Rev.*, **1899**, *6(4)*, 407–408 doi: 10.1037/h0071184.
16. L. Hall, P. Johansson, B. Tarning, S. Sikstrom, and T. Deutgen, *Cognition*, **2010**, *117*, 54–61.
17. H. Nienborg and B. G. Cumming, *Nature*, **2009**, *459*, 89–92 doi:10.1038/nature07821.
18. M. L. Demattè, R. Österbauer, and C. Spence, *Chem. Senses*, **2007**, *32*, 603–610 doi: 10.1093/chemse/bjm030.
19. D. E. Gaygen and A. Hedge, *Chem. Senses*, **2009**, *34*, 85–91 doi:10.1093/chemse/bjn057.
20. T. Kobayashi, N. Sakai, T. Kobayakawa, S. Akiyama, H. Toda, and S. Saito, *Chem. Senses*, **2008**, *33*, 163–171 doi: 10.1093/chemse/bjm075.
21. M. L. Demattè, D. Sanabria, R. Sugarman, and C. Spence, *Chem. Senses*, **2006**, *31(4)*, 291–300 doi: 10.1093/chemse/bjj031.
22. (a) A. S. Crisinel and C. Spence, *Atten. Percept. Psychophys.*, **2010**, *72*, 1994–2002; (b) A. S. Crisinel and C. Spence, *Chem. Senses*, **2012**, *37*, 151–158 doi:10.1093/chemse/bjr085.
23. A. S. Crisinel and C. Spence, *Acta Psychol.*, **2011**, *138(1)*, 155–161 doi: 10.1016/j.actpsy.2011.05.018.
24. B. Mesz, M. A. Trevisan and M. Sigman, *Perception*, **2011**, *40*, 209–219.
25. H. S. Seo, M. Guarneros, R. Hudson, H. Distel, B. C. Min, J. K. Kang, I. Croy, J. Vodicka, and T. Hummel, *Chem. Senses*, **2011**, *36(2)*, 177–187. doi: 10.1093/chemse/bjq112.
26. B. Derntl, V. Schöpf, K. Kollndorfer, and R. Lanzenberger, *Chem. Senses*, **2013**, *38*, 67–75 doi: 10.1093/chemse/bjs084.
27. C. Ferdenzi, S. C. Roberts, A. Schirmer, S. Delplanque, S. Cekic, C. Porcherot, I. Cayeux, D. Sander, and D. Grangjean, *Chem. Senses*, **2013**, *38(2)*, 175–186 doi: 10.1093/chemse/bjs083.
28. R. L. Doty and E. L. Cameron, *Physiol. Behav.*, **2009**, *97*, 213–228.
29. L. D. Grunebaum, J. Murdock, M. P. Castanedo-Tardan, and L. S. Baumann, *J. Cosmet. Dermatol.*, **2011**, *10(2)*, 89–93 doi: 10.1111/j.1473-2165.2011.00554.x.
30. L. A. Nguyen, B. A. Ober, and G. KL. Shenaut *Chem. Senses*, **2012**, *37(8)*, 745–754 doi: 10.1093/chemse/bjs060.
31. Autobiographical Odour Memory, Johan Willander, Ph. D. thesis, Stockholm University, **2007**.
32. G. M. Zucco, L. Aiello, L. Turuani, and E. Köster, *Chem. Senses*, **2012**, *37(2)*, 179–189 doi: 10.1093/chemse/bjr089.
33. Y. Masaoka, H. Sugiyama, A. Katayama, M. Kashiwagi, and I. Homma, *Chem. Senses*, **2012**, *37*, 379–388 doi: 10.1093/chemse/bjr120.
34. M. B. Toffolo, M. A. Smeets, and M. A. van den Hout, *Cogn Emot.*, **2012**, *26*, 83–92.
35. M. Matsunaga, T. Isowa, K. Yamakawa, Y. Kawanishi, H. Tsuboi, H. Kaneko, N. Sadato, A. Oshida, A. Katayama, M. Kashiwagi, and H. Ohira, *Act. Nerv. Super. Rediviva.* **2011**, *53(3)*, 114–120 ISSN 1337-933X.
36. J. Croy, S. Olgun, and P. Joraschky, *Emotion*, **2011**, *11*, 1331–1335.
37. J. Seubert, A. F. Rea, J. Loughead, and U. Habel, *Chem. Senses*, **2009**, *34*, 77–84 doi: 10.1093/chemse/bjn054.

38. J. A. Gottfried, R. Deichmann, J. S. Winston, and R. J. Dolan, *J. Neurosci.*, **2002**, *22(24)*, 10819–10828.

39. J. N. Lundstrom, J. A. Boyle, R. J. Zatorre, and M. Jones-Gotman, *Cereb. Cortex*, **2008**, *18(6)*, 1466–1474.

40. J. H. B. de Groot, M. A. M. Smeets, A. Kaldewaij, M. J. A. Duijndam, and G. Semin, *Psychol. Sci.*, **2012**, *23(11)*, 1417–1424 doi: 10.1177/0956797612445317.

41. M. Schredl, D. Atanasova, K. Hörmann, J. T. Maurer, T. Hummel, and B. A. Stuck, *J. Sleep Res.*, **2009**, *18*, 285–290 doi:10.1111/j.1365-2869.2009.00737.x.

42. D. C. Barnes, J. Chapuis, D. Chaudhury, and D. A. Wilson, *PLoS One*, **2011**, *6*, e18130.

43. M. Beaulieu-Lefebre, F. C. Schneider, R. Kupers, and M. Ptito, *Brain Res. Bull.*, **2011**, *84(3)*, 206–209 doi: 10.1016/j.brainresbull.2010.12.014.

44. P. Rombaux, C. Huart, A. G. De Volder, I. Cuevas, L. Renier, T. Duprez, and C. Grandin, *Neuroreport*, **2010**, *21(17)*, 1069–1073 doi: 10.1097/WNR.0b013e32833fcb8a.

45. R. Kupers, M. Beaulieu-Lefebre, F. C. Schneider, T. Kassuba, O. B. Paulson, H. R. Siebner, and M. Ptito, *Neuropsychologia*, **2011**, *49(7)*, 2037–2044 doi: 10.1016/j.neuropsychologia.2011.03.333.

46. M. Fukada, E. Kano, M. Miyoshi, R. Komaki, and T. Watanabe, *Chem. Senses*, **2012**, *37(4)*, 347–356 doi: 10.1093/chemse/bjr108.

47. C. R. Schubert, K. J. Cruickshanks, M. E. Fischer, G.-H. Huang, B. E. K. Klein, J. S. Pankow, and D. M. Nondahl, *Chem. Senses*, **2012**, *37(4)*, 325–334 doi: 10.1093/chemse/bjr102.

48. M. Guarneros, T. Hummel, M. Martínez-Gómez, and R. Hudson, *Chem. Senses*, **2009**, *34*, 819–826 doi:10.1093/chemse/bjp071.

49. A. S. Bomback and A. C. Raff, *Kidney Int.*, **2011**, *80*, 803–805.

50. T. D. Brisbois, I. H. de Kock, S. M. Watanabe, M. Mirhosseini, D. C. Lamoureux, M. Chasen, N. Macdonald, V. E. Baracos, and W. V. Wismer, *Ann. Oncol.*, **2011**, *22(9)*, 2086–2093 doi: 10.1093/annonc/mdq727.

51. T. D. Brisbois, I. H. de Kock, S. M. Watanabe, V. E. Baracos, and W. V. Wismer, *J. Pain Symptom. Manage.*, **2011**, *41*, 673–683.

52. P. Maurage, C. Callot, P. Philippot, P. Rombaux, and P. de Timary, *Biol. Psychol.*, **2011**, *88*, 28–36.

53. J. Lotsch, G. Geisslinger, and T. Hummel, *Trends Pharmacol. Sci.*, **2012**, *33(4)*, 193–199 doi: 10.1016/j.tips.2012.01.004.

54. P. M. Jenkins, D. P. McEwen, and J. R. Martens, *Chem. Senses*, **2009**, *34*, 451–464 doi:10.1093/chemse/bjp020.

55. J. C. McIntyre, E. E. Davis, A. Joiner, C. L. Williams, I.-C. Tsai, P. M. Jenkins, D. P. McEwen, L. Zhang, J. Escobado, S. Thomas, K. Szymanska, C. A. Johnson, P. L. Beales, E. D. Green, and J. C. Mulliken, NISC Comparative Sequencing Program, A. Sabo, D. M. Muzny, R. A. Gibbs, T. Attié-Bitach, B. K. Yoder, R. R. Reed, N. Katsanis, and J. R. Martens, *Nature Medicine*, **2012**, *18*, 1423–1428 doi:10.1038/nm.2860.

56. S. Steinbach, F. Proft, H. Schulze-Koops, W. Hundt, P. Heinrich, S. Schulz, and M. Gruenke, *Scand. J. Rheumatol.*, **2011**, *40(3)*, 169–177 doi: 10.3109/03009742.2010.517547.

57. L. Kalogjera and D. Dzepina, *Curr. Allergy Asthma Rep.*, **2012**, *12*, 154–162 doi: 10.1007/s11882-012-0248-5.

58. M. Larsson, D. Finkel, and N. L. Pedersen, *J. Gerontol.*, **2000**, *55B(5)*, 304–310.

59. B. A. Belluscio, L. Jin, V. Watters, T. H. Lee, and M. Hallett, *Mov. Disord.*, **2011**, *26(14)*, 2538–2543 doi: 10.1002/mds.23977.

60. J. E. Bergman, G. Bocca, L. H. Hoefsloot, L.C. Meiners, and C. M. van Ravenswaaij-Arts, *J. Pediatr.*, **2010**, *158(3)*, 474–479.

61. C. Segalas, J. Labad, P. Alonso, E. Real, M. Subira, B. Bueno, S. Jimenez-Murcia, and J. M. Menchon, *Anxiety*. **2011**, *28*, 932–940.

62. R. J. Stevenson, *Conscious. Cogn.*, **2011**, *20*, 1887–1898.

63. Y. Masaoka, M. Kawamura, A. Takeda, M. Kobayakawa, T. Kuroda, H. Kasai, N. Tsuruya, A. Futamura, and I. Homma, *Neurosci. Lett.*, **2011**, *503*, 163–166.

64. M. Shirasu, S. Nagai, R. Hayashi, A. Ochiai, and K. Touhara, *Biosci. Biotechnol. Biochem.*, **2009**, *73(9)*, 2117–2120 doi: 10.1271/bbb.90229.

65. G. Lippi and G. Cervellin, *Clin. Chem. Lab. Med.*, **2011**, *50(3)*, 435–439 doi: 10.1515/cclm.2011.672.
66. J.-N. Cornu, G. Cancel-Tassin, V. Ondet, C. Girardet O. Cussenot, *Eur. Urol.*, **2011**, *59*, 197–201 doi: 10.1016/j.eururo.2010.10.006.
67. A. Sturzu, S. Sheikh, H. Echner, T. Nägele, M. Deeg, C. Schwentner, M. Horger, U. Ernemann, and S. Heckl, *Invest. New Drugs*, **2013**, doi: 10.1007/s10637-013-9943-x.
68. A. Sturzu, H. Echner, U. Klose, S. Sheikh, T. Nägele, C. Schwentner, U. Ernemann, and S. Heckl, *Curr. Pharm. Biotechnol.*, **2012**, *13(2)*, 373–377.
69. H. Mohapatra and S. T. Phillips, *Angew. Chem. Int. Edn.*, **2012**, *124(44)*, 11307–11310 doi: 10.1002/ange.201207008.
70. M. S. Weinberg, A. P. Bhatt, M. Girotti, C. V. Masini, H. E. W. Day, S. Campeau, and R. L. Spencer, *Endocrinology*, **2009**, *150(2)*, 749–761 doi: 10.1210/en.2008-0958.
71. D. S. Phillips and G. K. Martin, *Physiol. Behav.*, **1971**, *7(4)*, 535–537.
72. A. Moussaieff, N. Rimmerman, T. Bregman, A. Straiker, C. C. Felder, S. Shoham, Y. Kashman, S. M. Huang, H. Lee, E. Shohami, K. Mackie, M. J. Caterina, J. M. Walker, E. Fride, and R. Mechoulam, *FASEB J.*, **2008** doi: 10.1096/fj.07-101865.
73. G. Goldzak-Kunik, R. Friedman, M. Spitz, L. Sandler, and M. Leshem, *Am. J. Clin. Nutr.*, **2012**, *95(2)*, 272–282 doi: 10.3945/ajcn.111.020131v1.
74. M. Schecklmann, C. Pfannstiel, A. J. Fallgatter, A. Warnke, M. Gerlach, and M. Romanos, *J. Neural. Transm.*, **2012**, *119*, 721–728 doi: 10.1007/s00702-011-0752-0.
75. N. Rapps, K. E. Giel, E. Sohngen, A. Salini, P. Enck, S. C. Bischoff, and S. Zipfel, *Eur. Eat. Disord. Rev.*, **2010**, *18(5)*, 385–389 doi: 10.1002/erv1010.
76. I. Dudova, J. Vodicka, M. Havlovicova, Z. Sedlacek, T. Urbanek, and M. Hrdlicka, *Eur. Child Adolesc. Psychiatry*, **2011**, *20*, 333–340.
77. T. Tavassoli, and S. Baron-Cohen, *J. Autism Dev. Disord.*, **2012**, doi: 10.1007/s10803-011-1321-y.
78. C. Hardy, M. Rosedale, J. W. Messinger, K. Kleinhaus, N. Aujero, H. Silva, R. R. Goetz, D. Goetz, J. Harkavy-Friedman, and D. Malaspina, *Bipolar. Disord.*, **2012**, *14*, 109–117.
79. A. G. Cumming, N. L. Matthews, and S. Park, *Eur. Arch. Psychiatry Clin. Neurosci.*, **2011**, *261*, 251–259, doi: 10.1007/s00406-010-0145-7.
80. J. Hidalgo, G. Chopard, J. Galmiche, L. Jacquot, and G. Brand, *Rhinology*, **2011**, 49, 513.
81. E. M. Garland, S. R. Raj, A. C. Peltier, D. Robertson, and I. Biaggioni, *Neurology*, **2011**, *76*, 456–460.
82. A. Bahar-Fuchs, S. Moss, C. Rowe, and G. Savage, *Chem. Senses*, **2010**, *35*, 855–862 doi: 10.1093/chemse/bjq094.
83. J. Wang, P. J. Eslinger, R. L. Doty, E. K. Zimmerman, R. Grunfeld, X. Sun, M. D. Meadowcroft, J. R. Connor, J. L. Price, M. B. Smith, and Q. X. Yang, *Brain Res.*, **2010**, *1357*, 184–19.
84. W. Li, J. D. Howard, J. A. Gottfried, *Brain*, **2010**, *133(9)*, 2714–2726.
85. S. Forster, A. Vaitl, S. J. Teipel, I. Yakushev, M. Mustafa, C. la Fougère, A. Rominger, P. Cumming, P. Bartenstein, H. Hampel, T. Hummel, K. Buerger, W. Hundt, and S. Steinbach. Functional representation of olfactory impairment in early Alzheimer's disease. *J. Alzheimer's. Dis.*, **2010**, *22*, 581–591.
86. I. Makowska, I. Kloszewska, A. Grabowska, I. Szatkowska, and K. Rymarczyk, *Arch. Clin. Neuropsychol.*, **2011**, *26(3)*, 270–279 doi: 10.1093/arclin/acr011.
87. D. Jimbo, M. Inoue, M. Taniguchi, and K. Urakami, *Psychogeriatrics.*, **2011**, *11*, 196–204.
88. C. Rouby, T. Thomas-Danguin, M. Vigouroux, G. Ciuperca, T. Jiang, J. Alexanian, M. Barges, I. Gallice, J. L. Degraix, and G. Sicard, *Int. J. Otolaryngology*, **2011**, 203805 doi: 10.1155/2011/203805.
89. D. W. Wesson, E. Levy, R. A. Nixon, and D. A. Wilson, *J. Neurosci.*, **2010**, *30(2)*, 505–514 doi: 10.1523/JNEUROSCI.4622-09.2010.
90. M. A. Babizhayev, A. I. Deyev, and Y. E. Yegorov, *Curr. Clin. Pharmacol.* **2011**, *6*, 236–259.
91. Z. B. Liu, W. M. Niu, X. H. Yang, Y. Wang, and W. G. Wang, *J. Tradit. Chin. Med.*, **2010**, *30*, 283–287.
92. J. Ruiz-Martinez, A. Gorostidi, E. Goyenechea, A. Alzualde, J. J. Poza, F. Rodríguez, A. Bergareche, F. Moreno, A. López de Munain, and J. F. Martí Massó, *Mov. Disord.*, **2011**, *26(11)*, 2026–2031 doi: 10.1002/mds.23773.

93. T. Parrao, P. Chana, P. Venegas, M. I. Behrens, and M. L. Aylwin, *Neurodegener. Dis.*, 2012, *10*, 179–182 doi: 10.1159/000335915.
94. C. H. Hawkes and R. L. Doty, *The Neurology of Olfaction*, Cambridge University Press, Cambridge, 2009.
95. T. Baba, A. Takeda, A. Kikuchi, Y. Nishio, Y. Hosokai, K. Hirayama, T. Hasegawa, N. Sugeno, K. Suzuki, E. Mori, S. Takahashi, H. Fukuda, and Y. Itoyama, *Mov Disord.*, 2011, *26*, 621–628.
96. T. Baba, A. Kikuchi, K. Hirayama, Y. Nishio, Y. Hosokai, S. Kanno, T. Hasegawa, N. Sugeno, M. Konno, K. Suzuki, S. Takahashi, H. Fukuda, M. Aoki, Y. Itoyama, E. Mori, and A. Takeda, *Brain*, 2012, *135*, 161–169.
97. M. F. Damholdt, P. Borghammer, L. Larsen, and K. Ostergaard, *Mov Disord.*, 2011, *26*, 2045–2050.
98. N. Sobel, M. E. Thomason, I. Stappen, C. M. Tanner, J. W. Tetrud, J. M. Bower, E. V. Sullivan, and J. D. Gabrieli, *PNAS*, 2001, *98*, 4154–4159.
99. J. Wang, H. You, J. F. Liu, D. F. Ni, Z. X. Zhang, and J. Guan, *Am. J. Neuroradiol.*, 2011, *32*, 677–681.
100. R. S. Wilson, L. Yu, J. A. Schneider, S. E. Arnold, A. S. Buchman, and D. A. Bennett, *Chem. Senses*, 2011, *36*, 367–373.
101. C. H. Hawkes, K. Del Tredici, and H. Braak, *Neuropath. Appl. Neruobiol.*, 2007, *33(6)*, 599–614 doi: 10.1111/j.1365-2990.2007.00874.x.
102. R. D. Prediger, A. S. Aguiar Jr., F. C. Matheus, R. Walz, L. Antoury, R. Raisman-Vozari, and R. L. Doty, *Neurotox. Res.*, 2011, doi: 10.1007/s12640-011-9281-8.
103. (a) J. F. Morley and J. E. Duda, *Biomark. Med.*, 2010, *4*, 661–670; (b) L. M. Chahine and M. B. Stern, *Curr. Opin. Neurol.*, 2011, *24(4)*, 309–317 doi: 10.1097/WCO.0b013e3283461723.
104. H. W. Berendse, D. S. Roos, P. Raijmakers, and R. L. Doty, *J. Neurol. Sci.*, 2011, *310*, 21–24.
105. W. Chen, S. Chen, W. Y. Kang, B. Li, Z. M. Xu, Q. Xiao, J. Liu, Y. Wang, G. Wang, and S. D. Chen, *J. Neurol. Sci.*, 2012, *316(1–2)*, 47–50 doi: 10.1016/j.jns.2012.01.033.
106. J. Deeb, M. Shah, N. Muhammed, R. Gunasekera, K. Gannon, L. J. Findley, and C. H. Hawkes, *QJM.*, 2010, *103*, 941–952 doi: 10.1093/qjmed/hcq142.
107. L. Kertelge, N. Brüggemann, A. Schmidt, V. Tadic, C. Wisse, S. Dankert, L. Drude, J. van der Vegt, H. Siebner, H. Pawlack, P. P. Pramstaller, M. I. Behrens, A. Ramirez, D. Reichel, C. Buhmann, J. Hagenah, C. Klein, K. Lohmann, and M. Kasten, *Mov Disord.*, 2010, *25*, 2665–2669 doi: 10.1002/mds.23272.
108. M. Suzuki, M. Hashimoto, M. Yoshioka, M. Murakami, K. Kawasaki, and M. Urashima. *Neurol.*, 2011, *11*, 157.
109. J. F. Morley, D. Weintraub, E. Mamikonyan, P. J. Moberg, A. D. Siderowf, and J. E. Duda, *Mov Disord.*, 2011, *26*, 2051–2057.
110. T. M. Rolheiser, H. G. Fulton, K. P. Good, J. D. Fisk, J. R. McKelvey, C. Scherfler, N. M. Khan, R. A. Leslie, and H. A. Robertson, *J Neurol.*, 2011, *258*, 1254–1260 doi: 10.1007/s00415-011-5915-2.
111. G. M. Zucco and F. Bollini, *Psychiatry Res.*, 2011, *190(2–3)*, 217–220 doi: 10.1016/j.psychres.2011.08.025.
112. B. Gopinath, K. J. Anstey, C. M. Sue, A. Kifley, and P. Mitchell, *Am. J. Geriatr. Psychiatry*, 2011, *19*, 830–834.
113. D. Yang, Q. Li, L. Fang, K. Cheng, R. Zhang, P. Zheng, Q. Zhan, Z. Qi, S. Zhong, and P. Xie, *Neurosci.*, 2011, *192*, 609–618.
114. D. R. Marks, K. Tucker, M. A. Cavallin, T. G. Mast, and D. A. Fadool, *J. Neurosci.*, 2009, *29(20)*, 6734–6751.
115. B. Palouzier-Paulignan, M.-C. Lacroix, P. Aimé, C. Baly, M. Caillol, P. Congar, A. K. Julliard, K. Tucker, and D. A. Fadool, *Chem. Senses*, 2012, *37(9)*, 769–797 doi: 10.1093/chemse/bjs059.
116. S. W. Brown and L. H. Goldstein, *Epilepsy Res.*, 2011, *97*, 236–242.
117. R. Efron, *Brain*, 1956, *79*, 267–281 doi: 10.1093/brain/79.2.267.
118. A. Saisu, M. Tatsumoto, E. Hoshiyama, S. Aiba, and K. Hirata, *Cephalalgia*, 2011, *31*, 1023–1028.

119. A. Stankewitz and A. May, *Neurol.*, **2011**, *77(5)*, 476–482 doi: 10.1212/WNL.0b013e318 227e4a8.
120. E. R. Coleman, B. M. Grosberg, and M. S. Robbins, *Cephalalgia*, **2011**, *31*, 1477–1489.
121. A. Lutterotti, M. Vedovello, M. Reindl, R. Ehling, F. Dipauli, B. Kuenz, C. Gneiss, F. Deisenhammer, and T. Berger, *Mult. Scler.*, **2011**, *17*, 964–969.
122. S. B. Dahlslett, O. Goektas, F. Schmidt, L. Harms, H. Olze, and F. Fleiner, *Eur Arch. Otorhinolaryngol.*, **2011**, doi: 10.1007/s00405-011-1812-7.
123. K. Erb, G. Bohner, L. Harms, O. Goektas, F. Fleiner, E. Dommes, F. A. Schmidt, B. Dahlslett, and L. Ludemann, *J. Neurol. Sci.*, **2012**, *316(1–2)*, 55–60 doi: 10.1016/j.jns.2012.01.031.
124. A. M. Silva, E. Santos, I. Moreira, A. Bettencourt, E. Coutinho, A. Goncalves, C. Pinto, X. Montalban, and S. Cavaco, *Mult. Scler. J.*, **2012**, *18(5)*, 616–621 doi: 10.1177/1352458511427156.
125. V. Tepavcevic, F. Lazarini, C. Alfaro-Cervello, C. Kerninon, K. Yoshikawa, J. M. Garcia-Verdugo, P. M. Lledo, B. Nait-Oumesmar, and A. Baron-Van Evercooren, *J. Clin. Invest.*, **2011**, *121(12)*, 4722–4734 doi: 10.1172/JCI59145.
126. O. Goektas, F. Schmidt, G. Bohner, K. Erb, L. Ludemann, B. Dahlslett, L. Harms, and F. Fleiner, *Rhinol.*, **2011**, *49*, 221–226.
127. F. A. Schmidt, F. Fleiner, L. Harms, G. Bohner, K. Erb, L. Ludemann, B. Dahlslett, and O. Goktas, *Rofo*, **2011**, *183*, 531–535.
128. D. Malaspina, R. Goetz, A. Keller, J. W. Messinger, G. Bruder, D. Goetz, M. Opler, S. Harlap, J. Harkavy-Friedman, and D. Antonius, *Schizophr. Res*, **2012**, *135(1–3)*, 144–151 doi: 10.1016/j.schres.2011.11.
129. V. Kamath, B. I. Turetsky, M. E. Calkins, C. G. Kohler, C.G. Conroy, K. Borgmann-Winter, D. E. Gatto, R.E. Gur, and P. J. Moberg, *World J. Biol. Psychiatry*, **2011**, doi: 10.3109/15622975.2011.
130. B. Crespo-Facorro, S. Paradiso, N. C. Andreasen, D. S. O'Leary, G. L. Watkins, L. L. B. Ponto, and R. D. Hichwa, *JAMA*, **2001**, *286(4)*, 427–435 doi: 10.1001/jama.286.4.427.
131. V. Kamath, B. I. Turetsky, and P. J. Moberg, *Psychiatry Res.*, **2011**, *187(1–2)*, 30–35 doi: 10.1016/j.psychres.2012.12.011.
132. R. J. Stevenson and R. Langdon, *Cogn. Neuropsychiatry*, **2012**, *17(4)*, 315–333, doi: 10.1080/13546805.2011.633748.
133. R. Langdon, J. McGuire, R. Stevenson, and S. V. Catts, *Br. J. Clin. Psychol.*, **2011**, *50*, 145–163.
134. A. Nguyen, P. E. Pelavin, M. E. Shenton, P. Chilakamarri, R. W. McCarley, P. G. Nestor, and J. J. Levitt, *Brain Imaging Behav.*, **2011**, *5*, 252–261 doi: 10.1007/s11682-011-9129-0.
135. J. L. Hellier, N.L. Arevalo, H. J. Blatner, A. K. Dang, A.C. Clevenger, C. E. Adams, and D. Restrepo, *Brain Res.*, **2010**, *1358*, 140–140 doi: 10.1016/j.brainres.2010.08.027.
136. S. Boesveldt, J. Frasnelli, A. R. Gordon, and J. N. Lundstrom, *Biol. Psychol.*, **2010**, *84*, 313–317.
137. P. J. Bulsing, M. A. M. Smeets, and M. A. van den Hout, *Chem. Senses*, **2009**, *34*, 111–119 doi: 10.1093/chemse/bjn062.
138. S. C. Kärnekull, F.U. Jönsson, M. Larsson, and J. K. Olofsson, *Chem. Senses*, **2011**, *36*, 641–648 doi: 10.1093/chemse/bjr028.
139. A. Ihrig, J. Hoffman, and G. Triebig, *Internat. Arch. Occupat. Environ. Health*, **2006**, *79(4)*, 332–338.
140. S. Yamazaki, K. Hoshino, and M. Kusuhara, *Anti-Aging Medicine*, **2010**, *7(6)*, 60–65.
141. R. L. Doty, P. Shaman, S. L. Appelbaum, R. Gilberson, J. Sikorsky, and L. Rosenberg, *Science*, **1984**, *226*, 1441–1443.
142. C. Murphy, C. R. Schubert, K. J. Cruickshanks, B. E. K. Klein, R. Klein, and D. M. Nondahl, *J. Amer. Med. Assocn.*, **2002**, *288*, 2307–2312.
143. J. C. Stevens, W. S. Cain, F. T. Schiet, and M. W. Oatley, *Perception*, **1989**, *18*, 265–276.
144. O. Almkvist, B. Berglund, and S. Nordin, 1992, Odor detectability in successfully aged elderly and young adults. Reports from the Dept. of Psychology, Univ. of Stockholm, No. 744.
145. R. J. Elsner, *Arch. Gerontol. Geriatr.*, **2001**, *33*, 81–94.
146. J. M. Pinto, S. Thanaviratananich, M. G. Hayes, R. M. Naclerio, and C. Ober, *Chemical Senses*, **2008**, *33(4)*, 319–329.
147. S.S. Schiffman, *JAMA*, **1997**, *278*, 1357–1362.
148. N. E. Rawson, *Sci. Aging Knowl. Environ.*, **2006**, *5*, pe6 doi: 10.1126/sageke.2006.5.pe6.

149. J. M. Pinto, S. Thanaviratananich, M. G. Hayes, R. M. Naclerio, and C. Ober, *Chem. Senses*, **2008**, *33(4)*, 319–329.

150. M. I. Griep, T. F. Mets, P. Vogelaere, K. Collys, M. Laska, and D. L. Massart, *Tidschr. Gerontol. Geriatr*, **1997**, *28(1)*, 11–17.

151. C. Adam, F. Beurier, C. Godefroy, and J.-M. Sieffermann, Poster at ISOT/AChemS, San Francisco, July 2008.

152. (a) J. C. Stevens, W. S. Cain, and R. J. Burke, *Chem. Senses*, **1988**, *13*, 643–653; (b) W. S. Cain and J. C. Stevens, *Ann. N. Y. Acad. Sci.*, **1989**, *561*, 29–38.

153. C. H. Hawkes, M. Shah, and A. Fogo, *Neurol*, **2005**, Abs. P01.147, *6 Supplement 1*.

154. C. J. Wysocki and A. N. Gilbert, in *Nutrition and the Chemical Senses in Aging*, C. Murphy, Ed., New York Academy of Sciences, New York, 1989, pp 12–28.

155. M. L. Pelchat, in *Compendium of Olfactory Research*, T. S. Lorig, Ed., Kendall/Hunt Publishing co., Dubuque, Iowa, 2001 pp 3–12.

156. D. Buschhütter, M. Smitka, S. Puschmann, J. C. Gerber, M. Witt, N. D. Abolmaali, and T. Hummel, *NeuroImage*, **2008**, *42(2)*, 498–502.

157. D. M. Yousem, R. J. Geckle, W. B. Bilker, and R. L. Doty, *Ann. N. Y. Acad. Sci.*, **1998**, *855*, 546–555.

158. C. Murphy, C. R. Schubert, K. J. Cruickshanks, B. E. K. Klein, R. Klein, and D. M. Nondahl, *JAMA.*, **2002**, *288*, 2307–2312.

159. M. I. Griep, P. van der Niepen, J. J. Sennesael, T. F. Mets, D. L. Massart, and D. L. Verbeelen, *Nephrol. Dialysis Transplantation*, **1997**, *12(10)*, 2093–2098.

160. B. N. Landis, C. G. Konnerth, and T. Hummel, *Laryngoscope*, **2004**, *114(10 I)*, 1764–1769.

161. T. L. Green, L. D. McGregor, and K. M. King, *Can. J. Neurosci. Nursing*, **2008**, *30(2)*, 10–13.

162. S. Nordin, O. Almkvist, B. Berglund, and L.-O. Wahlund, *Arch. Neurol.*, **1997**, *54(8)*, 993–998.

163. R. L. Doty, D. A. Deems, R. E. Frye, R. Pelberg, and A. Shapiro, *Neurol*, **1988**, *38(8)*, 1237–1244.

164. C. H. Hawkes, B. C. Shephard, and S. E. Daniel, *J. Neurol. Neurosurgery and Psychiatry*, **1997**, *62*, 436–446.

165. M. Ferry, *Nutr. Rev.*, **2005**, S22-S29.

166. T. Nakashima, C. P. Kimmelman, and J. B. Snow Jr., *Arch. Otolaryngol.*, **1984**, *110*, 641–646.

167. A. T. Loo, S. L. Youngentob, P. F. Kent, and J. E. Schwob, *Int. J. Dev. Neurosci.*, **1996**, *14*, 881–900.

168. Y. Rosli, L. J. Breckenridge, and R. A. Smith, *J. Electron Microscopy*, **1999**, *48*, 77–84.

169. K. Nagano, T. Katagiri, S. Aiso, H. Senoh, Y. Sakura, and T. Takeuchi, *Exp. Toxicol. Pathol.*, **1997**, *49*, 97–104.

170. T. Hirai, S. Kojima, A. Shimada, T. Umemura, M. Skai, and C. Itakura, *Neuropathol. Appl. Neurobiol.*, **1996**, *22*, 531–539.

171. A.C. Lee, H. Tian, X. Grosmaitre, and M. Ma, *Chem. Senses*, **2009**, *34*, 695–703 doi:10.1093/chemse/bjp056.

172. M.-B. Genter, and S. F. Ali, *Neurobiol. Aging*, **1998**, *19*, 569–574.

173. T. V. Getchell, X. Peng, A. J. Stromberg, K. C. Chen, G. C. Paul, N. K. Subhedar, D. S. Shah, M. P. Mattson, and M. L. Getchell, *Aging Res. Rev.*, **2003**, *2*, 211–243.

174. T. V. Getchell, N. S. Krishna, N. Dhooper, D. L. Sparks, and M. L. Getchell, *Ann. Otol. Rhinol. Laryngol.*, **1995**, *104*, 47–56.

175. N. E. Rawson, G. Gomez, B. J. Cowart, A. Kriete, E. Pribitkin, and D. Restrepo, *Neurobiol. Aging.*, **2012**, *33(9)*, 1913–1919 doi: 10.1016/j.neurobiolaging.2011.09.036.

176. J. K. Kalmey, J. G. Thewissen, and D. E. Dluzen, *Anatomical Record*, **1998**, *251*, 326–329.

177. E. Meisami, L. Mikhail, D. Baim, and K. P. Bhatnagar, *Ann. N. Y. Acad. Sci.*, **1998**, *855*, 708–715.

178. C. G. Smith, *J. Comp. Neurol.*, **1942**, *77*, 589–594.

179. J. W. Hinds and N. A. McNelly, *J. Comp. Neurol.*, **1981**, *203*, 441-453.

180. K. K. Wong, M. L. Muller, H. Kuwabara, S. A. Studenski and N. I. Bohnen, *Neurosci. Lett.*, **2010**, *484*, 163–167.

181. C. R. Schubert, K. J. Cruickshanks, B. E. Klein, R. Klein, and D. M. Nondahl, *Laryngoscope*, **2011**, *121(4)*, 873–878 doi: 10.1002/lary.21416.

182. N. E. Rawson and A. -S. LaMantia, *J. Neurobiol.* **2006**, *66(7)*, 653–676 doi: 10.1002/neu.20236.

183. N. E. Rawson and A. -S. LaMantia, *Experimental Gerontology*, **2007**, *42(1–2)*, 46–53.

184. K. K. Yee and N. E. Rawson, *Brain Res.*, **2000**, *124*, 129–132.

185. N. Etchamendy, V. Enderlin, A. Marighetto, R. M. Vouimba, V. Pallet, R. Jaffard, and P. Higueret, *J. Neurosci.*, **2001**, *21*, 6423–6429.

186. R. D. Prediger, L. C. Batista, and R. N. Takahashi, *Neurobiol. Aging*, **2005**, *26*, 957–964.

187. G. S. Chiu, D. Chatterjee, P. T. Darmody, J. P. Walsh, D. D. Meling, R. W. Johnson, and G. G. Freund, *J. Neurosci.*, **2012**, *32(40)*, 13945–13955 doi: 10.1523/JNEUROSCI.0704-12.2012.

188. M. J. Spinetta, M. T. Woodlee, L. M. Feinberg, C. Stroud, K. Schallert, L. K. Cormack, and T. Schallert, *Psychopharmacol.*, **2008**, *201(3)*, 361–371.

189. E. Marchetti, M. Jacquet, G. Escoffier, M. Miglioratti, A. Dumuis, J. Bockaert, and F. S. Roman, *Brain Res.*, **2011**, *1405*, 49–56.

190. W. C. Watt, H. Sakano, Z. Y. Lee, J. E. Reusch, K. Trinh, and D. R. Storm, *Neuron*, **2004**, *411*, 955–967.

191. Aging Well with Your Sense of Smell www.senseofsmell.org/resources/publications.php.

192. B. Gopinath, K. J. Anstey, A. Kifley, and P. Mitchell, *Maturitas*, **2012**, *72(1)*, 50–55 doi: 10.1016/j.maturitas.2012.01.009.

193. B. Gopinath, C. M. Sue, A. Kifley, and P. Mitchell, *J. Gerontol. A Biol. Sci. Med. Sci.*, **2012**, *67A(2)*, 204–209 doi: 10.1093/gerona/glr165.

194. C. Warren and S. Warrenburg, "Mood benefits of Fragrance" in "Aroma-chology: the impact of science on the future of Fragrance" Symposium, November 12 1991.

195. R. A. Baron, *Aroma-Chology Review*, **1997**, *6(1)*, 3,10-11.

196. S. S. Schiffman and E. A. Sattely-Miller, Pleasant odours improve mood of men as well as women at mid-life, in *Compendium of Olfactory Research*, A. Gilbert, Ed., Kendall/Hunt Publishing co., Dubuque, Iowa, 1995.

197. W. Redd and S. Manne, Using aroma to reduce distress during magnetic resonance imaging. In A. Gilbert, Eds, *Compendium of Olfactory Research 1982–1994*, Dubuque, Iowa: Kendall/Hunt Publishing Company, 1995, 47–52.

198. (Report on 54th Annual Fragrance Symposium of the American Society of Perfumers, Perfumer and Flavorist, 2008, *33*, 24–29).

199. Euromonitor, 4th September 2007.

200. K.-H. Plattig, G. Kobal, and W. Thumfart, *Zeitschr. Gerontol*, **1980**, *13(2)*, 149–157.

201. S. Sahadevan, S. M. Saw, W. Gao, L. C. S. Tan, J. J. Chin, C. Y. Hong, and N. Venketasubramanian, *J. Amer. Geriatrics Soc.*, **2008**, *56(11)*, 2061–2068.

202. B. J. Gurland, D. E. Wilder, R. Lantigua, Y. Stern, J. Chen, E. H. P. Killeffer, and R. Mayeux, *Int. J. Geriatric Psychiatry*, **1999**, *14(6)*, 481–493.

203. Soap, Perfum. Cosmet., October **2006**.

Chapter 5

The Scents of Nature

BASIC REQUIREMENTS FOR ODOUR

As discussed in the previous chapter, for humans, odorous substances must be volatile enough to allow sufficient numbers of their molecules to reach the olfactory epithelium. The two key parameters affecting volatility are vapour pressure and the balance between hydrophobicity and hydrophilicity, usually denoted by the log P value. Vapour pressure in turn depends partly on molecular weight. We therefore find that substances with a molecular weight over 300 Da are too poorly volatile to serve as odorants. Substances with a low log P, that is, those that are more water-soluble than oil-soluble, usually experience stronger inter-molecular forces such as hydrogen bonds, and this has the effect of reducing vapour pressure since more energy is required to escape from the solid or liquid phases. Since the odorants must cross the mucus layer in order to reach the receptors, their water solubility is an important factor in its own right. Putting all of this together means that the majority of odorant molecules are based on carbon, have no more than 18 carbon atoms in their structure and, more often than not, have only one functional group, most likely an oxygen-containing functionality. Of course, there are exceptions to this latter generalisation, such as the nitromusks or vanillin, but these are exceptions.

WHY NATURE MAKES VOLATILE CHEMICALS

Volatile organic compounds are produced in large variety and quantity in nature and for many different reasons. In the first chapter, the deliberate production of volatile chemicals for communication between insects was discussed. However, the bulk of volatile organic substances serve other purposes. Monoterpenes such as myrcene (**5.1**) and α-pinene (**5.2**) are produced by trees in response to injury. These molecules form part of the rosin that exudes from wounds in the tree and their susceptibility to autoxidation means that they can polymerise to form a protective

Chemistry and the Sense of Smell, First Edition. Charles S. Sell.
© 2014 John Wiley & Sons, Inc. Published 2014 by John Wiley & Sons, Inc.

barrier. Chemically reactive precursors such as isopentenyl pyrophosphate (**5.3**) (present in the exudates of pine and rubber trees) can also polymerise to form a protective barrier over a wound. Other molecules serve as chemical defence agents through anti-bacterial or anti-fungal properties, thus protecting the plant from pathogens. Many of the odorous components of oils such as cinnamon, frankincense and myrrh fall into this category. For example, myrrh is the resin of the shrub *Commiphora abyssinica* and contains a number of anti-bacterial or anti-fungal components such as the eudesmane derivative (**5.4**). Cineole (**5.5**) is produced in large quantity by many eucalyptus species, and it is thought that its role is to provide a highly flammable atmosphere around the trees so that, in the event of a forest fire, the oxygen in the air around the tree is depleted rapidly thus limiting damage to the tree. Isopentenyl pyrophosphate (**5.3**) is the key intermediate in terpenoid biosynthesis and so is produced in great quantity. Elimination of the pyrophosphate produces isoprene (**5.6**). Such production is possibly largely coincidental but it serves to release 100 million tonnes of isoprene into the atmosphere annually from trees such as poplar and it has been shown to be a contributing factor to smog in California (1). Some odorants are produced as a result of degradation processes. The carotenoids are important pigments in nature and are insufficiently volatile to produce odours. However, degradation of carotenoids gives us some of nature's finest odours. The ionones found, for example, in violets and boronia, the damascones found in roses and other sources and the theaspiranes produced by tobacco and tea are all carotenoid breakdown products. The principal odour components of ambergris and orris are degradation products of other terpenoids. Many malodorous chemicals are breakdown products. The foul smells associated with spoilt food and faeces alert us to bacterial contamination and a consequent health risk. The amines present in urine and rotting fish are another set of warning signals. The smell of sweat results from decomposition of human chemicals by bacteria. The odour of the cocktail of sweat chemicals has been shown to be characteristic of the individual producing the precursor secretion and so could possibly play (or have played) a role in recognition (Figure 5.1).

Although nature does not produce this vast array of volatile organic chemicals for the benefit of humans, we have adapted to use them to give us useful information about our environment and to delight us aesthetically.

The primary metabolites of nature, that is, those that are essential to all living organisms, can be grouped into four main classes: proteins, nucleic acids, carbohydrates and lipids. Degradation and reaction products of these are important in forming the aroma of many foodstuffs. For example, the aromas of coffee and roast meat are largely the result of the Maillard reaction between proteins and carbohydrates. Degradation of lipids gives us a variety of smells from the pleasant notes of fresh air (due to aliphatic aldehydes) to the foul stench of stale sweat arising from fatty acids in the C_4 to C_{10} range. However, it is the secondary metabolites, in particular the terpenoids, shikimates and polyketides, that provide us with most of nature's scents. Secondary metabolites are those that are found in certain organisms but not others. Variations in production of secondary metabolites can be used

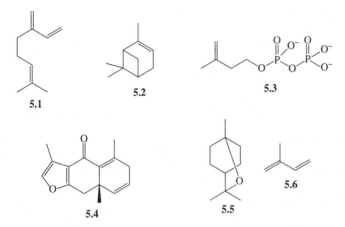

Figure 5.1 Isopentenyl pyrophosphate and some volatile terpenoids.

to classify organisms, especially plants, and this approach is known as *chemotaxonomy*. As far as fragrance is concerned, the terpenoids are, by far, the most important group, followed by the shikimates.

The principal routes by which all secondary metabolites are synthesised in nature all start from glucose (**5.7**). Glucose is formed from carbon dioxide and water by plants and photosynthetic algae. This is an unfavourable reaction and the energy required to drive it forward comes from sunlight which is trapped by the porphyrin chromophore of chlorophyll. Glycolysis, that is, the cleavage of the central C–C bond of glucose, produces phosphoenol pyruvate (**5.8**), which is one of the key starting materials for the shikimates, the other being erythrose 4-phosphate(**5.9**), another glucose breakdown product. Decarboxylation of phosphoenolpyruvate and subsequent esterification with coenzyme-A (CoA) (**5.10**) gives acetyl CoA (**5.11**), the direct precursor for the lipids and polyketides and also, via mevalonic acid (**5.12**), the terpenoids. These conversions are shown in Figure 5.2.

Decarboxylation and dehydration of mevalonic acid produces the terpenoid precursor isopentenyl pyrophosphate (**5.3**) in what is known as the *MVA route*. An alternative to this route is known as the *MEP route* and starts from pyruvate and glyceraldehyde 3-phosphate. The key intermediate is 2-C-methyl-D-erythritol 4-phosphate (MEP) (**5.13**) which is also converted to isopentenyl pyrophosphate (**5.3**). Green algae use the MEP pathway, whereas archea, fungi and animals use the MVA pathway and plants and bacteria use both.

The onward routes from the precursors illustrated in Figure 5.2 will be described below under the headings of terpenoids, shikimates and lipids and polyketides. All of these biochemical reactions use the same basic chemistry as that used by synthetic chemists in the laboratory, but in nature they are directed by enzymes which serve the triple role of reducing entropy by bringing the reagents together in the right configuration, lowering the enthalpic barrier to forming transition states and directing the stereochemical course of the reaction.

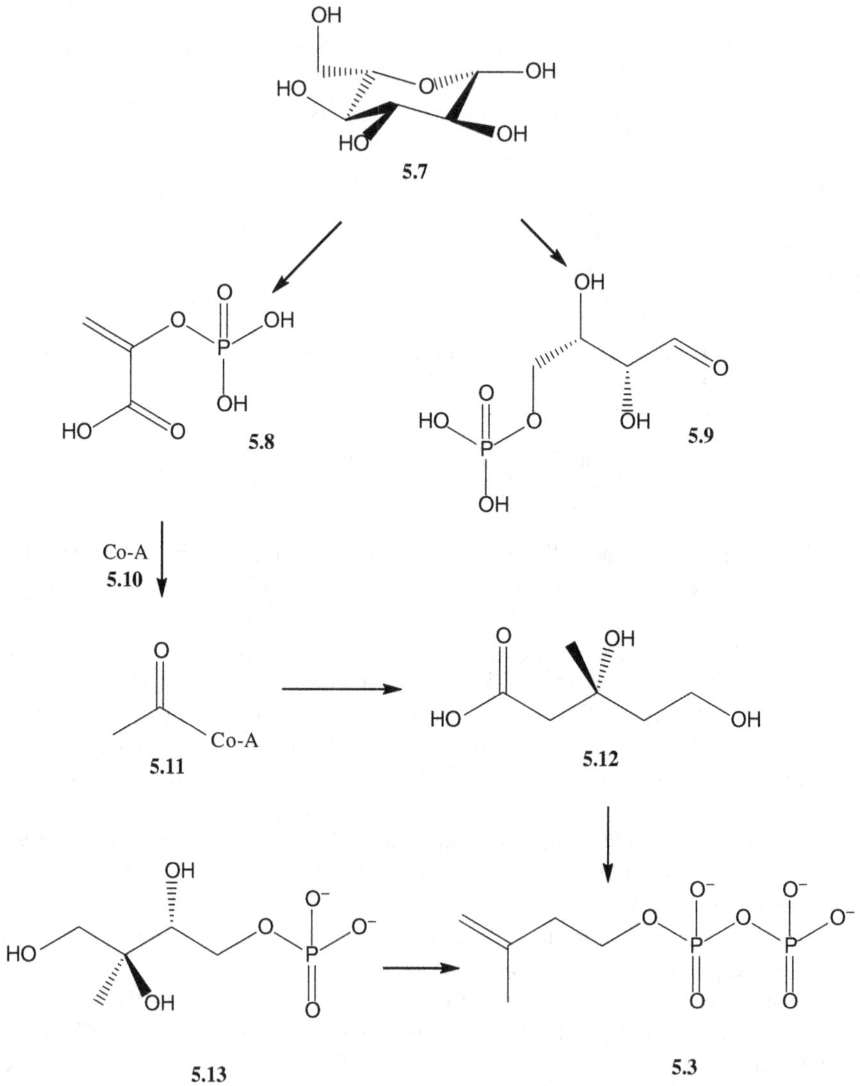

Figure 5.2   Glucose as starting material for secondary metabolites.

Thus the range of enzymes that are available to the plant determines the range of secondary metabolites and, since the presence of the enzymes is dependent on the plant's DNA, the link is established between plant genotype and the selection of secondary metabolites it produces. Co-factors, also known as *co-enzymes* are sometimes required by enzymes. Of particular importance in biosynthesis are CoA (**5.10**), which activates the carbonyl carbon atom of acids to nucleophilic attack and activates the adjacent hydrogen atoms to abstraction by base; nicotinamide adenine dinucleotide phosphate (NADP) (**5.14**) and its reduced form (NADPH) (**5.15**), which serve as redox reagents through hydride transfer; and adenosine triphosphate (ATP) (**5.16**), which, through phosphorylation of alcohols, makes them susceptible to nucleophilic attack. These co-enzymes are shown in Figure 5.3. Readers interested in further detail on biogenesis are referred to the books by Bu'Lock (2) and Mann et al. (3) and the review by Croteau (4) dealing with terpenoid biosynthesis.

TERPENOIDS

Terpenoids (sometimes also referred to as *isoprenoids*) are defined as materials whose molecular structures contain carbon backbones which are made up of iso-prene (2-methylbuta-1,3-diene) (**5.6**) units. Therefore, the number of carbon atoms in any terpenoid will be a multiple of five, although degradative and other metabolic processes can affect this. In such cases, the overall structure will still indicate the substance's terpenoid origins and they will still be considered to be terpenoids. The 'isoprene' skeletal feature stems from their biosynthesis from mevalonic acid. This is the key biosynthetic route to terpenoids although an alternative has also been identified. (5) Terpenoids serve a wide range of functions in nature includ-ing defensive resins, pheromones, anti-oxidants and the pigments responsible for vision. Volatile terpenoids are important as natural chemical signals, and it is not surprising that they have also reached prominence in the flavour and fragrance industry.

The first terpenoids to be studied in detail contained two isoprene units and were called *monoterpenoids*. Consequently, the basic nomenclature system is based on 10 carbon units rather than the 5 of isoprene, and this is shown in Table 5.1. The term *terpene* was originally applied to the hydrocarbons found in turpentine and similar oils but is sometimes applied more generally to any unsaturated terpenoid hydrocarbon. Usually, it will indicate an unsaturated monoterpenoid hydrocarbon, the higher non-oxygenated homologues being referred to as *sesquiterpenes, diter-penes*, and so on.

Carotenoids are a specific class of tetraterpenoids as will be explained below. Steroids are a subgroup of triterpenoids and are defined as those that produce Diels's hydrocarbon when distilled from zinc dust. Because of the issue of volatil-ity as described above, the volatile members of the terpenoid family are almost all found in the hemiterpenoid, monoterpenoid and sesquiterpenoid families. The prime exceptions are certain degradation products of larger terpenoids and a few odorous steroids. This text will therefore concentrate on the first three terpenoid classes.

The junction between isoprene units is not random but most often is formed through the so-called head-to-tail coupling as shown in Figure 5.4. In certain cases, a tail-to-tail coupling occurs, also shown in Figure 5.4. This coupling is a charac-teristic feature of the central coupling used to form the precursors for the steroids and carotenoids as will be seen later. The explanation of the coupling systems lies in the biosynthesis as described below.

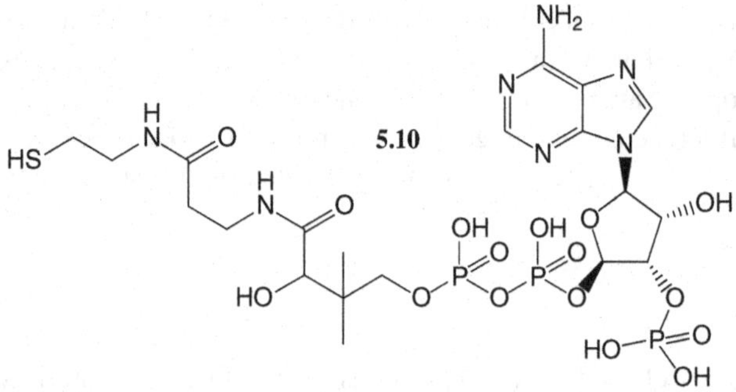

Figure 5.3   Key coenzymes in biosynthesis.

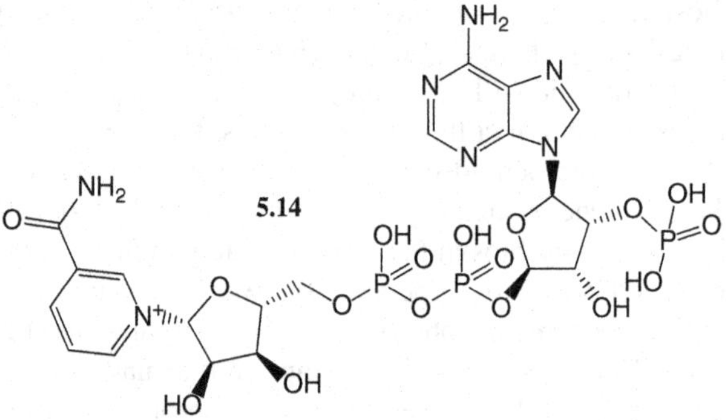

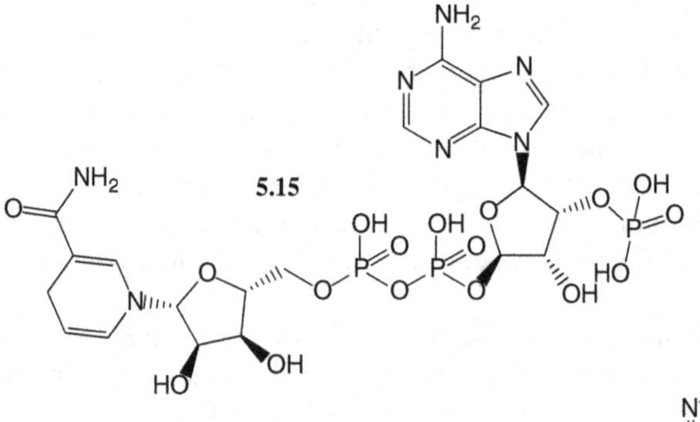

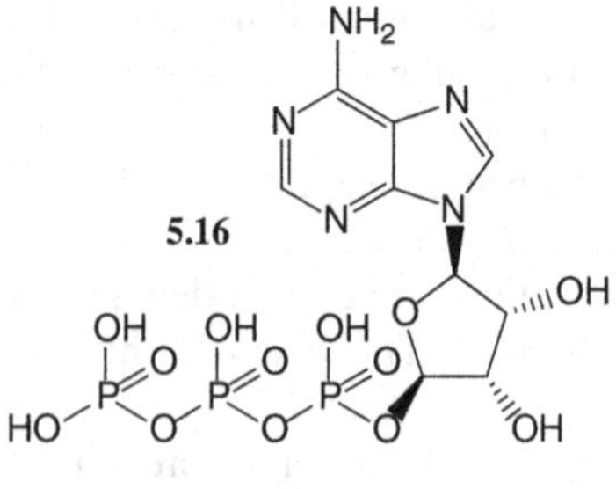

Table 5.1   Classification of Terpenoids

Name	No. of isoprene units	No. of carbon atoms
Hemiterpenoids	1	5
Monoterpenoids	2	10
Sesquiterpenoids	3	15
Diterpenoids	4	20
Sesterterpenoids	5	25
Triterpenoids	6	30
Tetraterpenoids	8	40
Carotenoids	8	40
Polyisoprenoids	> 8	> 40

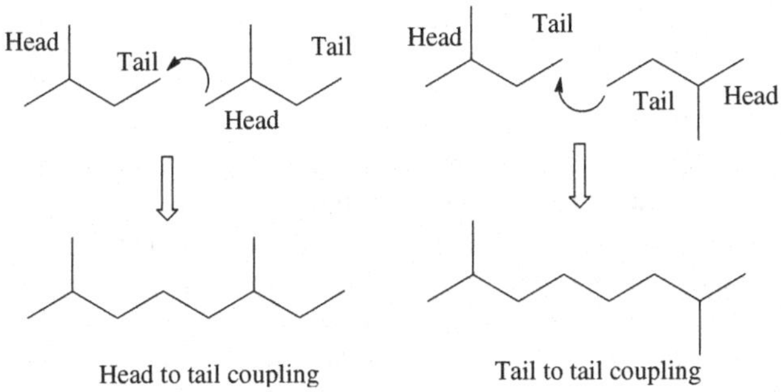

Figure 5.4   The two modes of coupling of isoprene units.

After formation of a linear skeleton, the chain may be cross-linked to produce rings, and further classification of terpenoids is based on the resulting ring systems. Some common ring systems of importance in the chemistry of fragrant terpenoids are shown in Figure 5.5. A clear listing of the known ring systems can be found in the book by Devon and Scott (6).

Both linear and cyclic structures can be functionalised by the introduction of oxygen or other heteroatoms. As the IUPAC names of terpenoids are often rather unwieldy, trivial names are in common use. The original trivial names usually indicate a natural source of the material, for example, pinene from pine species. Semi-systematic names are often based on the ring system and the oxygenation pattern. For example, 1-methyl-4-isopropylcyclohexane is referred to as *p-menthane* and numbered as shown in Figure 5.5. Greek letters are sometimes used to indicate the order in which the isomers were discovered or their relative abundance in an essential oil and sometimes to refer to the location of the double bond in isomeric olefins.

Terpenoids can be analysed by the usual methods. For the volatile members of the family, gas chromatography-mass spectrometry (GC/MS) is a particularly useful tool. In laboratories (e.g. those in the major fragrance companies) that are

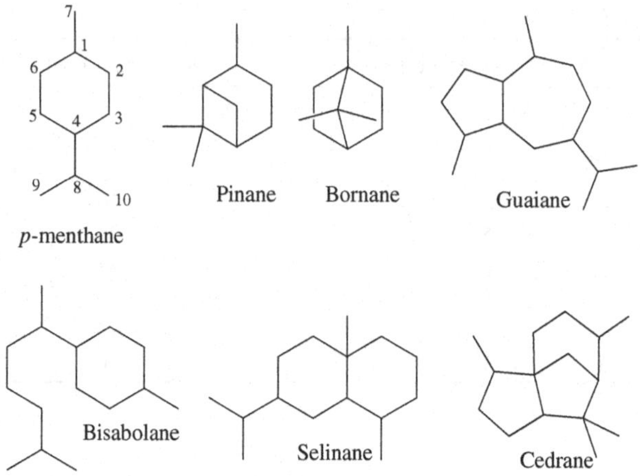

Figure 5.5   Some common terpenoid skeletons.

accustomed to analysing mixtures of volatile terpenoids, GC/MS is the major analytical technique employed, and such laboratories will have extensive libraries of the mass spectra of terpenoids to assist in this. However, the mass spectral fragmentation patterns of closely related terpenoids are often so similar as to render definitive identification by MS alone impossible. For these materials and those for which there is no reference (e.g. compounds newly isolated from nature), nuclear magnetic resonance (NMR) spectroscopy is the analytical tool of choice. Physical techniques such as density, refractive index and optical rotation are relatively inexpensive and prove useful in quality control.

Being secondary metabolites, individual terpenoids may be common to a number of species or may be produced by only one organism. Comparison of the terpenoids produced by a plant will give an indication of which metabolic pathways operate in it and can therefore be used to aid in the classification of species, a procedure known as *chemical taxonomy*.

Terpenoid Biosynthesis

The two key building blocks for terpenoids in nature are isopentenyl pyrophosphate (**5.3**) and prenyl pyrophosphate (**5.17**). Coupling of these two isomeric materials under enzymic control, as shown in Figure 5.6, gives geranyl pyrophosphate (**5.18**), the precursor for all monoterpenoids. Subsequent addition of isopentenyl pyrophosphate (**5.3**) to geranyl pyrophosphate (**5.18**) in a similar way produces farnesyl pyrophosphate (**5.19**) and hence sesquiterpenoids. Further repetition of the process leads on to the diterpenoids and higher.

The mechanism of the tail-to-tail coupling is shown in Figure 5.7 using generalised R and R_1 groups to represent two terpenoid chains. When both chains are C_{15} (one farnesyl and one nerolidyl), the product is squalene (**5.20**), which is the

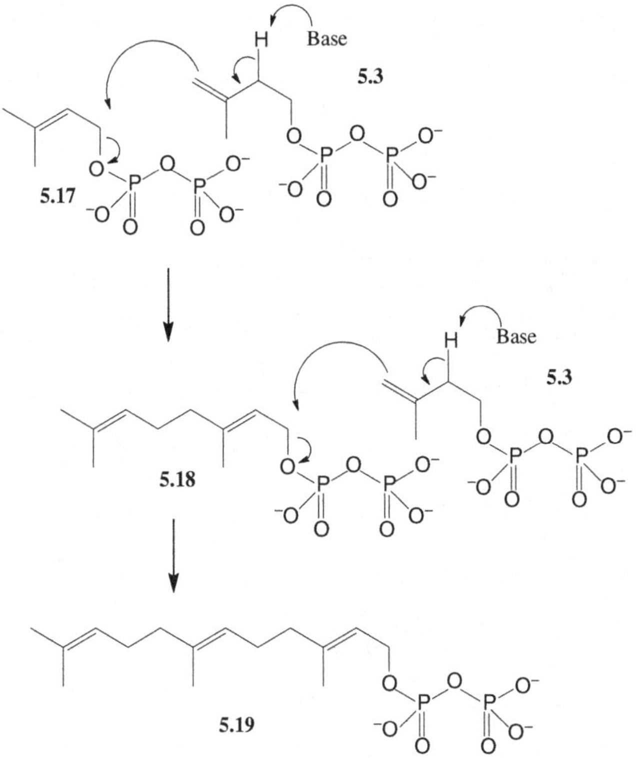

Figure 5.6   Biosynthesis of basic linear terpenoid skeletons.

biogenetic precursor for the steroids. When both chains are C_{20} (one geranylger-anyl and one geranyllinalyl), the product, after dehydrogenation, is lycopene (**5.21**), which is the precursor for the carotenoids. A few steroids and a larger number of carotenoid degradation products are important odorants in nature.

The linear terpenoid precursors can undergo many different cyclisation processes. The longer the chain, the more different possibilities there are for cyclised structures. The cyclisation processes are essentially carbocation reactions, with the initial cation being formed by loss of the pyrophosphate residue from the parent linear structure. Some cyclisation processes for monoterpenoids are shown in Figure 5.8 and for sesquiterpenoids in Figure 5.9.

Hemiterpenoids

As mentioned earlier, isoprene (**5.6**) is produced by many species of tree, but its volatility means that it is not often found in essential oils. Prenol (**5.22**), isoprenol (**5.23**) and esters of these are found in various oils such as ylang in which prenol,

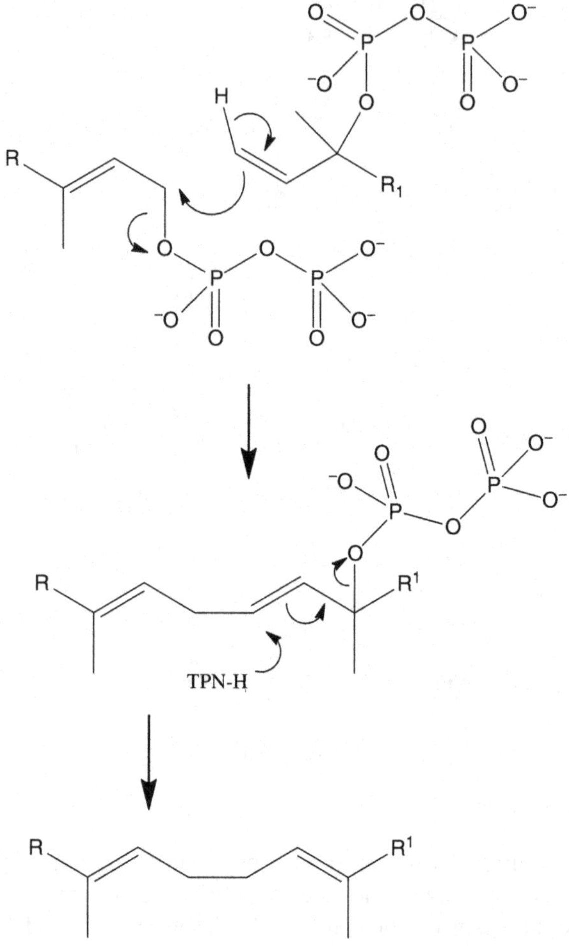

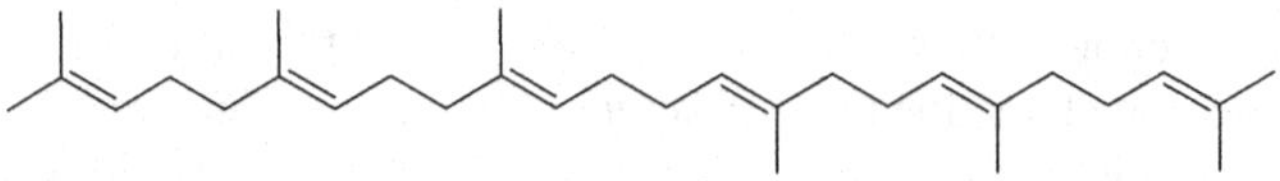

Figure 5.7   Tail to tail coupling in terpenoids.

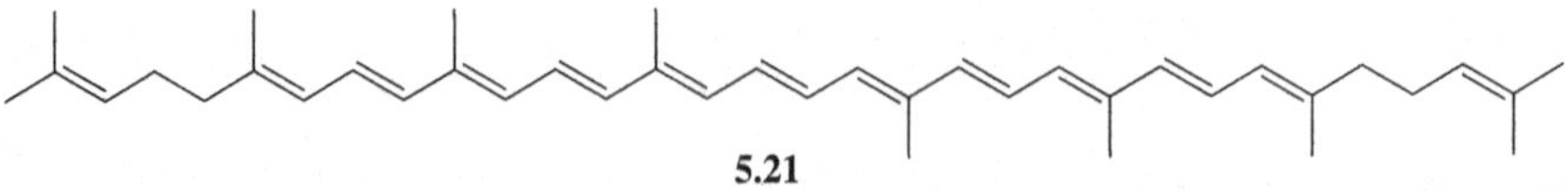

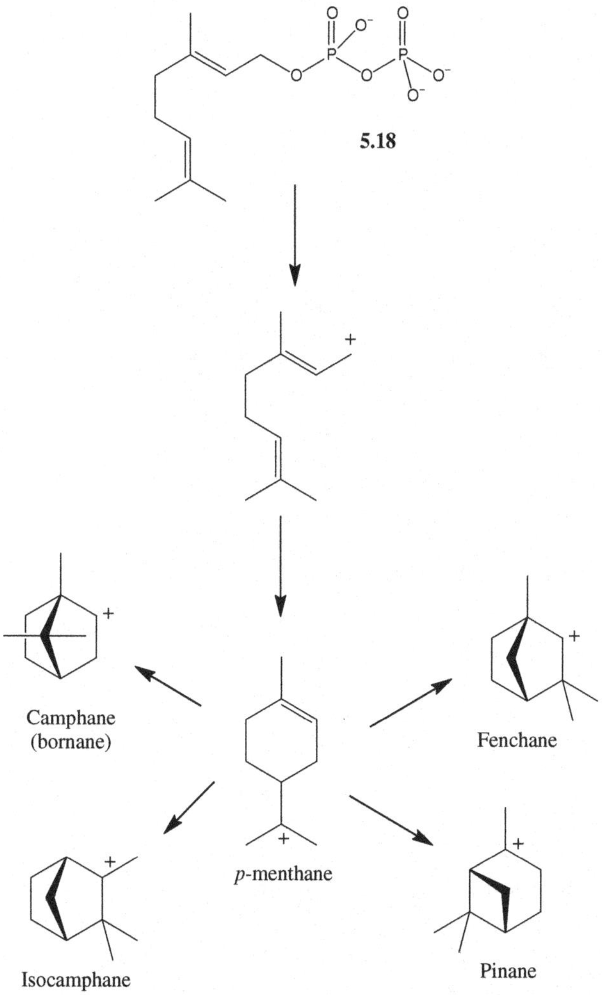

5.18

Figure 5.8   Biosynthetic routes to some monoterpenoid skeleton types.

its acetate and benzoate have all been found. Isoprenol and its acetate have been found in jonquil (daffodil) (Figure 5.10).

Monoterpenoids

Myrcene (**5.1**), 7-methyl-3-methylene-1,6-octadiene, is also known as β-*myrcene* and is very widespread in nature. This is not surprising as it is formed in nature by elimination from geranyl pyrophosphate, the precursor of all monoterpenoids. It can also be formed by elimination of water from alcohols such as geraniol or

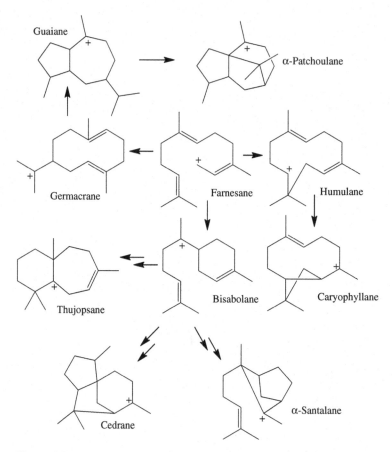

Figure 5.9   Biosynthetic routes to some sesquiterpenoid skeleton types.

5.22   **5.23**

Figure 5.10   Hemiterpenoid alcohols.

linalool and so its presence in natural extracts may be as an artefact (formed during the extraction process) rather than as a genuine plant metabolite.

Both geometric isomers of β-ocimene (**5.24, 5.25**) occur extensively in nature, whereas the isomers of α-ocimene (**5.26, 5.27**) and allo-ocimene (**5.28, 5.29**) occur in a more limited range of species. None of them is extracted or used commercially to a significant extent (Figure 5.11).

Limonene is very widespread in nature. The richest sources are the oils contained in the peel of citrus fruits, which contain levels up to 90%. The major

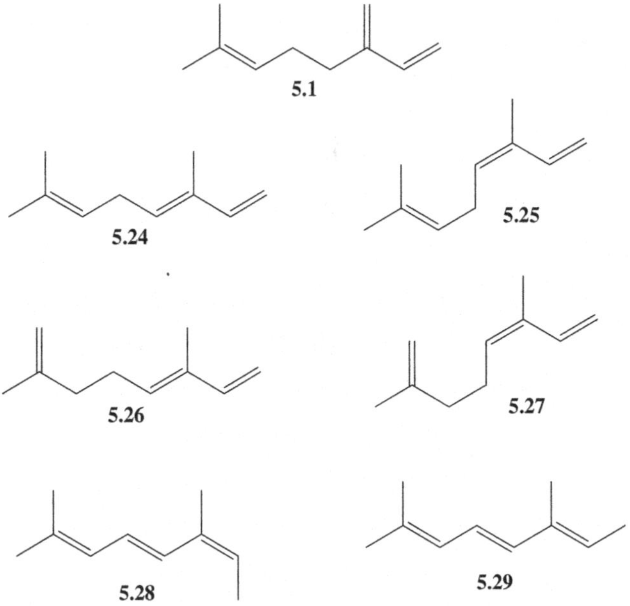

Figure 5.11 Linear monoterpenoid hydrocarbons.

source of limonene is indeed from citrus peel, largely as a by-product of the fruit juice industry, and production exceeds 50,000 tonnes per annum. Citrus fruit produce the (*R*)-(+)-enantiomer (**5.30**) and so the bulk of commercially available limonene is dextrorotatory. The laevorotatory enantiomer is available but in much more restricted supply and at a higher price. (*R*)-Limonene is used as such in perfumery. It has little odour value of its own but contains trace odoriferous impurities originating from the oil, mostly orange, from which it was extracted (7,8). The chemistry of limonene has been reviewed by Thomas and Bessière (9).

Both α-phellandrene (**5.31**) and β-phellandrene (**5.32**) occur widely in essential oils. (−)-α-Phellandrene can be isolated from *Eucalyptus dives* oil. A particularly rich source of (*S*)-(−)-β-phellandrene (**5.33**) is the lodgepole pine, *Pinus contorta*. It is also found at a level of about 2% in south-eastern U.S. turpentine, and processing the turpentine gives a fraction containing about 28% (−)-β-phellandrene and 62% (−)-limonene (**5.34**). *p*-Cymene (**5.35**) has been identified in over 1800 essential oils and plant extracts, particularly rich sources being oregano and thyme. Since it is at a free energy 'well' and is readily formed by autoxidation of monoterpenoid olefins, care must always be taken in deciding whether its presence is natural or if it is an artefact of the extraction process.

The terms *dipentene* or *p-menthadienes* are used to indicate a mixture of monoterpenoid hydrocarbons and ethers usually produced as by-products from processes for the manufacture of other terpenoids. The major source is as a by-product from the manufacture of synthetic pine oil. Important components of

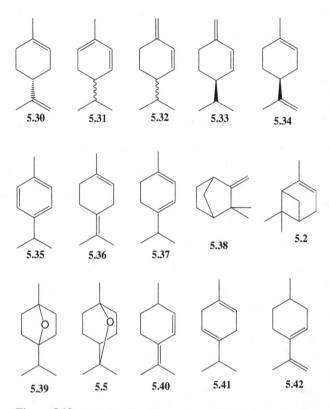

Figure 5.12   Some monocyclic monoterpenoid hydrocarbons and ethers.

dipentene are limonene (**5.30** and **5.34**), terpinolene (**5.36**), α-terpinene (**5.37**), camphene (**5.38**), 1,4-cineole (**5.39**), 1,8-cineole (**5.5**), α-pinene (**5.2**), *p*-cymene (**5.35**), 2,4(8)-*p*-menthadiene (**5.40**), α-phellandrene (**5.41**), β-phellandrene (**5.32**) and γ-terpinene (**5.41**). The *p*-menthadienes are isomerized to an equilibrium mixture by either strong acid or strong base. The equilibrium composition from acid-catalysed isomerisation has been determined as 53.3% α-terpinene (**5.36**), 13.8% γ-terpinene (**5.40**), 29.8% 2,4(8)-*p*-menthadiene (**5.39**) and 3.1% 3,8-*p*-menthadiene (**5.42**) (Figure 5.12) (10).

α-Pinene (**5.2**) is very widespread in nature. The most significant sources commercially are, of course the pine, fir and spruce species which are used to produce turpentine. Like the α-isomer, β-pinene occurs very widely in nature and, again like the α-isomer, the main source is from fractional distillation of turpentine. Camphene (**5.38**) is also widespread in nature. 3-Carene (also known as delta-3-carene or Δ-3-carene) (**5.43**) is less widespread in nature than the pinenes. However, it is found in many turpentines and can be extracted from them by fractional distillation. Turpentine from the western United States and Canada averages about 25% 3-carene; much of it is unutilised although it is obtained in high optical purity.

5.43   **5.44**   **Figure 5.13**   Isomerisation of 2-carene to 3-carene.

Turpentines from the Scandinavian countries, the CIS, Pakistan, and India all contain significant quantities of 3-carene. 3-Carene can be isomerised by base to give 2-carene (**5.44**), another natural compound although less widespread in occurrence that the 3-isomer (Figure 5.13).

Linalool (**5.45**) is more properly spelt linalöol and pronounced with two distinct o sounds, the first long and the second short. This gives an indication of one of its principal sources, linaloe oil, the essential oil of the Indian tree *Bursera delpechiana*, which contains levels of about 30% linalool and 45–50% of its acetate. However, it is common practice nowadays to omit the diaeresis and even to spell the name with a single letter o. Other synonyms include linalyl alcohol, Licareol (extract from rosewood) and Coriandrol (extract from coriander). It occurs very widely in nature. The richest source is Ho leaf oil from China and Taiwan, which typically contains over 90% linalool with levels as high as 97.5% having been reported (11). Rosewood oil will typically contain 75–85% linalool and it is a major component of many flower (e.g. about 80% in freesia and 75% in honeysuckle) and herb (e.g. 65–80% in coriander) oils. Linalyl acetate is also a frequently encountered component of plant oils. Oils in which it plays a particularly important organoleptic role include lavender (~50%) and citrus leaf oils (also ~50%).

Natural linalool is extracted mainly from three species. The largest is Brazilian rosewood of which about 100–150 tpa (tonnes per annum) was once produced. In recent years, there has been some concern about the endangering of the rosewood species by over-harvesting and consequently about the sustainability of production of the oil and so production has fallen. About 10 tpa is produced from Chinese and Taiwanese Ho leaf. Production from Linaloe oil is only a few tpa in India (12). Minor sources of natural linalool include shin and coriander oils (13). In the past, it was also extracted from lavender and bergamot.

The names of the isomeric alcohols geraniol and nerol are derived from those of the plants from which they were first isolated: geraniol (**5.46**) from geranium and nerol (**5.47**) from neroli (orange flower). Other names for geraniol-rich grades include Gerallol, Meranol, Reuniol, Rhodeanol, Rhodinol, Roseneone and Roseol, and for nerol-rich grades they include Allerol, Lorena, Neraniol, Nerodol, Nerolol and Nerosol (Figure 5.14).

These two materials are most easily dealt with together since they often occur together in nature and syntheses usually also produce both simultaneously. The equilibrium mixture is comprised of 60% geraniol and 40% nerol. The isomers can

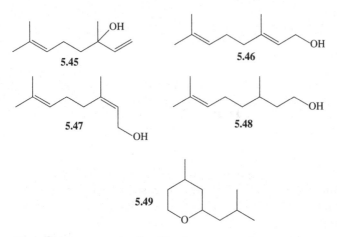

Figure 5.14   Major linear monoterpenoid alcohols and rose oxide.

be separated by efficient fractional distillation, and products of many compositions are available. The odour profile varies with the isomer ratio amongst other factors.

Both alcohols are widespread in nature, geraniol more so than nerol. The richest geraniol sources are *Monarda fistulosa* 93%, palmarosa (*Cymbopogon martini*) 80–85%, citronella (*Cymbopogon nardus, Cymbopogon winterianus*) 30%, lemongrass (*Cymbopogon citratus*) 30% and geranium 50%. Nerol is highest in rose, palmarosa, citronella and davana.

The most important commercial source of natural geraniol is citronella (especially *C. winterianus*) in which oil it is present at a level of about 30%. The oil also contains citronellol and citronellal, and the geraniol must be separated from these by fractional distillation. Production is estimated at about 5000 tpa (14). Other significant sources are geranium (150 tpa from an oil containing 15–80%), Jamrosa (*Cymbopogon jwarancusa*) (100 tpa from an oil containing 80–85%), palmarosa (*C. martini*) (80 tpa from an oil containing 75–85%) and Dhanrosa (*Cymbopogon flexuosus*) (10 tpa from an oil containing 80–85%). Nerol and linalool do not form complexes with calcium chloride, thus permitting isolation of a pure grade of geraniol from palmarosa. Esters of these alcohols also occur widely. Essential oils and citronella, palmarosa and geranium oils thus can also serve as sources of natural grades of the corresponding esters.

Geraniol possesses a mild, sweet floral odour which is distinctly rose in character and therefore is only used in rose perfumes. The odour of pure nerol is sweet, rosy refreshing and 'wet' seashore (15). The purer it is, that is, the less geraniol it contains, the more of the 'wet' character there is.

Citronellol (**5.48**) is widespread in nature in both enantiomeric forms, the richest sources being rose and geranium. It is one of the components responsible for the insect-repellent properties of citronella oil. Previously, it was extracted from various oils, especially citronella and geranium, or by hydrogenation of citronellal isolated from citronella oil. Nowadays, the major source, by far, is synthetic. Citronellol has a fresh rosy odour and finds extensive use in floral fragrances and also in flavours for citrus and other fruit notes.

Rose oxide (**5.49**) is found in a number of essential oils but particularly those of rose and geranium. It is used to give dry, green and rosy top-notes to fragrances. Racemates and optically pure (both laevo and dextro) forms are commercially available and used in fragrances.

α-Terpineol (**5.50**) is very widespread in nature, usually occurring at relatively low levels in oils. It is a major component of 'pine oil', a material well known from its use in pine disinfectants. Natural pine oil was a product derived from the extraction of aged pine stumps, and sulfate pine oil is a product separated from crude sulfate turpentine in about 5% yield.

The major component of pine oil is α-terpineol (**5.50**). Other components in the product include; β-terpineol (**5.51**), γ-terpineol (**5.52**), α-fenchol (**5.53**), borneol (**5.54**), terpinen-1-ol (**5.55**) and terpinen-4-ol (**5.56**), p-menthadienes (mainly limonene and terpinolene) and 1,4-cineole (**5.39**) and 1,8-cineole (**5.5**). The mechanisms of some of these reactions are shown in Figure 5.15. The ethers 1,4- and 1,8-cineole are also formed by cyclisation of the p-menthane-1, 4- and 1,8-diols. The bicyclic alcohols α-fenchol and borneol (endo-1,7,7-trimethyl-bicyclo[2.2.1]heptan-2-ol) are formed by the Wagner–Meerwein rearrangement of the pinanyl carbocation and subsequent hydration.

Pure α-terpineol has a delicate sweet floral, lilac-type odour (though only present at very low levels, if at all, in lilac). Its odour qualities are greatly affected by impurities, and many fragrance houses buy cheap grades and re-distill it to their perfumery quality. The monoterpenoid hydrocarbons tend to give a pine character, whereas the other alcohols and the phenols give a medicinal quality. Low cost, ready availability and stability to air, soap and household products make it a very useful ingredient.

α-Terpinyl acetate has a sweet, herbaceous, refreshing odour of the bergamot-lavender type similar to, but weaker than, that of linalyl acetate. It occurs in many oils such as cardamom, lovage and laurel.

There are eight isomeric forms of p-menthane-3-ol (**5.57**) of which (1R, 3R,4S)-(−)-menthol, usually known as l-menthol (**5.58**) is the commonest in nature and has the best organoleptic character of the eight. It has a good profile of clean, fresh and minty taste with a low contribution of the musty and bitter components that mar the profile of many of the other isomers. l-Menthol occurs primarily in plants of the genus *Mentha*. The two most important of these commercially are *M. arvensis* (cornmint) and *M. piperita* (peppermint). *M. arvensis* oil contains between 70% and 75% l-menthol and is by far the most important source of it. l-Menthol is known for its refreshing, diffusive odour characteristic of peppermint. However, its main uses stem from its physiological cooling effect. When applied to skin or mucus membranes, l-menthol creates the sensation of cooling independent of the actual temperature of the tissue concerned. This property makes it useful in toothpaste and other oral care products, in confectionery, tobacco, cosmetic products and in pharmaceuticals.

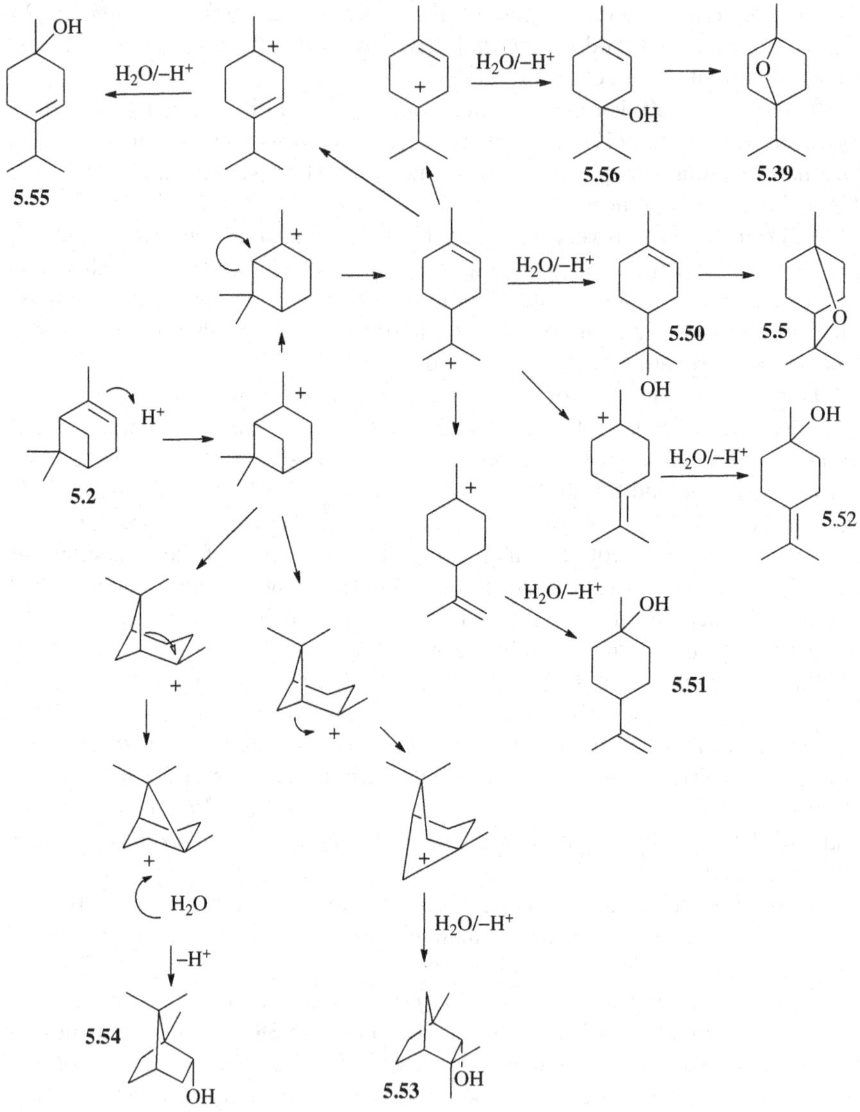

Figure 5.15   Formation of pine oil components.

Isopulegol occurs in some species including *Eucalyptus citriodora* and Citronella. It has been reported to have a minty herbaceous odour but Yamamoto has found that, when (−)-isopulegol (**5.59**) is very pure, both chemically and enantiomerically, it has virtually no odour. However, it does impart a feeling of freshness, crispness and coolness to citrus fragrances (16).

Other *p*-menthanes of significance in mint include the ketones menthone, isomenthone, pulegone, piperitone and carvone and the ether menthofuran.

Menthone is sometimes referred to as *trans-menthone* but the prefix is essentially redundant as the cis isomer is called *isomenthone*. Menthone and isomenthone can be interconverted by epimerisation, the equilibrium mixture containing about 70% menthone and 30% isomenthone. In fact, isolation of either in pure form is difficult because of the ease of interconversion. The direction of rotation of plane-polarised light switches on epimerisation so that *l*-menthone (**5.60**) isomerises to *d*-isomenthone (**5.61**) and *d*-menthone to *l*-isomenthone. Menthone is fairly widespread in nature, the *l*-isomer being commoner than the *d*-isomer. It is commonest in mints, pennyroyal and sages, which are also the oils containing the highest levels. Low levels are found in oils such as geranium and rose. Menthone is used in mint reconstitutions and to some extent in other essential oil reconstitutions and perfumes. The history of its uses and other properties have been reviewed by Clark (17).

(−)-Piperitone (**5.62**) is usually isolated from the oil of *E. dives* but some is also synthesised from limonene (13). It has a fresh minty, camphoraceous note and finds some use in perfumes (largely for masking) but more in flavours such as caraway and tarragon. It is also used as a mint flavour ingredient in oral care products. (+)-Pulegone (**5.63**) is present at about 75% in the oil of pennyroyal (*Mentha pulegium*), which is grown commercially in Southern Europe and North Africa (Figure 5.16).

Carvone, also known as *p-mentha-6,8-dien-2-one* and *carvol*, is a material of commercial importance and has been reviewed by Clark (18). There are two enantiomers, of which the (*R*)-(−)- is the commoner and is used in much greater quantities. If the stereochemistry is not specified, it is usually the (*R*)-(−)-enantiomer, commonly referred to as *l*-carvone, which is intended.

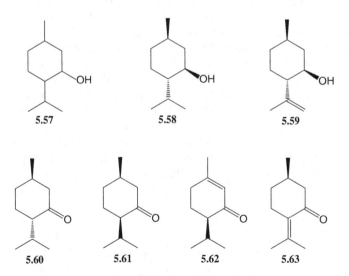

Figure 5.16   Some *p*-menthane alcohols and ketones.

Both isomers occur fairly widely in essential oils. The most significant natural sources of carvone are spearmint, dill and caraway. The term *spearmint* is applied to various *Mentha* species including *M. cardiaca, M. gracilis, M. spicata* and *M. viridis* and these usually contain 55–75% of the (*R*)-(−)-enantiomer (**5.64**). The (*S*)-(+)-enantiomer (**5.65**) is found in dill (*Anethum graveolens*) at levels of 30–65% and at 50–75% in caraway (*Carum carvi*). Carvone is used both as a synthetic material and in the form of essential oils. Very little is isolated and purified from natural sources. By far, the most important use of *l*-carvone is as a spearmint flavour in oral care preparations such as toothpastes and mouthwashes. Chewing gum is another significant use for *l*-carvone. The *d*-isomer is used in flavours, particularly for pickles, though usually in the form of dill or caraway essential oils. Carvone can be used to effect in small quantities in perfumery but this represents a very small fraction of total consumption.

Carvone is one of the classic examples of odour differences between enantiomers. In 1971, Friedman and Miller (19) interconverted the enantiomers using the reactions shown in Figure 5.17 and proved conclusively that the spearmint character of *l*-carvone and the dill/caraway character of *d*-carvone were intrinsic properties of the respective enantiomers and were not due to trace impurities. Thus epoxidation of *l*-carvone (**5.64**) gave (**5.66**), which could be reduced to *d-trans*-carveol (**5.67**) and that oxidised to *d*-carvone (**5.65**). Similarly, *d*-carvone could be converted via the epoxide (**5.68**) and *l-trans*-carveol (**5.69**) to *l*-carvone (**5.64**). In the same year, however, Leitereg et al. (20) carried out a well-constructed sensory evaluation of the carvone enantiomers and demonstrated that only about two-thirds of people can distinguish between them.

Menthofuran (**5.70**) occurs in a variety of oils, mostly in mints and it is one of the characteristic components of *M. piperita* (peppermint). It can be prepared from pulegone and is mainly used in reconstitutions of peppermint oil (13). The commonest isomer is the (+)-form. Menthofuran can also be prepared from pulegone by sulfonation with fuming sulfuric acid followed by pyrolysis of the resulting sultone. Use of (+)-pulegone as feedstock gives (+)-menthofuran (Figure 5.18).

Thymol is the trivial name for 2-isoproyl-5-methylphenol. Its natural sources, history, world consumption, substitutes, analogues, isomers and derivatives have been reviewed by Clark (21). Thymol (**5.71**) is found in a number of species, mostly from the *Thymus, Ocimum* and *Monarda* families. It takes its name from thyme (*T. vulgaris*) of which it is an organoleptically important component. The levels present vary widely not only from species to species but also from plant to plant within a species. As it is a phenol, it can be extracted from herb oils using aqueous sodium hydroxide and subsequent acidification. Such techniques were used to produce thymol in the past, particularly from thyme, oregano and basil. Material isolated in this way tended to contain some carvacrol (**5.72**). This is a disadvantage because the medicinal, phenolic and tarry odour of carvacrol spoils the sweeter, herbal and medicinal odour of thymol. Thymol has anti-bacterial, anti-fungal and anti-parasitic properties. It is less toxic than phenol, the LD_{50} of thymol being 980 mg/kg for rats in contrast to 530 mg/kg for phenol. Its antibacterial properties mean that it inhibits plaque formation and therefore it finds use in

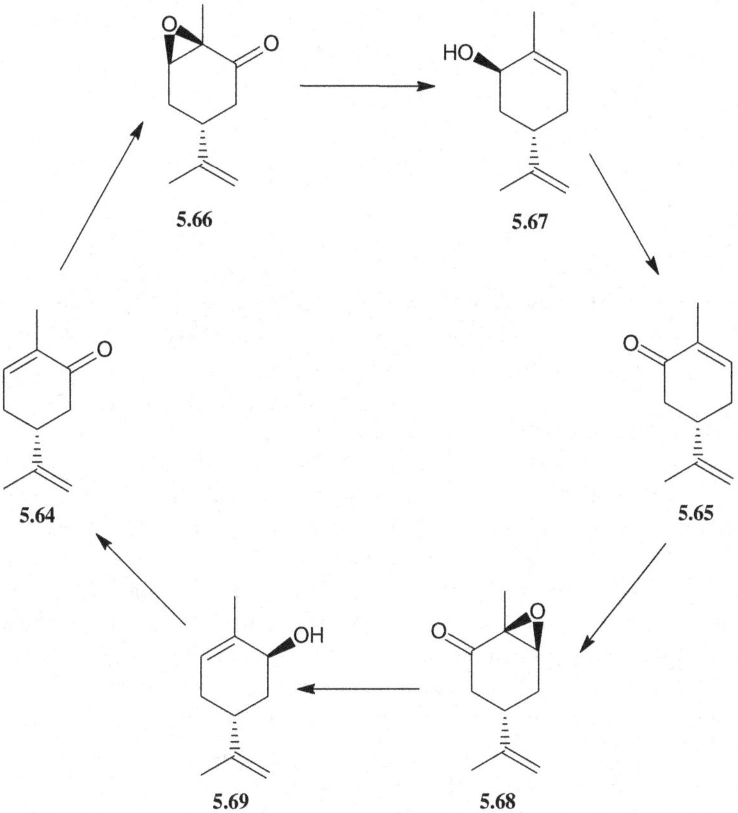

5.66

5.67

5.65

5.64

5.69

5.68

Figure 5.17   Interconversion of carvone enantiomers.

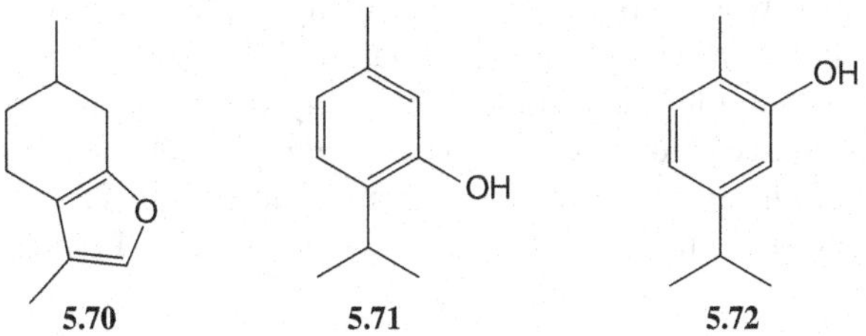

5.70   **5.71**   **5.72**

Figure 5.18   Menthofuran and the *p*-menthane phenols.

oral care applications. It has been used as a fungicidal treatment for fabrics and as an anthelmintic for both humans and animals. Synthetic thymol finds relatively little use in perfumes and flavours, oils such as thyme and basil being used in preference.

Borneol (*endo*-1,7,7-trimethylbicyclo[2.2.1]heptan-2-ol) (**5.54**), isoborneol (*exo*-1,7,7-trimethylbicyclo[2.2.1]heptan-2-ol) (**5.74**) and their acetates occur in a

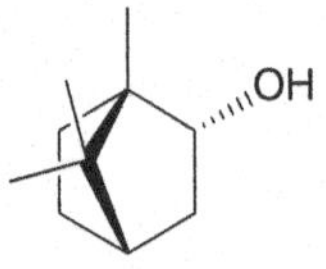

Figure 5.19 Borneols.

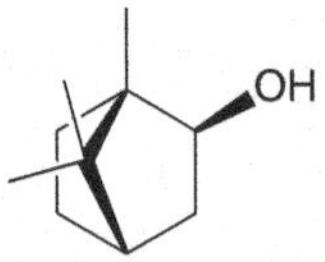

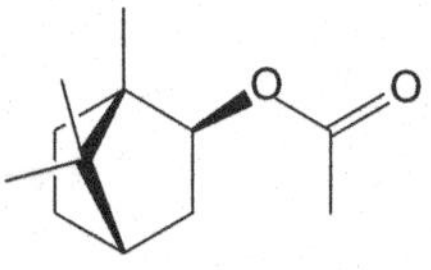

wide variety of herbs and other plants. They are used in perfumes for soaps and detergents for woody, camphor and pine notes with a relatively low cost. Two mono-ethers of isoborneol with glycols are used as fixatives in perfumery. The major product in perfumery, by far, is isobornyl acetate (**5.75**), borneol being used at about 1/10th the volume of the former (Figure 5.19).

1,8-Cineole (**5.5**) is also known as *eucalyptol, cajeputol, cajuputol, kajeputol,* 1,8-epoxy-*p*-menthane, 1,8-oxido-*p*-menthane and the anhydride of *p*-menthane-1,8-diol. It is very widespread in nature, which is not surprising since it is easily formed by trapping of the terpinyl carbocation by water or from acid-catalysed cyclisation of α-terpineol or similar reactions. Such reactions also explain its presence in pine oils and dipentene. Its natural sources, history, producers and capacity, pricing, imports, and substitutes, analogues and derivatives have been reviewed by Clark (22). Cineole has a camphoraceous odour strongly reminiscent of eucalyptus. It has some use in fragrance and in oral care preparations but the largest part is used in paramedical applications because of its anti-bacterial and decongestant properties. Since natural 1,8-cineole is readily available and inexpensive, commercial production is entirely from natural sources, primarily *Eucalyptus globulus* (Figure 5.20).

Citral exists as two geometric isomeric forms, geranial (**5.76**) (also known as *citral a*) which is the (*E*)-isomer and neral (**5.77**) (also known as *citral b*) which has the (*Z*)-configuration. When the word 'citral' is used, it usually implies a mixture of the two geometric isomers. Citral is widespread in nature, both isomers usually being present and in ratios varying considerably but usually within the 40 : 60 to 60 : 40 range. The richest sources are lemongrass (*C. citratus*), which contains 70–90% citral, and the fruit of *Litsea cubeba*, which typically contains 60–75%. It also occurs in lemon balm, ginger, basil, rose and, of course, citrus species. Lemons

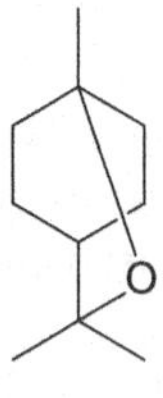

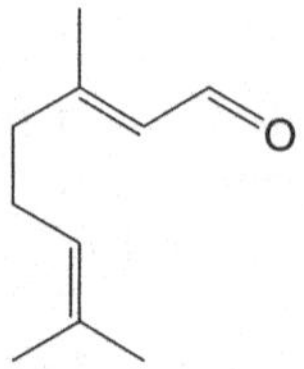

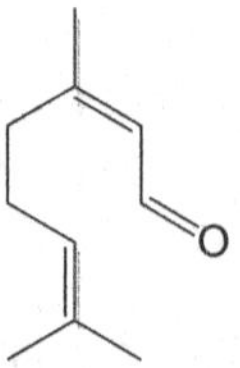

Figure 5.20 1,8-Cineole and citral.

usually only contain a few percent of citral, but it is the principal component responsible for their characteristic odour. Special grades are produced by extraction from lemongrass and *Eucalyptus staigeriana*, but the only natural source that competes economically as a source for citral as a feedstock is Chinese *L. cubeba*. Citral is used in flavours owing to its lemon character, but its use in fragrance is limited by its skin sensitisation potential. The major use of citral is an intermediate for production of other terpenoids, including vitamins.

The odour of citral is very desirable, as there is a strong association between lemon odour and cleanliness in the mind of the consumer. However, in addition to issues with skin safety, citral is not a very robust molecule. For example, even attempts to purify it by distillation are likely to lead to isomerisation (23) and addition of a mild acid such as ascorbic acid is needed to prevent this (24). In functional products, citral is subject to both acid and base reactions and to oxidation by air or by other components such as hypochlorite bleach. Consequently, a great deal of research has been conducted in the search for materials with the odour of citral but with much better stability.

Some of the first citral analogues to appear were its dimethyl acetal (**5.78**) and diethyl acetal (**5.79**). The acetal group is more stable than the aldehyde but the improved stability is bought at the price of a poorer match to the odour of lemon. A major breakthrough came with the discovery of geranyl nitrile (**5.80**). Geranyl nitrile has also been known under such names as Citraldia®, Citralva®, Citranile®, Citrylone®, geranionitrile, Geranitrile®, geranonitrile, Lemonitrile®, LRG 183, LRT 7 or LRT 8, although some of these have fallen out of use. Geranyl nitrile is an example of the many instances in which the nitrile function plays a similar role to the aldehyde group as far as odour is concerned, yet it is much more stable to acids, bases and oxidants. Thus, geranyl nitrile has an odour that is distinctly lemon in character and therefore became an important fragrance ingredient. However, more recently, concerns over its safety have led to its withdrawal as a fragrance ingredient. Double bonds are susceptible to attack by oxidants and also undergo acid-catalysed reactions, and so citronellyl nitrile (**5.81**) and tetrahydrogeranyl nitrile (**5.82**) have also been developed as fragrance ingredients. The latter is also known under the trade names Hypo-Lem® (IFF) and Virixal Nitrile® (PCAS). Citronitrile® (**5.83**) is an analogue of citral in which the isopropenyl group has been replaced by a benzene ring and the aldehyde by a nitrile (Figure 5.21).

Citronellal (**5.84**) occurs in a number of essential oils. The richest sources are *E. citriodora* (up to 85% citronellal content), some chemotypes of *Litsea cubeba* and citronella (*Cymbopogon nardus*) (typically 30–40% of the (+)-enantiomer). Swangi Leaf Oil (*Citrus hystrix*) is rich in (−)-citronellal, as it accounts for 60–80% of the oil obtained from the leaves (25). *Backhousia citriodora* contains up to 80% of the (−)-enantiomer. Natural grades of citronellal are commercially available from *E. citriodora* and citronella, but the major sources of commercial material are synthetic.

Hydroxycitronellal (**5.85**) is important in perfumery for its floral, muguet green and sweet odour which finds use in a wide range of floral fragrances. It

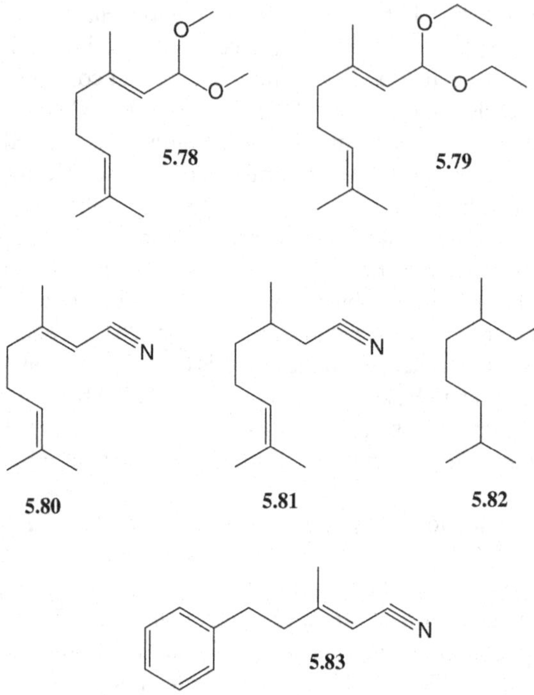

Figure 5.21   Citral analogues.

has been sold under many trade names such as Anthosal®, Centaflor®, Cyclalia®, Cyclia®, Cyclodor®, Cyclohydronal®, Fixol®, Fixonal®, Hycelea®, Hylea®, Majal®, Muguet synthetique®, Muguettine principe®, Storine®, Tilleul® and a special grade prepared from citronellal ex citronella was called *Laurine®*. Annual consumption is estimated at about 1000 tonnes (26) and the price would, typically be in the $12–15/kg range. It is valued for its floral odour which is strongly reminiscent of muguet and also has some lime blossom character. The dimethyl acetal and methyl anthranilate Schiff's base of hydroxycitronellal are also important fragrance ingredients. The latter (**5.86**), is available under trade names such as Anthralal®, Arangol®, Arerantae®, Aurangeol®, Auranol®, Aurantea®, Aurantein®, Aurantion®, Aurantine®, Aurantol®, Aurantolin®, Aurantorcol®, Aurentol®, Auriol®, Bigariol® and Bigaradia® (Figure 5.22).

Camphor (**5.87**) is widespread in nature in both enantiomeric forms. The richest source is the oil of camphor wood, *Cinnamomum camphora*, from which the (+)-enantiomer is extracted commercially (27). It is also an important contributor to the odour of lavender and of herbs such as sage and rosemary. Most synthetic camphor is produced from α-pinene (**5.2**) via camphene (**5.38**), and production runs to many thousands of tonnes/annum. It is used in perfumes but, more importantly, as a plasticiser, preservative, disinfectant and in paramedical applications. Its use

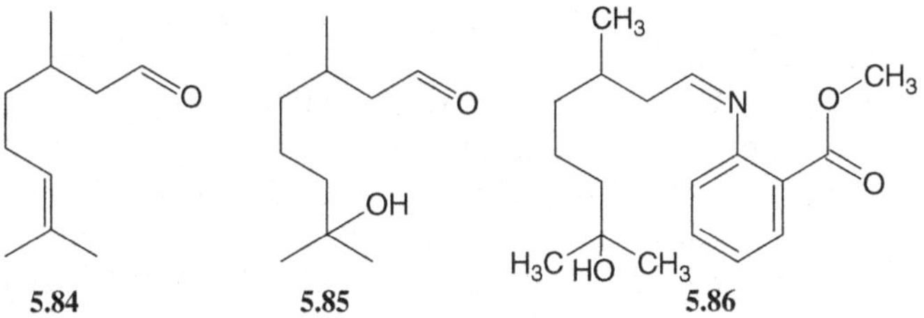

Figure 5.22 Citronellal and derivatives.

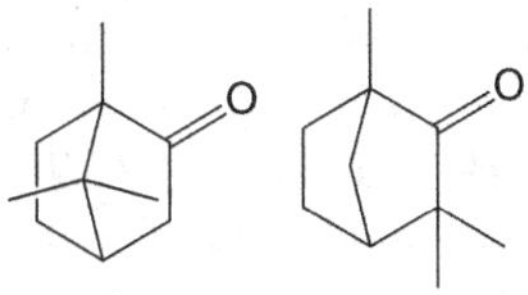

5.87   5.88

Figure 5.23 Camphor and fenchone.

in religious ceremonies in Asian countries is also a significant part of the total consumption (28).

Fenchone (**5.88**) occurs in a number of essential oils such as cedar leaf and lavender, and the (−)-isomer is particularly important in fennel. Synthetic material is made from pinene. It is used mainly in reconstitution of fennel oils (Figure 5.23).

Sesquiterpenoids

Sesquiterpenoids contain three isoprene units, and the precursor for them all, in nature, is farnesyl pyrophosphate (**5.19**) as shown in Figure 5.6. Because there are now three double bonds in the molecule, as opposed to the two of monoterpenoids, the variety of possible cyclic structures is much greater. Skeletal rearrangements, migrations of methyl groups and even loss of carbon atoms to produce norsesquiterpenoids all contribute further to the variety. There are probably over 3000 sesquiterpenes that have been isolated and identified in nature. A large number of sesquiterpenoids possess interesting biological activities but most are of academic interest only and have no commercial application outside folk medicine.

As a consequence of their higher molecular weight, sesquiterpenoids are less volatile than their monoterpenoid counterparts. This means that a smaller percentage find use in perfumery. However, some of those which do have an odour have low thresholds and/or high intensities. They often have fixative properties and are generally used as base notes, in particular as woody notes, in perfumes. The use of sesquiterpenoids in perfumery was reviewed by McAndrew (29).

Three of the most important sesquiterpenoid rich oils are sandalwood, patchouli and vetiver.

Sandalwood oil is obtained by distillation of the wood of the parasitic tree *Santalum album*. The isomeric alcohols α-santalol (**5.90**) and β-santalol (**5.91**) account for about 90% of the oil. Over-harvesting has taken the species to the brink of being endangered and supply is, consequently, now very short. Fortunately, the odour can be recreated using synthetic materials such as the isobornylcyclo-hexanols and the campholenic aldehyde derivatives syntheses of which will be described in Chapter 6. Details of sandalwood chemistry can be found in Chapter 6 of the book by Sell (30).

Patchouli oil is distilled from the fermented leaves of *Pogostemon cablin*. Its main constituent is patchouli alcohol (**5.92**), also known as *patchoulol*. However, this and the other major components have relatively little odour and the character-istic scent arises from some of the minor compounds present. Vetiver oil is distilled from the roots of the tropical grass *Vetivera zizanoides*. It contains a very large num-ber of sesquiterpenoids, α-vetivone (**5.93**), β-vetivone (**5.94**) and khusimol (**5.95**) accounting for about 35% but, as with patchouli, it is a number of minor compo-nents that are responsible for its valuable odour. Details of patchouli and vetiver chemistry can be found in Chapter 7 of the book by Sell (Figure 5.24) (30).

Caryophyllene (**5.96**), often referred to as β-*caryophyllene*, is very widespread in nature, occurring in such diverse plants as pepper, Melissa, ylang–ylang, marigold and the herbs sage and basil. The richest source is clove bud oil (*Eugenia*

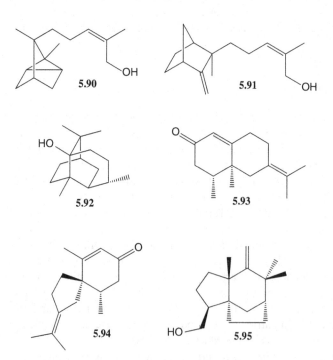

Figure 5.24   Sandalwood, patchouli and vetiver components.

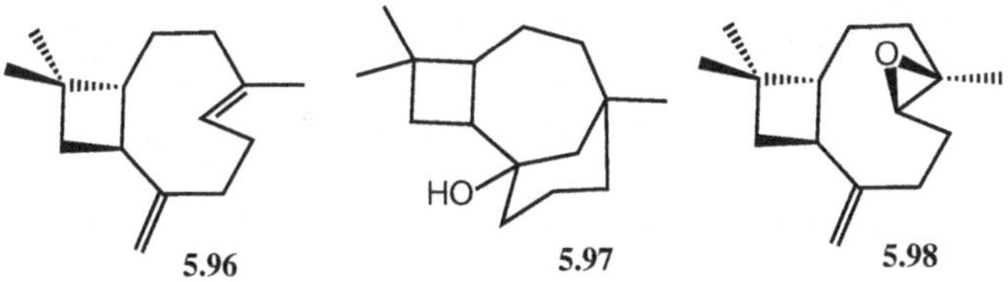

Figure 5.25   Caryophyllene and derivatives.

caryophyllata) from which it is extracted commercially, albeit as a by-product of eugenol production. It is used to produce a variety of materials that are used in perfumery. Examples include caryophyllene alcohol or caryolanol (**5.97**) (from acid catalysed hydration) and the epoxide (**5.98**) (from treatment with a per-acid) which has a woody odour with a hint of amber. The epoxide also occurs widely in nature and is useful in reconstituting essential oils. The caryophyllene molecule is very strained, having a trans-double bond in a nine-membered ring to which a four-membered ring is fused. Thus, many of its reactions proceed with rearrangement. For example, acetylation gives a mixture of ketones known as *acetylcaryophyllene*. A description of its chemistry can be found in Chapter 7 of the book by Sell (Figure 5.25) (30).

Longifolene (**5.99**) occurs in and is commercially extracted from the oil of *Pinus longifolia*. It has a very strained ring system, and exposure to acids causes an exothermic rearrangement to isolongifolene (**5.100**). Epoxidation of isolongifolene gives the corresponding epoxide (**5.101**), which undergoes acid-catalysed rearrangement to isolongifolanone (**5.102**). This is valued in perfumery for its warm woody, amber odour and is produced under trade names such as Valanone B®, Isolongifolanone, Timberone®, and Piconia® (31). The Prins reaction of isolongifolene with formaldehyde has also produced a number of useful products (32). The most significant of these is the mixture of the acetates (**5.103**) and (**5.104**), which is known under the trade name of Amboryl Acetate®. All of these reactions are shown in Figure 5.26. Further chemistry of longifolene is described in Chapter 7 of the book by Sell (30).

The name 'cedar' is used to describe a variety of trees of the *Cedrus*, *Juniperus*, *Cupressus* and *Thuja* families. In chemical terms, they fall into two main categories, those that are rich in cedrene, cedrol and thujopsene, and those that are rich in atlantones. The second category is used *per se* in perfumery, for example, in the form of Atlas Cedarwood Oil. The first category, which includes Texan, Virginian and Chinese cedar, has a much larger production and the oils are used both *per se* and as feedstocks for a range of fragrance ingredients. Over-harvesting is becoming an issue with cedarwood, even though the situation is not yet as severe as it is for sandalwood.

The main components of these latter cedarwood oils are α-cedrene (**5.105**), cedrol (**5.106**) and thujopsene (**5.107**). The oils also contain a variety of minor components (33). Fractional distillation gives hydrocarbon and alcohol fractions. Cedrene and thujopsene are separated from the former, and recrystallisation of the

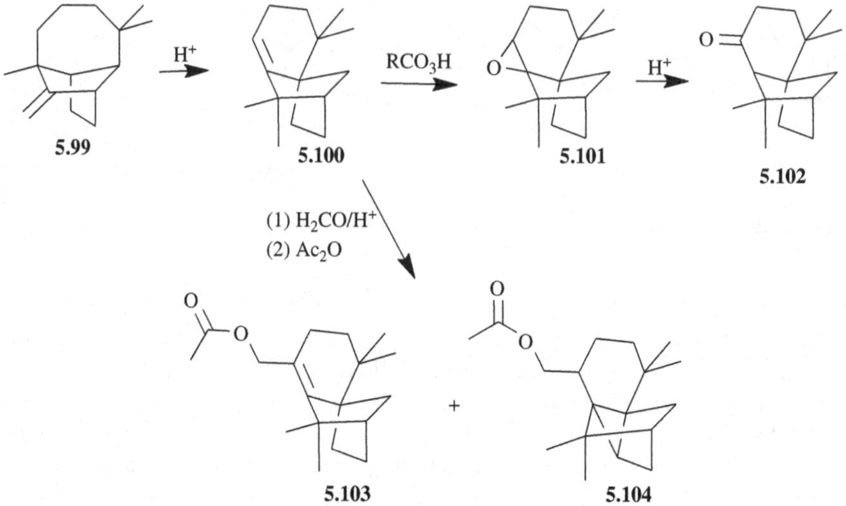

Figure 5.26   Longifolene and derivatives.

latter gives cedrol. A crude alcoholic fraction containing both cedrol and isomers such as widdrol (**5.108**) together with some ketonic material is sold as cedrenol. Thujopsene is also isolated from Hiba Wood Oil (26).

Acetylation of the alcohol mixture yields cedryl acetate (**5.109**). Cedryl methyl ether (**5.110**), also known as Cedramber®, is readily made by the methylation of cedrol (34). Both the ester and the ether have woody/amber odours and are therefore useful in perfumery. Cedrene oxide (**5.111**) also has a woody/amber smell of use in perfumery, and Ambrocenide® (**5.112**), another amber material, can be prepared from it by hydrolysis and reaction with acetone. These reactions and products are shown in Figure 5.27.

The most important cedarwood derivative is the mixture produced by acylation of the oil as shown in Figure 5.27. This product has a rich woody odour with cedar, amber and musky character (35). Either the hydrocarbon fraction or the entire oil can be used as feedstock because, during the acylation process, the alcohols are dehydrated to the corresponding olefins and these then react together with the original hydrocarbon oil components. The product is known under a range of trade names such as acetylcedrene, Lignofix®, Lixetone®, methyl cedryl ketone and Vertofix®. The major component of the mixture is acetylcedrene (**5.113**) but the major contributor to the odour is the acetylation product (**5.114**) of rearranged thujopsene (36–38). The Friedel–Crafts reaction is normally carried out with sulfuric or polyphosphoric acids as catalysts but, interestingly, when titanic chloride was used, the acylation product of cedrene was not acetylcedrene but the odourless vinyl ether (**5.115**) (39).

Guaiol (**5.116**) and bulnesol (**5.117**) are the major components of Guaiacwood Oil, which is distilled from the South American tree *Bulnesia sarmienti*.

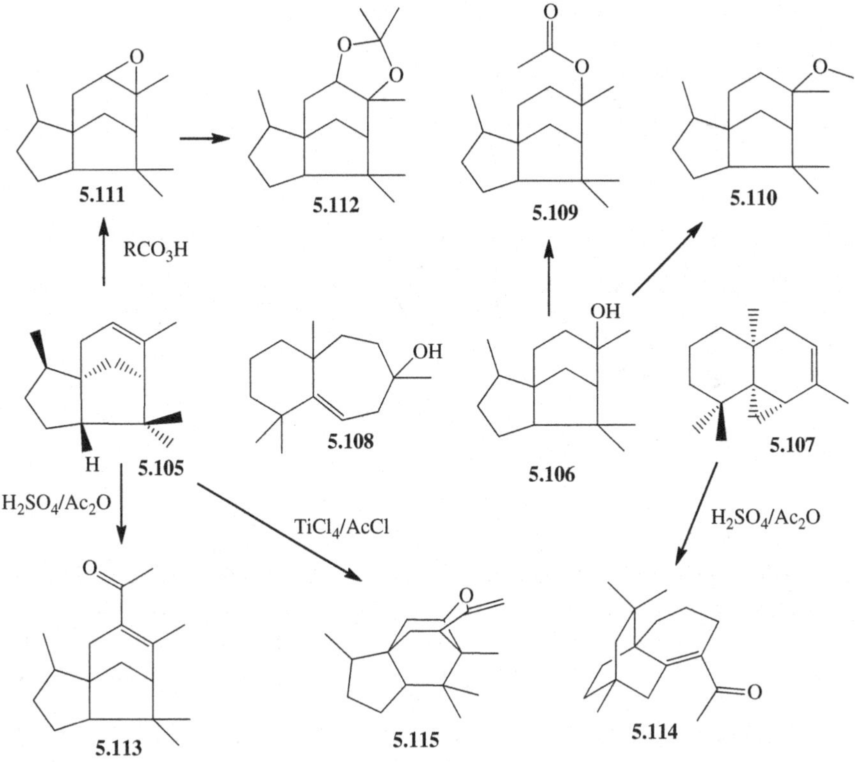

Figure 5.27   Cedarwood chemistry.

Both the oil and guaiol extracted from it are used as fixatives in perfumery. The acetylated oil and guaiyl acetate are also used in this way. Guaiazulene (**5.118**) is formed from guaiol by dehydration and dehydrogenation. It occurs in a number of essential oils and is used as an anti-inflammatory agent. Guaiazulene is also prepared from α-gurjunene (**5.119**), which is the major component of Gurjun Balsam. This is obtained from South-East Asian trees of the *Dipterocarpus* species and is used in a similar way to Guaicwood Oil. Azulenes are blue in colour and chamazulene (**5.120**) is the dye responsible for the blue colour of chamomile oil (Figure 5.28).

Farnesol (**5.121**), also known as *farnesyl alcohol*, occurs in many blossom oils. The trans-trans isomer is the commonest but all four possible isomers do occur in nature and all share the same odour, reminiscent of muguet and linden blossom. A mixture of *trans-trans*- and *trans-2-cis-*6-isomers occurs, together with nerolidol (**5.122**), in Cabrueva Oil (from *Myrocarpus frondosis*). Natural farnesol can be extracted from the oil but the yield is only 2–3%. Synthetic material is therefore the most important commercially. Farnesol provides an excellent background note in floral accords such as muguet and lilac and is also used in floral

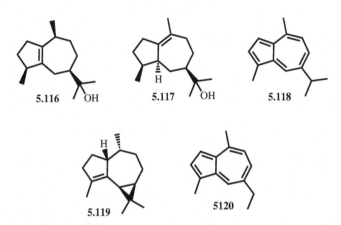

Figure 5.28   Some guaiane sesquiterpenoids.

and oriental fragrances. It is also valued as a blender and fixative. Its acetate has a faint green-floral odour and is used as a fixative for rose bases. Nerolidol (**5.122**) is also known as *peruviol* and exists as both enantiomeric forms of each of the cis and trans isomers. It occurs in many essential oils. For example, the (+)-*trans*-isomer is found in Cabreuva Oil and the (−)- in the oil of *Dalbergia parviflora* and a special grade of nerolidol is available from the former.

α-Bisabolol (**5.123**) occurs in chamomile (40) and lavender (41) oils. Chamomile is recognised as a European medicinal plant, and bisabolol contributes to its healing and soothing effects. The richest natural source is the New Caledonian shrub *Myoporum crassifolium* Forst. Original reports found that the wood oil contained up to 80% of a diastereomer of α-bisabolol, which was named *anymol* (42,43). This has now been shown to be (−)-*epi*-α-bisabolol (**5.124**) in work, which places the level at about 65% (44). All four stereoisomers are known in nature. For example, in addition to chamomile, (−)-α-bisabolol (**5.125**) occurs in the heartwood of *Vanillosmopsis erythropappa* (45); (+)-α-bisabolol (**5.126**) is a constituent of poplar bud extract *Populus balsamifera*; and (+)-*epi*-α-bisabolol (**5.127**) is the main constituent (~30%) of the essential oil from the South African sage *Stevia stenophylla* (46). α-Bisabolol has been found to reduce infections (47), have a spasmolytic effect similar to that of papaverine (48), have a beneficial effect on ultraviolet (49) and heat (50) burns, anti-inflammatory properties (51), and inhibit the growth of *Corynebacterium sp.* and *Staphylococcus epidermis*, the bacteria mainly responsible for the decomposition of human perspiration (52). It is available both as a homochiral isolate from plant sources and as a racemate from synthesis, through the acid-catalysed cyclisation of nerolidol (53). It is an almost colourless oil with a faint, sweet floral odour and has fixative properties in perfumery. However, its main use is as an anti-phlogistic agent in cosmetics. (−)-Cubebol (**5.128**) occurs in a number of essential oils. The richest source is Cubeb Oil (*Piper cubeba*) in which it is present at about 10% (54). Patent

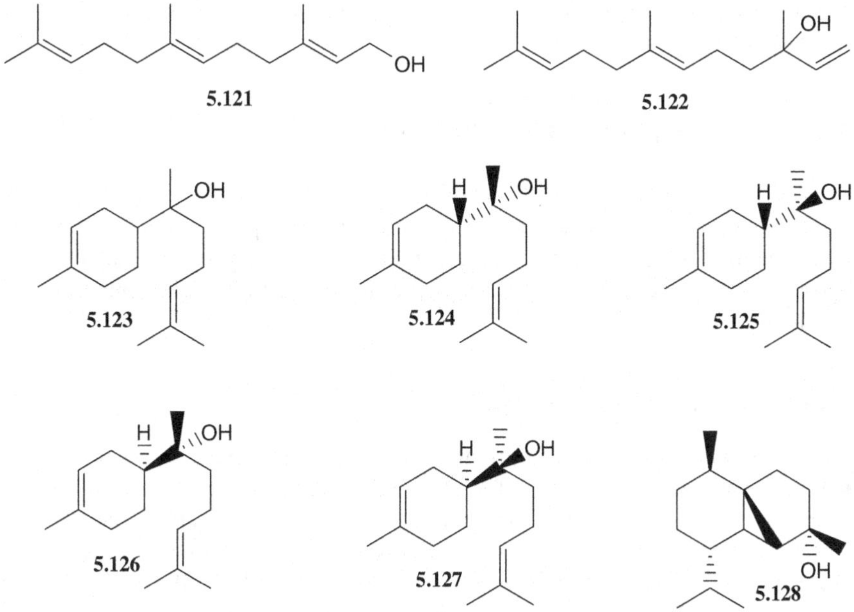

Figure 5.29 Some sesquiterpenoid alcohols.

applications have been filed by Firmenich claiming it as a cooling and refreshing agent in combination with other cooling agents (Figure 5.29) (55–57).

(+)-Nootkatone (**5.129**) was identified in grapefruit and found to be an important part of its taste (58). It has also been found in oranges and lemon, and can be prepared by oxidation of either valencene (**5.130**) or nootkatene (**5.131**) as shown in Figure 5.30. Its usage is restricted largely by its intensity rather than its price, although the latter is in thousands of dollars. Valencene has been isolated from orange juice and orange peel oil and is also found in lemon and grapefruit oils. However, the level of valencene in these oils is low. Nootkatene is readily obtainable from the wood of *Chamaecyparis nootkatensis* by steam distillation and can be converted to nootkatone by hydrochlorination and subsequent oxidation with Jones' reagent (59). Some chemotypes of the tree produce nootkatone, which can therefore be extracted from them. Nootkatone has a powerful sweet and citrus odour

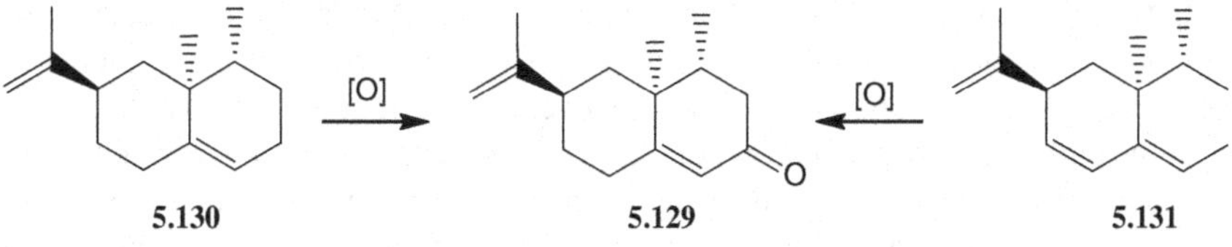

Figure 5.30 Nootkatone and its preparation from valencene and nootkatene.

and is of importance in grapefruit flavours, contributing to both the aroma and the bitter taste (60).

Terpenoid Degradation Products

It is obvious that degradation of natural products will occur as a result of decay but it also can be the product of metabolic processes that produce materials of use to the organism. The three groups of terpenoid degradation products that are most important in terms of odorous materials are those of carotenoids, ambreine and iripallidal.

Carotenoid Degradation Products

The carotenoids are the most important group of the tetraterpenoids. As mentioned earlier, the biogenetic precursor for the carotenoids is lycopene (**5.21**), which is formed by an initial tail-to-tail fusion of geranylgeranyl and geranyllinalyl pyrophosphates (in an analogous way to that shown for the formation of squalene in Figure 5.4) followed by dehydrogenation. It is responsible for the distinctive red colour of tomatoes. Cyclisation at one or both ends of the chain then produces the mono- and bi-cyclic members of the family. Oxidation can occur at the ends of the chain to produce further carotenoids, and oxidative degradation produces materials with fewer than the starting 40 carbon atoms. The geometrical configuration of the double bonds is usually trans. The prefix 'neo' is often used to designate isomers containing at least one cis-configuration. The prefix 'apo' indicates carotenoids that are oxidative degradation products retaining more than half of the carotene structure. About 600 naturally occurring carotenoids have been identified and characterised (61,62). Carotenoids are widely distributed in both terrestrial and marine plant and animal life. It has been estimated that nature produces about 100 million tpa of carotenoids, whereas synthetic production amounts to only several hundred tpa (63,64).

The best known carotenoids are α-carotene (**5.132**), β-carotene (**5.133**) and γ-carotene (**5.134**), which occur widely in nature. One important role of these in nature is as precursors for 11-*cis*-retinal (**5.135**), which is the pigment used in the light-sensitive rod and cone cells of the visual system. The structures of these carotenoids are shown in Figure 5.31, and a comparison of the structures of the degradation products shown in Figures 5.32–5.35 will illustrate how the odorous derivatives are related to their precursors.

In terms of odour, the most significant terpenoid degradation products are the ionones, damascones and safranic acid derivatives. The molecular structures of these materials all contain a 2,2,6-trimethylcyclohexyl fragment, derived from the ring at the end of the original carotenoid. Safranic acid derivatives have a single carbon attached to the 1-position of the ring, whereas the ionones and damascones have a chain of four carbon atoms at this position. The ionones and damascones differ in that the ionones are oxygenated at the third carbon away from the ring, whereas

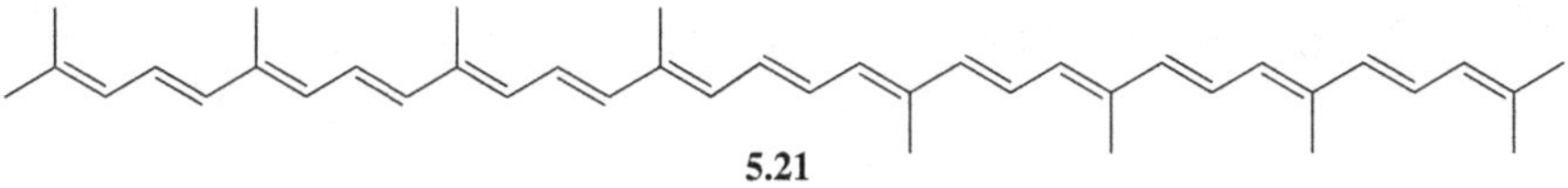

5.21

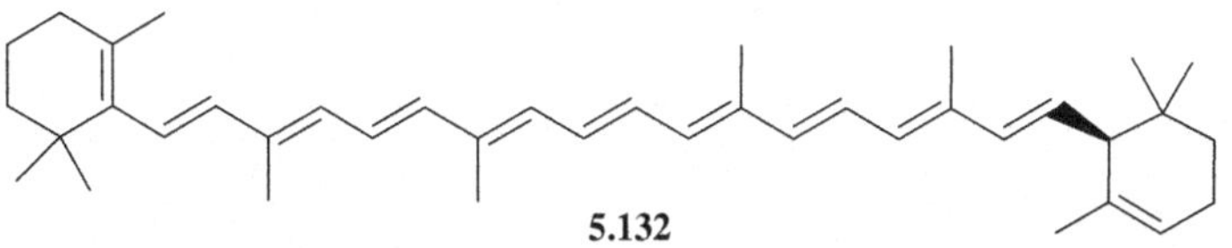

5.132

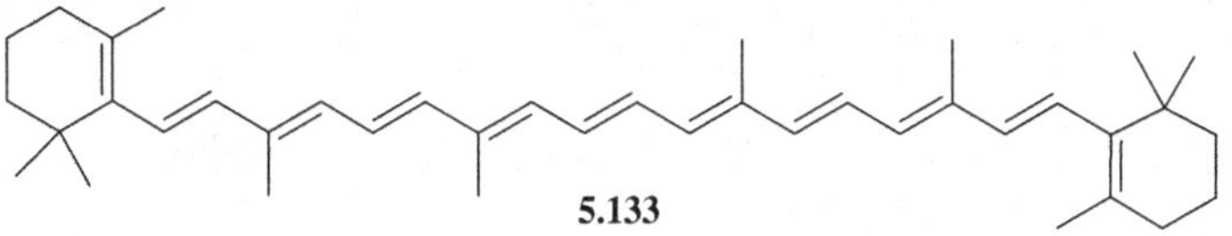

5.133

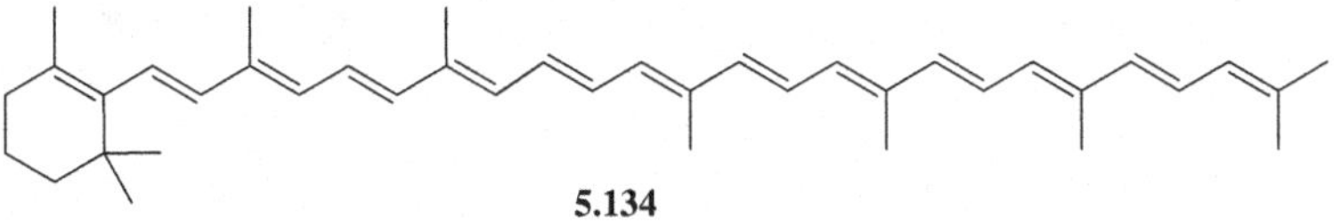

5.134

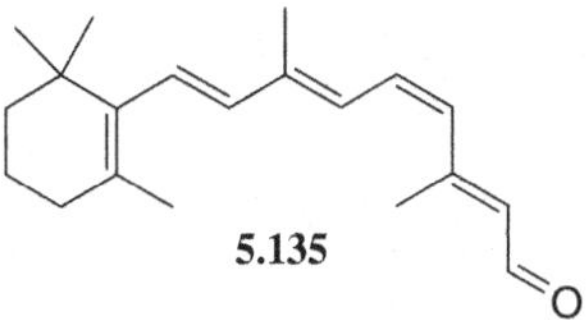

5.135

Figure 5.31 Some carotenoids.

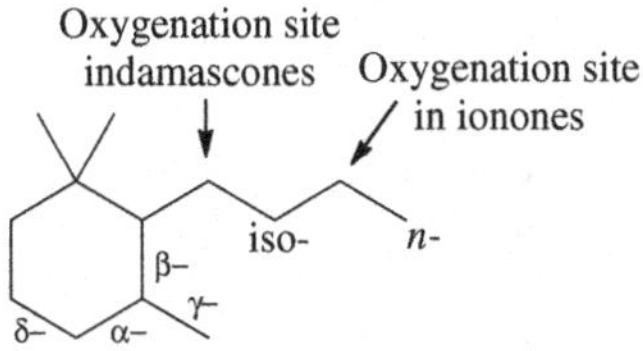

Figure 5.32 Nomenclature in ionones and damascones.

the damascones are oxygenated at the carbon attached to the ring. The nomenclature system for double bonds and substituents is shown in Figure 5.32. Double bond positions are identified by Greek letters, and the location of substituents in the ionone family by the prefixes '*n*' and '*iso*'. One potential source of confusion is that α-*iso*-methylionone is often referred to as γ-*methylionone*.

The ionones are materials of major importance in perfumery. They occur naturally in a variety of flowers, fruits, leaves, such as tobacco, and even roots, such as carrots. In odour terms, the ionones are associated with violet and, indeed,

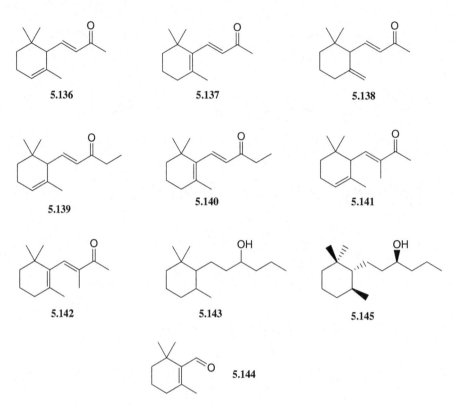

5.136   **5.137**   **5.138**

5.139   **5.140**   **5.141**

5.142   **5.143**   **5.145**

5.144

Figure 5.33   Some ionones and related compounds.

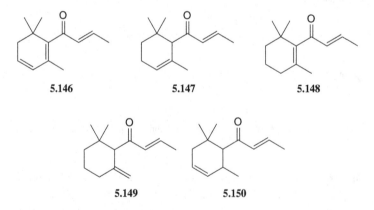

5.146   **5.147**   **5.148**

5.149   **5.150**

Figure 5.34   Some damascones.

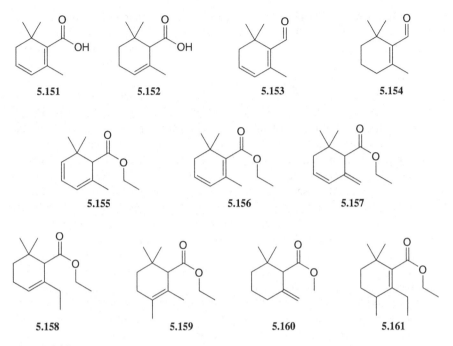

Figure 5.35 Safranic acid and derivatives.

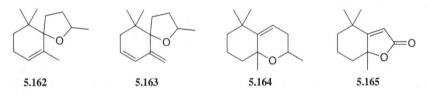

Figure 5.36 Some other carotenoid degradation products.

α- and β-ionones account for 57% of the volatile components of violet flowers (*Viola odorata*). β-Ionone also has a woody odour character and is of use not only in perfumery but also as a key intermediate in the synthesis of vitamins A, E and K. Annual production of ionones for perfumery use runs into hundreds of tonnes but that of β-ionone is much higher because of its use as an intermediate for vitamin synthesis. The methylionones do not occur in nature but their odour properties have made them valuable analogues of the natural materials. In perfumery, α-*iso*-methylionone is far more important than the ionones and its tonnage is about 10 times that of α- and β-ionones (13). As a result of the manufacturing route, ionones and methylionones for fragrance use are often mixtures of various isomers and each producer will have its own composition signature(s).

α-Ionone (**5.136**) occurs widely in nature, for example, in violets, blackberries, plums and tobacco. It has a warm, woody, floral odour with balsamic and sweet tones and is strongly reminiscent of violet flowers. It is used widely in perfumery. β-Ionone (**5.137**) is also very widespread in nature being found in, amongst others, rose, osmanthus, raspberries, cherries, tobacco, carrots and capsicums. It has a warm woody, dry and fruity odour and is greener than α-ionone. However, it is less useful than the latter and is used particularly in woody perfumes. In production terms, it is the most important of all the ionones because of its use in vitamin manufacture. γ-Ionone (**5.138**) is not observed in nature, and is of minor importance in perfumery. α-n-Methylionone (**5.139**) is also known as α-*cyclocitrylidenebutanone* and *Cetone Alpha*. It has a floral, sweet-oily odour of moderate tenacity and is used as a blender in perfumery. β-n-Methylionone (**5.140**) is also known as β-*methylionone, Cetone Beta*®, *Iraldeine Beta*® and *Raldeine Beta*®. It has a warm woody odour of the same type as β-ionone and is the least important of the methylionones. α-*iso*-Methylionone (**5.141**) is also available under trade names such as Iraldeine Gamma®, Raldeine Gamma® and Noviraldiol® and is also, somewhat confusingly, called γ-*methylionone*. It has a sweet floral odour with woody and tobacco nuances and is a very versatile ingredient. This makes it the most important of the entire ionone family as far as perfumery is concerned, and annual production runs into thousands of tonnes. β-*iso*-Methylionone (**5.142**) is available under trade names such as δ-methylionone, Rhodione Methyl Delta®, Ironal Methyl Delta®, Raldeine Delta® and Iraldeine Delta®. The use of delta in the name is potentially confusing, as it could also signify a different position of the double bond which, in this case, it does not. It has a light warm, woody floral character with a distinct animalic/ambergris note. This makes it of great interest to perfumers but its high price limits it to a very low level of use (Figure 5.33).

One ionone analogue that has become an important perfumery ingredient is that which is known by the trade names of Timberol® and nor-Limbanol® (**5.143**). It is produced by reaction of citral with 2-pentanone, then cyclisation in a manner exactly analogous to that of the ionones and then hydrogenation (65). It has a highly diffusive powdery-woody-amber odour and is used in a wide range of perfume formulations. There are two centres of asymmetry in the molecule and, hence, four stereoisomers. It was found that a mixture of the isomers with a trans relationship around the ring had a more interesting odour than a mixture of all four isomers (66). A mixture high in trans isomers can be produced from β-cyclocitral (**5.144**) by hydrogenation followed by aldol condensation with 2-pentanone and then hydrogenation (67).

Later, it was found that the most interesting individual isomer is the (+)-*trans*-nor-Limbanol (**5.145**) which has the lowest odour threshold of the four with a powerful, elegant woody-amber character and is very radiant (68–72). This material is also now also being used in fragrances.

The first member of the damascone family to be identified was β-damascenone (**5.146**). It was isolated from the Damask rose, *Rosa damascena*, the oil of which contains about 0.05% of it (73) and the structure was confirmed by synthesis (74). Since then, damascones have been identified in a wide variety of natural products.

Table 5.2   Odour Thresholds of Damascones

Compound	Threshold (ng/l air)
α-Damascone	0.14
β-Damascone	0.19
γ-Damascone	4.5
δ-Damascone	0.021
β-Damascenone	0.013

They are usually present at very low levels, but their powerful odours mean that they contribute much to the odour profiles of the plants and oils containing them. As an indication of this, the thresholds in air of some of the damascones, as given by Williams (75) in his review of their history, are shown in Table 5.2. Just as their occurrence in nature has been found to spread from rose to other plant sources, so their use in perfumery has expanded from rose to a wide variety of applications. Their success in perfumery has resulted in a great deal of activity in the search for analogues (75–77) (Figure 5.34).

The allylic transposition of the α,β-unsaturated ketone function in the side chain makes the damascones a much more challenging synthetic target than the corresponding ionones. Consequently, the damascones are much more expensive, prices being in the $250–1500/kg range. However, their intense odours mean that they can be used even at these prices. The major producer is Firmenich, who make many hundreds of tonnes in collaboration with DRT in France.

α-Damascone (**5.147**) has a floral, fruity, rose and apple odour character. The (S)-(−)-enantiomer is found in tea and is over 100 times stronger than its antipode (78–80). β-Damascone (**5.148**) is found in rose, osmanthus, tea, rum and tobacco. It has a fruity, floral, blackcurrant, plum, rose, honey and tobacco odour profile. γ-Damascone (**5.149**) is fruitier than α- or β-damascone. However, its threshold of detection is an order of magnitude higher than either of these. It has a powerful floral, rosy and fruity note with pine and green character. δ-Damascone (**5.150**) is not found in nature. It is also known as *Dihydrofloriffone*®. It is very diffusive with a threshold similar to that of damascenone and has a fruity, blackcurrant, floral and woody odour character. β-Damascenone (**5.146**) is present in Bulgarian rose oil at a level of only 0.05% but is a major contributor to the overall odour of the oil. Not surprisingly, it is used in reconstitutions of rose oil. It is also found in rose, apricot, beer, grape, mango, tomato, wine, rum, raspberries, passionfruit and blackberries. Its detection threshold is 1/10th of those of α- or β-damascone.

Safranic acid (**5.151**) and cyclogeranic acid (**5.152**) and related materials are formed by the degradation of the central chain of carotenoids right back to the carbon next to the cyclohexane ring at the end. The corresponding aldehydes safranal (**5.153**) and β-cyclocitral (**5.154**) both make important olfactory contributions to the oils containing them. Safranal is important to the odour of saffron and accounts for up to 70% of the volatiles in it. It also occurs in osmanthus, tea, grapefruit and paprika, amongst others. β-Cyclocitral is found in rum, tea, tomato, melon, paprika, peas and broccoli. These aldehydes are not particularly stable in application and so their importance as top-notes has led to a search for more stable materials with similar odour types (Figure 5.35).

The first synthetic material in the family to be commercialised was ethyl safranate, which was found to have the same limit of detection as β-damascenone (81). It is actually a mixture of three isomers: α- (**5.155**), β- (**5.156**) and γ- (**5.157**), the ratio between them depending on the reaction conditions. The α-isomer has a woody-ionone like odour, the β- is herbal and spicy and the γ- is fruity, reminiscent of apple and plum. Another speciality in this area is Givescone® (82–84), which is a mixture of two isomers (**5.158**) and (**5.159**), and has a rosy, spicy, fruity and woody odour. More recent introductions in the family include Romascone® (**5.160**) (85) and Myrascone® (**5.161**). The former is damascene-like, fruity and blackberry-like, whilst the latter is dry herbal and fruity.

There are a number of other groups of volatile carotenoid degradation products that occur in nature and contribute to the odours of the plants in which they are found. Many of these are commercially available in small quantities at high price. Examples include the theaspiranes (**5.162**), which are found in tea, passionfruit and tobacco; vitispiranes (**5.163**) found in some oils such as lemon balm; edulans (**5.164**) found in passionfruit, osmanthus and tobacco; and dihydroactindiolide (**5.165**) found in tea, osmanthus, tomato, ambergris and tobacco (Figure 5.36).

Ambreine Degradation Products

The sperm whale, *Physeter catodon* (formerly *Physter macrocephalus*) produces, in its intestinal tract, a material known as *ambergris*. Ambergris is found washed up on beaches and in former days was also removed from whales that had been killed. Never a secure source of supply, the decline in whale numbers has made natural ambergris an increasingly rare commodity. Ambergris was found to be comprised of up to 46% cholestanol-type steroids, principally epicoprosterol (**5.166**), and 25–45% of the triterpene ambreine (**5.167**) (86). Lumps of ambergris, usually about 20 cm in diameter but on occasions weighing up to 400 kg, are excreted into the sea where they undergo a series of degradative reactions in the presence of air, salt water and light. As this happens, the ambergris fades from dark brown to pale grey and develops an odour that is highly prized in perfumery (87,88). There are many degradation products of ambreine, quite a number of which contribute to the characteristic animalic, briny, ozonic and faecal character of the odour. The degradative transformations have been reproduced *in vitro* (89–92), and this chemistry has been reviewed by Sell (30). Three degradation products of importance to perfumery are the naphthofuran (**5.168**), α-ambrinol (**5.169**) and dihydro-γ-ionone (**5.170**). As might be imagined from its relationship to the ionones, the last of these contributes to the tobacco notes of ambergris. It also serves as a starting material for the synthesis of α-ambrinol (**5.169**), to which it is easily converted by means of the Prins reaction. The naphthofuran is by far the most important of all the ambreine degradation products (Figure 5.37).

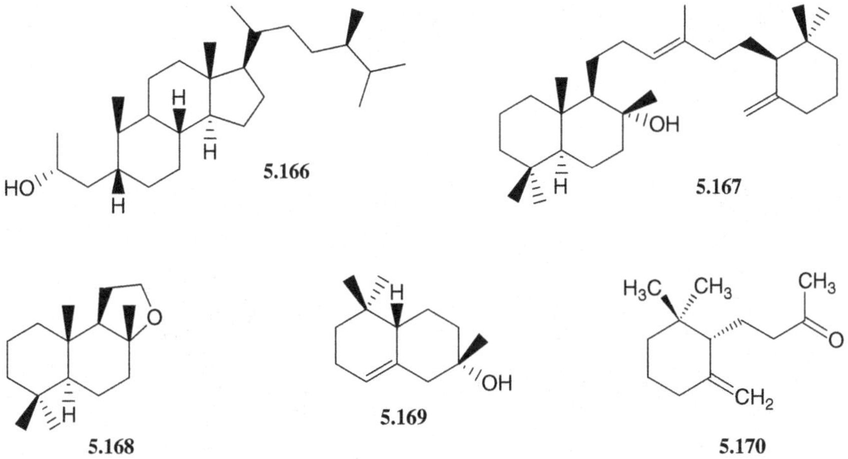

Figure 5.37   Some key ambergris components.

This naphthofuran (3a,6,6,9a-tetramethyldodecahydronaphtho[2,1-b]furan) (**5.168**) is one of the most expensive fragrance ingredients typically costing well over \$500/kg for enantiomerically pure grades and over \$350/kg for racemic material. Despite its high price, it is used extensively in perfumery, albeit at low levels, not only for its powerful ambergris odour but also for its enriching effect on fragrance compositions. In view of its value to perfumery, a great deal of effort has been, and continues to be, invested into the search for efficient synthetic routes to the naphthofuran. Annual worldwide production runs to several tens of tonnes and a variety of qualities is available under trade names such as Amberlyn®, Ambermore®, Ambertone®, Ambrofix®, Ambrox®, Ambroxan®, Ambroxid®, Ambroxid Rein®, Ambroxide®, Fixateur 404® and Sylvamber® for enantiopure materials derived from partial synthesis; Ambrox DL®, Cetalox® and Synambran® for racemic materials by total synthesis and Cetalox-laevo® for enantiopure product from total synthesis. Currently, the most important synthetic routes are based on sclareol (**5.171**) ex Clary Sage (*Salvia sclarea*) as feedstock, as will be described in Chapter 6. α-Ambrinol (**5.169**) is a pale yellow liquid with a powerful animalic, ambergris odour and it is used at low levels in a wide variety of fragrances.

In view of the value of these nature-identical ingredients and the difficulties in synthesising them, it is not surprising that many analogues have been developed and are used in perfumery. The structures of some are shown in Figure 5.38. Ambra Oxide® (**5.172**) is the pyran equivalent of (**5.168**), and Grisalva® (**5.173**) is another close analogue. Some other ambergris materials in use, such as Karanal® (**5.174**) (93), are not closely related to the natural materials but do elicit similar odours. Polywood® (**5.175**) is related to ambrinol but has a woody odour rather than being animalic. Jeger's ketal. (**5.176**), named after its discoverer but also called

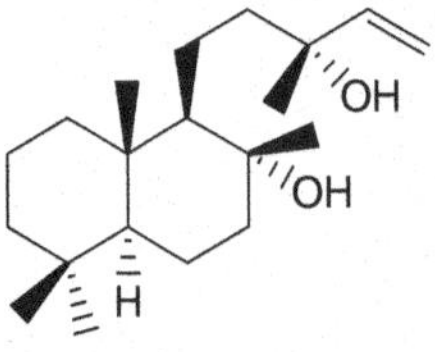

Figure 5.38   Sclareol and some amber analogues.

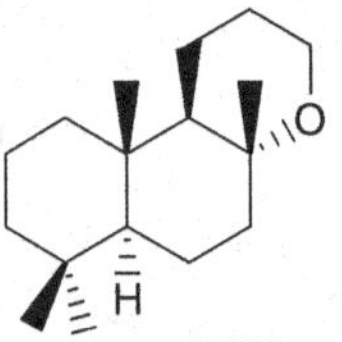

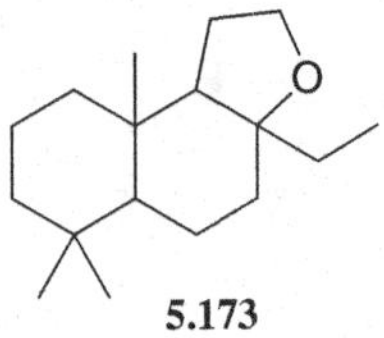

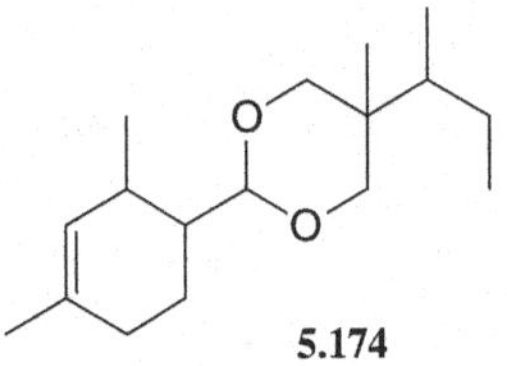

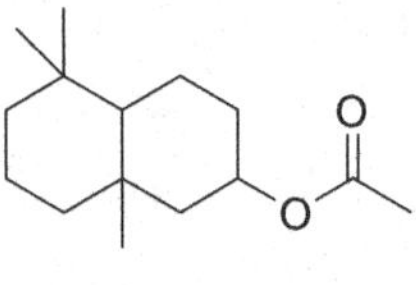

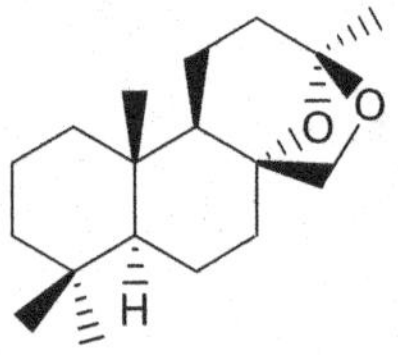

Amberketal®, *Ambraketal®* and *Ketamber®*, is another material with a powerful ambergris odour.

Iripallidal Degradation Products

The triterpenoid iripallidal (**5.177**) occurs in the rhizomes of the iris *Iris pallida*, and this degrades to give, amongst other products, the irones. The natural extract is known as *orris* and owes its odour largely to the irones. These have the same nomenclature system as the ionones and are therefore known as α-*irone* (**5.178**), β-*irone* (**5.179**) and γ-*irone* (**5.180**) (Figure 5.39).

SHIKIMIC ACID DERIVATIVES

The second largest group of natural odorants are the shikimates. Their biosynthesis is outlined in Figure 5.40. Addition of phosphoenol pyruvate (**5.8**) to erythrose 4-phosphate (**5.9**) leads to shikimic acid (**5.181**), the key intermediate from which this family of natural products takes its name. The shikimates are characterised by having a benzene ring with a C_1 to C_3 substituent and oxygenation usually at positions 3 and/or 4 and/or 5 relative to it. These basic features are all present in shikimic acid itself, though, through the course of biosynthesis, all of the original oxygen atoms in the ring are lost and those appearing in the final products are introduced later by oxidation. Loss of the ring oxygen atoms and aromatisation gives benzoic acid (**5.182**), which can be oxidised to *p*-hydroxybenzoic acid (**5.183**) or *o*-hydroxybenzoic acid, better known as *salicylic acid* (**5.184**). Addition of a second molecule of phosphoenol pyruvate (**5.8**) to shikimic acid gives chorismic acid (**5.185**) and an oxy-Cope-like rearrangement then gives the phenylpyruvic acid

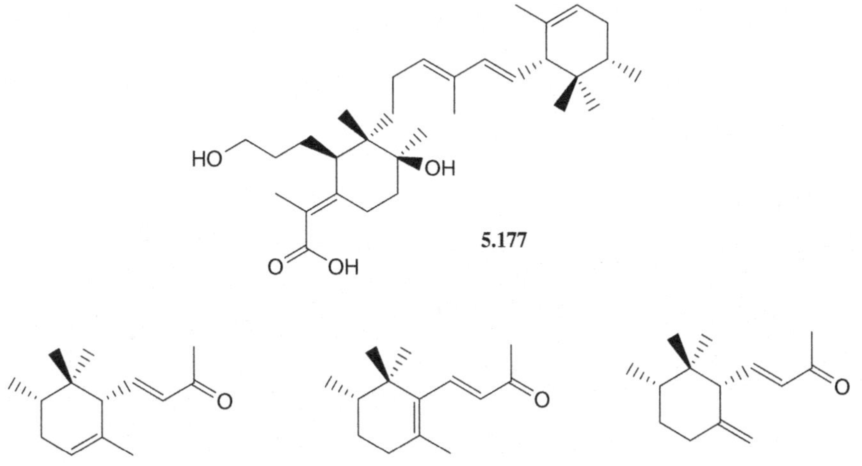

5.177

5.178   **5.179**   **5.180**

Figure 5.39 Iripallidal and the irones.

derivative prephenic acid (**5.186**). This acid serves as the precursor for phenylala-nine (**5.187**). Deamination of phenylalanine (**5.187**) gives cinnamic acid (**5.188**). Like benzoic acid (**5.182**), cinnamic acid (**5.188**) can be hydroxylated in either the ortho or para positions of the ring, giving *o*-coumaric acid (**5.189**) or *p*-coumaric acid (**5.190**), respectively. Further hydroxylation of the latter gives caffeic acid (**5.191**), and etherification of the hydroxyl group in the meta position gives fer-ulic acid (**5.192**), a key ingredient for some significant odorants. Oxidation of the methyl ether of ferulic acid (**5.192**) and cyclisation gives methylenecaffeic acid (**5.193**).

The shikimic acid pathway is of vital importance to plants, as it leads to the monomers from which lignin is constructed. The flavonoid family of plant pigments and antioxidants are also shikimate derivatives. This pathway also provides both plants and animals with three of the essential amino acids, namely phenylalanine (**5.187**), tryptophane (**5.194**) and tyrosine (**5.195**) (Figure 5.41).

Various oxidations, reductions and side-chain modifications are involved in transforming the basic intermediates of Figure 5.40 in the array of shikimates that are found in nature. These reactions can occur in various sequences, and many shikimates can be produced by the same reactions occurring in a different order, even in the same plant. For example, anisaldehyde (**5.196**) could be biosynthesised from shikimic acid (**5.181**) either via hydroxylation of cinnamic acid (**5.188**) to give *p*-coumaric acid (**5.190**) followed by methylation of the phenolic group and degradation of the side chain, or via benzoic acid (**5.182**) with hydroxylation to *p*-hydroxybenzoic acid (**5.183**) followed by methylation of the phenolic group and reduction to the aldehyde, as shown in Figure 5.42. The following discussion of odorous shikimates will group them primarily according to their chemical struc-tures rather than exact biosynthetic routes.

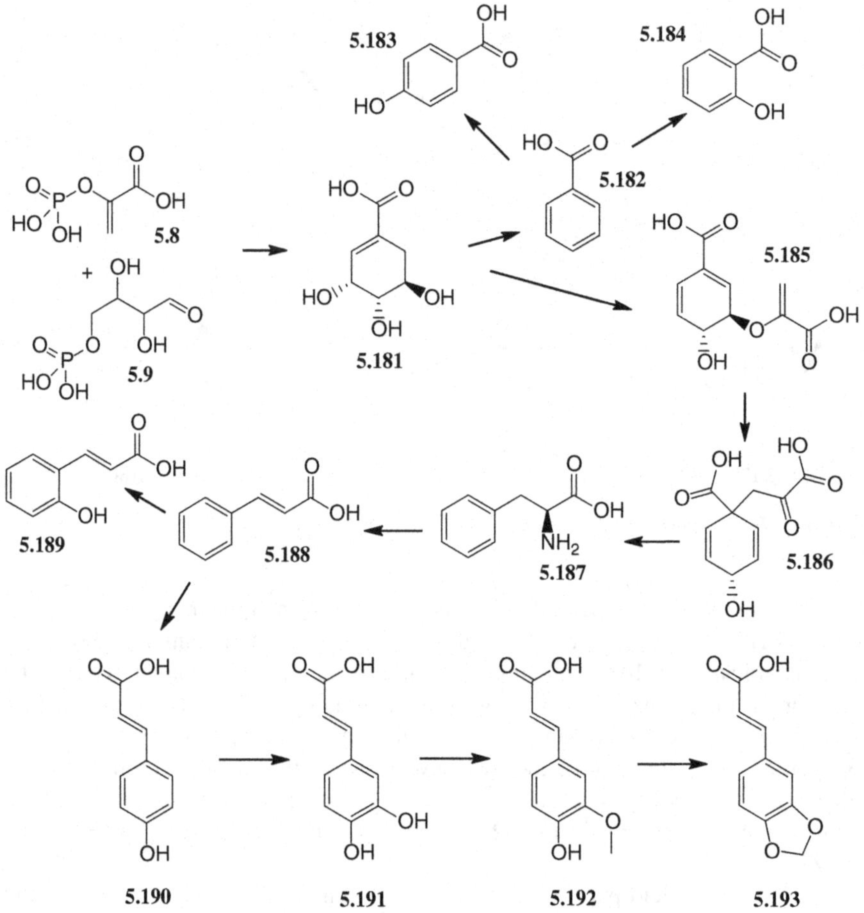

Figure 5.40   Basic flow of shikimic acid pathway.

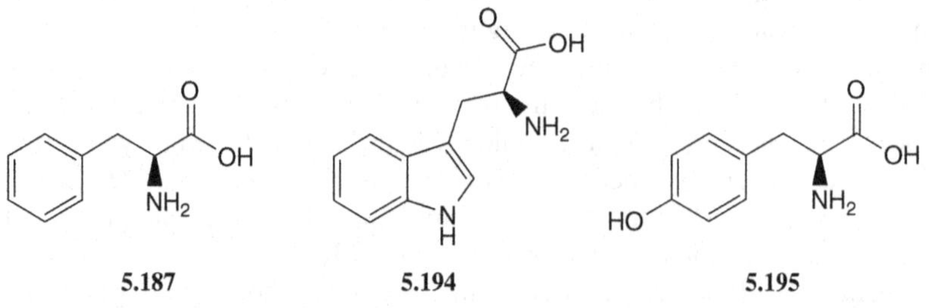

Figure 5.41   Shikimate derived amino acids.

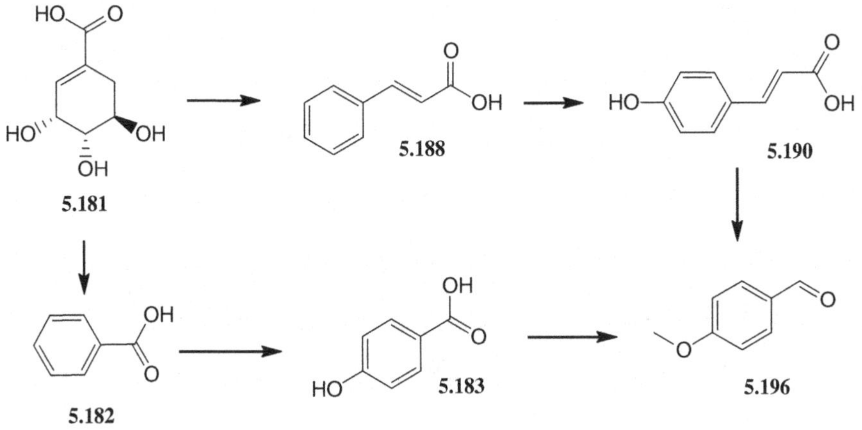

Figure 5.42 Possible routes to anisaldehyde.

Benzyl alcohol (**5.197**), benzoic acid (**5.182**) and their esters occur in a wide range of flowers and hence essential oils. For example, benzyl acetate (**5.198**) is a major component of jasmine, and methyl benzoate (**5.199**) is found in ylang. Benzaldehyde (**5.200**) is found in various kernels and in some flowers and other sources but is usually present as amygdalin (**5.201**), a glycoside of its cyanohydrin. Hydrolysis of amygdalin produces benzaldehyde and hydrogen cyanide, and the author believes that it is this association in natural sources rather than any effect at olfactory receptors that accounts for the similarity of their odours (Chapter 2). Salicylic acid (**5.184**) derives its name from the willow (*Salix*) in which it is found. Its esters are widespread in nature. For example, methyl salicylate (**5.202**) is a major component of wintergreen and an important odour contributor to ylang. Methyl anthranilate (**5.203**) is yellow in colour and has an intense sweet odour which contributes to the bouquets of many flowers (Figure 5.43).

Anisaldehyde (**5.196**) occurs in quite a number of flowers and other botanical sources. The corresponding alcohol (**5.204**) and acid (**5.205**) as well as esters of these do occur but less frequently. Vanillin (**5.206**) is of considerable commercial importance since it is the principal odour component of vanilla, but it does occur in a variety of other natural sources. Heliotropin (**5.207**) is also known as *piperonal* and is the principal component of heliotrope flowers (Figure 5.44).

Cinnamaldehyde (**5.208**) and cinnamyl acetate (**5.209**) are important components of cinnamon and also occur in other oils. Cinnamyl alcohol (**5.210**) and cinnamic acid (**5.188**) are less widespread in occurrence. Methyl cinnamate (**5.211**) is widespread and is a major component of basil (Figure 5.45).

Indole (**5.212**) and scatole (**5.213**) are degradation products of tryptophane (**5.194**), and 2-phenylethanol (**5.214**) of phenylalanine (**5.187**). Indole is important to the odour of jasmine but associations with its presence in faeces and manure mean that it is not always well received when smelt *per se*. 2-Phenylethanol, also known as *phenethyl alcohol*, is one of the key components of rose oil (together with

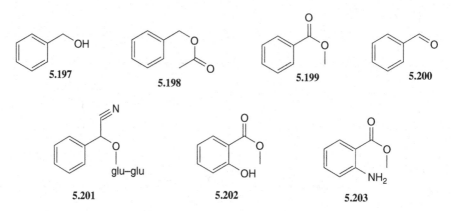

Figure 5.43 Some shikimates related to benzoic acid.

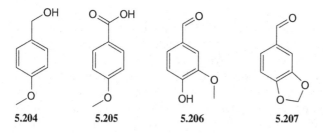

Figure 5.44 Some shikimates related to *p*-hydroxybenzoic acid.

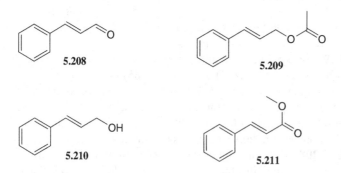

Figure 5.45 Some shikimates related to cinnamic acid.

geraniol and citronellol) and it and various of its esters, contribute significantly to a wide range of floral odours (Figure 5.46).

Coumarin (**5.215**) contributes a sweet base note to many natural scents, including new mown hay and lavender. The furocoumarin bergaptene (**5.216**) is found in

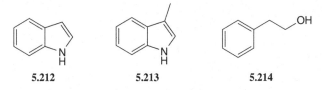

5.212   **5.213**   **5.214**

Figure 5.46 Some shikimates from degradation of amino acids.

various sources, most notably bergamot. It must be removed from essential oils before use in perfumery because of its activity as a photosensitiser (Figure 5.47).

Both isomers of anethole, (*E*)-anethole (**5.217**) and (*Z*)-anethole (**5.218**), occur in a variety of herbs and spices. (*E*)-anethole (**5.217**) is commoner and is usually found at higher levels. Estragole (**5.219**) is also known as *methyl chavicol* and occurs in various herbs, especially tarragon. All three of these *p*-coumaric acid derivatives have been found to have levels of toxicity that restrict perfumery use of both the compounds *per se* and oils containing them. Use of the herbs as food flavours is not affected by this regulation (Figure 5.48).

Eugenol (**5.220**), isoeugenol (**5.221**), methyleugenol (**5.222**) and both isomers of methylisoeugenol (**5.223**) and (**5.224**) occur in a wide range of herbs (e.g. bay, tarragon and basil), spices (e.g. clove, pimento and nutmeg) and flowers (e.g. carnation and ylang). The richest source of eugenol is in cloves and it is an important part of the odour of carnation (Figure 5.49).

Safrole (**5.225**) is the major component of Sassafras Oil. The isomeric isosafrole (**5.226**) occurs less widely in nature. Safety restrictions have made both substances obsolete in perfumery (Figure 5.50).

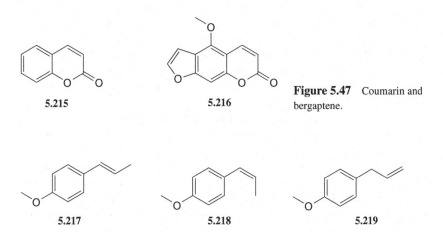

5.215   **5.216**   **Figure 5.47** Coumarin and bergaptene.

5.217   **5.218**   **5.219**

Figure 5.48 Some shikimates derived from *p*-coumaric acid.

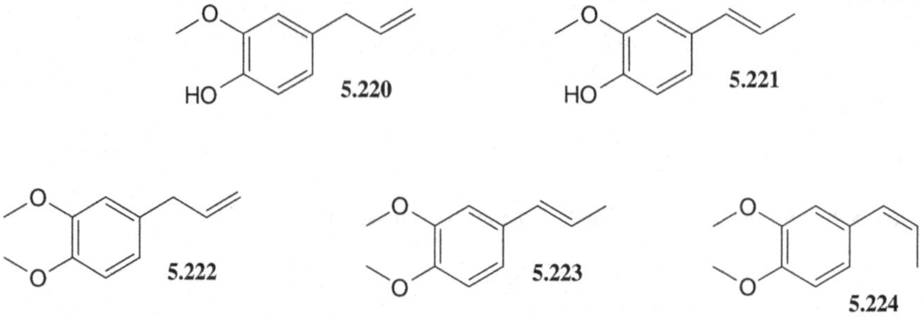

Figure 5.49   Ferulic acid derivatives.

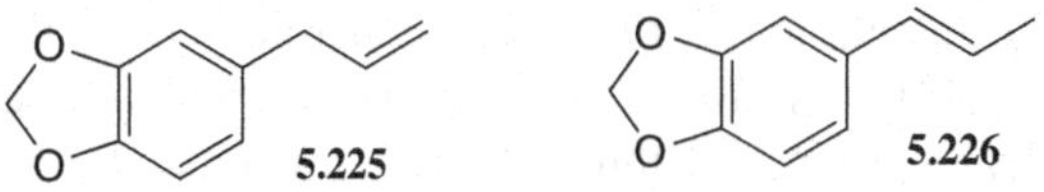

Figure 5.50   Methylenecaffeic acid derivatives.

LIPIDS AND POLYKETIDES

Addition of acetyl CoA (**5.11**) units to each other in a reaction akin to the Claisen ester condensation leads to straight chains containing multiples of two carbon atoms, as shown in Figure 5.51. If the ketone functions are left intact, intramolecular cyclisation reactions result in molecules such as orsellinic acid (**5.227**). Examination of the structure of polyketides usually reveals evidence of the presence, or former presence (e.g. a double bond formed by dehydration of an alcohol), of oxygenation at alternate carbon atoms in the backbone of the molecular structure. The most important odorants derived from orsellinic acid are those derived from the lichen *Evernia prunastri*, commonly known as *oakmoss* or *treemoss*. Its most organoleptically significant components are methyl 3-methylorsellinate (**5.228**) and ethyl everninate (**5.229**), the former in particular. Natural extracts are now restricted because of minor components such as chloratranol (**5.230**) and the depside atranorin (**5.231**) which are skin sensitisers. For this reason, in perfumery the natural moss extracts have now been replaced to a considerable extent by the purer, safer synthetic methyl 3-methylorsellinate (**5.228**) (Figure 5.52).

If during biosynthesis the ketone functions of the polyketide chain are reduced to methylene groups, the pathway constitutes homologation of acetic acid to give the natural fatty acids. Addition of side chains can occur through incorporation of propionic acid or methylation of a straight-chain acid. The shorter chain acids are volatile and have quite potent odours. They can be formed *per se* or as the result of degradation reactions. The most important degradation route is via β-oxidation, which basically constitutes the reverse of the biosynthesis process and leads to successive reduction of the chain by two carbon atoms with concomitant production of acetic acid. Various enzymic oxidations, reductions and reactions involving chain

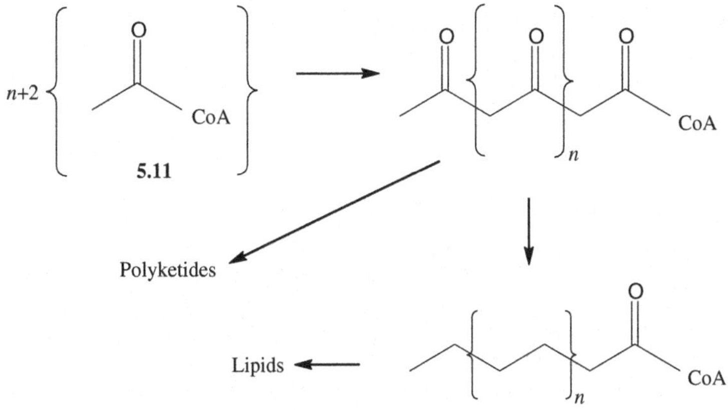

Figure 5.51   Lipid and polyketide biosynthesis.

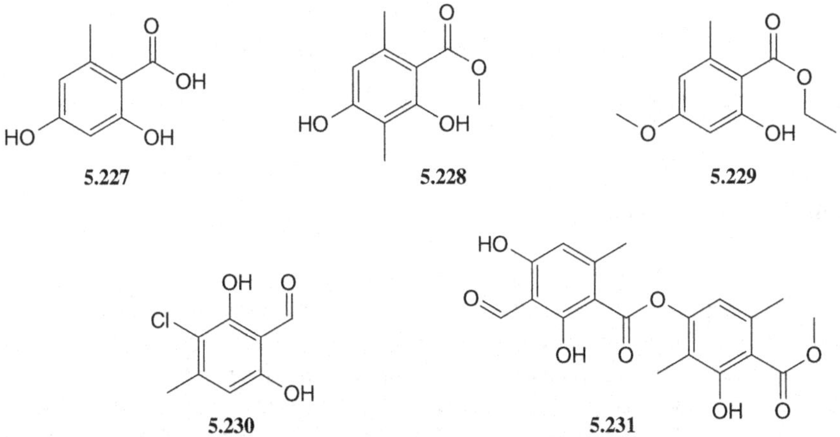

Figure 5.52   Orsellinic acid and derivatives.

breaking can occur, leading to a wide variety of odorous lipid derivatives. In terms of production of odorants, another important degradation is the autoxidation of unsaturated lipids to give homoallylic hydroperoxides, which are reduced to the corresponding alcohols and then undergo cleavage as shown in Figure 5.53.

The large variety of saturated and unsaturated fatty acid precursors, combined with the many different reactions available for modification of lipid molecules, means that there is an enormous range of derivatives with sufficient volatility to reach the olfactory receptors. The distinctive aromas of most fruits owe much to fatty esters such as ethyl acetate (**5.232**), ethyl butyrate (**5.233**), amyl acetate (**5.234**) and suchlike, whilst the characteristic odour of freshly cut grass is due to *cis*-3-hexenol (**5.235**) (Figure 5.54).

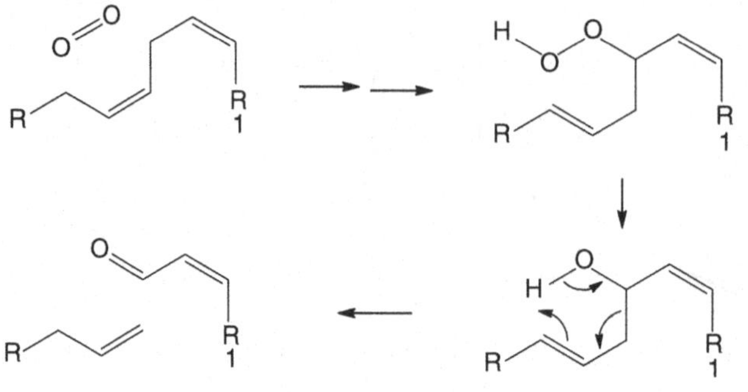

Figure 5.53   Autoxidative cleavage of lipids.

Figure 5.54   Some odorous lipid derivatives.

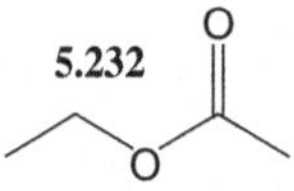

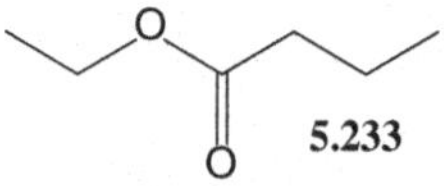

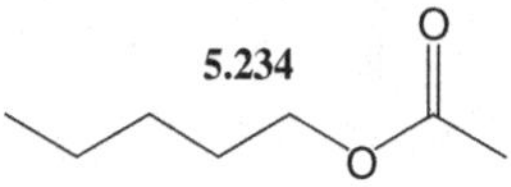

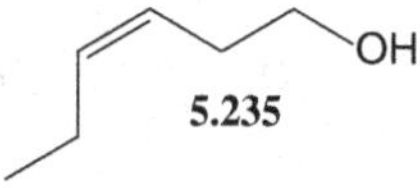

As an illustration of the variety of lipid derivatives found in nature, we can consider some of the many decane derivatives that occur in plants. Decane (**5.236**) itself can be found in osmanthus (*Osmanthus fragrans*), *Narcissus tazetta* and at levels up to 10% in the flowers of *Acacia praecox*. 1-Decene (**5.237**) occurs in the flowers of *Nicotiana rustica* and 1-decyne (**5.238**) in *Eremocharis triradiata*. 1-Decanol (**5.239**) occurs in coriander leaf (also known as *cilantro*) (*Coriandrum sativum*) and its acetate (**5.240**), propionate (**5.241**) and butyrate (**5.242**) in ambrette seed (*Hibiscus abelmoschus*). 2-Decanol (**5.243**) is found in *Boronia megastigma* and its acetate (**5.244**) in rue (*Ruta graveolens*). The seeds of the Vietnamese plant *Alpinia katsumodal* contain both 2-decenol (**5.245**) and 3-decenol (**5.246**), and *Phaeomeria speciosa* contains 9-decenol (**5.247**), this last having become an important synthetic rose ingredient. Decanal (**5.248**) is an olfactorily important component of orange oil (*Citrus aurantium*). Analyses of coriander leaf give decanal (**5.248**) and various unsaturated analogues as components. These include (*E*)- and (*Z*)-2-decenal, (**5.249**) and (**5.250**), respectively, (*Z*)-4-decenal (**5.251**), 5-decenal {stereochemistry undefined drawn as (*E*)-} (**5.252**) and 9-decenal (**5.253**). 2-Decanone (**5.254**) has been found in rue (*Ruta graveolens*), 3-decanone (**5.255**) in cornmint (*Mentha arvensis*) and dec-3-ene-2-one (**5.256**) in *Acronychia pedunculata*. Decanoic acid (**5.257**) occurs in the fruits of guava (*Psidium guajava*) and *Zizyphus jujuba*.

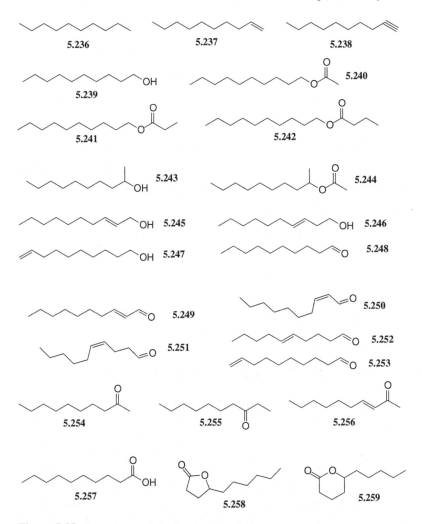

Figure 5.55 Some decane derivatives found in plants.

γ-Decalactone (**5.258**) and δ-decalactone (**5.259**) both occur in osmanthus (*Osmanthus fragrans*), whilst the former is found in tuberose (*Polyanthes tuberosa*) and the latter in honeysuckle (*Lonicera caprifolium*) (Figure 5.55).

   Two organoleptically important components of jasmine (*Jasminum officinale var. grandifolium*) are jasmone (**5.260**) and methyl jasmonate (**5.261**). These are derived from arachidonic acid (**5.262**) by a series of reactions including autoxidation, cyclisation and degradation (Figure 5.56) (94).

   Another group of lipid-derived natural products are the musks. In all of these, the lipid chain is cyclised to give a macrocyclic odorant. Muscone (**5.263**) is found in the anal glands of the musk deer (*Moschus moschiferus*), civetone (**5.264**) in

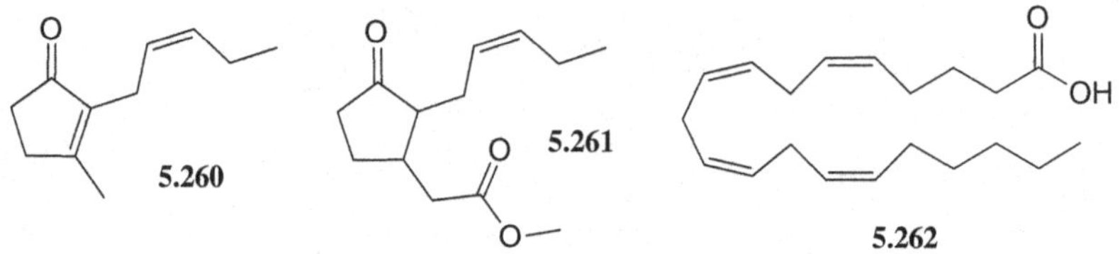

Figure 5.56   The jasmonoids and their biosynthetic precursor.

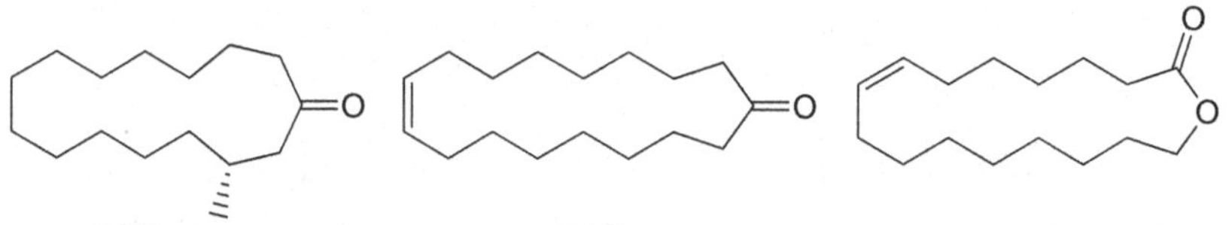

Figure 5.57   The natural musks.

the anal glands of the civet cat (*Civettictus civetta*) and ambrettolide (**5.265**) in the seeds of ambrette (Ab*elmoschus moschatus*) (Figure 5.57).

HUMAN USE OF ODOROUS PLANT EXTRACTS

Humans use plant odorants to find food and to assess its quality, but they also use them for aesthetic purposes and perfumery is one of the oldest industries. The earliest method used to concentrate the odorous components of plants was that of enfleurage in which the plant material was mixed with fat and, when the odorants had been absorbed into it, the fat was melted to allow it to be removed from the plant debris by filtration. This technique was used into the twentieth century. With the development of distillation in the Middle Ages, alcohol became available as a solvent. This allowed separation of the odorous components from the fats and waxes which enfleurage also extracts. The product of enfleurage is known as a *concrete* and the ethanolic extract as *absolute*. Distillation of plant material also has a history of use, usually in the form of steam distillation in which the presence of water limits the temperature to 100 °C and therefore reduces the degree of degradation of the plant volatiles. The oils extracted in this way are known as *essential oils*. The ancients believed that the oils represented the spirit of the plant. In Aristotelian cosmology, spirit was considered to be the fifth element (the first four being fire, air, earth and water) and was known as the *quintessence*, from which is derived the term *essential* in this context. In most cases, the oils are removed by simple separation of the organic phase from the aqueous phase after condensation. The main exception is rose because 2-phenylethanol (**5.214**), one of the major components of rose oil, is moderately water-soluble and remains in the aqueous phase giving the product known as *rose water*. Chemical technology provided a variety of solvents from the mid-ninetenth century onwards and these solvents could be used to replace non-volatile animal fats and the solvent could be removed by distillation to give the concrete. Benzene was the commonest solvent used for this purpose in the early part of the twentieth century until its toxicity was recognised and alternatives were sought. The commonest alternative was the azeotrope of ethyl acetate and hexane, which has similar solvation and distillation properties to those of benzene. Liquid carbon dioxide is another alternative and one that is growing in popularity. Citrus oils are separated from the peel of the fruit by physical expression. For bergamot, this was once done manually by squeezing the fruit with a special tool. Nowadays, mechanical expression is used, and orange, lemon and grapefruit oils are obtained as by-products of the fruit juice industry. More details of these various processes and a table of typical products can be found in the book by Sell (95).

DIVERSITY OF NATURAL ODORANTS

Essentially, all natural odours are complex mixtures of chemicals with different functional groups and often from more than one biosynthetic route. The variety of skeletal types and the range of chemical functionality found in odorous natural products are astounding, and new discoveries are being made all the time. However, care must be taken with modern chemical analysis of plant material because of the very high sensitivity of modern analytical techniques and the consequent possibility of trace contamination of samples either before harvesting or during isolation (e.g. trace impurities in solvents used for extraction) leading to false reports of substances occurring in plants. Most analyses of plant volatiles nowadays are done using GC/MS and this can also lead to misidentifications due to software interpreting a close match of a mass spectrum as a positive identification. The exact pattern of secondary metabolites is used by botanists to classify plants, a process known as *chemotaxonomy*. However, there are some general patterns that can be seen, and the experienced natural products chemist or fragrance chemist will learn to anticipate certain classes of chemicals in any given botanical source. Flower oils often contain monoterpenoids, especially alcohols, aldehydes and esters, carotenoid degradation products and/or shikimates, usually with 0 to 3 carbon atoms in the side chain. Fruits are well known for their lipid-derived volatiles (such as aliphatic esters) and many also contain carotenoid degradation products (such as damascones and theaspiranes). Citrus fruits owe much of their character to aldehydes including terpenoid aldehydes (e.g. decanal (**5.248**) in orange and citral (**5.76** and **5.77**) in lemon). The key components of herbs are often terpenoid hydrocarbons (e.g. pinenes in rosemary) and/or ketones (e.g. menthones in mint) or shikimates (e.g. estragole (**5.219**) in tarragon), usually with a three-carbon side chain. Spices are usually characterised by shikimates, usually with a three-carbon side chain (e.g. anethole (**5.217** and **5.218**) in anise and eugenol (**5.220**) in clove). Woody scents are usually rich in cyclic terpenoids, monoterpenoids in pine and related species and sesquiterpenoids in precious woods and woody scents such as vetiver and patchouli.

MALODOURS IN NATURE

It is not surprising that not everything in nature smells attractive to us humans since we use our sense of smell to alert us to danger and consequently develop aversions to substances that we judge to constitute a threat, whether that threat is real or not.

One real threat that we all learn early in life is that decaying matter indicates the presence of harmful bacteria. Bacterial degradation of proteins leads to the formation of volatile nitrogen and sulfur-containing substances and, so, again, it is not surprising that amines, thiols and sulfides are amongst the molecules considered most offensive by most people. Indole (**5.212**) and scatole (**5.213**) are degradation products of tryptophane (**5.194**) but their presence in some floral scents might explain why they do not induce quite such an adverse reaction as do ammonia (**5.266**), hydrogen sulfide (**5.267**) or dimethyl trisulfide (**5.268**). We learn at an early age to avoid bacterial contamination from sources where bacterial degradation of protein is rife, such as faeces (hence sewage), decomposing flesh and sepsis. In addition to the substances mentioned above, key warnings of bacteria come from trimethylamine (**5.269**), putrescine (**5.270**), cadaverine (**5.271**), methanethiol (**5.272**), dimethyl sulfide (**5.273**), dimethyl disulfide (**5.274**) and diallyl sulfide (**5.275**) (Figure 5.58).

Lipid oxidation products are also warning of bacterial action, perhaps the best known example being that of butyric acid (**5.276**) as a warning of milk that has 'gone off'. However, bacterial degradation of lipids also provides us with a wide variety of odorants that can be put to use for the benefit of animals through social recognition. Thus, across the range of mammals, body odours can be and are used

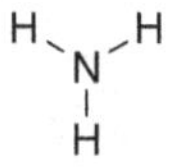

5.266

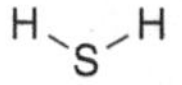

5.267

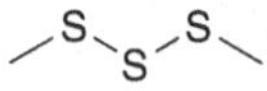

5.268

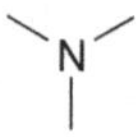

5.269

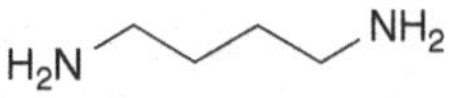

5.270

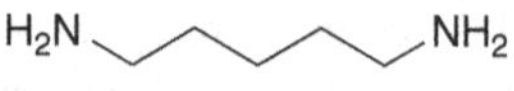

5.271

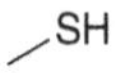

5.272

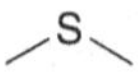

5.273

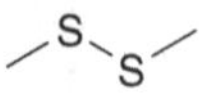

5.274

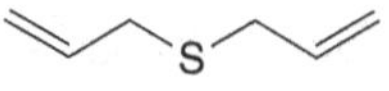

5.275

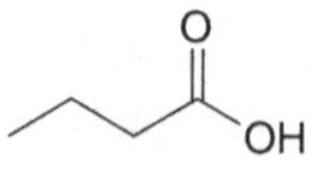

5.276

Figure 5.58 Some malodorants found in nature.

for species identification, individual recognition and determination of such states as sex, reproductive status and hierarchical status as discussed in Chapter 1.

Human Sweat Malodour

In terms of human malodours, the most studied is axillary sweat. Fresh human sweat is odourless and it is bacterial action on their components that is responsible for the malodour that develops (96,97).

The precursors are produced in the apocrine sweat glands of the axilla and production starts at puberty (98). It has been shown that the flora responsible for production of the odour belong to the family of *Coryneform* bacteria (98,99). The volatiles released by aged sweat comprise very complex mixtures of chemicals. Penn et al. found about 5000 chemicals in human skin headspace but did not discriminate between exogenous and endogenous substances. Of the materials they identified, 373 were consistent over time and statistical analysis of the results showed distinctive individual patterns (100). The ability to distinguish between the body odour of different people represents an amazing feat of pattern recognition by the brain because the differences between the odours of individual humans lie in the relative proportions of components rather than the identity of those components. Despite Freud's assertion that humans are microsmatic, there is plenty of evidence (see the examples in Chapter 1) that we are capable of performing this remarkable analytical task.

Three interesting questions arise from this. Firstly, what are the components of human sweat volatiles? Secondly, how are they formed? Thirdly, is the variation dependent on the human, their skin flora or both? The key odorous chemical constituents of human sweat odour fall into three main groups, namely fatty acids, sulfanyl alcohols and steroids.

Preti et al. (101) identified 25 of the acids and some γ-lactones. The acids were a mixture of straight- and branched-chain acids, both saturated and unsaturated, mostly in the C6–C11 range. They confirmed the presence of both (*E*)- and (*Z*)-3-methyl-2-hexenoic acids (**5.277** and **5.278**, respectively) and the importance of the former to the characteristic axilla malodour (Figure 5.59). They also reported a number of 4-ethyl-substituted acids and 4-ethyloctanoic acid (**5.279**) in particular. This acid is commonly known as *goat acid* since it is found in male goats. Its odour is very characteristic of the odour of goats, and it has a very low odour threshold. Despite a previous report (102) claiming that isovaleric acid is an important component of axilla malodour, they did not report finding it in their samples. Isovaleric acid is, however, known to be a key contributor to the odour of sweaty feet where it is produced by degradation of leucine by *Staphylococcus epidermis* (103). Natsch et al. (104,105) reported finding a variety of carboxylic acids, hydroxy acids and dicarboxylic acids in sweat malodour. They showed that these are present in fresh sweat in the form of glutamine conjugates, and the free acids are released by a zinc-dependant N_α-acyl-glutamine aminocyclase (N-AGA) which is produced by *Corynebacteria* species. This reaction is shown in Figure 5.60 using the glutamine

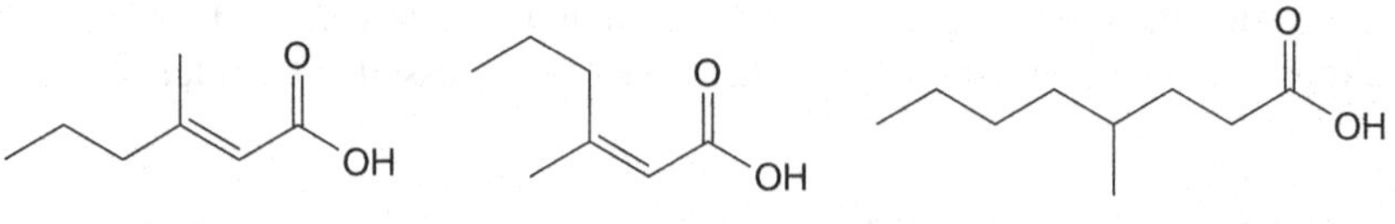

Figure 5.59   Some key sweat acids.

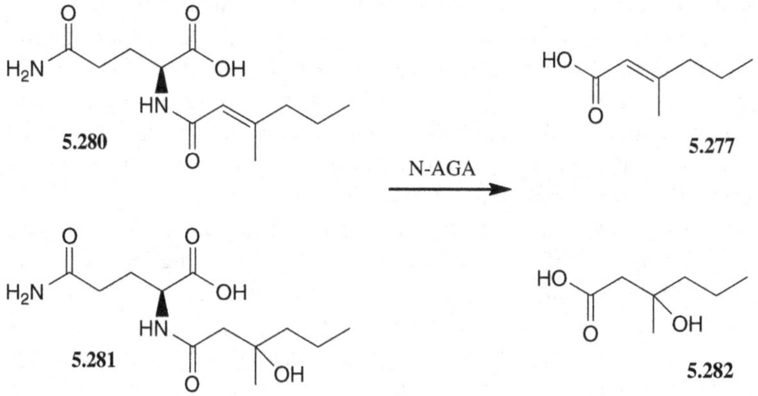

Figure 5.60   Release of sweat acids from their glutamine conjugates.

conjugates (**5.280** and **5.281**, respectively) of (E)-3-methyl-2-hexenoic acid (**5.277**) and 3-hydroxy-3-methylhexanoic acid (**5.282**). These two acids are generally considered to be the most significant in terms of axillary malodour.

Starkenmann et al. (106) identified the cysteine-glycine dipeptide S-conjugate (**5.283**) of 3-methyl-3-sulfanyl-1-hexanol (**5.284**) in fresh human sweat and suggested that the malodorous sulfanyl alcohol might be released from it by a β-lyase from *Staphylococcus haemolyticus*. However, Natsch et al. (107) showed that two enzymes are involved. Firstly, the peptide link is cleaved by a peptidase to give (**5.285**), and then the sulfanyl alcohol is released by cystathione- β-lyase from *Corynebacteria*. These authors also reported the occurrence of three other sulphanyl alcohols in sweat malodour: 2-methyl-3-sulphanyl-1-butanol (**5.286**), 3-sulphanyl-1-pentanol (**5.287**) and 3-sulphanyl-1-hexanol (**5.288**). All of these materials have very intense odours. For example, the detection threshold in air of (**5.287**) is 2 pg/l and that of (**5.284**) is 1 pg/l (Figure 5.61).

Starkenmann et al. (108) showed that the ratio of the precursors for 3-hydroxy-3-methylhexanoic acid (**5.282**) to 3-methyl-3-sulphanyl-1-hexanol (**5.284**) was about three times higher in men than in women. Women have much greater potential to produce the latter, which is generally considered to be the more unpleasant of the two, in their axillary malodour.

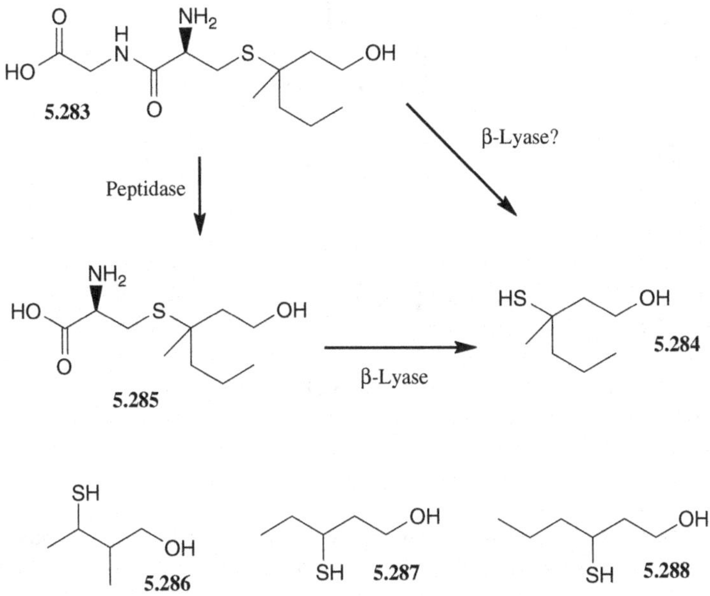

Figure 5.61   Release of sweat sulphanyl alcohols from their conjugates.

The biochemistry of sweat production by these routes has been reviewed by Gautschi et al., who also discuss the possibility of inhibition of the enzymes involved (109).

5-α-Androstenone (androstenone) (**5.289**) was first identified as a component of human sweat by Gower (110). It is of particular importance in the sweat of males. Previously, it had been well known as a metabolite of male pigs and was linked to the distinctive off-note (boar taint) in the taste of their meat (111,112). It is also present in truffles (*Tuber melanosporum*) and this possibly accounts for the use of pigs in finding these (113). The possible role of androstenone as a sex pheromone in pigs was discussed in Chapter 1. That role and the relationship of androstenone to androgens such as testosterone have led to considerable interest in its role in humans despite its lack of any androgenic hormonal activity.

5-α-Androstenone has a urinous odour, whereas 3-α-androstenol (**5.290**) (another sweat component) (i.e. OH group in 3-position and pointing down) has a more pleasant odour, often described as musky or sandalwood. Anosmias to androstenone are discussed in Chapter 4. For those who can detect it, it has a very low odour threshold, which Baydar et al. estimated to be 0.0008 ppb (Figure 5.62) (114).

Rennie et al. showed that it is *Coryneform* rather than other bacteria that are responsible for steroid metabolism, just as for the acids and sulfanyl alcohols (115). Gower et al. showed that the two odorous steroids are produced from androstadienone (**5.291**) and androstadienol (**5.292**), both of which have much higher odour detection thresholds and are therefore usually present at levels below

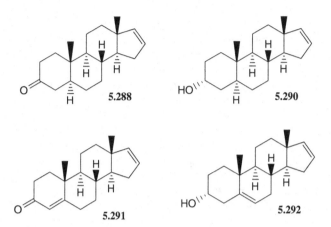

Figure 5.62   Steroids in sweat.

those detectable by smell (116). They suggested that low-odour individuals have lower *Coryneform* populations on skin of axilla rather than lower levels of the precursors.

Sulfate esters of some steroids were discovered in sweat, and it was suggested that these might be precursors for the odorous steroids (117); but this hypothesis has little experimental support and the necessary precursors are not amongst those which have been found.

As mentioned above, human axillary malodour is a very complex mixture of chemicals. Kuhn and Natsch (118) have shown that, whilst there are normally considerable differences in sweat odour composition between individuals, the differences in odour between twins is of the same order as that between one individual on different days. There has been much speculation in the literature that differences in sweat volatiles composition might be linked to the major histocompatibility complex, but Natsch et al. (119) have found no evidence to support this hypothesis.

REFERENCES

1. A. Guenther, T. Karl, P. Harley, C. Wiedinmyer, P. I. Palmer, and C. Geron, *Atmos. Chem. Phys. Discuss.*, **2006**, *6(1)*, 107–173.
2. J. D. Bu'Lock, *The Biosynthesis of Natural Products*, McGraw-Hill, New York, 1965.
3. J. Mann, R. S. Davidson, J. B. Hobbs, D. V. Banthorpe, and J. B. Harbourne, *Natural Products: Their Chemistry and Biological Significance*, Longman, London, 1994.
4. R. Croteau, *Chem. Rev.* **1987**, *87*, 929.
5. M. Rohmer, *Pure Appl. Chem.*, **2003**, *75(2-3)*, 375–387.
6. T. K. Devon and A. I. Scott, *Handbook of Naturally Occurring Compounds*, Vol. 2, The Terpenes, Academic Press, New York, 1971.
7. C. S. Sell, *Chem. Biodivers.*, 2004, *1*, 1899.
8. M. H. Boelens, H. Boelens, L. J. van Gemert, *Perfumer & Flavorist*, 1993, *18*, 1–15.
9. A. F. Thomas and Y. Bessière, *Nat. Prod. Rep.*, **1989**, *6(3)*, 291.
10. R. B. Bates, E. S. Caldwell, and H. P. Klein, *J. Org. Chem.* **1969**, *34*, 2615.

11. S. T. Ohashi et al., *Perfumer & Flavorist*, 1997, *22(2)*, 1.
12. G. S. Clark, *Perfumer & Flavorist*, 1988, *13(4)*, 49–54.
13. H. U. Däniker, *Flavors and Fragrances (Worldwide)*, SRI International, Stanford, 1987.
14. G. S. Clark, *Perfumer & Flavorist*, 1998, *23(3)*, 19.
15. S. Arctander, *Perfume and Flavor Chemicals (Aroma Chemicals)*, Vols. I–II, Montclair, N.J., 1969.
16. T. Yamamoto, US 5,773, 410, 1998, to Takasago.
17. G. S. Clark, *Perfumer & Flavorist*, 1994, *19(3)*, 41–45.
18. G. S. Clark, *Perfumer & Flavorist*, 1989, *14(3)*, 35–40.
19. L. Friedman and J. G. Miller, *Science*, **1971**, *172*, 1044.
20. T. J. Leitereg, D. G. Guadagni, J. Harris, T. R. Mon, and R. Teranishi, *J. Agric. Food Chem.*, **1971**, *19*, 785.
21. G. S. Clark, *Perfumer & Flavorist*, 1995, *20(1)*, 41.
22. G. Clark, *Perfumer & Flavorist*, 2000, *25*, 3.
23. G. Ohloff, *Tetrahedron Lett.*, **1960**, *1(32)*, 10.
24. D. E. Sasser, US 5,094,720, 1992, to Union Camp Corp.
25. (a) C. Moreuil, and R. Huet, *Fruits*, **1973**, *28*, 703. (b) A. Sato, K. Asano, and T. Sato, *J. Essent. Oil Res.*, **1990**, *2*, 179.
26. L. P. Somogyi, B. Rhomberg, and N. Takei, *Flavors and Fragrances*, SRI International, Stanford (CA), 1995.
27. E. Guenther, *The Essential Oils*, Vol. IV, D. Van Nostrand Co., Inc., New York, 1950, pp. 256–328.
28. E. Klein and W. Rojahn, 6th International Congress on Essential Oils, paper No. 163, San Francisco, Calif., 1974.
29. B. A. McAndrew, *Perfumer & Flavorist*, 1992, *17(4)*, 1.
30. C. S. Sell, *A Fragrant Introduction to Terpenoid Chemistry*, Royal Society of Chemistry, Cambridge, 2003, ISBN 0 85404 681 X.
31. T. S. Santhanakrishnau, R. Sohti, U. R. Nayak, and S. Dev, *Tetrahedron*, **1970**, *26*, 65.
32. (a) G. Ferber, *Perf. Cosmet.* **1978**, *68*, 18. (b) H. R. Ansari, N. Unwin, and H. R. Wagner, US 4,100,110, 1978, to Bush Boake Allen.
33. G. C. Kitchens, J. Dorsky, and K. Kaiser, *Givaudanian* , **1971**, *1*, 3.
34. J. H. Blumenthal, US 3,373,208, 1968, to International Flavors & Fragrances.
35. T. F. Wood, *Givaudanian*, **1970**, *1*, 3.
36. H. U. Däniker, A. R. Hochstetler, K. Kaiser, and G. C. Kitchens, *J. Org. Chem.*, **1972**, *37*, 1.
37. H. U. Däniker, A. R. Hochstetler, K. Kaiser, and G. C. Kitchens, *J. Org. Chem.*, **1972**, *37*, 6.
38. W. G. Dauben, L. E. Friedrich, P. Oberhänsli, and E. I. Aoyagi, *J. Org. Chem.*, **1972**, *37*, 9.
39. B. A. McAndrew, S. E. Meakins, C. S. Sell, and C. Brown, *J. Chem. Soc. Perkin Trans. 1*, **1983**, 1373, doi: 10.1039/P19830001373
40. F. S. M. Zaoral and V. Herout, *Coll. Czech. Chem. Comm.*, **1953**, *16*, 626.
41. C. F. Seidel, P. H. Müller, and H. Schinz, *Helv. Chim. Acta*, **1944**, *27*, 738.
42. K. G. O'Brien, A. R. Penfold, and R. L. Werner, *Aust. J. Chem.*, **1953**, *6*, 166.
43. K. G. O'Brien, A. R. Penfold, M. D. Sutherland, and R. L. Werner, *Aust. J. Chem.*, **1954**, *7*, 298.
44. C. Menut, P. Cabalion, E. Hnawia, H. Aganiet, J. Waikedre, and A. Fruchier, *Flavour Fragr. J.*, **2005**, *20(6)*, 621.
45. E. Fleikamp, G. Nonnenmacher, and O. Isaac, *Z. Naturforsch. (B)*, **1981**, *36*, 114–118.
46. E.-J. Brunke, F.-J. Hammerschmidt, *Dragoco Report*, 1984, *2*, 37.
47. M. Hava and J. Janku, *Rev. Czech. Med.*, **1957**, *3*, 130.
48. G. C. Dull, J. L. Fairley, R. Y. Gottshall, and E. H. Lucas, *Antibiot. Anmerkungen*, **1956-7**, 682.
49. J. V. Janku and C. Zita, *Cs. Farmacie*, **1954**, *3*, 93.
50. C. Zita and B. Steklova, *Cs. lek Ces*, **1955**, *94*, 204.
51. V. Yakoviev and A. von Schlichtergral, *Arzneim. Forschung*, **1969**, *19*, 615.
52. E. Klein, W. Rojahn, *Dragoco Report*, 1977, *24 (4)*, 79.
53. C. D. Gutsche, J. R. Maycock, and C. T. Chang, *Tetrahedron*, **1968**, *24*, 859.
54. B. M. Lawrence, Perfumer & Flavorist, 1980, *5(6)*, 27.
55. M. I. Velazco, L. Wünsche, and P. Deladoey, US 6,214,788, 2001, to Firmenich.

56. M. I. Velazco, L. Wünsche, and P. Deladoey, EP Application 1,541,039 A1, 2005, to Firmenich.
57. M. I. Velazco, L. Wünsche, and P. Deladoey, CH 60699, 1999, to Firmenich.
58. G. L. K. Hunter and W. B. Brogden, *J. Food Sci.*, **1965**, *30*. 876
59. G. Ohloff, DE 1,948,033, 1970, to Firmenich.
60. G. Ohloff, *Scent and Fragrances*, transl. by W. Pickenhagen and B. M. Lawrence, Springer-Verlag, Berlin, 1994, p. 136.
61. http://www. Leffingwell.com. Accessed 2013 November 27 and references contained therein.
62. T. W. Goodwin, *Chemistry and Biochemistry of Plant Pigments*, 2nd Edn., Academic Press, Inc., New York, 1976.
63. O. Isler, R. Rügg, and U. Schwieter, *Pure Appl. Chem.*, **1976**, *14*, 245.
64. O. Isler, *Pure Appl. Chem.*, **1979**, *51*, 447.
65. E. Klein and W. Rojahn, DE 2,807,584, 1978, to Dragoco Gerberding.
66. K. H. Schulte-Elte, W. Giersch, B. Winter, H. Pamingle, and G. Ohloff, *Helv. Chim. Acta*, **1985**, *68(7)*, 1961.
67. K. H. Schulte-Elte, G. Ohloff, B. L. Müller, and W. Giersch, EP 118,809, 1984, to Firmenich.
68. K-H Schulte-Elte, C. Margot, C. Chapuis, D. P. Simmons, and D. Reichlin, EP 0 457,022 B1, 1991, to Firmenich.
69. K-H. Schulte-Elte and G. Ohloff, US 4,623,750, 1983, to Firmenich.
70. K-H. Schulte-Elte, US 4,626,381, 1983, to Firmenich.
71. K-H. Schulte-Elte and G. Ohloff, US 4,711,875, 1983, to Firmenich.
72. K-H. Schulte-Elte and H. Pamingle, EP 121,828, 1983, to Firmenich.
73. E. Kovats, *J. Chromatogr.*, **1987**, *406*, 185.
74. E. P. Demole, P. Enggist, U. Säuberli, M. Stoll, and E. Kovats, *Helv. Chim. Acta*, **1970**, *53*, 541.
75. A. Williams, *Perfumer & Flavorist*, 2002, *27(2)*, 18.
76. P. Weyerstahl and K. Licha, *Liebigs Ann. Chem.*, **1996**, *(5)*, 809–814, doi: 10.1002/jlac. 199619960527
77. K. S. Ayyar, R. C. Cookson, and D. A. Kagi, *J. Chem. Soc. Chem. Commun.*, **1973**, 161.
78. C. Fehr and J. Galindo, *J. Am. Chem. Soc.*, **1988**, *110*, 6909.
79. C. Fehr and J. Galindo, EP 326869, 1989, to Firmenich.
80. C. Fehr and J. Galindo, *J. Org. Chem.*, **1988**, *53*, 1828.
81. H. J. Wille, W. M. B. Könst, and J. Kos, GB 1456152, 1974, to Naarden International.
82. P. A. Ochsner and H. Schenk, FR 2,296,611, 1976, to Givaudan.
83. H. Schenk, FR 2,327,226, 1977, to Givaudan.
84. P. A. Ochsner and H. Schenk, US 4,006,108, 1977, to Givaudan.
85. C. Fehr and J. Galindo, US 5,015,625, 1991, to Firmenich.
86. E. J. Lederer, *J. Chem. Soc.*, **1949**, 2115–2125. doi: 10.1039/JR9490002115
87. G. Ohloff, The Fragrance of Ambergris, in *Fragrance Chemistry*, Ed. E. T. Theimer), Academic Press, New York, 1982.
88. C. S. Sell, *Chem. Ind.*, **1990**, *16*, 520.
89. B. D. Mookherjee and R. R. Patel, Proceedings 7th International Conference of Essential Oils, Kyoto, **1977**, paper 137, 479. Chem Abstr, **1980**, *92*, 203412h.
90. E. Lederer, *Fortschritte der chemie organischer Naturstoffe*, Springer-Verlag, 1950, 6, p. 120.
91. G. Ohloff, K-H. Schulte Elte, and B. L. Muller, *Helv. Chim. Acta*, **1977**, *60*, 2763.
92. K-H. Schulte Elte, B. L. Muller, and G. Ohloff, *Nouv. J. Chim.*, **1978**, *2*, 247.
93. C. P. Newman, K. J. Rossiter, and C. S. Sell, EP 276998, 1988, to Unilever.
94. J. Mann, R. S. Davidson, J. B. Hobbs, D. V. Banthorpe, and J. B. Harbourne, *Natural Products: Their Chemistry and Biological Significance*, Longman, Harlow, 1994.
95. Sell, C. S. *The Chemistry of Fragrances From Perfumer to Consumer*, 2nd Edn, Royal Society of Chemistry, Cambridge, 2006.
96. J. Hurley and W. Shelley, *The Human Apocrine Gland in Health and Disease*, Charles C Thomas, Springfield, Illinois, 1960.
97. N. Shehadeh and A. M. Kligman, *J. Invest. Dermatol.*, **1963**, *41*, 1–5.
98. D. B. Gower *Biochemistry of Steroid Hormones*, 2nd Edn., Ed. H. L. J. Makin, Blackwell, Oxford, 1984, pp. 170–206.

99. J. J. Leyden, K. J. McGinley, E. Hoelzle, J. N. Labows, and A. M. Kligman, *J. Invest. Dermatol.*, **1981**, *77*, 413–416.

100. D. J. Penn, E. Oberzaucher, K. Grammer, G. Fischer, H. A. Soini, D. Wiesler, M. V. Novotny, and R. G. Brereton, *J. R. Soc. Interface*, **2007**, *4(13)*, 331–340.

101. X.-N. Zeng, J. L. Leyden, H. J. Lawley, K. Sawano, I. Nohara, and G. Preti, *J. Chem. Ecol.*, **1991**, *17(7)*, 1469–1492.

102. J. N. Labows, Odour detection, generation and etiology in the axilla, in *Antiperspirants and Deodorants*, Eds. C. Felden and K. Laden, Marcel Dekker, New York, 1988, pp. 321–343.

103. K. Ara, M. Hama, S. Akiba, K. Koike, K. Okisaka, T. Hagura, T. Kamiya, and F. Tomita *Can. J. Microbiol.*, **2006**, *52*, 357–364.

104. A. Natsch, H. Gfeller, P. Gygax, J. Schmid, and G. Acuna, *J. Biol. Chem.*, **2003**, *278(8)*, 5718–5727.

105. A. Natsch, S. Derrer, F. Flachsmann, and J. Schmid, *Chem. Biodivers.*, **2006**, *3*, 1–19.

106. C. Starkenmann, Y. Niclass, M. Toccaz, and A. J. Clark, *Chem. Biodivers.*, **2005**, *2*, 705–716.

107. A. Natsch, J. Schmid, and F. Flachsmann, *Chem. Biodivers.*, **2004**, *1*, 1058–1072.

108. M. Troccaz, G. Borchard, C. Vuilleumier, S. Raviot-Derrien, Y. Niclass, S. Berucci, and C. Starkenmann, *Chem. Senses*, **2009**, *34*, 203–210.

109. M. Gautschi, A. Natsch, and F. Schröder, *Chimia*, **2007**, *61*, 27–32 doi: 10.2533/chimia.2007.27.

110. D. B. Gower, *J. Steroid Biochem.*, **1972**, *3*, 45–103.

111. D. B. Gower and B. A. Ruparelia, *J. Endocrinol.*, **1993**, *137*, 167–187.

112. V. Prelog and L. Ruzicka, *Helv. Chim. Acta*, **1944**, *27*, 61.

113. P. Z. Margalith *Steroid Microbiology*, Charles C Thomas, Springfield, IL 1986.

114. A. Baydar, M. Petrzilka, and M.-P. Schott, *Chem. Senses*, **1993**, *18*, 661–668.

115. P. J. Rennie, D. B. Gower, K. T. Holland, A. I. Mallet, and W. J. Watkins, *Int. J. Cosmet. Sci.*, **1990**, *12*, 197–202.

116. D. B. Gower, K. T. Holland, A. I. Mallet, P. J. Rennie, and W. J. Watkins, *J. Steroid Biochem. Mol. Biol.*, **1994**, *48*, 409–418.

117. J. N. Labows *Antiperspirants and Deodorants*, Eds. C. Felgen and K. Laden, Marcel Dekker, New York, 1988.

118. F. Kuhn and A. Natsch, J. R. , *Soc. Interface*, **2009**, *6*, 377–392 doi: 10.1098/rsif.2008.0223.

119. A. Natsch, F. Kuhn, and J.-M. Tiercy, *J. Chem. Ecol.* **2010**, *36*, 837–846. doi: 10.1007/S10886-010-9826-y.

Chapter 6

Manufacture of Fragrance Ingredients

INTRODUCTION

In his book *Histories*, the Greek writer Herodatus included a section on the perfumes of Arabia (1). He asked Arabian perfumers about the source of their raw materials and was told a pack of lies. Manufacturing perfume is relatively easy, if you know the formula and if you have the ingredients. In order to protect their business, the perfumers of Herodatus's era relied on secrecy to protect their formulae and their raw material sources. Nowadays, modern analytical chemistry means that formulae can no longer be kept secret, and information such as trade statistics and necessary disclosures regarding chemical manufacture make raw material sources a matter of public knowledge. So, modern fragrance houses rely on patented materials and superior manufacturing technology as the best methods of protecting their business. The design and development of patentable ingredients will be discussed in Chapter 7. This chapter deals with the manufacture of fragrance ingredients.

Top flavour and fragrance (F&F) companies must have strong ingredient portfolios to survive. Those that do not quickly move down the league table. They need a continual flow of new molecules in order to cope with changing market needs such as changes in consumer product chemistry, changing cost (feedstock, technology), changing environmental constraints, changing safety constraints and so on, as will be discussed in the next chapter. They also need to be the lowest cost producer of a reasonable percentage of their ingredient palette in order to protect themselves from competitors and others. For example, the largest scale products can become targets for major chemical companies whose scale and technological strength can result in them being able to manufacture a product at lower cost than can a fragrance house. If all of the ingredients in a perfume formula can be purchased at the same cost or at lower cost than available to the fragrance house, then that piece of business will soon be lost to a competitor or a contract blending operation and the fragrance house will have squandered its perfume creation and development investment. One problem for the F&F company is to know where its production cost stands relative to the market since a co-producer never knows the real market price. A competing supplier will never disclose their real selling price but will always imply a lower figure in an attempt to lead the F&F company to cease production. Similarly, a customer company will always imply that it can source an ingredient more cheaply elsewhere in order to lead the F&F company to reduce its selling price. Strong nerves and good technical assessment of competitors' likely production costs are needed when setting prices and deciding whether to continue production of any fragrance ingredient.

One advantage that the F&F companies have is the understanding of odour quality. Fragrance ingredients are produced to high standards of consistent chemical purity but that is only part of the story. Odour quality is also vital. To take a ridiculously extreme example, one part per billion of hydrogen sulfide will have a greater effect on the odour of a product than would 20% by weight of propylene glycol. The role of the quality control perfumer is essential in fragrance chemical production, and the production chemist must have the necessary skills in purification in order to meet the exacting demands of the former. I remember one instance where a salesman from a chemical company presented a sample of their latest proposed product to a quality control perfumer and was told, just as the stopper was removed from the bottle, that the odour quality was totally unacceptable. A chemical purity of 99.5% was woefully inadequate in view of the intense off-notes emanating from the bottle. The inability to produce consistently acceptable odour quality is one key reason why chemical companies have always failed to break into the fragrance ingredients market without the help of an F&F company. The other key factors relate to the economic realities of the F&F industry.

To put fragrance in context, the total world volume of fragrance is of the order of 300,000 tonnes, about enough to half-fill one supertanker. The single largest volume fragrance/flavour ingredient is menthol with an annual consumption of about 12,000 tonnes. Such volumes are dwarfed by those of other industries. For instance, the annual production of methyl methacrylate, the monomer for making poly (methyl methacrylate) (commonly known as *Perspex* or *Plexiglass*) is about 2,200,000 tonnes (2) and that of ethylene is about 16,000,000 tonnes. In terms of environmental impact, the world fragrance volume is even tinier when compared to the release of volatile organic chemicals by plants. For example, it is estimated that trees liberate 100,000,000 tonnes of isoprene into the atmosphere annually (3). The low tonnage of fragrance ingredients makes production at an acceptable cost a significant challenge for the process chemist, especially for those ingredients with more complex molecular structures. In general, there is an exponential relationship between price and volume of fragrance ingredients. Some essential oils have prices in the region of thousands of dollars per kilogram, but their use is limited to fine fragrance and can be measured in tens or hundreds of kilograms per annum. A very few synthetic or semi-synthetic chemicals command prices in the hundreds of dollars per kilogram. The large volume fragrance materials such as geraniol, linalool or phenylethanol have prices under \$10/kg. Just as the tonnage of fragrance ingredients is dwarfed by those of petrochemicals, so their value is dwarfed by that of pharmaceuticals. The anti-cholesterol drug Lipitor is worth \$11,900,000,000 per annum which is 5–50 times the value of the high volume fragrance ingredients, and sales of that one drug alone are considerably more than those of the entire fragrance industry. In 2011, the five largest fragrance companies, with their market (%) share in brackets, were as follows: Givaudan (19.1), Firmenich (13.6), IFF (12.6), Symrise (9.4) and Takasago (6.8) (4). These five companies dominate the fragrance ingredients market.

Perhaps the oldest written perfume formula is that described in the book of Exodus (Exodus, 30, 22–38). It consists of myrrh (5.75 kg), cinnamon (2.875 kg), calamus (2.875 kg) and cassia (5.75 kg) in olive oil (4 l). Such a formula is typical of ancient perfumery. The ingredients are all botanical extracts, and the solvent is olive oil. Perfumery remained much the same until the Middle Ages when Arabs began to apply distillation as a technology. This meant that essential oils came into use, and also ethanol became available as a solvent for extraction of ingredients from plant material and a replacement for fixed oils, such as olive oil, in the final perfume. Distillation completely revolutionised perfumery and so, by way of example, all of the ingredients of the famous Esterhazy bouquet (Table 6.1) (5) of eighteenth

Table 6.1   The Esterhazy Bouquet

- Extrait de fleur d'orange (from pomade), 1 pint

- Esprit de rose (triple), 1 pint

- Extract of vetivert, 1 pint

- Extract of vanilla, 1 pint

- Extract of orris, 1 pint

- Extract of tonquin, 1 pint

- Esprit de neroli 1 pint

- Extract of ambergris, 1/2 pint

- Otto of santal, 1/2 drachm

- Otto of cloves, 1/2 drachm.

century Austria were products of either distillation (esprit and otto) or extraction using a distilled solvent (extract or extrait).

The second major technological innovation of the perfume industry was the introduction of synthetic organic chemistry in the nineteenth century. The first synthetic materials to be used as fragrance ingredients were compounds such as coumarin, heliotropin and vanillin. These chemicals are all components of natural oils and extracts, and so are nowadays referred to as 'nature-identical'. The three mentioned were all present in the landmark fragrance Jicky which was launched in 1889. The first reported syntheses of the compounds were in 1868, 1869 and 1877, respectively. By the early 1920s, perfumers had begun to experiment with totally novel entities, and the most celebrated success was the use of non-nature-identical aldehydes providing a novel top note to complement the natural rose and jasmine oils of Chanel 5. Ernest Beaux, the perfumer who created this most famous of fragrances, remarked: 'One has to rely on chemists to find new aroma chemicals creating new, original notes. In perfumery, the future lies primarily in the hands of chemists'. Beaux's belief in the role of chemistry was certainly correct, though chemistry has proved to be of fundamental importance in many more ways than that foreseen by Beaux. Process chemistry is one discipline that has come to be increasingly vital for modern perfumery.

NATURAL FRAGRANCE INGREDIENTS

The fragrance industry still uses plant materials as ingredients. Volatile odorants can be extracted from plant material by expression, distillation or solvent extraction.

The term 'expression' refers to the use of physical pressure to crush plant material and force oils out of the matrix, which usually is a combination of lignin and cellulose. This is an important process for non-volatile or 'fixed' oils such as those of olive, sunflower, rapeseed or walnut. The only volatile oils that are still extracted in this way are the citrus peel oils. Originally, these oils were extracted by hand using a metal tool to press the peel and squeeze out the oil. Nowadays, a large proportion of commercial citrus oils are produced as by-products of the fruit juice industry. For example, whole oranges are crushed mechanically, filtered or centrifuged to remove the solids and the peel oil separated from the juice. Specialist equipment is used to extract oils such as bergamot in those cases where the juice is not used as food.

Extraction of oils by distillation almost always means steam distillation. The presence of water in the system limits the temperature in the still and, therefore, reduces thermal decomposition of more delicate chemicals in the oil. The yield of essential oil is usually of the order of only a few percent of the dry weight of the plant material and, therefore, distillation is usually carried out close to the site where the plants are grown in order to reduce transportation costs. Water is becoming an increasingly scarce resource, especially in many of the regions (such as North Africa) where the largest oil production is found, and so the water is recycled in the still by means of return of the aqueous phase of the distillate to the still pot. The original device for achieving this was known as a *Florentine flask*. At the end of the distillation, the aqueous phase, known as the *waters of cohobation*, is usually discarded but, in the case of rose oil, which has a significant solubility in water, the water is used as a separate ingredient. Rose water is the aqueous phase from rose distillation and is a traditional ingredient of Turkish delight. Otto and Attar are terms, derived from Arabic, used in perfumery for distilled oils. The French term 'ésprit de' and the English 'spirits of' give a clue to the ancient Aristotelian thinking regarding distillation. Aristotle postulated four basic physical elements, namely, fire, air, earth and water, with a fifth element, quintessence, to account for the life force or spirit. Distillation was thought to be the process of removing this spirit from the plant, hence quintessential oil, nowadays abbreviated to essential oil.

Hydrodiffusion is a technique similar to distillation except that the steam is introduced at the top of the vessel containing the plant material and the water/oil mixture is removed from the bottom. The main function of the steam is to burst the oil-bearing glands in the plant tissue, rather than to vaporise the oil. It, therefore, uses less energy than distillation and is a gentler process but, since the product is not distilled, the oil can contain non-volatile materials.

Essential oils usually contain terpenoid hydrocarbons. These can be components in the plant or can be formed by dehydration during processing. Removal of the monoterpene hydrocarbons can be achieved by fractional distillation and the products, which are enriched in the more odorous oxygenated components, are known as *deterpenated* or *folded oils*.

Ancient perfumery did use solvent extraction, though the available solvents were limited to triglyceride oils and fats. Enfleurage is a process that was used from ancient times up to the early twentieth century. Plant material was pressed into a layer of fat and left to stand while the volatile plant oils diffused into the fat. The fat was then melted and filtered to remove the solid plant material. Re-solidification gave a product known as a *concrète*. The ancient Egyptians placed cones of concrète on their heads to perfume their bodies as the fat melted in the heat. More recently, the concrète would be extracted with ethanol to separate the more ethanol-soluble oxygenated materials from the fat. The resultant extract is known as an 'absolute ex concrète'. It is also possible to produce an essential oil by distillation of the concrete. The term 'absolute' is used generally to describe an oil that has been solvent-extracted from either a concrete or directly from plant material. The major solvent of the nineteenth and early twentieth centuries was benzene. Nowadays, because of the toxicity of benzene, other solvents are used in its place. One is the azeotrope of ethyl acetate and hexane, which has similar boiling and solvation properties to those of benzene. Of increasing importance is carbon dioxide, either as a liquid or in the super-critical state.

The term 'tincture' refers to an ethanolic extract. Tinctures are less common nowadays, those of ambergris and vanilla perhaps being the most significant. Ambergris is animal-derived rather than plant-derived. It is produced in the intestinal tract of the sperm whale (*Physeter catodon*) in response to ingested irritants such as shells. The material excreted into the sea by the whale is dark brown in colour and relatively low in odour. Exposure to light, air and salt water enables various chemical reactions to occur, and the ambergris becomes lighter in colour, eventually turning pale grey, and much more highly odorous. An account of the chemistry can be found in the book by Sell (6). Ambergris is collected from beaches in various locations such as New Zealand. It has always been expensive and, because of the decline in whale numbers and the time required for beachcombing, the price has risen even higher. Some natural ambergris is used in the most expensive fragrances, but synthetic routes from plant or mineral sources nowadays supply almost all of the perfumery requirements via the key odour component.

Nowadays, many essential oils must go through a subsequent process to remove harmful components before they can be used in perfume. The psoralene bergaptene, for example, must be removed from bergamot oil because of its phototoxicity. This is relatively easy since bergaptene is less volatile than the odorous components of the oil. Some oils tend to contain metals, the best known example being patchouli which usually contains iron. The iron does not constitute a health hazard but does cause discoloration of the oil and products containing it. Distillation or treatment with a chelating agent can be used in this case.

Natural extracts are complex mixtures of chemicals and their composition varies with country of production, weather and many other factors. This makes analysis much more difficult than in the case of synthetic materials which are usually either single, pure chemicals or simple mixtures. Production of natural materials is labour intensive and low yielding, and so their costs are high. All of these factors combine to make adulteration commercially attractive, and fragrance companies must use the best analytical equipment and skilled analysts in order to avoid purchasing adulterated oils. The analytical methods required for this include such advanced technology as isotopic analysis. An account of this topic and a list of the commonest plant extracts can be found in Chapter 3 of the book edited by Sell (7).

In the flavour industry, there is a strong market for 'natural' and 'organic' ingredients. Some of this rubs off into the fragrance industry, but the cost implications are much more significant in fragrance than in flavour. Foe example, synthetic vanillin costs only a few dollars per kilogram, whereas the cost of one kilogram of natural vanillin will be in thousands of dollars. Because of consumer perception, the flavour industry can tolerate such a differential, but the fragrance industry cannot. To produce an 'organic' essential oil is more costly than the regular one, and there is no means of positive verification by chemical analysis. There is a market but it tends to be a niche one because of the cost.

If a fragrance ingredient can be extracted from a vegetable source, it is renewable but that does not necessarily make it sustainable. Over-harvesting of Indian sandalwood (*Santalum album*) has led to the tree becoming an endangered species and, therefore, production has essentially ceased. Similarly, rosewood oil is no longer used. But sustainability goes far beyond that. For example, large modern fragrance houses will not use any oil or extract that has been obtained using child labour. Leading companies nowadays tend to work with local growers to ensure that their agricultural and extraction practices are sustainable and, often, the companies will also contribute to social programmes in the communities. Examples would include the establishment of sustainable Australian sandalwood oil through Givaudan working with an environmental organisation and local government; R. C. Treatt helping provide wells in Indian villages to support the local agriculture (including essential oil production); and Givaudan working with Madagascan villagers to ensure sustainable vanilla production, and providing a local school in addition.

SYNTHETIC FRAGRANCE INGREDIENTS

Sustainability of agricultural practices affects natural feedstocks for chemical synthesis in the same way that it does essential oil production. So, a fragrance company seeking to move from petrochemical feedstocks to renewable alternatives must add sustainability into the equation. For instance, at present turpentine is an important raw material for the synthesis of fragrance ingredients, but most of the turpentine used for this purpose is sulfate turpentine, which is a by-product of the paper industry. As recycling of paper has increased, the need for fresh wood pulp has fallen and thus the supply of sulfate turpentine has become tighter. If all offices were truly to become paperless and newspapers were to be replaced by radio, television and electronic communication, then the supply of sulfate turpentine would be jeopardised. On the other hand, research into sustainable bio-fuels has led to the potential for sustainable production of raw materials by fermentation of biomass or readily available carbohydrates. An example is the production of farnesene (**6.1**) by the Californian biotech company Amyris. Some miscellaneous terpenoids are shown in Figure 6.1.

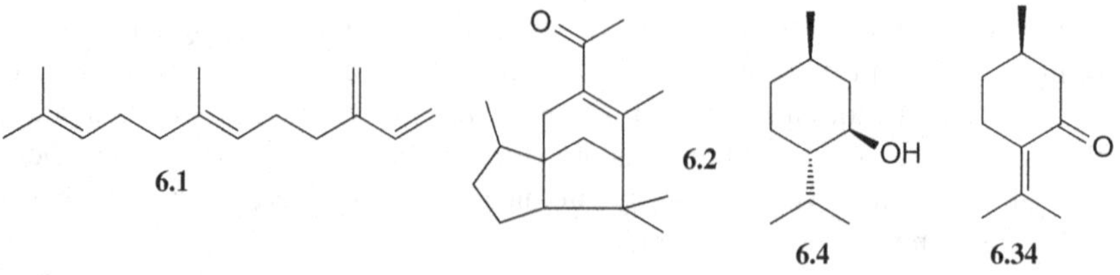

6.1   **6.2**   **6.4**   **6.34**

Figure 6.1   Miscellaneous terpenoids.

Some fragrance ingredients, such as cedrane derivatives like acetylcedrene (**6.2**), are prepared only from natural feedstocks because the cost of chemical synthesis would be prohibitive in view of the complexity of their molecular structure (leading to high cost) and their (relatively low) value in perfumery. Other ingredients are much more readily produced from basic petrochemicals. However, most fragrance ingredients are fairly equally available from either petrochemical- or plant-derived feedstocks. The balance between sources then depends on other factors such as technology, availability, cost and environmental issues. The section on terpenoid production below provides some good examples of the balance between plant-based and petrochemical-based routes to perfume ingredients.

Basic Principles of Modern Fragrance Chemicals Manufacturing

Whilst manufacturers of fine chemicals work to chemical and physical specifications, organoleptic specifications are also of crucial importance in the F&F industry. Tolerances around analytical parameters are often greater than those around odour standards. The organoleptic quality is important in determining price, and so, in most instances, various grades of a material will exist with very different prices. To the individual purchaser, the price will vary depending on the amount bought. The overall volume of use is determined by a combination of price and quality. For this reason, both prices and volumes vary depending on prevailing economic conditions.

When synthetic fragrance ingredients were first introduced, they offered the advantages over natural materials of lower cost, consistent quality and more reliable availability. These advantages had such a dramatic effect on the industry that issues such as chemical feedstock availability, energy consumption and waste were given lower priority. Nowadays, sustainability is the key word, and feedstocks, energy, environmental performance and process safety are strong driving forces of chemical development. No process will be perfect in all respects, and a balance giving the best overall performance will be sought. The long-term aim of the development chemist is to continually improve that balance to bring the process as close to perfection as possible. Catalytic methods have received particular attention over the last few decades; recent advances in their use for terpenoid manufacture have been reviewed by Swift (8). A more detailed account of the basic principles of chemical process development can be found in the book by Sell (6). A few examples will serve to illustrate how process chemistry has changed and improved since the beginning of the modern industry in the mid-nineteenth century and how the various sustainability drivers work together in design of new processes.

Synthesis of Citral

One of the best examples is the production of citral (**6.3**) and, from it, (−)-menthol (**6.4**). Atom efficiency (also known as *atom utilisation*) is a useful tool for the rapid

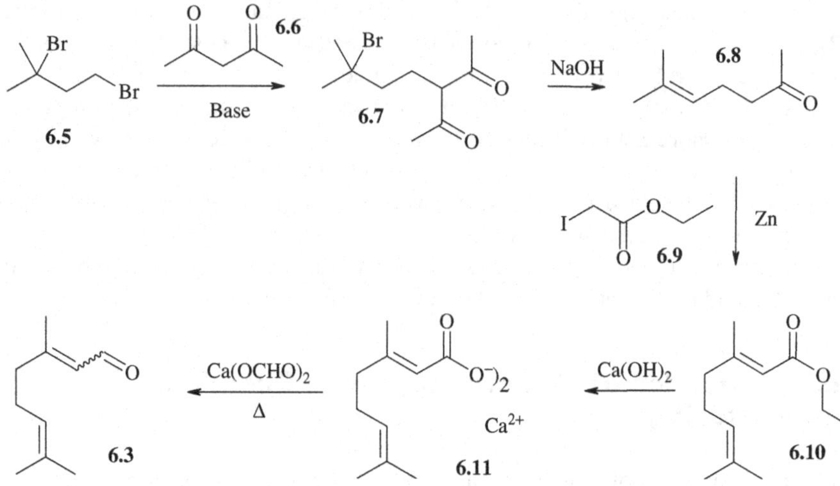

Figure 6.2   The Barbier–Bouveault–Tiemann synthesis of citral.

assessment of the environmental performance of a synthetic route. It is defined as the molecular weight of the desired product of a reaction divided by the total of the molecular weights of all species produced in the reaction, multiplied by 100 to give a percentage figure. Thus, different routes to a target can be compared (assuming 100% yield at every stage) to see how they compare. Estimates can then be refined by making realistic guesses about achievable yields. All of this can be done before carrying out any practical work, and so the process chemist can select only the most promising routes for development. We can also use the concept to see how the process chemistry industry has progressed in terms of environmental performance, as illustrated by the following account of processes for the manufacture of citral (**6.3**).

The earliest synthesis of citral was that of Barbier, Bouveault and Tiemann as shown in Figure 6.2. It was started by Barbier and Bouveault who synthesised ethyl geranate (**6.10**) in 1896 and completed by Tiemann in 1898 with the reduction of ethyl geranate (**6.10**) to citral (**6.3**) (9). The reaction of 1,3-dibromo-3-methylbutane (**6.5**) with acetylacetone (**6.6**) gave the bromoketone (**6.7**), which was dehydrobrominated and deacylated to give 6-methylhept-5-en-2-one (**6.8**). Reformatsky reaction with zinc and ethyl iodoacetate (**6.9**) gave ethyl geranate (**6.10**), which was hydrolysed to give calcium geranate (**6.11**), the pyrolysis of which with calcium formate gave citral (**6.3**). Granted that this is an academic route intended to establish the structure of citral (**6.3**) rather than to produce it in quantity, it still serves to illustrate that the use of stochiometric quantities of reagents leads to poor overall atom efficiency. In this case, a mere 20% of the mass produced would be the desired product, even if every step were to proceed with 100% yield.

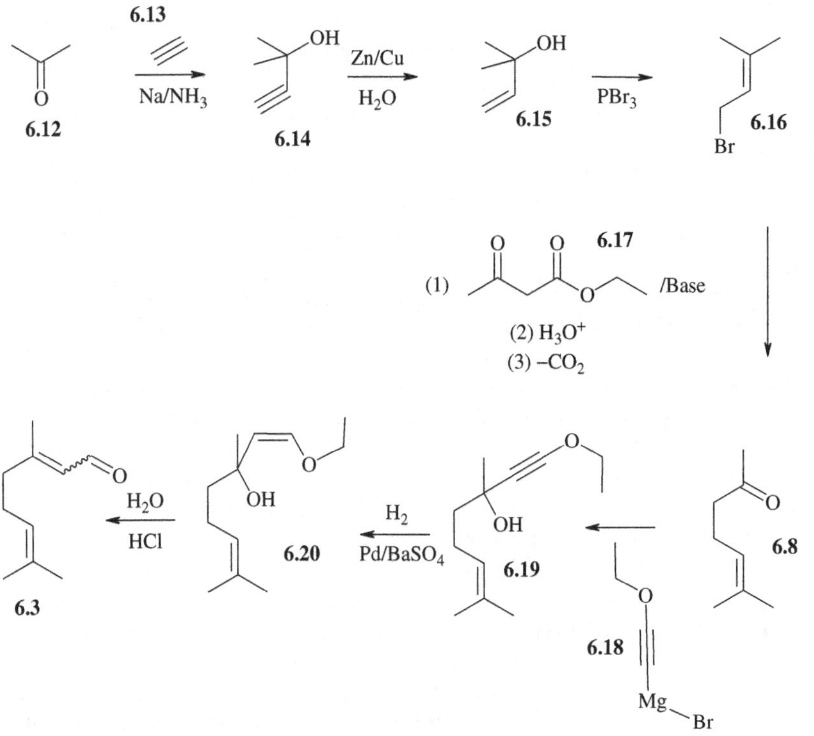

Figure 6.3 The Arens–van Dorp synthesis of citral.

In 1948, Arens and van Dorp achieved a new synthesis route of citral, which is shown in Figure 6.3. It still gave only 20% atom efficiency but paved the way for much more efficient processes because their strategy essentially used acetone (**6.12**) and acetylene (**6.13**) as starting materials, though the second molecule of each to be incorporated into the product was disguised by activating and/or protecting groups. This basic strategy was to prove key to later, much more efficient syntheses. Addition of acetylene to acetone gave methylbutynol (**6.14**), which was hydrogenated to methylbutenol (**6.15**). Treatment of this with phosphorus tribromide gave prenyl bromide (**6.16**), which could be alkylated with ethyl acetoacetate (**6.17**) (an activated equivalent of acetone) followed by hydrolysis and decarboxylation to give methylheptenone (**6.8**). The second acetylene unit carried an ethoxy function and was used as its magnesium salt (**6.18**). Lindlar hydrogenation of (**6.19**) thus produced gave the hydroxyether (**6.20**), which yielded citral (**6.3**) on hydrolysis.

The discovery of the Carroll reaction enabled methylbutenol (**6.15**) to be treated directly with ethyl acetoacetate (**6.17**) to give methylheptenone. Direct addition of acetylene (**6.13**) gave dehydrolinalool (**6.21**) and an acid-catalysed

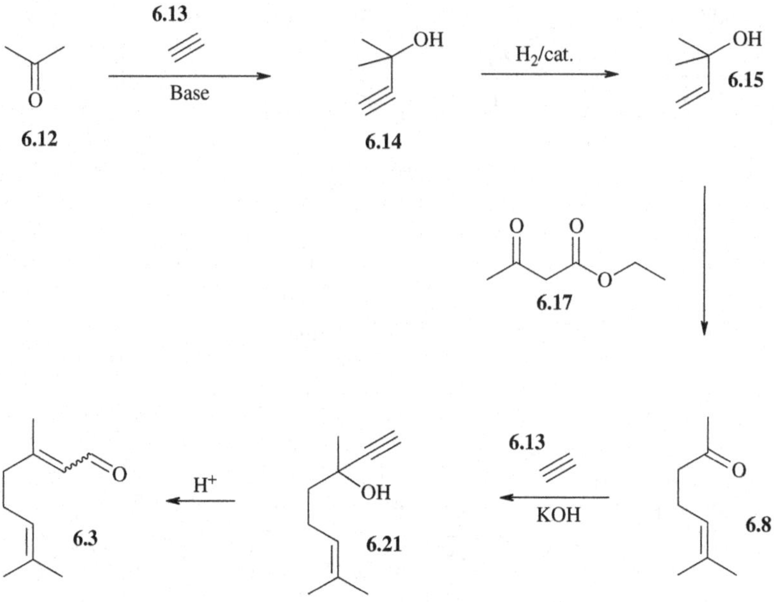

Figure 6.4   The Carroll improvement.

rearrangement gave citral (**6.3**). This improvement dates from about 1950 and improved the atom efficiency to 45% as shown in Figure 6.4.

Use of the Claisen rearrangement of the product of the reaction of methylbutenol (**6.15**) and the methyl enol ether of acetone (**6.22**) in the process developed in 1957 further improved the atom efficiency to 55% (Figure 6.5). The addition of acetylene to acetone under Favorski–Babayan conditions gives methylbutynol (**6.14**), which can be hydrogenated to methylbutenol (**6.15**) using a Lindlar catalyst (10–14).

When this is treated with methyl propenyl ether (the vinyl ether of acetone), trans-etherification takes place followed by a Claisen-Cope rearrangement to give methylheptenone (**6.8**) (15). Addition of acetylene produces dehydrolinalool (**6.21**), a key intermediate for subsequent conversions. Citral can be produced from dehydrolinalool via a Meyer-Schuster rearrangement (16, 17), rearrangement using a vanadate catalyst (18), or by rearrangement of its acetate in the presence of copper salts (19, 20), trisilylorthovanadates (21) or vanadium catalysts in the presence of silanols (22), and yields of up to 90% can be obtained (23, 24).

In about 1977, BASF developed the process shown in Figure 6.6, which gives 80% atom efficiency and the only by-product is water. So, although 20% of the weight is lost as water, this is environmentally benign. The ene reaction of isobutene (**6.23**) with formaldehyde (**6.24**) gives isoprenol (**6.25**) (25, 26). This

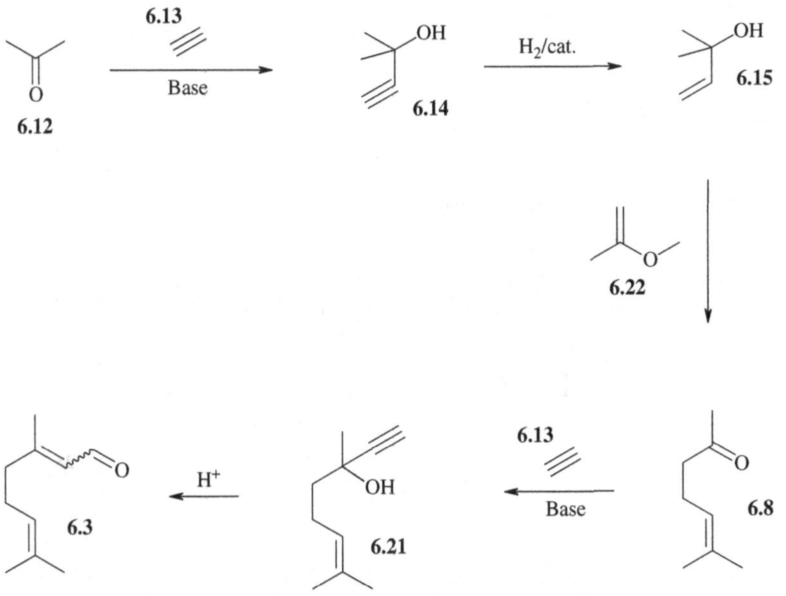

Figure 6.5   The Claisen improvement.

can be hydrogenated to give prenol (**6.26**) (27) or oxidised and isomerised to give prenal (**6.27**) (28–30). Formation of the prenyl enol ether of prenal (**6.28**) is followed by Claisen rearrangement to (**6.29**) (31, 32), which then undergoes a Cope rearrangement to give citral (**6.3**) (33–35).

Synthesis of (–)-Menthol

BASF have since extended this synthesis to (–)-menthol as shown in Figure 6.7 (36). The absolute stereochemistry of menthol is of crucial importance since one of the eight possible stereoisomers, namely, (–)-menthol, possesses a much stronger physiological cooling effect than the others and it is this property, rather than its odour, which is sought after in flavours and cosmetics. Any synthesis of menthol must, therefore, achieve high stereoselectivity to be commercially valuable. The substance known as *citral* (**6.3**) and produced by routes such as those described above consists of a mixture of (*E*)- and (*Z*)-isomers, known as *geranial* (**6.30**) and *neral* (**6.31**), respectively. These can be separated by distillation, and each hydrogenated using the appropriate enantiomer of a chiral catalyst gives (*R*)-citronellal (**6.32**). This can be cyclised using an acidic catalyst, and the one chiral centre of the starting (**6.32**) directs the stereochemistry of formation of the two new centres to give (–)-isopulegol (**6.33**), which is easily hydrogenated to (–)-menthol (**6.4**).

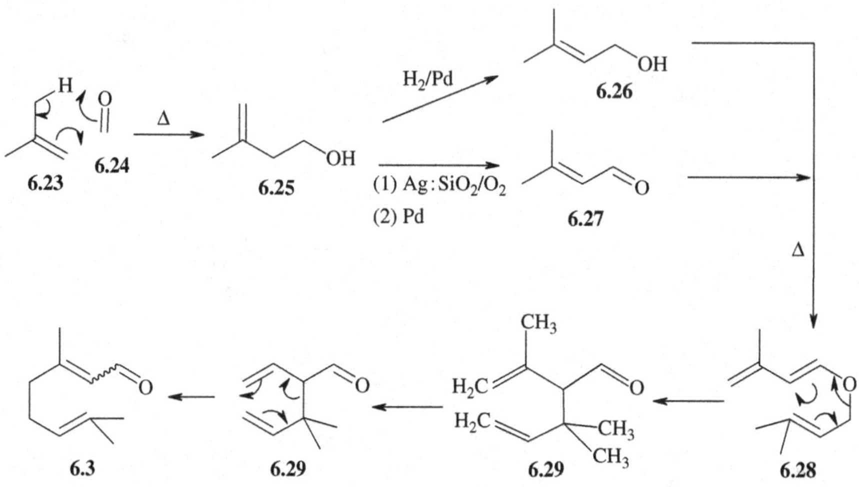

Figure 6.6   The BASF route to citral.

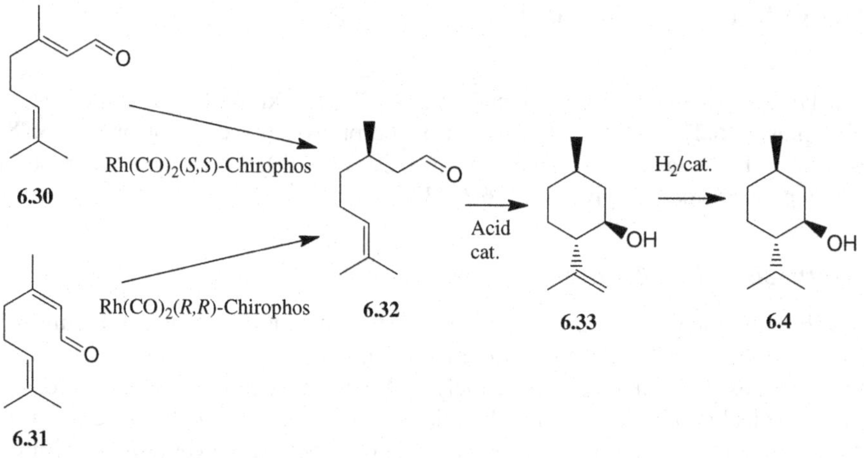

Figure 6.7   The BASF synthesis of menthol.

This represents a very elegant synthesis of menthol using only catalytic reactions and with only water as by-product.

Menthol provides an interesting example of competing production routes that currently exist in economic balance. It will be interesting to see how sustainability factors shift the balance over the coming years. The BASF feedstocks, isobutylene and formaldehyde, are currently sourced from petrochemicals but could, in principle, also be produced by biotech processes and thus make the synthesis sustainable.

The three other key routes all have strengths and weaknesses, and those who operate them are continually seeking to build on the former and address the latter. There are also a number of other routes to (−)-menthol (**6.4**), many starting from natural homochiral precursors such as (+)-pulegone (**6.34**) (Figure 6.1) which is present at about 75% in the oil of pennyroyal (*Mentha pulegium*). Details of these can be found in the *Kirk–Othmer Encyclopaedia* (37). None of them currently competes economically and their sustainability might also be poorer than the current major routes if feedstocks for those could be produced using biotechnology.

Production of menthol from mint plants, such as *Mentha piperita* and *Mentha arvensis*, are clearly renewable but their sustainability is less certain. Mint competes with food crops for arable land and, as the world's population grows, this competition will become more severe. The need for fertiliser is one factor that also adversely affects mint production because the Haber process is very energy intensive and lifecycle analysis shows that it is a significant factor in mint production.

The Symrise menthol process is shown in Figure 6.8. The first two steps are straightforward but are currently reliant on petrochemical feedstocks. Friedel–Crafts addition of propene to *m*-cresol (**6.35**) produces thymol (**6.36**) (38, 39) and this is hydrogenated, using a catalyst containing Co (49.7%), Mn (22.9%) and Cu (0.2%), to a mixture of all eight stereoisomers of 2-isopropyl-5-methylcyclohexanol (**6.37**) (39–42). Fractional distillation is then used to separate racemic menthol, the other six isomers being recycled via the hydrogenation stage. The racemic menthol thus prepared was originally resolved through its 3,5-dinitrobenzoate ester, but it was later found that the benzoate served equally well. A super-saturated solution of the racemic benzoates is seeded with crystals of one enantiomer inducing that isomer to crystallise out (43). The opposite enantiomer is, therefore, concentrated in the mother liquor. The resolved esters can then be saponified to the corresponding alcohols and the benzoic acid recycled. It is possible to equilibrate any menthol isomer to the same equilibrium mixture (44), but in practice the other diastereomers are most easily recycled via the hydrogenation stage, along with the *d*-menthol. The recycles are so efficient that the overall yield of *l*-menthol from this route is over 90%. The disadvantage of the process lies in terms of the labour, time, plant capacity and energy consumed in operating all of the separation and recycle processes. Biochemical methods for the resolution of menthol are now being developed. Symrise have patented a method for the selective hydrolysis of racemic menthyl benzoate using a lipase (45). A lipase from, for example, *Candida rugosa*, will hydrolyse the benzoate of *l*-menthol but not that of *d*-menthol. Thus treating the racemic benzoate with the lipase gives a mixture of *l*-menthol and *d*-menthyl benzoate which are easily separated without recourse to fractional crystallisation.

The Takasago process, shown in Figure 6.9, starts from β-pinene (**6.38**) which is pyrolysed to give myrcene (**6.39**). Addition of diethyl amide to myrcene (**6.39**) gives geranyl diethyl amine (**6.40**) and this can be isomerised stereoselectively to the diethyl enamine of (*R*)-citronellal (**6.41**). This key step, in which chirality is introduced, was developed by Professor Noyori of Nagoya University and which is part of the work for which he was awarded the Nobel Prize for chemistry in

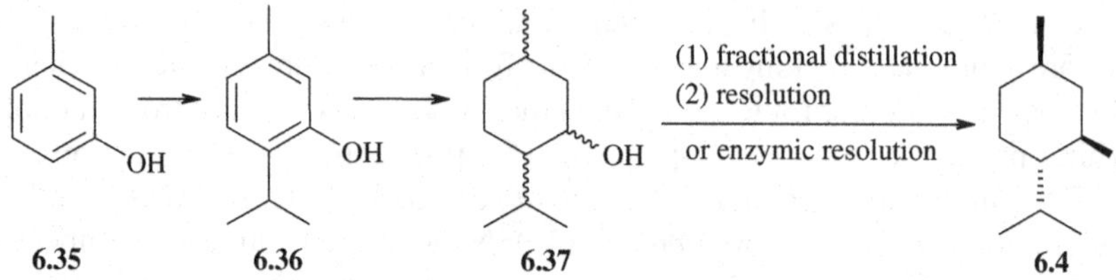

Figure 6.8 The Symrise menthol process.

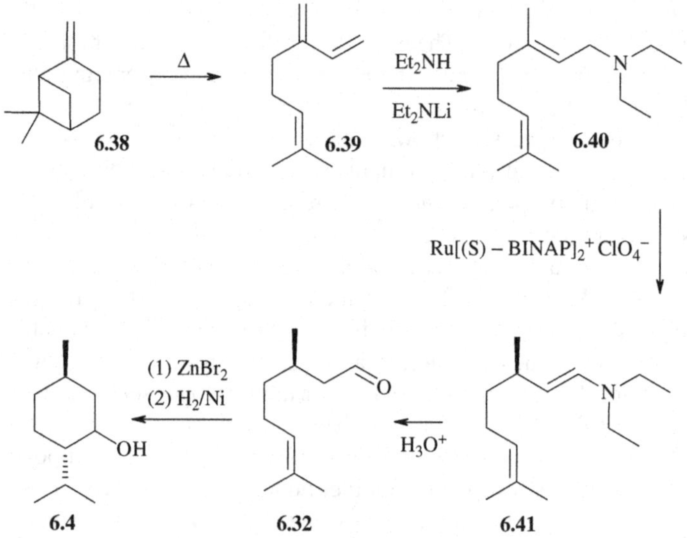

Figure 6.9 The Takasago menthol process.

2001 (46–49). It proceeds with an enantiomeric excess (ee) of more than 96% and the catalyst turn-over number is 400,000, which makes the process very efficient. Hydrolysis to (R)-citronellal (**6.32**) followed by acid-catalysed cyclisation to (−)-isopulegol (**6.33**) and subsequent hydrogenation gives (−)-menthol (**6.4**). A significant improvement to the cyclisation process was made when it was discovered that zinc salts such as the chloride and bromide, gave much higher selectivity for (−)-isopulegol, the ratio of the desired isomer to the others being about 94 : 6 (50–52). Further development has improved selectivity even further to a ratio of 99.7 : 0.3 through use of the tris-(2,6-diarylphenoxy)aluminium catalyst (53).

    As an example of the continuing effort to provide ever more efficient syntheses, Takasago reported investigations into the possibility of a dual catalyst system for selective reduction of citral (**6.3**) to (*R*)-citronellal (**6.32**) without the need to separate geranial (**6.30**) from neral (**6.31**). Their attempt to do this used a palladium on barium sulfate catalyst to maintain equilibrium between the two citral isomers and an organocatalyst that selectively reduced only one of them (54). Although the process is not yet commercially competitive, it is a significant research success and demonstrates how process chemistry is continuing to advance.

Synthesis of Coumarin

The Reimer–Tiemann reaction was important in early fragrance chemistry since it allowed the synthesis of aromatic aldehydes from the corresponding hydrocarbons. The Reimer in question went on to form a fragrance company, Haarmann and Reimer, which is now part of Symrise. However, the Reimer–Tiemann reaction does present a safety issue in terms of the generation of a carbene intermediate, and there is also an effluent problem in terms of inorganic waste and the use of chloroform. An example of how the chemical industry started greening chemical processes before the subject became popular in academic circles is presented by the development, by Haarmann and Reimer, of a catalytic process to produce saligenin (**6.42**) from phenol (**6.43**), thereby providing salicylaldehyde (**6.44**) as an intermediate for coumarin (**6.45**) synthesis (Figure 6.10). Some para isomer (**6.46**) is formed alongside the saligenin (**6.42**), and this can be used to produce *p*-anisaldehyde (**6.47**) via oxidation, using oxygen and a lead-doped platinum on carbon catalyst, and methylation. The ratio between o- and p-substitution depends on the catalyst employed and varies from 99 : 1 o:p using magnesium acetate to 50 : 50 using KOH. The balance can, therefore, be selected to suit commercial demand.

Synthesis of (–)-Carvone

Another example of greening of processes is provided by the synthesis of (–)-carvone (**6.48**). For many years, carvone was synthesised from (+)-limonene (**6.49**) by the addition of nitrosyl chloride (formed *in situ* from isopropyl nitrite and hydrochloric acid), rearrangement of the adduct (**6.50**) to carvoxime (**6.51**) and hydrolysis, as shown in Figure 6.11. This process was developed by Norda, which became part of Quest International and is now part of Givaudan. (+)-Limonene (**6.49**) is the major component of orange peel oil and is readily available as a by-product of orange juice production. Millennium chemicals later developed a process, also based on (+)-limonene (**6.49**), but using cleaner catalytic route as shown in Figure 6.12. This route involves epoxidation of (+)-limonene (**6.49**), rearrangement of the epoxide (**6.52**) using a zinc octoate and 2-aminophenol complex as a catalyst and subsequent oxidation of the (–)-carveol (**6.53**) thus produced to (–)-carvone (**6.48**) (55) Millennium Chemicals is now known as Renessenz.

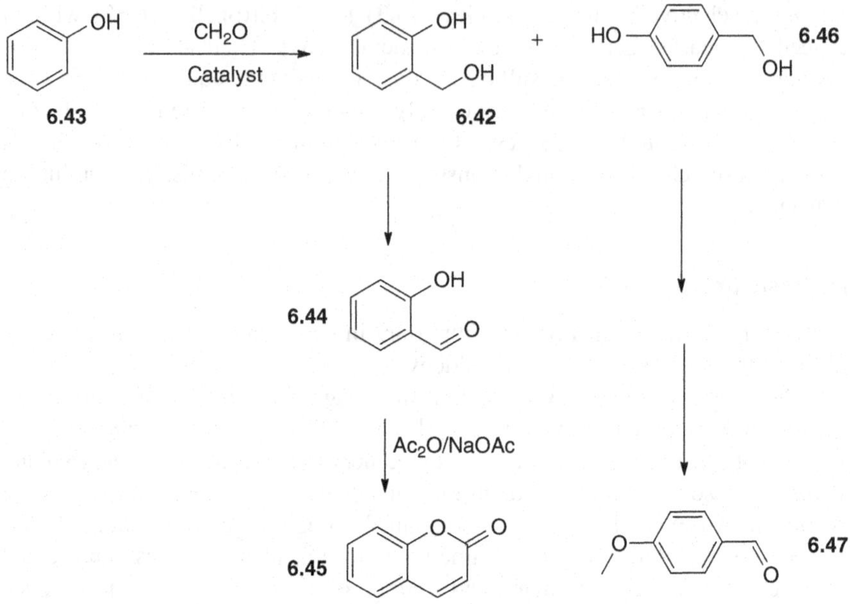

Figure 6.10   The Haarmann–Reimer route to coumarin and anisaldehyde.

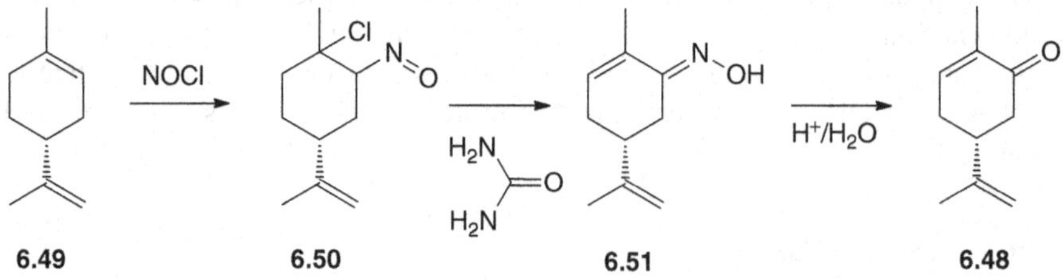

Figure 6.11   The Norda carvone process.

Biotechnology

In recent years, there has been a lot of interest in the potential use of enzymes as catalysts for chemical processes. The attractive features of enzymes include their high selectivity and the fact that they function at or close to ambient temperature, factors that are environmentally favourable in terms of waste production and energy consumption, respectively. A good example of the use of enzymes is the resolution of menthol as described above. However, enzymes usually work best in aqueous media, in which fragrance ingredients are usually not very soluble and, even more disadvantageously, many enzymes require expensive co-factors in order to function. The simplest solution to the latter problem is to use intact organisms so that co-factors are produced and recycled by the organism. The disadvantage of this

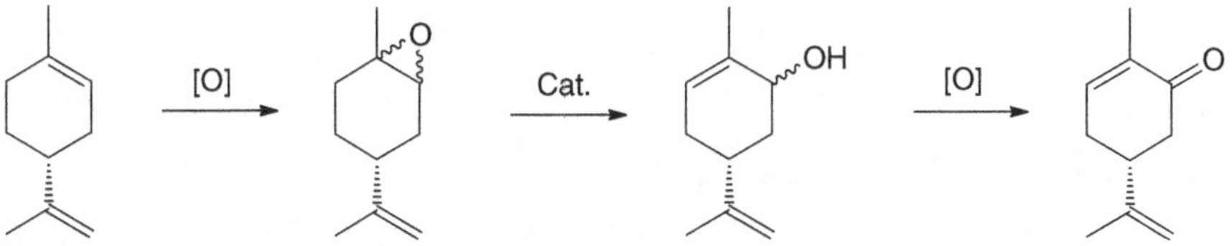

Figure 6.12   The Millennium carvone process.

approach is that the organism also produces a plethora of other chemicals and uses feedstock as an energy source to drive its cellular chemistry. This results in low yield and production of biomass as a by-product. Recovery of the desired product from the dilute aqueous solution (necessary because of the low solubility of reagent and product) can be very energy intensive, thereby defeating one of the objectives.

An example of these difficulties is the use of fermentation to produce the musk hexadecanolide (**6.54**) from palmitic acid. The use of the yeast *Torulopsis bombicola* was first patented by Pfizer (56), and further development was published by Jeffcoat and Willis (57). The basic scheme, as shown in Figure 6.13, looks straightforward and suggests an environmentally friendly process. The yeast oxidises the terminal carbon atom of palmitic acid (**6.55**) to give juniperic acid (**6.56**), which can be cyclised using the Carothers technique (58) to give hexadecanolide (**6.54**). One problem is that the hydroxylation is not entirely specific and some 15-hydroxyhexadecanoic acid (**6.57**) is formed alongside the juniperic acid. This is not necessarily a major difficulty because 15-methylpentadecanolide (**6.58**) also elicits a musky odour. So, as long as regulatory clearance is obtained for it, the mixture of the two hydroxy acids can be lactonised to give a mixture of the two musks. A more serious issue is the fact that the two hydroxy acids are not produced as such but are bound to a disaccharide, sophorose, by a mixture of ester and/or glycoside links and the remaining alcoholic groups of the sugar are partially acetylated, giving a complex mixture. Freeing the hydroxy acids using acidic hydrolysis generates a waste stream of sophorose-derived material which is difficult to handle and adds to the biomass waste. The Carothers technique produces the desired lactones, but much of the glycerol (which is used as a solvent for the reaction but also enables azeotropic removal of the lactones from the equilibrium mixture of polyesters) is lost by polymerisation, thus adding to the total waste produced. The overall E-factor (ratio of desired product to total mass produced) of the process is, therefore, very low.

An example of the successful use of fermentation is that of the oxidation of sclareol (**6.59**) to sclareolide (**6.60**) as shown in Figure 6.14. The former is produced by clary sage (*Salvia sclarea*) and is a by-product during the production of the essential oil from the plant. The latter is an intermediate in the synthesis of the napthofuran (**6.61**) as will be described later. Previous methods for oxidation of sclareol (**6.59**) involved the use of heavy metal oxidants, permanganate and/or

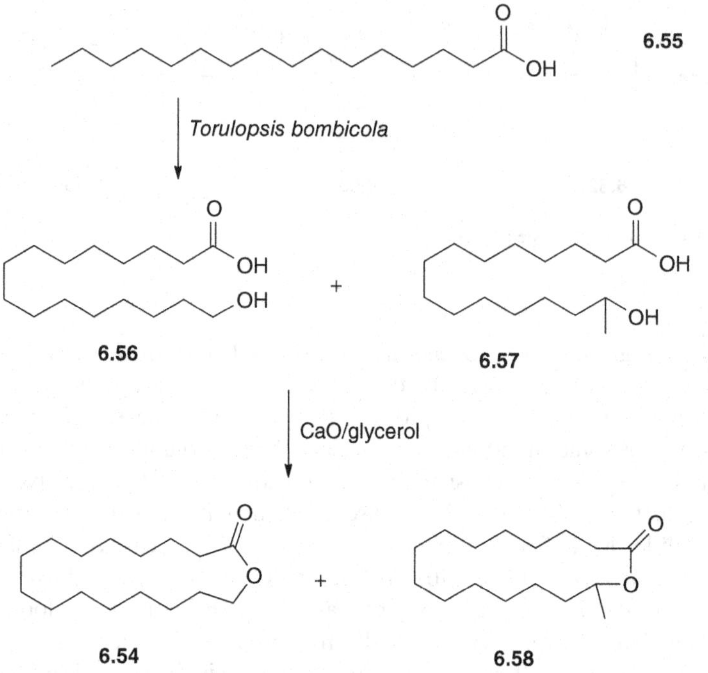

Figure 6.13   Synthesis of hexadecanolide from palmitic acid.

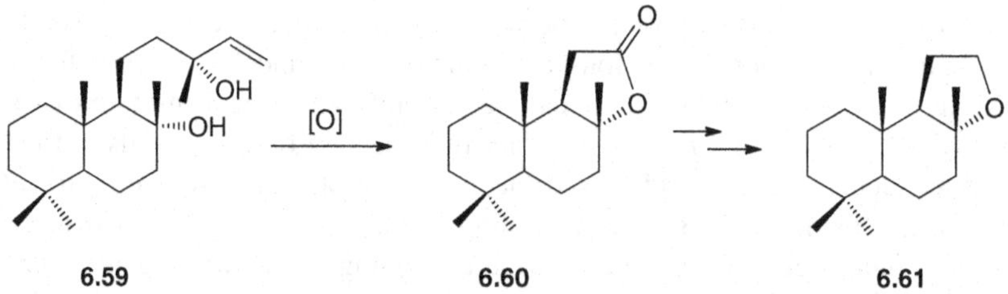

Figure 6.14   Oxidation of sclareol.

chromate and therefore presented a serious environmental problem. The oxidation is now achieved by use of fermentation with *Cryptococcus laurentii* (59).

MAJOR INDUSTRIAL SYNTHETIC ROUTES TO TERPENOIDS

The so-called rose alcohols [geraniol (**6.62**), citronellol (**6.63**) and linalool (**6.64**)] together with the corresponding aldehydes citral (**6.3**) and citronellal (**6.65**) lie at the heart of the major production routes in terpenoid chemistry, as shown in

Figure 6.15. (Note that absolute stereochemistry is shown only for (−)-menthol (**6.4**) and, obviously, the intermediates in its synthesis must, where appropriate, possess the correct absolute stereochemistry as explained elsewhere in the text.) In the figure, the basic feedstocks are contained in the ellipses and it can be seen that currently there is a balance between natural (hence renewable) and petrochemical feedstocks. The major natural feedstocks at present are turpentine and limonene from citrus oils. Sulfate turpentine is currently more significant than gum turpentine but, as discussed above, it is entirely dependent on paper manufacture. These sources are renewable, but this does not necessarily mean that they are sustainable. The major petrochemical feedstocks are not currently sustainable but, again, as explained previously, with advances in fermentation and synthetic biology they might become available from sustainable sources in the future. Thus the various feedstocks and synthetic routes will continue to co-exist in techno-economic balance in the foreseeable future.

Terpenoids from Petrochemical Feedstocks

The major petrochemical routes to the rose alcohols and the corresponding aldehydes are directed towards linalool and citral. This is partly for historic reasons because the earliest syntheses involved 6-methylhept-5-en-2-one (**6.8**) as a key intermediate and the early industrial syntheses were developed from these academic routes. The other reason is that the major companies manufacturing terpenoids from petrochemical feedstocks are all producers of vitamins A, E and K, and citral is the key to the synthesis of these.

The Swiss pharmaceutical company Hoffmann-La Roche started the first major synthetic production of terpenoids in 1957. The plant and processes now belong to DSM. The basic process has been refined over the last four decades and the current scheme is as shown in Figure 6.5 (60).

BASF use the synthesis of citral (**6.3**) shown in Figure 6.6, and a similar process is used by the Japanese rubber company, Kuraray. BASF have patented other routes to the key monoterpenoids such as those shown in Figure 6.16. In the first chemical step, formaldehyde (**6.24**) and acetone (**6.12**) react under the influence of a basic catalyst to produce methyl vinyl ketone (**6.66**). This then undergoes an ene reaction with isobutylene (**6.23**) to give the methylheptenone isomer 6-methylhept-6-en-2-one (**6.67**). Both reaction steps can be carried out in a single process operation, lending further elegance to the overall scheme (25, 61, 62). Isomerisation of this material to methylheptenone (**6.8**) is straightforward (63, 64), and the unconverted isobutylene (**6.23**) and acetone (**6.12**) can be recycled to the process, thus making it commercially feasible (65, 66). Addition of acetylene (**6.13**) to (**6.8**) and (**6.67**) gives the isomeric alcohols (**6.21**) and (**6.68**), respectively (67). Partial hydrogenation of (**6.21**) gives linalool (**6.64**). When these materials are used for the preparation of ionones and vitamins, *iso*-dehydrolinalool (**6.68**) offers an advantage in that it is more easily protonated than dehydrolinalool (**6.21**) and this helps in the cyclisation reaction (31,68,69).

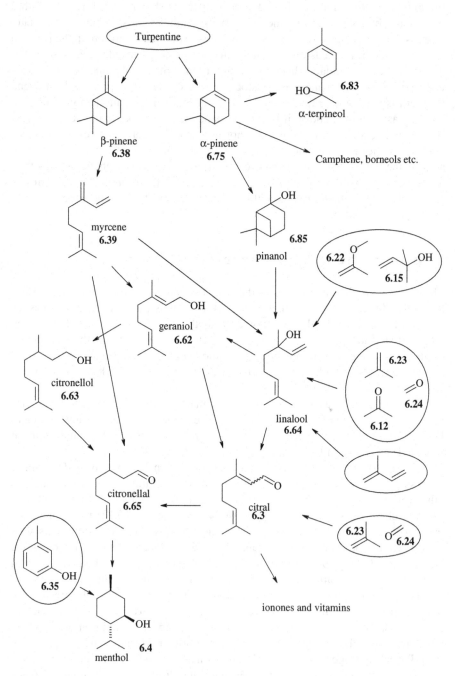

Figure 6.15   The key terpenoid manufacturing routes.

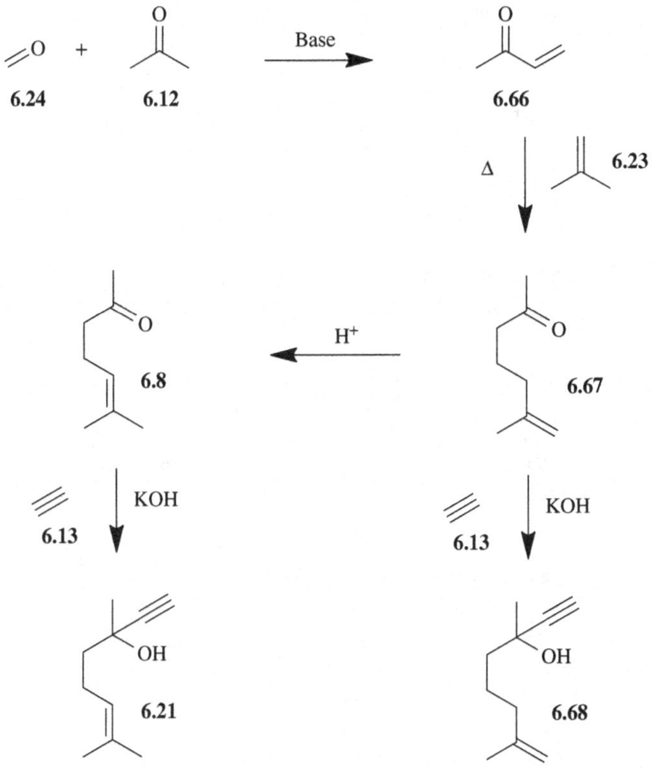

Figure 6.16 Key features of a BASF synthesis of linalool and vitamins.

Hydrochlorination of isoprene (**6.69**) produces prenyl chloride (**6.70**) and some of the isomeric 3-chloro-3-methylbut-1-ene (**6.71**), the ratio between the two depending on reaction conditions. The former undergoes S_N2 reactions while the latter prefers S_N2', hence both alkylate preferentially at the primary carbon atom. Therefore, treatment of the chlorides with acetone (**6.12**) in the presence of a base gives methylheptenone (**6.8**), as shown in Figure 6.17 (70). This is the basis of a process developed by Rhone-Poulenc in which a phase-transfer catalyst (PTC) is used to assist in the alkylation of acetone (71–73). A similar process is operated by Kuraray (74). Linalool produced in this way can be isomerised to geraniol using an orthovanadate catalyst (75).

Terpenoids from Turpentine

By far, the most important natural feedstock for terpenoid synthesis is turpentine. It can be produced by tapping suitable conifers, a process that involves making

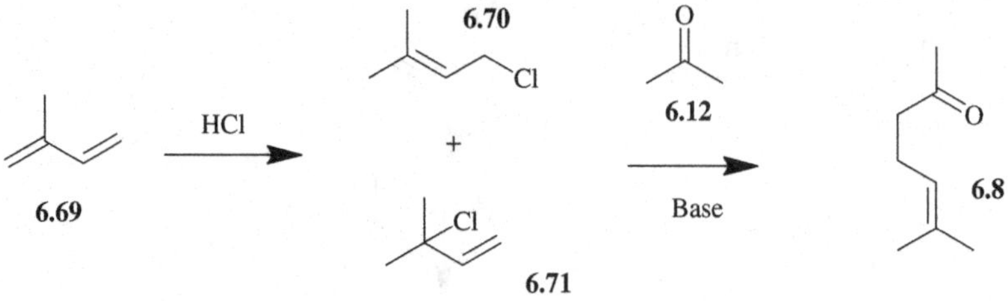

Figure 6.17 Manufacture of terpenoids from isoprene via prenyl chloride.

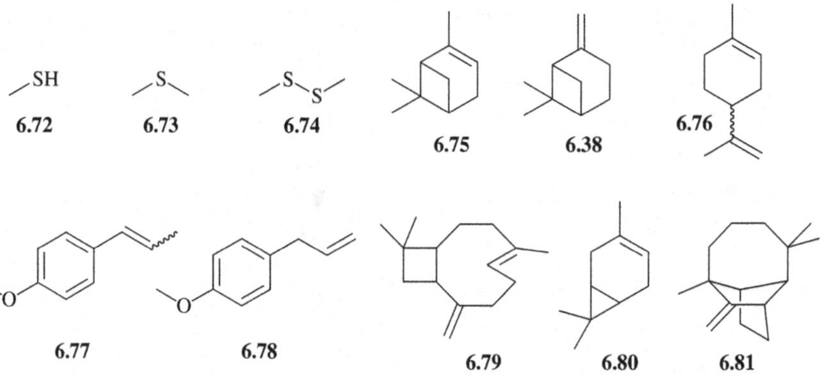

Figure 6.18 Turpentine components.

an incision in the bark and collecting the exudate in cups. Such turpentine is called *gum turpentine*. Wood turpentine is extracted mainly from tree stumps but the major source of turpentine is known as *crude sulfate turpentine* (CST) as it is a by-product of the Kraft paper process.

The composition of turpentine varies depending on the species of tree from which it is produced, and plantations may contain a variety of species and chemotypes, therefore the composition of crude turpentine is subject to variation. The largest production is sulfate turpentine from the south-eastern United States, which amounts to well over 100,000 tonnes/annum and in 2005 was priced under \$2 gallon^{-1}. The second most important is Chinese gum turpentine which is produced in about half that volume. Other significant sources include gum turpentine from Russia, Brazil, Portugal, India and Mexico and sulfate turpentine from Canada, these last few each being of the order of 10,000 tonnes/annum. Scandanavia and Chile are now also significant suppliers of CST (Figure 6.18).

Fractional distillation of CST gives a 'lights' fraction (1–2%) which contains *inter alia* some lower boiling sulfur compounds such as methyl mercaptan (**6.72**), dimethyl sulfide (**6.73**), and dimethyl disulfide (**6.74**). The distillation removes much of the sulfur-containing impurity originating from the Kraft process, but for many applications further desulfurisation is necessary. This 'lights' fraction is followed by the α-pinene (**6.75**) fraction (60–70%) and then the β- pinene (**6.38**) (20–25%). The next fraction is referred to as 'dipentene' (3–10%) and contains racemic limonene (**6.76**) together with other *p*-menthadienes. Then, after the pine oil fraction (3–7%) comprising ethers and alcohols, comes a fraction (1–2%) containing the shikimates anethole (**6.77**) and estragole (also known as *methylchavicol*) (**6.78**), and the sesquiterpene hydrocarbon, β-caryophyllene (**6.79**).

Turpentine from the western United States contains 12–43% of 3-carene (**6.80**). Indian turpentine contains about 60% of 3-carene together with about 15% of the sesquiterpene longifolene (**6.81**). Turpentine from Sweden, Finland, the CIS, and Austria also all contain 3-carene.

The two isomeric pinenes are, by far, the most important of these natural feedstocks as far as conversion to other terpenoids for the fragrance industry is concerned. The ethers anethole and estragole find use as F&F ingredients, and many of the other components are used as feedstocks for fragrance ingredients. Obviously, there are many uses apart from the F&F industry to which turpentine derivates are put.

Fragrance Ingredients Produced from α-pinene

The most significant syntheses starting from α-pinene (**6.75**) are those of linalool (**6.64**), geraniol (**6.62**), nerol (**6.82**) and α-terpineol (**6.83**). These and some others are shown in Figure 6.19.

Linalool (**6.64**) and geraniol (**6.62**) are produced from α-pinene (**6.75**) by the Renessenz Company at their site in Colonel's Island, Georgia (76). Their process starts with hydrogenation of α-pinene (**6.75**) using a special catalyst which gives a high selectivity for *cis*-pinane (**6.84**). This is necessary because the trans isomer is relatively unreactive in the subsequent reaction step. Autoxidation followed by catalytic hydrogenation of the intermediate hydroperoxide gives pinanol (**6.85**) with a cis/trans ratio of 75 : 25. The pinanols are distilled and then pyrolysed to give linalool (**6.64**). This pyrolysis, first reported by Ohloff and Klein (77), is run at relatively low conversion in order to minimise the formation of plinols (alcohols formed by ene cyclisation of linalool). Before conversion to geraniol, the linalool must be freed of impurities boiling close to the former. This isomerisation is carried out over a vanadate catalyst (78), the process of which is improved by first converting the linalool to its borate ester (79). A description of the mechanism of the reaction has been published. This gives a mixture of geraniol (**6.62**) and nerol (**6.82**) with a purity of 99% and a geraniol/nerol ratio of 68 : 32. Geraniol is the preferred isomer and can be separated from nerol by distillation, though many commercial grades of

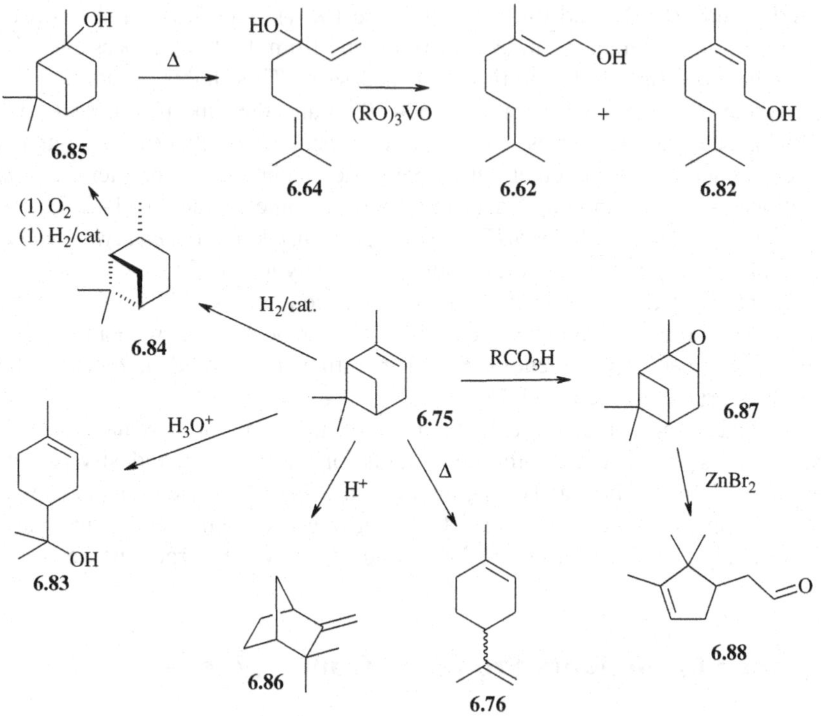

Figure 6.19   Major products from α-pinene.

'geraniol' are actually geraniol/nerol mixtures, the ratio varying from 98% of one isomer to 50 : 50.

Treatment of α-pinene with aqueous acid gives α-terpineol (**6.83**), whereas anhydrous acid gives camphene (**6.86**). The former is used as pine oil, a disinfectant and a fragrance ingredient. The use of β- and USY-zeolites as catalysts for the hydration of α-pinene and camphene leads to more selective and less polluting processes (80, 81). Pyrolysis of α-pinene gives a mixture containing dipentene (**6.76**) and ocimenes. Epoxidation of α-pinene gives pinene oxide (**6.87**), which is isomerised to give campholenic aldehyde (**6.88**), an important feedstock for a variety of sandalwood materials.

Fragrance Ingredients Produced from β-pinene

Figure 6.20 shows some of the major routes to terpenoids starting from β-pinene (**6.38**). The pyrolysis of β-pinene to produce myrcene (**6.39**) was first introduced

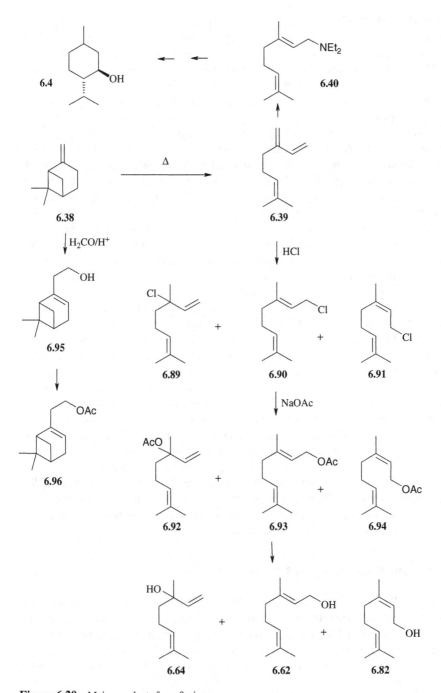

Figure 6.20 Major products from β-pinene.

by Glidden (SCM) in 1958 (82). The process is carried out at 500 °C and gives a product containing 75–77% myrcene (**6.39**). Hydrochlorination of myrcene (**6.39**) in the presence of a copper catalyst gives initially predominately linalyl chloride (**6.89**), which then isomerises to give a mixture of linalyl (**6.89**) (2–4%), geranyl (**6.90**) (50–55%) and neryl (**6.91**) (40–50%) chlorides (83, 84). The crude product from this reaction also contains bornyl and α-terpinyl chlorides as a consequence of traces of unchanged β-pinene in the pyrolysis product. These two, together with traces of other chlorinated impurities, make purification of the ultimate downstream products difficult. Bush, Boake and Allen (BBA) and Union Camp both developed processes similar to those of Glidden and these are now owned and operated by the fragrance house International Fragrances and Flavours (IFF) at their site in Jacksonville, Florida. Direct hydrolytic conversion of the halides to the rose alcohols is complicated by a side reaction giving unacceptably high levels of α-terpineol (**6.83**); so, instead, the halides are converted to the corresponding acetates (**6.92**)– (**6.94**), respectively, (or formates) by the addition of sodium acetate or sodium formate with a PTC (85, 86). Saponification of the acetates or formates gives the alcohols and sodium acetate or formate for recycling. Fractionation of the crude alcohol mixture gives both geraniol (**6.62**) and nerol (**6.82**) as products, usually as mixtures; high purity products are made by further distillation. This overall scheme suffers from three disadvantages. First, the levels of impurities create difficulties in preparing high quality rose alcohols. The trace chlorinated compounds are a particular issue as they must be removed to below ppb levels. Second, the wastewater generated by the process must be treated and this adds to cost. Third, β-pinene is less abundant than its α-isomer and hence less readily available and more expensive. The two isomers can be interconverted (87) but the thermodynamic equilibrium composition contains 96% α-pinene and only 4% β-pinene. Furthermore, α-pinene boils at 156 °C at 760 mmHg and β-pinene at 165 °C at 760 mmHg. This would make any attempt to produce β-pinene by equilibration of α-pinene, separation of the two by distillation and recycling of the α-pinene very expensive in energy terms.

Prins reaction of β-pinene with formaldehyde gives the alcohol nopol (**6.95**), the acetate (**6.96**) of which is used as a fragrance ingredient. Base-catalysed addition of diethylamine to myrcene gives geranyl diethyl amine (**6.40**), which is converted to menthol (**6.4**) as described above.

Fragrance Ingredients from Myrcene

Diels–Alder addition of myrcene (**6.39**) to acrolein (**6.97**) gives the aldehyde (**6.98**), which is known under trade names such as Myrac Aldehyde®, Empetal®, Acropal® and Vernaldehyde®, and has a natural green and aldehydic odour. The related hydroxy aldehyde Lyral® (**6.99**) has a sweet, light, floral (muguet) odour with excellent tenacity and radiance. Direct hydration of Myrac Aldehyde® (**6.98**) to give Lyral® is not possible because of competing reactions such as an internal

Prins reaction; so Lyral® is produced by addition of sulfur dioxide to Myrac Aldehyde® (**6.98**) to give (**6.100**) followed by hydration to (**6.101**), pyrolytic elimination of sulfur dioxide to give myrcenol (**6.102**) and then addition of acrolein (**6.97**). The aluminium chloride-catalysed addition of 3-methylpentan-3-ene-2-one (**6.103**) (the aldol product of 2-butanone and acetaldehyde) gives the monocyclic ketone (**6.104**). Cyclisation of this ketone using 85% phosphoric acid gives a mixture containing the isomeric bicyclic ketones (**6.105**)–(**6.107**), which was first commercialised by IFF under the trade name Iso E Super® (88–90). The success of the material was such that it is now a major fragrance ingredient and, following expiry of the IFF patents, is being produced by a number of different companies under different trade names. Later research (91) showed that the most organoleptically important component of the mixture was not one of the major components but a minor, rearranged product (**6.108**) and this led to the development of an analogue known as Georgywood® (**6.109**) (92). All of these reactions starting from myrcene are shown in Figure 6.21.

Fragrance Ingredients from Citronellene

Hydrogenation of either α-pinene (**6.75**) or β-pinene (**6.38**) gives pinane (**6.84**), and pyrolysis of this produces citronellene (**6.110**) (93) which is also commonly known as *dihydromyrcene*. Any optical activity in the pinane is transferred to the citronellene (94). Typically, the pyrolysis is carried out at 550–600 °C and the crude product contains about 50–60% citronellene which can then be purified by fractional distillation.

The highest tonnage product from citronellene (**6.110**) is dihydromyrcenol (**6.111**), which has a powerful fresh lime-like odour and rose to prominence after its use to impart a new masculine freshness to the fragrances Drakkar Noir and Cool Water. Dihydromyrcenol is prepared by acid-catalysed hydration of citronellene, as shown in Figure 6.22 (95). The more electron rich tri-substituted double bond reacts preferentially with acids, and selectivity for the desired product is high. In some processes, formic acid is used as the catalyst, in which case the intermediate product is a mixture of dihydromyrcenol (**6.111**) and its formate (**6.112**). The formate is easily hydrolysed by a base (96). The mixture of the alcohol and formate is also used in perfumery under the name Dimyrcetol®. Sulfuric acid is also used as a catalyst for the hydration (97). Concentrated sulfuric acid adds to the double bond, and subsequent dilution with water causes hydrolysis of the intermediate sulfate ester. This gives a simpler process and a higher yield of dihydromyrcenol. The fully saturated alcohol, tetrahydromyrcenol (**6.113**), is also a useful fragrance ingredient.

When citronellene (**6.110**) is treated with formic acid at higher temperatures (50 °C) than that required to produce dihydromyrcenol and its formate, an unexpected rearrangement occurs to produce α,3,3-trimethylcyclohexane methanol (**6.114**), known as *cyclodemol*, and its formate (**6.115**) (98). The product

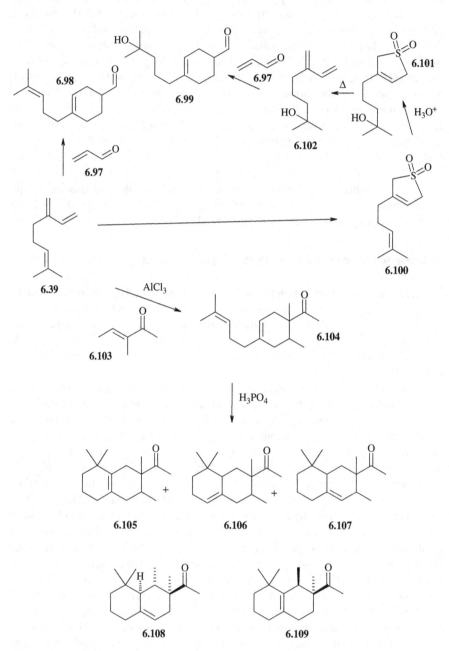

Figure 6.21   Fragrance ingredients from myrcene.

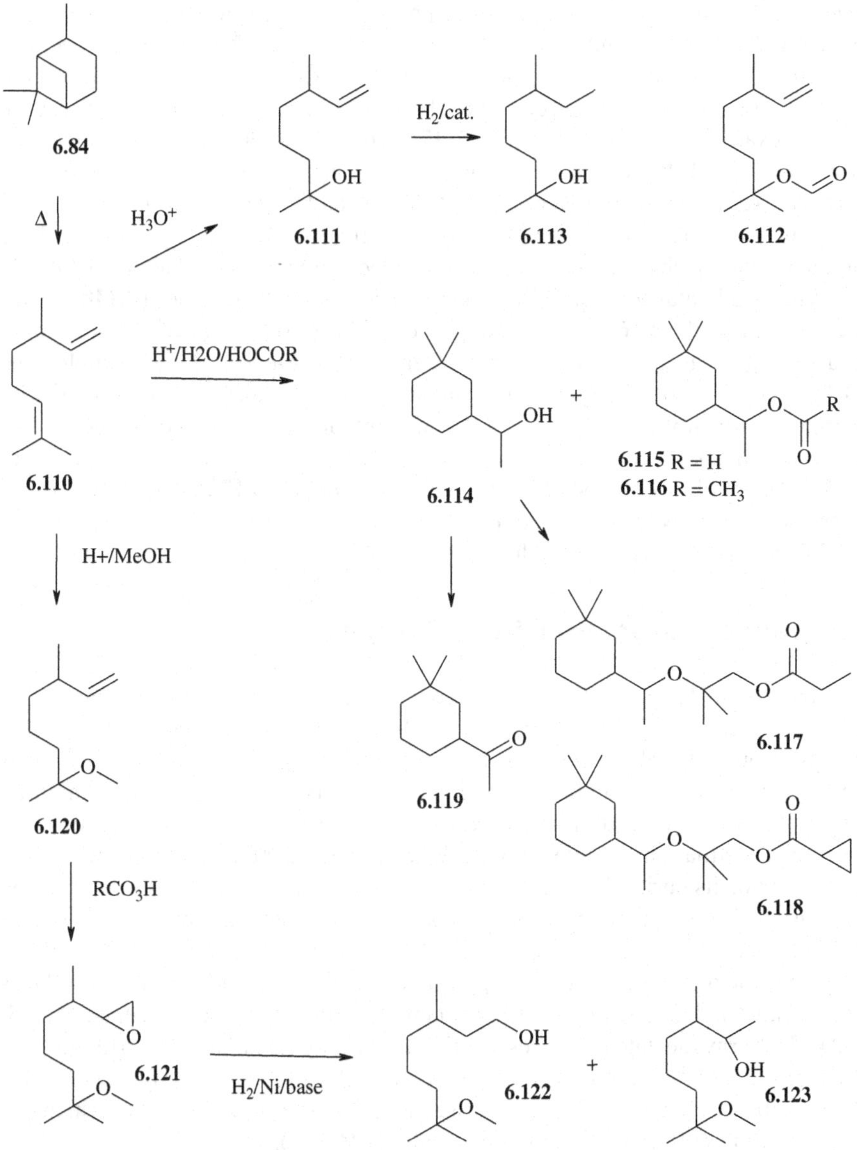

Figure 6.22   Fragrance ingredients from citronellene.

is formed by cyclization of dihydromyrcene to the cycloheptyl carbonium ion, which rearranges to give the more stable cyclohexyl compound (99). The formate ester, α,3,3-trimethylcyclohexane methanol formate (**6.115**), is a commercially available product known as Aphermate®, used to impart herbal, woody, ozone and fruity notes in fragrance. Similarly, treatment of citronellene with sulfuric acid in refluxing acetic acid for 6–8 h produces the acetate (**6.116**), which has both floral and musky notes and is known as Rosamusk® or Cyclocitronellene Acetate®. A series of musk compounds have been found based on Rosamusk but with an ethereal spacer between the cyclodemol fragment and the ester function. Typical examples are Helvetolide® (**6.117**), which is produced by Firmenich, and Serenolide® (**6.118**), which is produced by Givaudan. Oxidation of cyclodemol gives the ketone known as Herbac® (**6.119**), which has herbal, woody, minty and fruity notes. Stereochemistry has a marked effect on the odour of all of these products, and mechanistic studies have included stereochemical factors (100).

The acid-catalyzed addition of methanol to citronellene (**6.110**) gives methoxycitronellene (**6.120**) in good yield (101). Epoxidation of the remaining double bond with peracetic acid gives 2-methoxy-7,8-epoxy-2,6-dimethyloctane (**6.121**), which, on hydrogenation using a nickel catalyst, gives primarily methoxycitronellol (**6.122**). If a small amount of base is added to the hydrogenation, the product is a mixture 60% of the secondary alcohol Osyrol® or methoxyelgenol (**6.123**) and 40% of methoxycitronellol (**6.122**) (102, 103). Osyrol possesses a woody, floral odour characteristic of sandalwood and has become established as a fragrance ingredient.

Fragrance Ingredients from Camphene

Camphene (**6.124**) is produced commercially by the reaction of α-pinene (**6.75**) with a TiO_2 catalyst. Preparation of the catalyst has a great influence on the product composition and yield. Tricyclene is formed as a co-product but its reactivity is very similar to that of camphene; thus the product is generally used as a mixture. The p-menthadienes and dimers produced as by-products are easily removed by fractional distillation and the camphene has a melting point range of 36–52 °C, depending on its purity. The main use of camphene is as a feedstock for preparation of a variety of fragrance compounds, and some of the reactions employed are shown in Figure 6.23.

Addition of acids such as acetic, propionic, isobutyric, and isovaleric produces useful isobornyl esters, the most important of which is isobornyl acetate (**6.125**) (104). Isobornyl acetate possesses a fruity and woody odour and its perfumery use runs into thousands of tonnes per annum. Saponification of isobornyl acetate produces isoborneol (*exo*-1,7,7-trimethylbicyclo[2.2.1]heptan-2-ol (**6.126**), which can be oxidised or dehydrogenated to give camphor (**6.127**).

An important use of camphene is for the production of synthetic sandalwood materials. When camphene reacts with guaiacol (**6.128**) in the presence of a Brønsted or Lewis acid, a mixture of terpenylphenols is formed. Hydrogenation of the mixture results in hydrogenolysis of the methoxy group and gives a complex mixture of about 130 terpenylcyclohexanols such as 3-(2-isocamphyl)cyclohexanol (105, 106). The yield of the desired isomers in the final product can be improved by

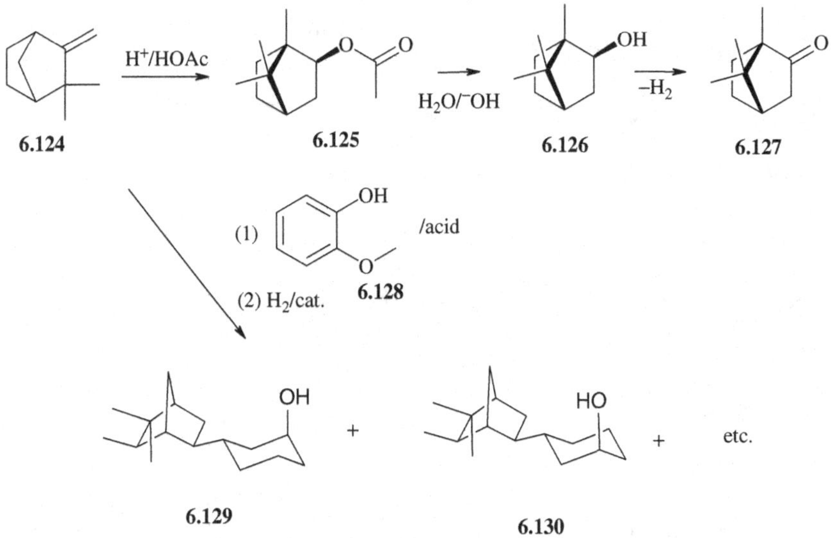

Figure 6.23 Fragrance ingredients from camphene.

reversing the etherification pattern of the camphene/guaiacol adduct (107–111). These materials are available under various trade names such as Candelum®, Indisan®, isobornylcyclohexanol (IBCH), Nardosandol®, Sandel®, Sandela®, Sandenol®, Sandeol® and Santalex®. Each of these has a unique isomer distribution and odour profile, and in several cases different qualities exist under the same trade name. Total annual consumption is in hundreds of tonnes. In a brilliant piece of work, Demole showed that the active isomers were those with an *exo*-isocamphane ring attached in the 3-position relative to an axial hydroxy group in the cyclohexanol, that is, structures (**6.129**) and (**6.130**) (112).

Sandalwood Odorants from Campholenic Aldehyde

Synthetic sandalwood odorants are of increasing importance because the supply of natural sandalwood is insufficient to meet demand and over-harvesting has put the future of the species (*S. album*) at risk as discussed previously. In addition to the isobornylcyclohexanols of the preceding paragraph, the sandalwood ingredients produced from campholenic aldehyde (**6.88**) have risen to major importance in perfumery. As shown in Figure 6.19, campholenic aldehyde (**6.88**) is produced from α-pinene by epoxidation and acid-catalysed rearrangement of the epoxide (113, 114). Zinc bromide is the most effective catalyst for the rearrangement but the chloride can also be used. As shown in Figure 6.24, the epoxidation and rearrangement both proceed stereoselectively, so that if the starting material is (+)-α-pinene

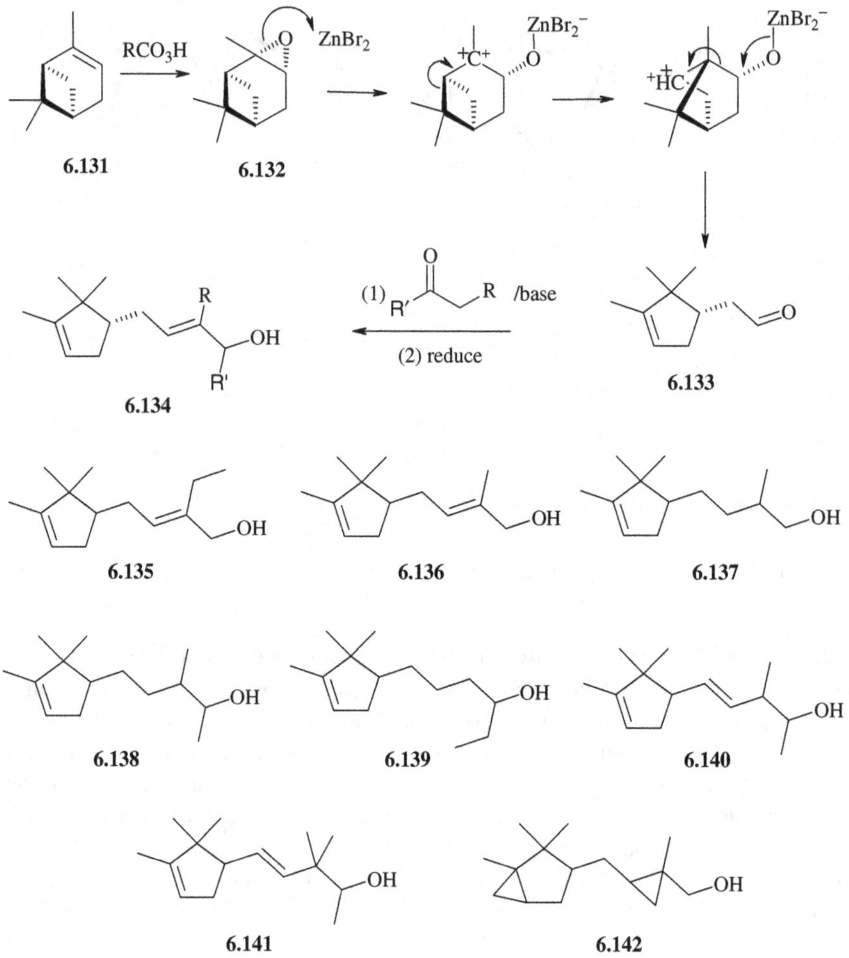

Figure 6.24   Sandalwood odorants from campholenic aldehyde.

(**6.131**) the epoxide is (+)-*cis*-α-pinene epoxide (**6.132**) and rearrangement gives (−)-campholenic aldehyde (**6.133**). The specific rotation of the aldehyde is very low and so it is difficult to check its enantiomeric purity, minor contaminants with high specific rotations proving particularly troublesome. Many commercial sources of pinene contain both enantiomers (usually not in equal proportions) and so the cheapest sources of campholenic aldehyde are usually of low enantiomeric purity. This has some significance as will be seen below.

The campholenic aldehyde-derived sandalwood materials were discovered by the East German company VEB Miltitz in the late 1960s (115). They found that, if campholenic aldehyde undergoes aldol condensation with an aldehyde or ketone and the resultant unsaturated ketone is reduced to an allylic alcohol (**6.134**) as shown in Figure 6.24, the products possess a fine odour reminiscent of sandalwood. The first of the series to be commercialised was the product (**6.135**) derived from an initial aldol with butanal. This material is now available under the trade names Anandol®, Bacdanol®, Balinol®, Bangalol®, Madrol®, Radjanol®, Sandolene®, Sandranol®, Santalinol® and Sriffol®. Versions with high enantiomeric excess are known as Dartanol® and Laevosandol®. In general, in this series, it has been found that the best sandalwood character is associated with those enantiomers prepared from (−)-campholenic aldehyde (**6.133**). Using propanal in place of butanal gives (**6.136**) which is known as Sandacore®, Sandalmysore Core®, Santacore®, Santalaire®, Santalice® and Santaliff®. A high enantiomeric excess version of this material is sold under the name Hindinol®. The saturated analogue of (**6.136**) is known as Brahmanol® (**6.137**) (116). If the double bond in the cyclopentene ring is also reduced, the resulting product does not have a sandalwood odour. When a ketone such as 2-butanone is used in the aldol reaction, a mixture of products results because the ketone can react on either side of its carbonyl group. When both the aldehyde and side-chain double bond of the product mixture from 2-butanone are reduced, a mixture of saturated alcohols is formed. Only those isomers with the basic structure (**6.138**) rather than (**6.139**) possess a sandalwood odour (117, 118). The former are the basis of the product known as Sandalore®. One minor component of Sandalore is the unsaturated material (**6.140**) in which the double bond has moved rather than being hydrogenated. This minor component contributes a disproportionate amount to the odour of Sandalore and is now manufactured independently and sold under the names Ebanol® and Ebalore® (119, 120). Alkylation of the intermediate ketone before reduction gives Polysantol® (**6.141**), also known as Nirvanol® and Suprasantol® (121, 122). Cyclopropanation of (**6.136**) gives Javanol® (**6.142**), which is the most powerful of all sandalwood chemicals in this series (123). The discovery of Javanol is a good example of the use of structure activity relationships (SARs) and molecular modelling in fragrance ingredient discovery (124). Prices of these materials start in the \$20–40/kg range and rise to over \$200 kg^{-1}. Many of them are produced and used internally by fragrance companies, so it is difficult to estimate volumes but the total figure is likely to be well over 1000 tpa.

Fragrance Ingredients from Citronellol

Hydrogenation of geraniol (**6.62**) and/or nerol (**6.82**) over a copper chromite catalyst gives a high yield of citronellol (**6.63**) (125).

Fractional distillation of the crude product produces a perfumery-quality citronellol. Geraniol can also be hydrogenated selectively to citronellol, without forming tetrahydrogeraniol, through use of a Raney cobalt catalyst (126). However, partial hydrogenation using a nickel catalyst does give some tetrahydrogeraniol. The product contains nerol, geraniol and tetrahydrogeraniol and has odour properties closer to materials derived from citronella oil and is therefore useful as a perfumery material (127). Optically active citronellol can be prepared in high optical purity by asymmetric hydrogenation of geraniol using a rhodium [(S)-BINAP]$_2$$^+$ catalyst (BINAP = 2,2′-bis(diphenylphosphino)-1,1′-binaphthyl) (128–131). The odour of the laevorotatory isomer is much finer than that of the racemate and so it commands a higher price than the racemate. Citronellol can also be made by the selective hydrogenation of citral (**6.3**) and citronellal (**6.65**) using a chromium-promoted Raney nickel catalyst (132, 133) or from citral using a palladium/ruthenium catalyst in the presence of trimethylamine (134).

Citronellol is important as a fragrance ingredient in its own right and also an intermediate in the synthesis of a number of other aroma chemicals, including its formate, acetate, propionate, butyrate and isobutyrate. Dehydrogenation of

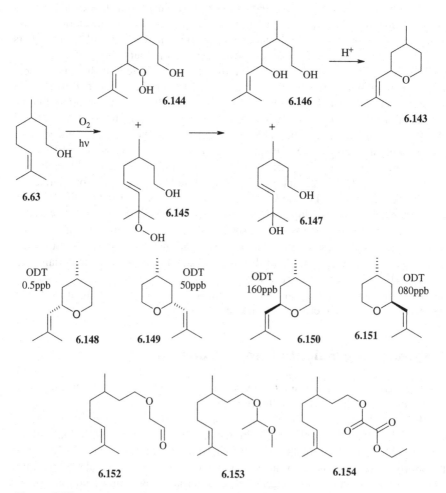

Figure 6.25 Fragrance ingredients from citronellol.

citronellol over a copper chromite catalyst produces citronellal **6.65** in good yield (101). If the dehydrogenation is done under distillation conditions in order to remove the lower boiling citronellal as it is formed, polymerisation or cyclisation of citronellal is prevented.

Citronellol is the key feedstock for the synthesis of rose oxide (**6.143**), as shown in Figure 6.25 (135). Dye-sensitised photoxidation of citronellol gives a mixture of two allylic hydroperoxides (**6.144**) and (**6.145**). These are reduced using bisulfite to give the corresponding diols (**6.146**) and (**6.147**), respectively, which are then treated with dilute sulfuric acid. Diol (**6.146**) cyclises to rose oxide whilst diol (**6.147**) remains unchanged, and the two are separated by distillation. Diol (**6.147**) is dehydrohydroxycitronellol and is used under trade names such as Allofixine®, Hydroxyol® and citronellol hydrate. It has a floral, sweet, green odour and is used as a fixative. Use of (−)-citronellol in the above synthesis produces a mixture, in about equal proportions, of (−)-*cis* and (−)-*trans* rose oxides, (**6.148**) and (**6.150**), respectively. Rose oxide provides a good example of the growing focus on chirality in the fragrance industry. It exists in four isomeric forms, two enantiomers each of two geometric isomers as shown in Figure 6.25 as structures (**6.148**)–(**6.151**), together with their odour detection thresholds (ODTs). These thresholds were determined by Yamamoto et al., who found that the absolute stereochemistry at C4 is more important than that at C2 (136, 137). The commonest of these in nature is the laevorotatory enantiomer of *cis*-rose oxide (**6.148**), which is found, for example, in rose and geranium oils. Not surprisingly, therefore, this is also the isomer with the finest rose character and the lowest odour threshold. The other isomers have more herbal character in their odour profile. Racemates and optically pure (both laevo and dextro) forms are commercially available and used in fragrances. The homochiral products, especially the laevorotatory ones, command much higher prices than does the racemate.

Some other aroma chemicals prepared from citronellol include citronellyl oxyacetaldehyde (**6.152**), which has a muguet note, and citronellyl methyl acetal (**6.153**), the mixed acetal with methanol of acetaldehyde, which has a floral, fruity, leafy and green character. Surprisingly, ethyl citronellyl oxalate (**6.154**) has a musky character in addition to rosy and fruity notes.

Production of Ionones

The synthetic routes to the ionone family of perfume ingredients are intimately linked to those of vitamins A, E and K because citral and β-ionone are intermediates in the latter. Ionones and methylionones for fragrance use are often mixtures of various isomers as a consequence of the manufacturing route, and each producer will have its own composition signature(s).

Aldol condensation of citral (**6.3**) with a ketone leads to materials known as ψ-*ionones*, as shown in Figure 6.26. Obviously, if acetone is used as the ketone, only one product, ψ-ionone (**6.155**), is produced. If an asymmetric ketone such as 2-butanone is used, then two products will result, the so-called *n*-methyl-ψ-ionone

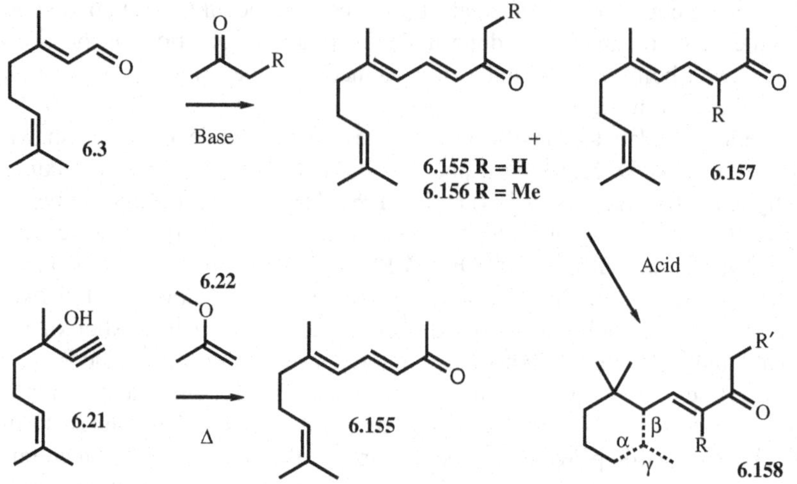

Figure 6.26   Synthesis of ionones.

(**6.156**) and the *iso*-methyl-ψ-ionone (**6.157**). Treatment of these ψ-ionones with an acid catalyst then gives a mixture of ionones, represented by structure (**6.158**) in which R and R' represent either H or Me as appropriate and one of the three dotted lines represents a double bond. (The Greek letters in the figure indicate the nomenclature for the double-bond isomers, as already mentioned in Chapter 5.) An alternative route to ionones (as distinct from methylionones) is to treat dehydrolinalool (**6.21**) with an acetone equivalent such as 2-methoxypropene (**6.22**) to yield ψ-ionone directly without going through citral (138). Diketene (139) and acetoacetate esters (139) have also been used. Similarly, the methyl enol ether of 2-butanone can react with dehydrolinalool to give *iso*-methyl-ψ-ionone (140).

Sodium or potassium hydroxides are the catalysts usually employed in the aldol condensation. Excess ketone is normally used and recovered and recycled. The exact conditions employed will affect the *n*/*iso* ratio of the products. For example, if the reaction temperature is kept at 0–10 °C, higher yields of the *iso*-methylpseudoionones, which are the more thermodynamically stable isomers, are obtained. The aldol intermediates have more time to equilibrate to the more stable isomers at the lower temperature. Sodium and potassium hydroxides tend to favour *n*-isomers, whereas quaternary ammonium hydroxides favour the *iso*-materials (141). Co-solvents such as methanol also affect the isomer ratio and can be very important in getting a high yield of the *iso*-methyl-ψ-ionones (142). Each producer has invested considerable effort into optimising ratios and guaranteeing consistency of product mix for each specified quality.

If phosphoric acid is used as the cyclisation catalyst, the α-isomer predominates, whereas with sulfuric acid the β-isomer is the major product. Use of boron trifluoride etherate in dimethylformamide gives predominantly the γ-isomer (143).

The reactions are normally carried out at atmospheric pressure, but higher temperatures and pressures can be employed (144). Prolonged acid treatment will lead to the thermodynamically favoured β-isomer.

Partially and totally hydrogenated ionones are also used in perfumery, and hydrogenation of the ketone also gives the ionols (145). Hydrogenation improves the stability of the materials but usually at the cost of odour quality.

Production of Damascones

As mentioned in Chapter 5, the damascones present a more difficult synthetic target than the ionones. The original production method is shown in Figure 6.27 using α-damascone (**6.159**) as an example (146) Allylation of dehydrolinalool (**6.21**) with allyl choloride (**6.160**) gives the acetylenic alcohol (**6.161**), which is rearranged to (**6.162**) and cyclised to α-damascone (**6.159**). The product of this synthesis contained a tiny trace of another compound with a very intense odour. This material was isolated and characterised, and it was found to have been formed by dehydration of alcohol (**6.161**) to the corresponding olefin (**6.163**), which then cyclised to (**6.164**) before hydrating to (**6.165**). This reaction by-product has now become an important perfumery ingredient in its own right and is sold under the trade name Dynascone® (147).

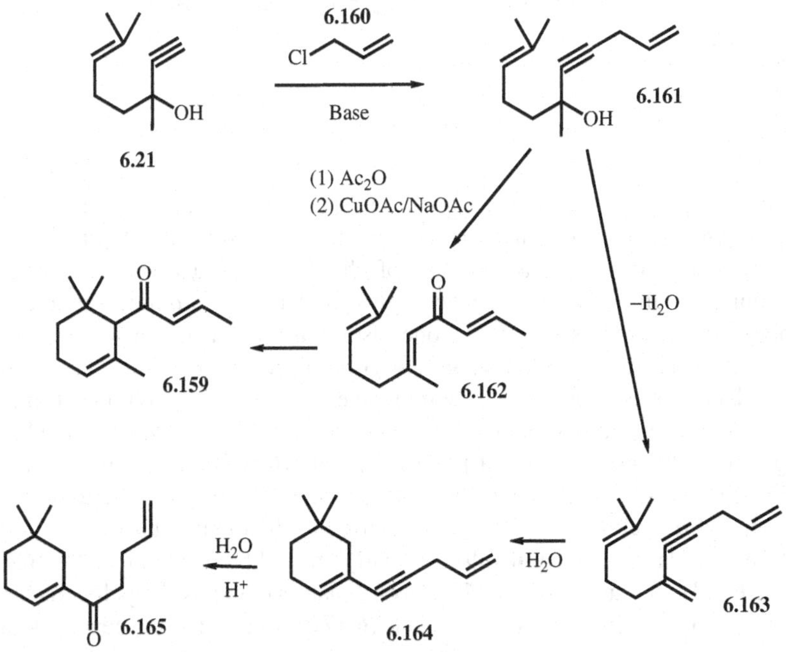

Figure 6.27   The original damascones production route.

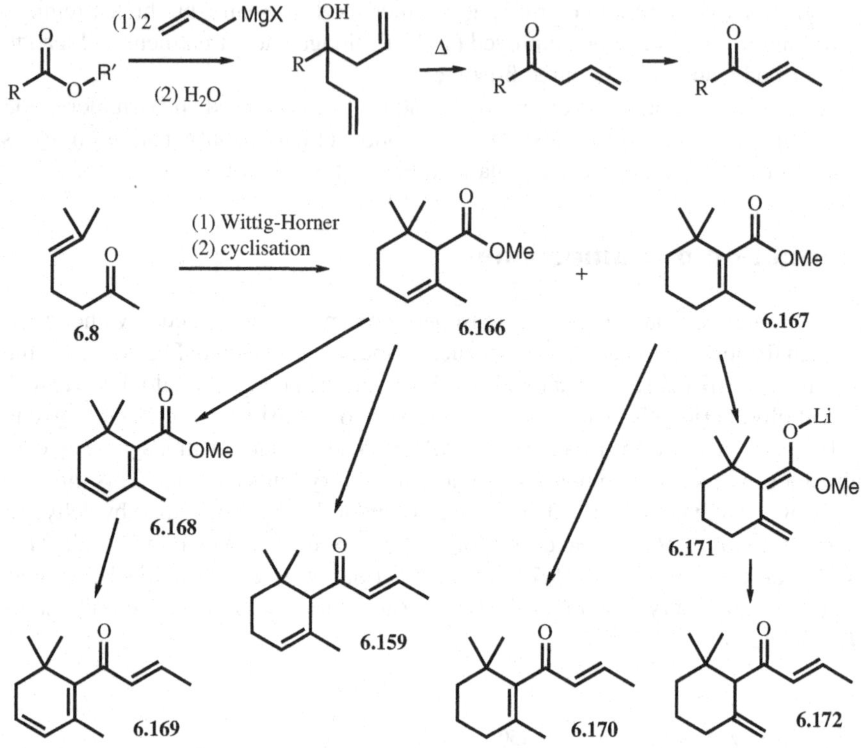

Figure 6.28   The current damascones production route.

A number of other routes to damascones have been devised (148, 149) but the one which is currently used in production is that shown in Figure 6.28. The key sequence is shown at the top of the figure. Reaction of an ester with 2 equiv. of allyl Grignard reagent gives a tertiary carbinol which can be pyrolysed to a β,γ-unsaturated ketone. Isomerisation using either acid or base then produces the corresponding α,β-unsaturated ketone (150). The synthesis of damascones starts with methylheptenone (**6.8**). A Wittig–Horner reaction followed by acid-catalysed cyclisation gives a mixture of methyl α-cyclogeranate (**6.166**) and methyl β-cyclogeranate (**6.167**). Application of the allyl Grignard–ene–isomerisation reaction sequence to (**6.166**) gives α-damascone (**6.159**). Methyl α-cyclogeranate can be converted to the diene ester (**6.168**) and this, in turn, to β-damascenone (**6.169**). Similarly, methyl β-cyclogeranate (**6.167**) can be converted to β-damascone (**6.170**). Deprotonation of methyl β-cyclogeranate (**6.167**) with butyl lithium gives the enolate (**6.171**), which can be treated with allyl Grignard reagent and subsequently isomerised to give γ-damascone (**6.172**) (151).

Production of 3a,6,6,9a-tetramethyldodecahydro-naphtho[2,1-b]furan

The importance of the eponymous naphthofuran (**6.59**) in perfumery has resulted in an ongoing search for efficient synthetic routes for its commercial production.

The labdane family of diterpenoids offers an attractive source of feedstock for the synthesis, as the members share the same substitution pattern and stereochemistry around the naphthalene ring system. The residues from distillation of the essential oil of clary sage (*S. sclarea*) contain about 50% by weight sclareol (**6.59**) and this is the major starting material for 3a,6,6,9a-tetramethyldodecahydronaphtho[2,1-b]furan synthesis. The production routes are shown in Figure 6.29.

The key intermediate in this synthesis is sclareolide (**6.60**), which can be obtained by chromic acid oxidation of sclareol (**6.59**). Alternatively, permanganate oxidation of sclareol gives the naphthopyran (**6.173**), which can be ozonised to the acid ester (**6.174**) and can then be saponified and lactonised to give sclareolide (**6.60**). As discussed above, fermentation routes are now also available enabling sclareol to be converted to sclareolide without the need for heavy metal oxidants (59). Reduction of sclareolide with either lithium aluminium hydride or borane gives the diol (**6.175**), which can be cyclised to the desired ether. This synthesis suffers the disadvantages of the use of either chromium or manganese oxidants, which creates an effluent problem and of over-oxidation with a subsequent requirement for a vigorous reductant to return the material to the desired oxidation level. Consequently, much research has been done to find alternative routes, and one ingenious example is that of Barton et al. (152, 153) and further developed by

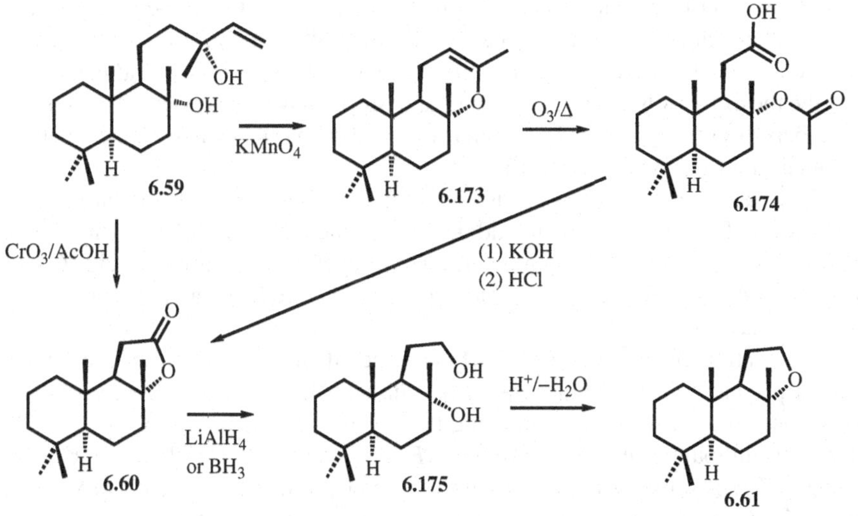

Figure 6.29   Production of 3a,6,6,9a-tetramethyldodecahydronaphtho[2,1-b]furan from sclareol.

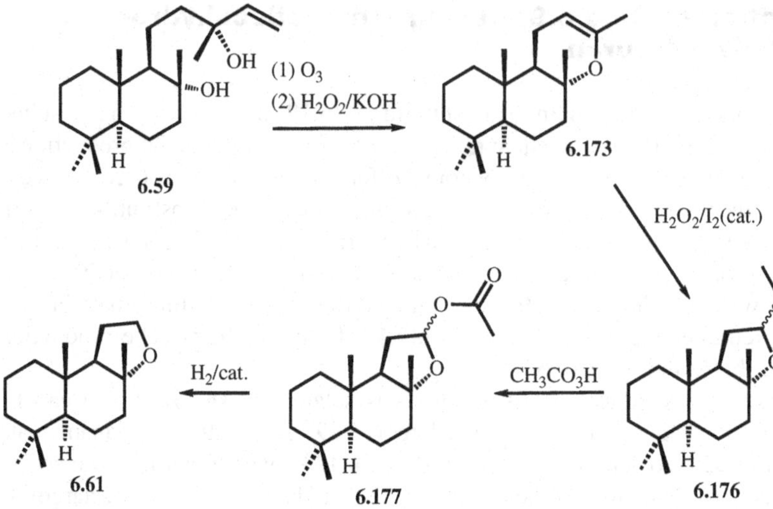

Figure 6.30   Alternative route to 3a,6,6,9a-tetramethyldodecahydronaphtho[2,1-b]furan from sclareol.

Tse, Davey and Payne (154, 155) as shown in Figure 6.30. Ozonolysis of sclareol (**6.59**) with an oxidative work-up gives norlabdane oxide (**6.173**). Treatment of this pyran with hydrogen peroxide in the presence of a catalytic amount of iodine gives the ketone (**6.176**). Bayer–Villiger oxidation of the latter using peracetic acid gives the acetal acetate (**6.177**), which can be hydrogenated to the target naphthofuran.

Many other diterpenoids and also monoterpenoids and sesquiterpenoids have been used as starting materials for naphthofuran (**6.59**), but none of these syntheses is of commercial importance. An excellent review of these will be found in Chapter 2 of the thesis by Bolster (156). Chauffat and Morris have also reviewed the history of ambergris development (157). Both reviews cover partial syntheses from natural products and total syntheses.

One basic synthetic approach towards total synthesis involves the biomimetic cyclisation of homofarnesic acid and derivatives thereof. The earliest syntheses of this type used acids such as stannic chloride to cyclise homofarnesic acid (**6.178**) to sclareolide (see Figure 6.31), which could then be reduced and cyclised in the conventional manner (158–166).

Cyclisation of homofarnesol (**6.179**) rather than the acid offers the advantage of producing the naphthofuran (**6.61**) directly rather than sclareolide (**6.60**). This was achieved by Vlad et al. using fluorosulfonic acid in nitropropane (167). However, although they started from (E,E)-homofarnesol (**6.179**), isomerisation to the (3Z,7E)-isomer was fast enough that their product contained a mixture of the desired target (**6.61**) and the isomeric cis-fused furan (**6.180**). A mixture of these two isomers can also be prepared from the monocyclic precursor (**6.181**)

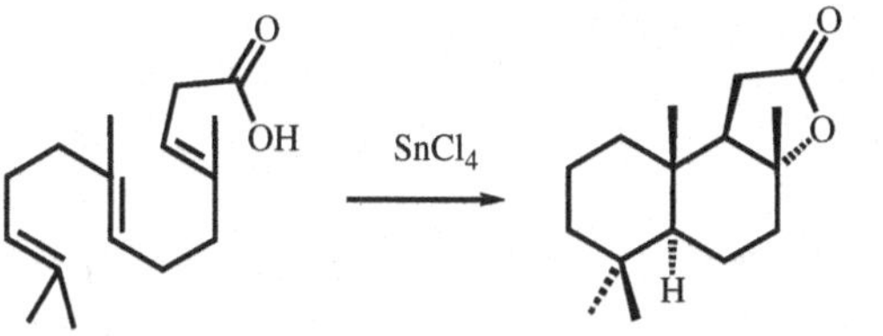

Figure 6.31   Acid-catalysed biomimetic cyclisation of homofarnesic acid.

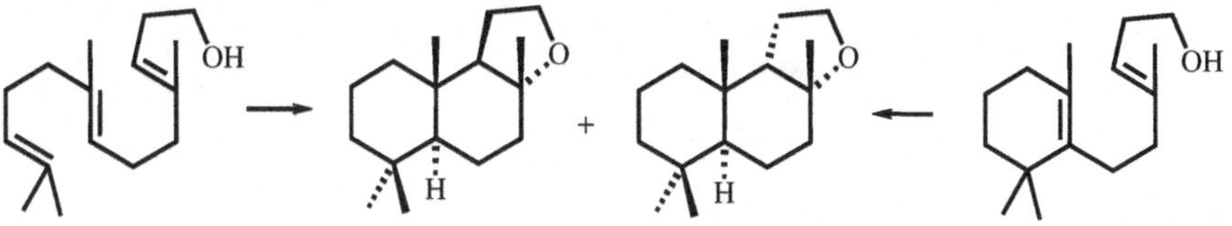

Figure 6.32   Acid-catalysed biomimetic cyclisation of homofarnesol.

(168–172). Both these reactions are shown in Figure 6.32. The latter was the subject of many years of development work which eventually led to processes for the production of high purity racemic and enantiomerically pure products (157).

NON-TERPENOID-RELATED FRAGRANCE INGREDIENTS

Apart from the musks, outside the field of terpenoid-related materials, fragrance ingredients families tend to be related to readily available feedstocks and, to a lesser extent, technologies. The most important feedstocks are benzene (**6.182**), toluene (**6.183**) and phenol (**6.43**). Benzene (**6.182**) and toluene (**6.183**) are obtained from petroleum. Phenol (**6.43**) was once isolated from coal tar but is now made from benzene (**6.182**) as shown in Figure 6.33. Friedel–Crafts addition of propene (**6.184**) to benzene (**6.182**) gives cumene (**6.185**), autoxidation of which gives the hydroperoxide (**6.186**), which can be cleaved to give phenol (**6.43**) and acetone (**6.12**).

Perfume Ingredients from Phenol

The use of phenol (**6.43**) to produce coumarin (**6.45**) and anisaldehyde (**6.47**) is shown in Figure 6.10. Alkylation of phenol (**6.43**) with isobutene (**6.23**) produces a mixture of o-t-butylphenol (**6.187**) and p-t-butylphenol (**6.188**). These phenols can be reduced and acetylated to give the esters (**6.189**) and p-t-butylphenol (**6.190**) respectively, both of which are important fragrance ingredients. Reaction of phenol (**6.43**) with carbon dioxide gives salicylic acid (**6.191**), a key intermediate in

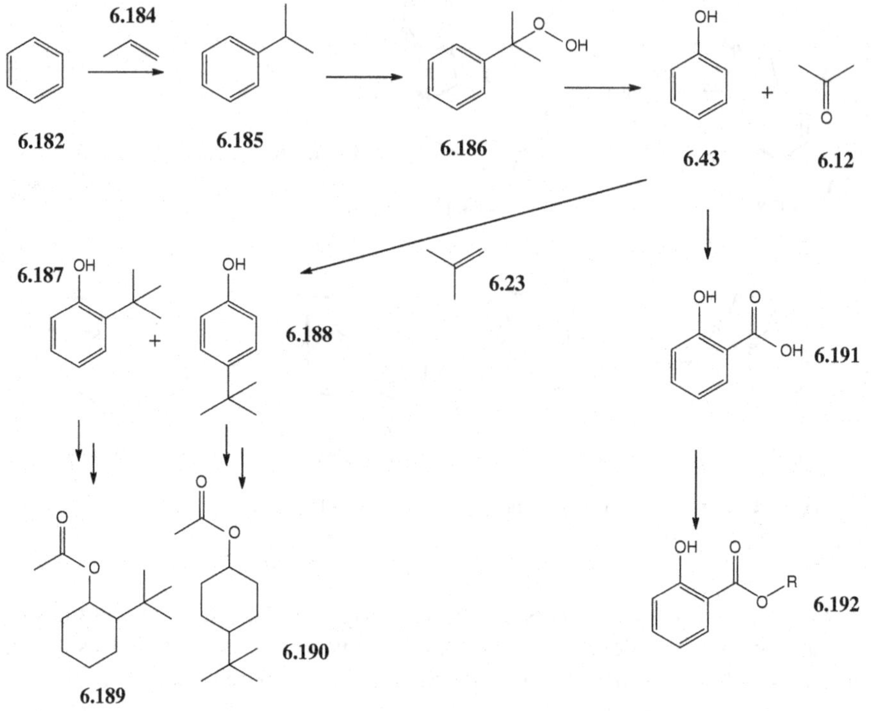

Figure 6.33   Manufacture and use of phenol.

the synthesis of aspirin. Various salicylate esters (**6.192**) are used in perfumes. The methyl ester (**6.192**, R = Me) is a major component of Wintergreen, and elicits an odour strongly reminiscent of it. Higher salicylate esters such as amyl, hexyl and benzyl provide useful background notes in fragrances and also have a fixative effect.

Routes to 2-Phenylethanol

2-Phenylethanol (**6.193**) is one of the highest volume perfume ingredients. It occurs widely in nature and is particularly important in rose oils. There are several routes to it from benzene and, although these exist in commercial equilibrium, the byproduct of the SMPO process is the most important because it is the lowest cost product currently. The three routes are shown in Figure 6.34. SMPO stands for styrene monomer/propylene oxide, both of which are important chemical feedstocks. Alkylation of benzene (**6.182**) with ethylene (**6.194**) produces ethyl-benzene (**6.195**), which can be autoxidised to the hydroperoxide (**6.196**). Cleavage of the hydroperoxide gives 1-phenylethanol (**6.197**) and propylene oxide (**6.198**).

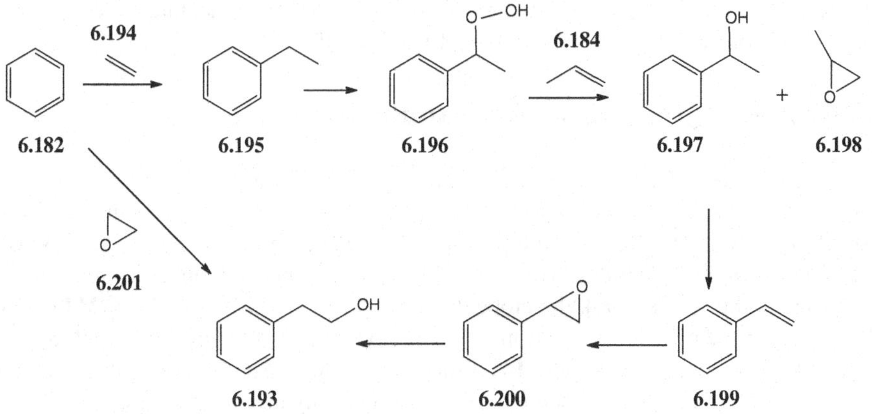

Figure 6.34   Manufacture of 2-phenylethanol.

Dehydration of the former gives styrene (**6.199**). Many minor products are formed together with 1-phenylethanol (**6.197**) in the cleavage reaction, and one of these is 2-phenylethanol (**6.193**). Separation of this alcohol and distillation to perfumery quality is far from trivial, but the fact that it is a by-product means that it is still relatively inexpensive and the scale of the SMPO process is so large that the volume of 2-phenylethanol (**6.193**) produced in this way meets most of the needs of the fragrance industry. Epoxidation of styrene (**6.199**) to its oxide (**6.200**) and subsequent hydrogenation gives 2-phenylethanol (**6.193**), as does addition of ethylene oxide (**6.201**) to benzene.

Perfume Ingredients from Toluene

Oxidation of toluene (**6.183**) gives benzaldehyde (**6.202**), which can undergo aldol condensation with other aldehydes to give the appropriate cinnamaldehydes (**6.203**), as shown in Figure 6.35. Of course, the major product of air oxidation of toluene (**6.203**) is the acid rather than the aldehyde, but the small amount of aldehyde in the product mixture is sufficient to supply the needs of the fragrance industry. The corresponding alcohol and its esters are also important fragrance ingredients, as are the alcohol and esters derived from cinnamaldehyde (**6.203**, R = H). Other cinnamaldehydes used extensively in perfumery are those derived from condensation of benzaldehyde (**6.202**) with heptanal and octanal, namely, amylcinnamaldehyde (ACA) (**6.203** R = C_5H_{11}) and hexylcinnamaldehyde (HCA) (**6.203** R = C_6H_{13}), respectively. Another example of routes in economic competition is also shown in Figure 6.35. Alkylation of toluene (**6.183**) with isobutene gives *p-t*-butyltoluene (**6.204**), which can be oxidised to the corresponding aldehyde (**6.205**). Condensation of this with propionaldehyde followed by hydrogenation, gives Lilial® (**6.206**). Alternatively, *t*-butylation of benzene gives (**6.207**), which undergoes Friedel–Crafts reaction with acrolein diacetate (**6.208**), and hydrolysis of the resultant adduct gives Lilial® (**6.206**).

Perfume Ingredients from Adipic Acid

Adipic acid (**6.209**) is a key intermediate in the production of Nylon 66 and is, therefore, available in large quantities at low price. Pyrolysis of its calcium or barium salts produces cyclopentanone (**6.210**), from which a variety of jasmine odorants are made (Figure 6.36). Aldol condensation and subsequent hydrogenation gives 2-alkylcyclopentanones such as Jasmatone® (**6.211**) and Heptone® (**6.212**). Isomerisation of 2-pentylidenecyclopentanone (**6.213**) to 2-pentylcyclopent-2-enone (**6.214**) followed by Michael addition of dimethyl malonate (**6.215**) and subsequent dealkoxycarbonylation gives methyl dihydrojasmonate (**6.216**). The most sought after jasmonate is (+)-(1*R*,2*S*) methyl Z-epijasmonate (**6.217**), but there is a cost associated with producing this in high stereochemical purity and so the cheaper methyl dihydrojasmonate (**6.216**) has a much higher volume of use.

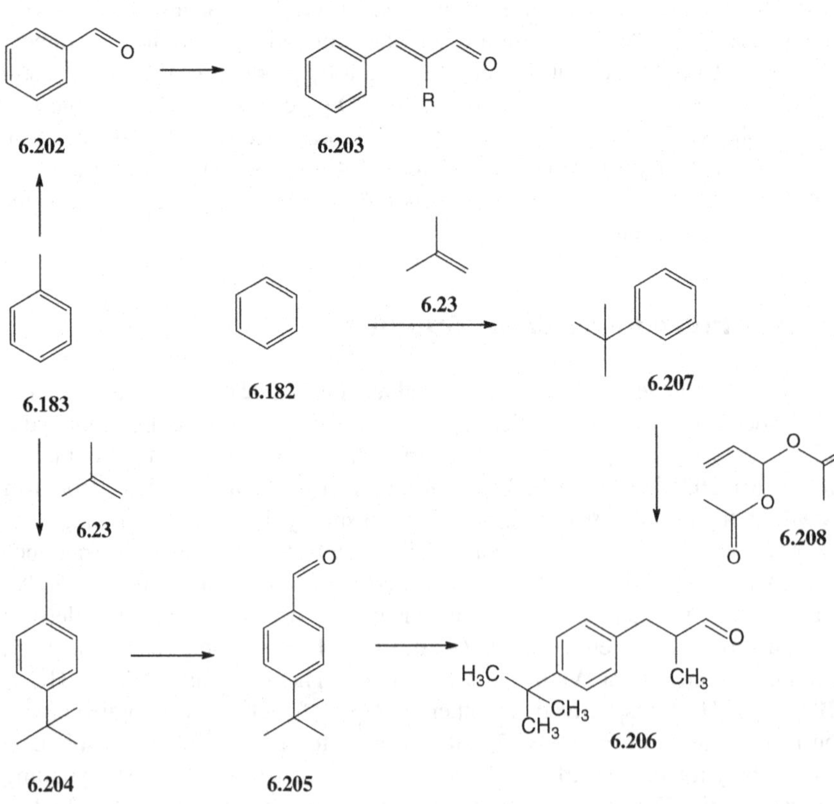

Figure 6.35   Manufacture of Lilial and cinnamic derivatives.

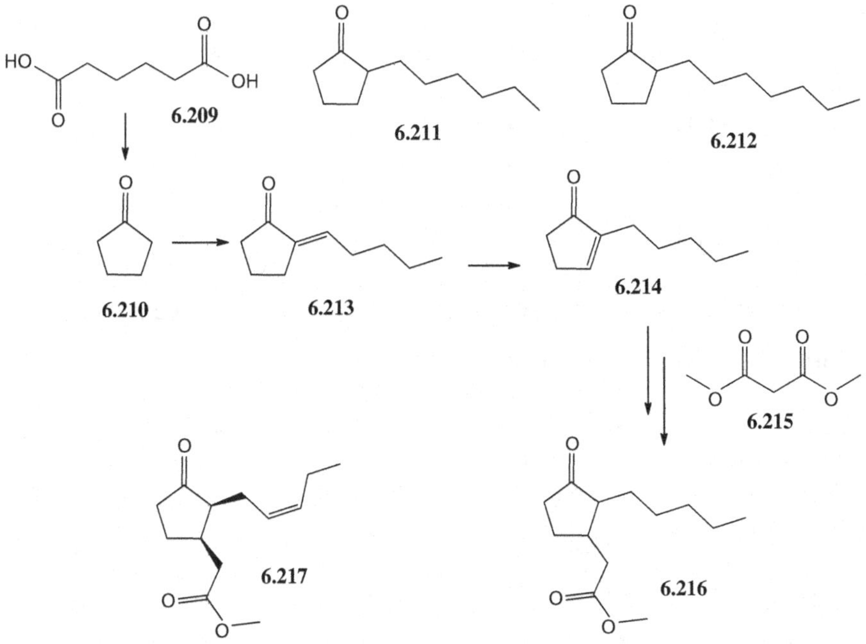

Figure 6.36   Cyclopentanone derivatives.

Perfume Ingredients from Dicyclopentadiene

Dicyclopentadiene (**6.218**) is another petrochemical intermediate that finds application in fragrance chemistry. Pyrolysis to cyclopentadiene (**6.219**) provides a potent dienophile for Diels–Alder reactions, as will be mentioned later. Ring strain in the dimer makes the double bond in the bridged system more reactive than the other and thus acid-catalysed addition of alcohols and acids occurs selectively giving various products such as Jasmacyclene® (**6.220** R = Me), Florocyclene® (**6.220** R = Et), Gardocyclene® (**6.220** R = ⁱPr), Pivacyclene® (**6.220** R = ᵗBu), Verdalia® A (**6.221** R = Me) and Fleuroxene® (**6.221** R = CH₂CH=CH₂). Longer reaction sequences give products such as Fruitate® (**6.222**), Dupical® (**6.223**), Scentenal® (**6.224**) and Vigoflor® (**6.225**) (Figure 6.37).

Perfume Ingredients from Naphthalene

Naphthalene (**6.226**) also provides several fragrance ingredients, and examples are shown in Figure 6.38. Acylation gives methyl naphthyl ketone (**6.227**), and hydroxylation followed by etherification gives Yara Yara® (**6.228** R = Me) and Nerolin Bromelia® (**6.228** R = Et). Oxidative degradation of naphthalene (**6.226**) to phthalic acid followed by Hoffmann rearrangement of the corresponding imide produces

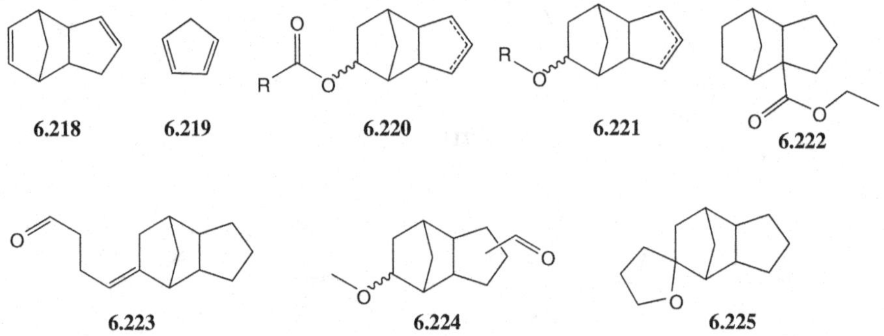

6.218   **6.219**   **6.220**   **6.221**   **6.222**

6.223   **6.224**   **6.225**

Figure 6.37   Dicyclopentadiene derivatives.

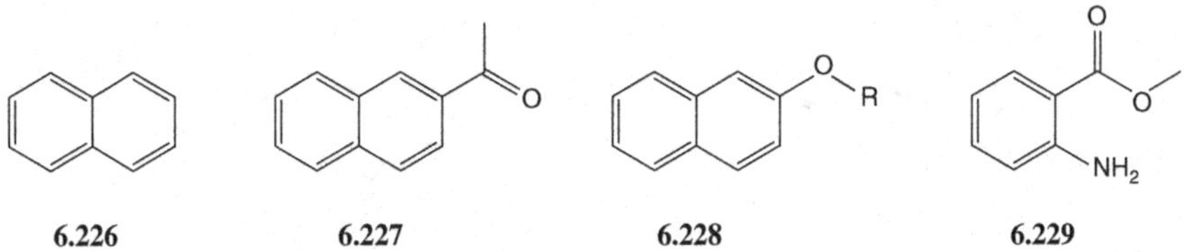

6.226   **6.227**   **6.228**   **6.229**

Figure 6.38   Naphthalene and derivatives.

methyl anthranilate (**6.229**). This amine is an ingredient in its own right, eliciting a heavy sweet floral odour, but it is also important because of its ability to form Schiff's bases with aldehydes. The Schiff base formation helps to protect aldehydes from oxidation and other reactions and also reduces their volatility, thus improving substantivity. Hydrolytic release in use then provides the aldehydic note, albeit always in combination with the heavy odour of methyl anthranilate. Some Schiff's bases of methyl anthranilate are manufactured and sold as such, but more usually perfumers will add methyl anthranilate *per se* to a fragrance, knowing that spontaneous Schiff's base formation will occur.

Perfume Ingredients from Vegetable Oils

There is an accent nowadays on renewable feedstocks. These have always been part of fragrance chemistry and it is expected that their role will increase in the future. Erucic acid (**6.330**) is obtained from rapeseed oil and ozonolysis of it produces brassylic acid (**6.331**), the ethylene ester (**6.332**) of which is a musk. Pyrolysis of ricinoleic acid (**6.333**) from castor oil gives undecylenic acid (**6.334**) (Figure 6.39), which serves as a feedstock for various fragrance ingredients. Reduction gives 10-undecenol (**6.335**) and undecanol (**6.336**), which can be dehydrogenated to 10-undecenal (**6.337**) and undecanal (**6.338**), respectively. Anti-Markownikov bromination of methyl undecylenate (**6.339**) and reaction with 1,4-butandiol (**6.340**) gives the musk Cervolide® (**6.341**).

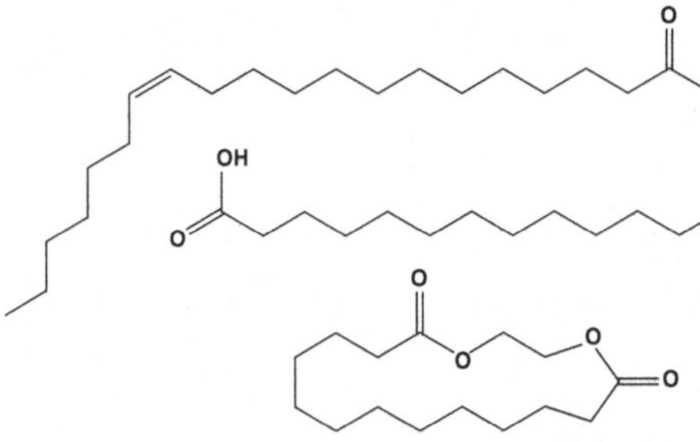

6.330

6.331

6.332

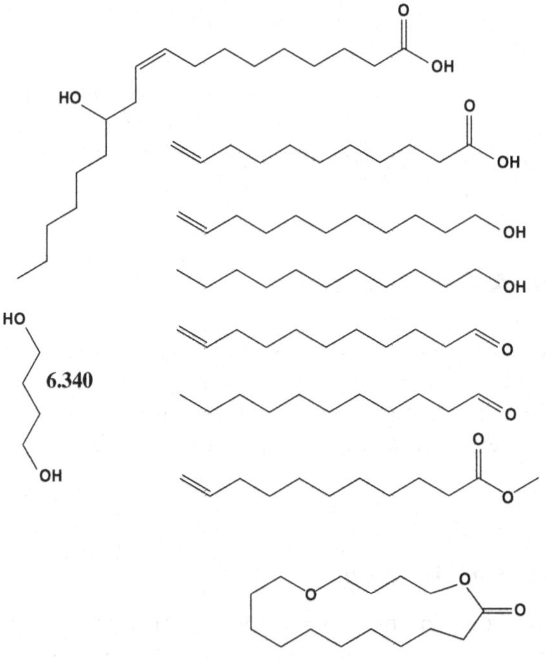

6.333

6.334

6.335

6.336

6.337

6.338

6.339

6.340

6.341

Figure 6.39 Fragrance ingredients from rapeseed and castor oils.

Perfume Ingredients from Clove and Sassafras oils

Eugenol (**6.342**) is obtained from clove buds, and safrole (**6.343**) from sassafras oil (Figure 6.40). Isomerisation, alkylation and ozonolysis can be used to produce various derivatives from these, including methyl isoeugenol (**6.344**), heliotropin (**6.345**) and vanillin (**6.346**). These products can all be obtained also by straightforward chemistry from phenol (**6.43**), via catechol (**6.347**). As in so many cases,

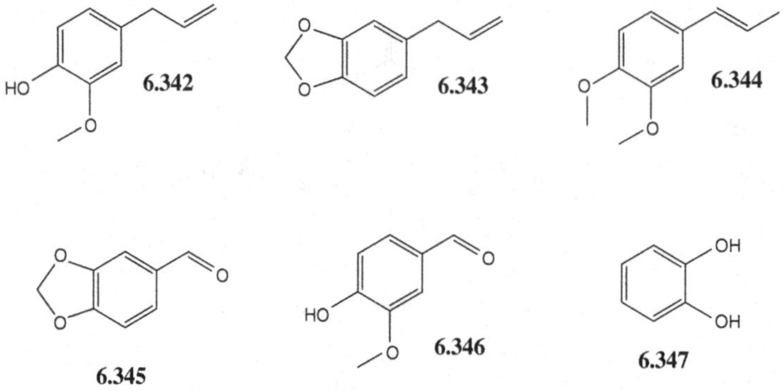

Figure 6.40   Shikimate-derived fragrance ingredients.

an economic balance exists that varies with availability of the natural, changes in process chemistry and such factors (especially environmental considerations). The most significant products from these natural sources are eugenol (**6.342**), its isomers and alkylated derivatives and heliotropin (**6.345**).

Use of Prins and Diels–Alder Reactions

The need for low cost and environmentally clean processing also makes some reactions more attractive than others for fragrance ingredient manufacture. Reactions such as the Prins and Diels–Alder are thermal or catalytic and do not involve stoichiometric reagents which lead to waste issues, as for example the Friedel–Crafts acylation does.

Perfume Ingredients via the Prins Reaction

Some typical Prins reactions are shown in Figure 6.41. Addition of formaldehyde (**6.24**) to 1-octene (**6.348**) in the presence of acetic acid, acetic anhydride and an acid catalyst produces a complex mixture of products one of which is the acetoxypyran (**6.349**). Various mixtures, each with distinctive ratio of components, are sold commercially. When homoallylic alcohols are used in the Prins reaction, the intermediate carbocation is trapped by the alcohol function to produce a pyran. The last step in the reaction sequence is elimination of a proton and this usually occurs in all possible directions, giving a mixture of isomers. Prins reaction of 4-methylpent-4-en-2-ol (**6.350**) with pentanal (**6.351**) gives a product mixture known as Gyrane® (**6.352**), whereas with benzaldehyde (**6.353**) it gives Pelargene® (**6.354**). Similarly, isoprenol (**6.25**) reacts with benzaldehyde (**6.353**) to give a mixture called Rosyrane® (**6.355**) and hydrogenation of this gives an important floral alcohol known as Mefrosol® (**6.356**).

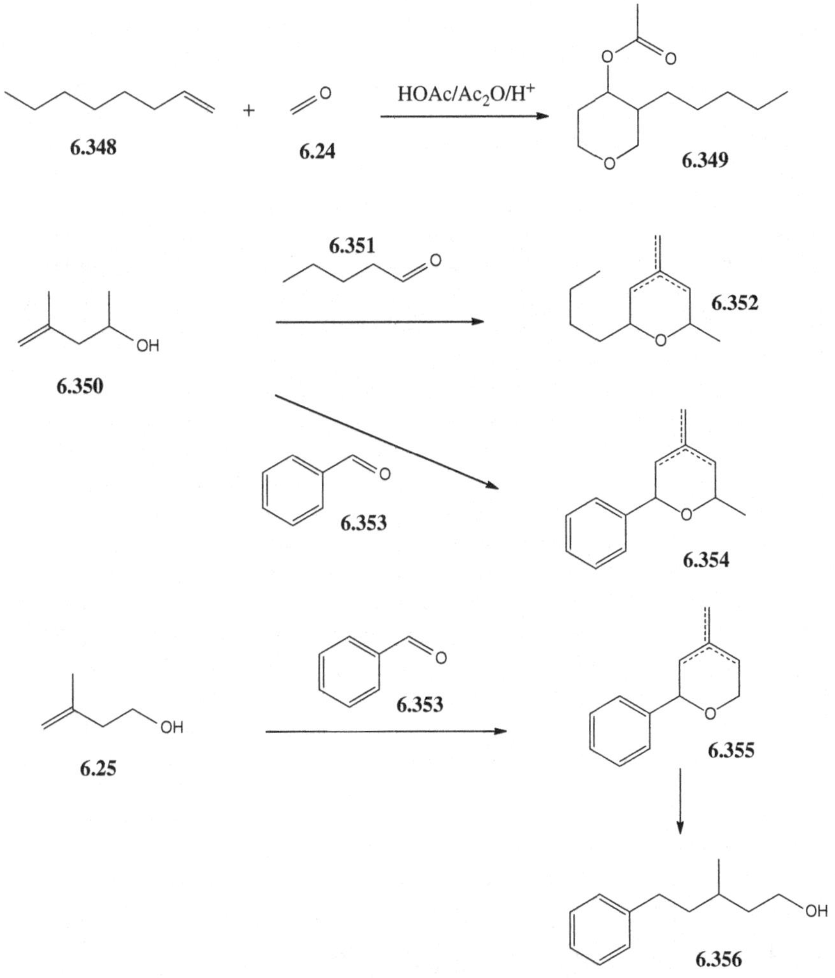

Figure 6.41 Some typical Prins reaction products used in perfumery.

Perfume Ingredients via the Diels–Alder Reaction

Diels–Alder reactions of myrcene (**6.39**) and myrcenol (**6.102**) are shown in Figure 6.21. Some other Diels–Alder reactions are shown in Figure 6.42. Addition of acrolein (**6.97**) to 2-methylpenta-1,3-diene (**6.357**) gives an aldehyde (**6.358**) known under various trade names such as Cyclal C.® Cyclopentadiene (**6.219**) reacts with 2-hexenal (**6.359**) to give Chrysanthal® (**6.360**) and α-terpinene (**6.361**) adds to acrolein (**6.97**) to give Maceal® (**6.362**). Butadiene (**6.363**) adds to methyl methacrylate (**6.364**) to give Tachrysate® (**6.365**). Hetero-Diels–Alder reactions can also be used, and a Lewis acid-catalysed addition of benzaldehyde (**6.353**) to isoprene (**6.69**) gives one (**6.366**) of the isomers of Rosyrane® (**6.355**) (173, 174).

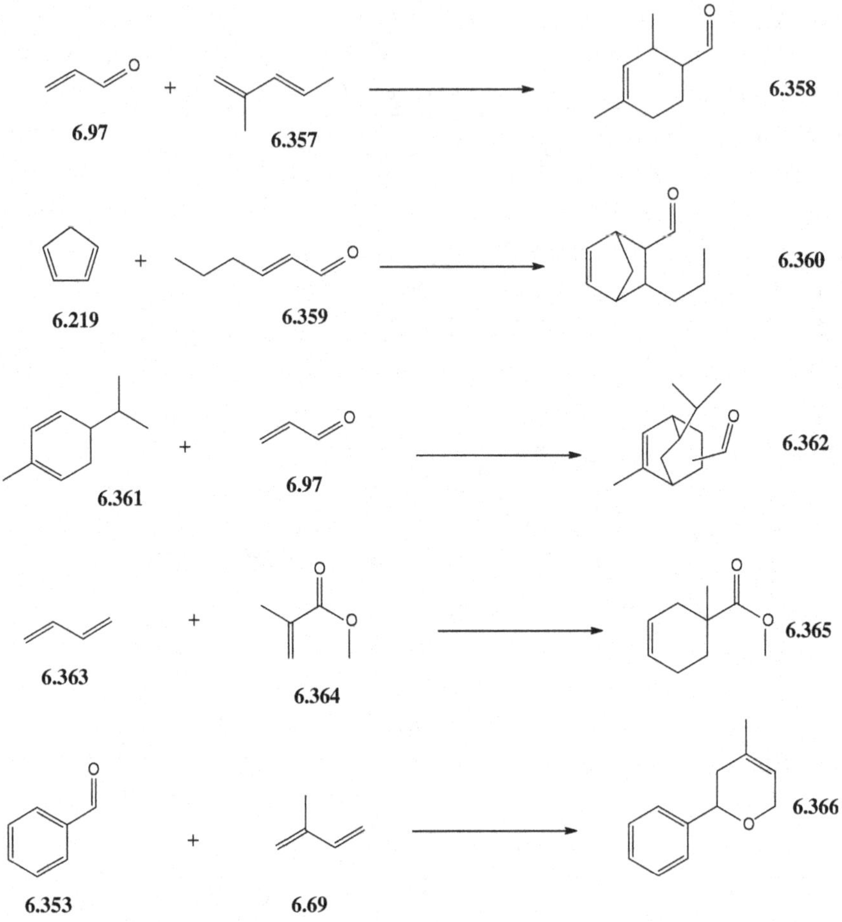

Figure 6.42   Some Diels–Alder reactions used in perfumery.

Synthesis of Musks

The first synthetic musk to be prepared was Musk Baur (**6.367**) in 1888. Baur was investigating explosives when he chanced upon this family of aromatics which elicit a musk odour. For many years, Musk Ambrette (**6.368**) was a favourite of perfumers, but recent concerns over the safety and environmental performance of the nitro musks has led to their demise. Figure 6.43 shows a typical synthesis of a nitro musk, Musk Ketone (**6.369**) in this case. The synthesis starts with *m*-xylene (**6.370**),

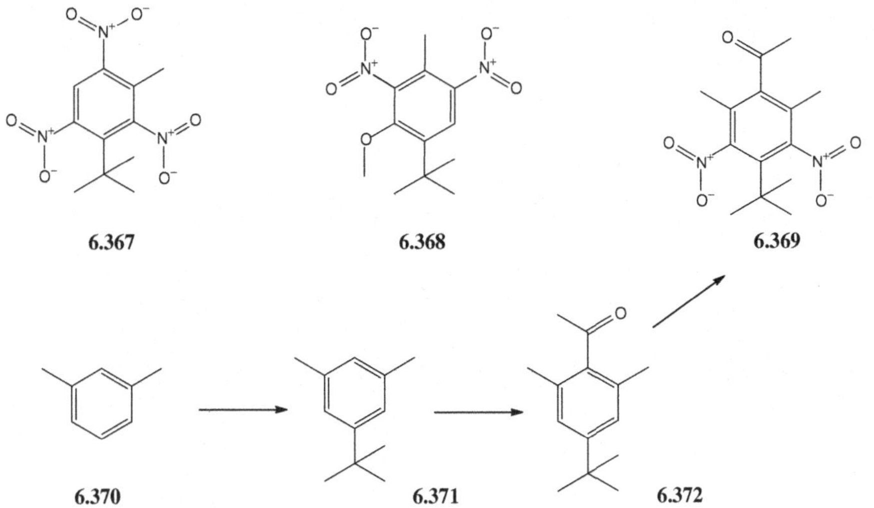

6.367   **6.368**   **6.369**

6.370   **6.371**   **6.372**

Figure 6.43   Nitro musks.

which is first alkylated by isobutene (**6.23**) to give (**6.371**), which is acetylated to give the ketone (**6.372**), and treatment of this with a mixture of nitric and sulfuric acids gives Musk Ketone (**6.369**).

The major musks of the second half of the twentieth century belonged to the polycyclic musk family. The two largest volume products in this group were Tonalid® (**6.373**) and Galaxolide® (**6.374**) and their syntheses are shown in Figure 6.44. The environmental performance of polycyclic musks is also being questioned, and so their tonnages are falling as they begin to be replaced by other musks. Tonalid® (**6.373**) is produced from *p*-cymene (**6.375**), which is first cyclialkylated with neohexene (**6.376**) followed by acetylation of the resulting tetralin (**6.377**). The starting material for Galaxolide® (**6.374**) is cumene (**6.185**), which is cyclialkylated with isoamylene (**6.378**). Acid-catalysed addition of propylene oxide (**6.198**) to the resultant indane (**6.379**) gives the alcohol (**6.380**), and reaction of this with formaldehyde (**6.24**) produces Galaxolide® (**6.374**).

The musk chemicals found in nature are macrocyclic: muscone (**6.381**), civetone (**6.382**) and ambrettolide (**6.383**) (Figure 6.45). Entropy presents a serious obstacle when synthesis of macroyclic molecules from linear precursors is attempted. The reactive groups at either end of the chain are more likely to react with a second molecule than with the opposite end of the same chain. Ruzicka solved the problem by using high dilution to prevent polymerisation and thus favour intra-molecular reaction. He was awarded the Nobel Prize for this work in 1926. However, high dilution is expensive on manufacturing scale because of the

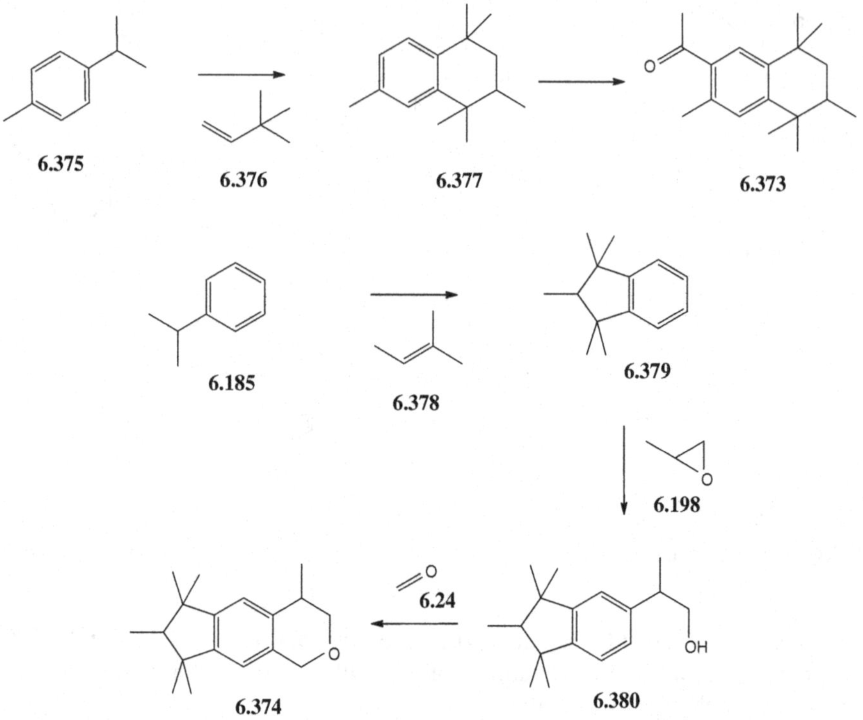

6.375

6.376

6.377

6.373

6.185

6.378

6.379

6.198

6.24

6.374

6.380

Figure 6.44   Representative syntheses of polycyclic musks.

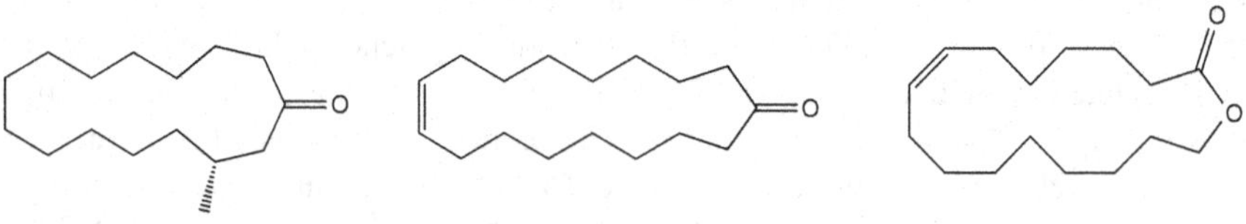

6.381

6.382

6.383

Figure 6.45   Nature's musks.

capital cost of the reactor and the cost of removal and recycle of large volumes of solvent. A more commercially practical solution to the problem was devised by Carothers in the 1950s but it only works for lactones (58).

Carothers' idea was to allow a hydroxyl acid to polymerise but to keep it under equilibrium conditions and remove the monocyclic lactone by azeotropic distillation. Glycerol proves to be a good solvent for this approach. In the presence of base, its hydroxyl groups are deprotonated and serve to keep the reaction moving; and its boiling point is close to those of macrocyclic musks so it distils with them but

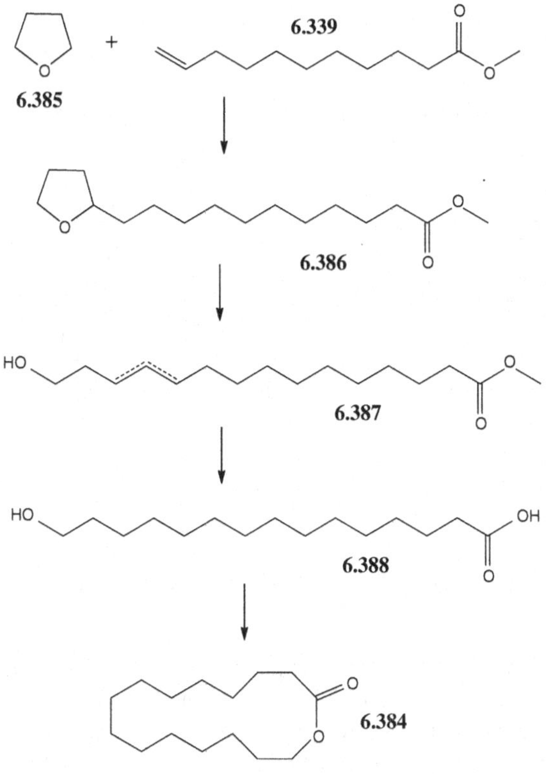

Figure 6.46 Synthesis of cyclopentadecanolide from methyl undecylenate.

is not miscible with them when the vapours condense, thus allowing easy separation of the product and recycling of the solvent. One disadvantage is that glycerol is not very stable in the presence of base at high temperature, resulting in loss of solvent and generation of effluent. Nonetheless, this technique or variants of it are used to produce various macrocyclic lactones. The examples of ethylene brassylate (**6.332**) and Cervolide® (**6.341**) are described above. Another example (shown in Figure 6.46) is that of cyclopentadecanolide (**6.384**) which, with ethylene brassylate, is currently one of the highest volume macrocyclic musks. In this synthesis, radical addition of tetrahydrofuran (**6.385**) to methyl undecylenate (**6.339**) gives the tetrahydrofuran (**6.386**). Pyrolysis of this opens the ring to give a mixture of two unsaturated alcohols (**6.387**). Hydrogenation and hydrolysis of this mixture gives 15-hydroxypentadecanoic acid (**6.388**), which can be cyclised under polymerisation/depolymerisation conditions to produce cyclopentadecanolide (**6.384**).

In 1959, Wilke (175) found that butadiene (**6.363**) could be trimerised to give cyclododecatriene (**6.389**), and hence cyclododecanone (**6.390**) became available. Eschenmoser then used the fragmentation reaction, which bears his name, to pioneer a new approach to macrocyclic musk synthesis, namely, that of fragmentation of the bridge between two smaller rings to produce a larger one (176). The outline

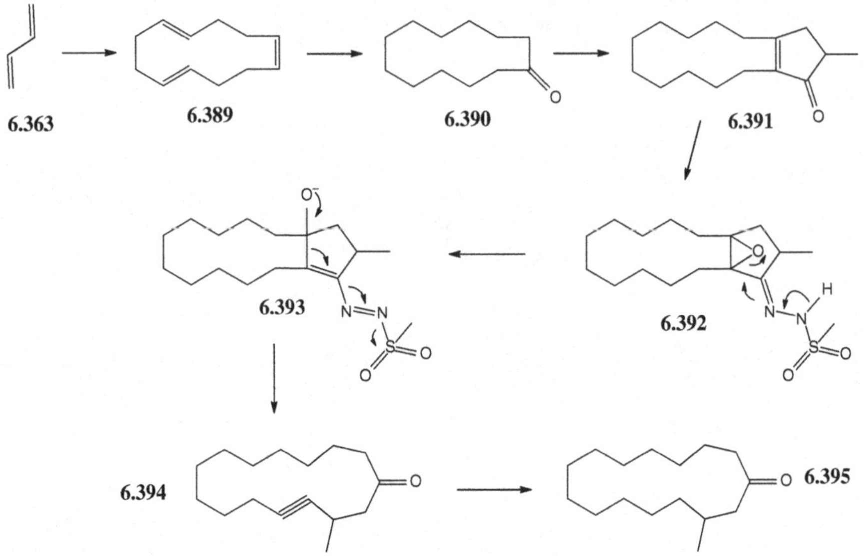

Figure 6.47   Eschenmoser's synthesis of racemic muscone.

of his reaction scheme is shown in Figure 6.47. He first built a cyclopentenone ring onto the side of the starting cyclododecanone (**6.390**) to give the bicyclic ketone (**6.391**). Use of the Eschenmoser fragmentation reaction via intermediates (**6.392**) and (**6.393**) then gave the cyclopentadecynone (**6.394**), which could be hydrogenated to racemic muscone (**6.395**).

A more commercially feasible fragmentation reaction is shown in Figure 6.48. Radical addition of allyl alcohol (**6.396**) to cyclododecanone (**6.390**) gives the hydroxyl ketone (**6.397**), which cyclises to the pyran (**6.398**). Acid-catalysed addition of hydrogen peroxide to this gives the peroxy acetal (**6.399**) and pyrolysis of this in the presence of a copper catalyst gives the mixture of isomeric olefinic ketones (**6.400**). This mixture can be hydrogenated to cyclopentadecanolide (**6.384**).

Olefin metathesis allows yet another strategy to be employed as evidenced by the synthesis of Animusk® (**6.402**) from cyclooctene (**6.401**), as shown in Figure 6.49. Dimerisation of the starting olefin produces cyclohexadecadiene (**6.403**), and mono-epoxidation and acid-catalysed isomerisation of this gives Animusk® (**6.402**).

As mentioned previously, a new family of musks has been introduced and is rising in significance, and typical examples are shown in Figure 6.22.

CONCLUDING REMARKS

The examples given in this chapter are illustrative of the types of processes used in the manufacture of fragrance ingredients and the principles behind them. More

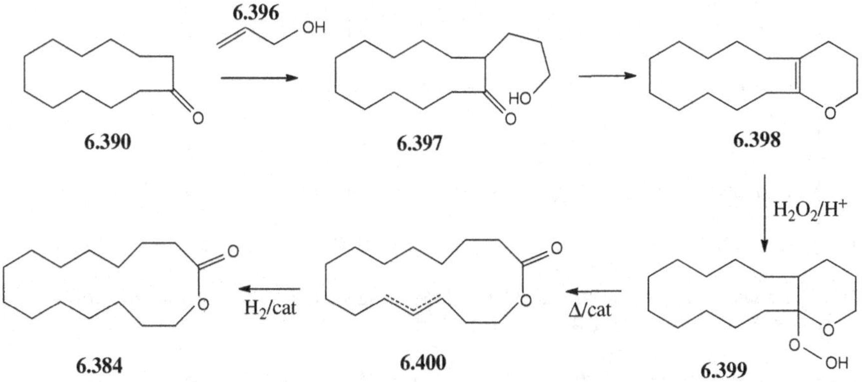

6.390 6.397 6.398

6.384 6.400 6.399

Figure 6.48 Synthesis of cyclopentadecanolide using a fragmentation reaction.

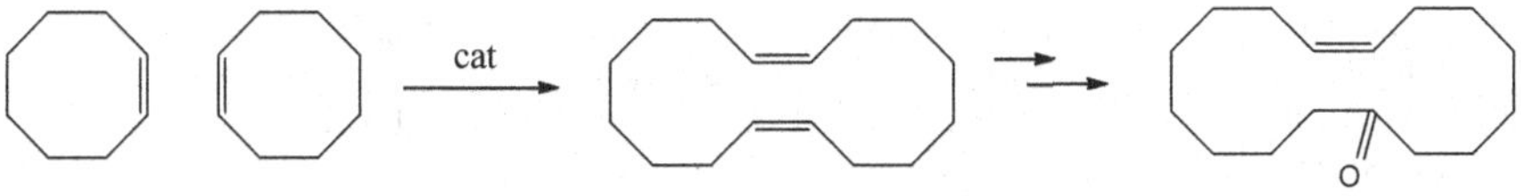

6.401 6.401 6.403 6.402

Figure 6.49 Musk synthesis using olefin metathesis.

comprehensive coverage can be found in the following sources. The chapter on terpenoids in the *Kirk–Othmer Encyclopedia* (37) and the book by Sell (6) are useful references for terpenoids and related materials. For both terpenoid and non-terpenoid ingredients, the books by Surburg and Panten (177) and by Sell (7) and *Ullmann's Encyclopedia of Industrial Chemistry* (178) are good sources of further information.

REFERENCES

1. *Herodatus, The Histories*, translator A. de Sélincourt, Penguin Books, Harmondsworth, 1968.
2. K. Nagai, *Appl. Catal. A: Gen.*, **2001**, *221*, 367–377.
3. F. Paulot, J. D. Crounse, H. G. Kjaergaard, A. Kürten, J. M. St. Clair, J. H. Seinfeld and P. O. Wennberg, *Science*, **2009**, *325(5941)*, 730–733 doi: 10.1126/science.1172910.
4. www.leffingwell.com/top_10.htm
5. D. G. Williams, *Perfumes of Yesterday*, Micelle Press, Weymouth, 2004, pp 144–145.
6. C. S. Sell, *A Fragrant Introduction to Terpenoid Chemistry*, Royal Society of Chemistry, Cambridge, 2003, ISBN 0 85404 681 X.
7. C. S. Sell, Ed., *The Chemistry of Fragrances: From Perfumer to Consumer* 2nd Edn., Royal Society of Chemistry, Cambridge, 2006.

8. K. A. D. Swift, *Topics in Catalysis*, **2004**, *27(1–4)*, 143.
9. (a) P. Barbier and L. Bouveault. *Compt. Rend.*, **1896**, *2(22)*, 393–395; (b) F. Tiemann and R. Schmidt, *Ber.*, **1896**, *29*, 913.
10. W. Kimel, J. D. Surmatis, J. Weber, G. O. Chase, N. W. Sax, and A. Ofner, *J. Org. Chem.* **1967**, *22*, 1611.
11. H. Lindlar, *Helv. Chim. Acta*, **1952**, *35*, 446.
12. H. Lindlar, US 2,681,938, 1954, to Hoffmann-LaRoche, Inc.
13. M. Derrien and J. F. Le Page, US 3,674,888, 1972, to Institut Francais du Petrole, des Carburants et Lubrifiants.
14. W. Kimel, N. W. Sam, S. Kaiser, G. G. Eichmann, G. O. Chase, and A. Ofner, *J. Org. Chem.* **1958**, *23*, 153.
15. (a) G. Saucy and R. Marbet, BE 634,738, **1964**, assigned to Hoffmann-La Roche; (b) G. Saucy and R. Marbet, *Helv. Chim. Acta.* **1967**, *50*, 2091.
16. K. H. Meyer and K. Schuster, *Ber.*, **1922**, *55*, 819.
17. H. Rupe and E. Kambli, *Helv. Chim. Acta*, **1926**, *9*, 672.
18. P. Charbardes, DE 1,811,517, 1969, to Rhône-Poulenc.
19. P. Chabardes and Y. Querou, GB 1,204,754, 1970, to Rhône-Poulenc SA.
20. P. Chabardes and Y. Querou US 3,920,751, 1975, to Rhône-Poulenc SA.
21. N. Hindley and D. A. Andrews, DE 2,353,145, 1973, to Hoffmann-La Roche.
22. T. Hosogai, T. Nishida and K. Itoh, JP 51,048,608, 1974, to Kuraray.
23. H. Pauling, D. A. Andrews, and N. C. Hindley, *Helv. Chim. Acta*, **1976**, *59*, 1233.
24. H. Pauling, US 3,981,896, 1976, to Hoffmann-LaRoche Inc.
25. H. Müller, H. Overwien and H. Pommer, DE-AS 1,618,098, 1967, to BASF.
26. P. R. Stapp, *Ind. Eng. Chem., Prod. Res. Dev.*, **1976**, *15*, 189.
27. H. Overwien and H. Müller, DE-AS 1,901,709, 1969, to BASF.
28. C. Dudeck et al., DE-AS 2,715,209, 1977, to BASF.
29. B. Meissner et al., DE-AS 2,715,208, 1977, to BASF.
30. W. F. Hölderich, in *"Studies in Surface Science and Catalysis"*, Eds., L. Guczi, F. Solymosi and P. Tetenyi, 1993, Vol. *75*, part A, p. 27.
31. H. Müller and H. Overwien US 3,686,321, 1972, to Badische Anilin- and Soda-Fabrik A-G.
32. A. Nissen and W. Aquila, EP 21,074, 1989, to BASF.
33. N. Götz and R. Fischer, DE 2,157,035, 1971.
34. P. Chabardes and J. Chasal, EP 344,043, 1988.
35. J. Paust, *Pure Appl. Chem.*, **1991**, *63*, 45.
36. C. Jäkel and R. Paciello, US 7534921 (2009) to BASF.
37. C.S. Sell, Terpenoids, in *The Kirk-Othmer Encyclopedia of Chemical Technology*, 5th Edn, John Wiley & Sons, Inc, New York.
38. (a) GB 776,204, 1957, to Bayer; (b) R. Stish et al., *Aug Chem* **1957**, *69*, 699, (Chem. Abstr., 1958, 52, 6244e).
39. A. W. Biedermann, DE 2,314,813, 1974 to Bayer.
40. N. E. Kologivova et al., *Zh. Prikl. Khim.*, **1963**, *36(12)*, 2740 (Chem. Abstr., 1964, 60, 9314f).
41. H. Terada and O. Koby, *Kagaku Zasshi*, **1962**, *65*, 1841 (Chem. Abstr., 1963, 59, 1685b).
42. W. Hickel, H. Feltkamp and S. Giger, *Ann.*, **1960**, *37*, 1 (Chem. Abstr., 1961, 55, 6401i).
43. J. Fleischer, K. Bauer and R. Hopp, DE 2,109,456, 1972, to Haarmann and Reimer.
44. A. B. Booth, US 2,843,636, 1959, to Glidden.
45. U. Bornscheuer, I.-L. Gatfield, E.-M. Hilmer, S. Vorlova and R. Schmidt, EP 1,223,223, 2002, to Haarmann and Reimer.
46. S. Akutagawa, *Appl. Catal. A*, **1995**, *128*, 171 and refs cited therein.
47. S. Inoue, H. Takaya, K. Tani, S. Otsuka, T. Sato and R. Noyori, *J. Amer. Chem. Soc.*, **1990**, *112(12)*, 4897.

48. K. Tani, T. Yamagata, S. Otsuka, S. Akutagawa, H. Kumobayashi et al., *J C S Chem Commun.*, **1982**, *11*, 600.
49. S. Akutagawa, *Topics in Catalysis 4*, **1997**, *3(4)*, 271.
50. Y. Nakatani and K. Kawashima, *J C S Chem. Commun.*, **1978**, 147.
51. Y. Nakatani and K. Kawashima, *Synthesis*, **1978**, 149.
52. JP 116,348, 1978, to Takasago (Chem. Abstr., 1979, 90, 87694g).
53. H. Yoji, I. Takesji and O. Yoshiki, EP 1,225,163, 2002, to Takasago.
54. H. Maeda, S. Yamada, H. Itoh and Y. Hori, *Chem. Commun.*, **2012**, *48*, 1772–1774 doi: 10.1039/c2cc16548a.
55. G. G. Koloeyer and J. S. Oyloe, US, 2001 899518 assigned to Millennium Chemicals.
56. I. M. Goldman and M. A. Perret, Fr. Pat. 1406122 19650716, 1965, to Pfizer.
57. R. Jeffcoat and B. J. Willis, *Dev. Food Sci.*, **1988**, *18*, 743–751.
58. E. W. Spanagel and W. H. Carothers, *J. Amer. Chem. Soc.*, **1935**, *57(5)*, 929–934 doi: 10.1021/ja01308a046.
59. M. I. Farbood, J. A. Morris and A. E. Downey, US 5,155,029, 1992, to IFF.
60. J. Dorsky, *Perfumer and Flavorist*, **1978**, *3(6)*, 51.
61. H. Pommer, H. Müller, and H. Overwien, DE 1,286,020, 1969, to Badische Anilin- and Soda-Fabrik A-G.
62. H. Pommer, H. Müller, and H. Overwien, DE 1,259,876, 1968, to Badische Anilin- and Soda-Fabrik A-G.
63. H. Müller, H. Köhl, and H. Pommer, US 3,670,028, 1972, to Badische Anilin- and Soda-Fabrik A-G.
64. H. Pommer, H. Müller, H. Köhl, and H. Overwien, DE 1,643,668, 1971, to Badische Anilin- and Soda-Fabrik A-G.
65. H. Pommer and A. Nurrenbach, *Pure Appl. Chem.* **1975**, *43(3–4)*, 527.
66. W. Reif and H. Grassner, *Chem. Ing. Tech.* **1973**, *45(10a)*, 646.
67. H. Pasedach, W. Hoffmann and W. Himmele, DE 1,643,710, 1967, to Badische Anilin and Soda-Fabrik A-G.
68. W. Friedrichsen, DE 973,089, 1959, to Badische Anilin- and Soda-Fabrik A.G.
69. H. Pommer, H. Müller, and H. Overwien, DE 1,268,135, 1968, to Badische Anilin- and Soda-Fabrik A.G.
70. Y. Tamai et al., DE 2,356,866, 1974, assigned to Kuraray.
71. M. Pichou, FR 1,548,516, 1968, to Societe des Usines Chimiques Rhone-Poulenc.
72. W. C. Meuley and P. Gradeff, CA 766,787, 1967, to Rhodia, Inc.
73. W. C. Meuly and P. Gradeff, FR 1,384,137, 1965, to Rhône-Poulenc SA.
74. Y. Tamai et al., US 3,983,175, 1976, to Kuraray Co., Ltd.
75. O. Yoshiaki et al., JP 75,58 004, 1973, to Kuraray.
76. J. Dorsky, p 399 in K Bauer and D Garbe, Common Fragrance and Flavour Materials, VCH Verlag, Weinheim, 1985.
77. G. Ohloff, E. Klein, and G. Schade, US 3,240,821, 1966, to Studiengesellschaft Kohle.
78. P. Chabardes and C. Grard, US 3,925,485, 1975, to Rhône-Poulenc.
79. B. J. Kane, US 4,254,291, 1978, to SCM Corp.
80. J. C. van der Waal, H. van Bekkum and J. Vital, *J. Mol. Catal.*, **1996**, *105*, 185.
81. H. Valente and J. Vital, Proc. 4th Int. Symposium on Heterogenous Catalysis and Fine Chemicals, Basel, Switzerland, 1996, p 214.
82. L. A. Goldblatt and S. Palkin, US 2,420,131, 1947, to Glidden.
83. R. Weiss, US 2,882,323, 1959, to Ameringen Häbler Inc.
84. P. W. Mitchell, L. T. McElligott and D. E. Sasser, EP 132,544, 1984, to Union Camp Corp.
85. R. L. Webb, US 3,031,442, 1958, to The Glidden Co.
86. R. L. Webb, US 3,076,839, 1963, to The Glidden Co.

87. J. M. Derfer, US 3,278,623, 1966, to The Glidden Co.
88. J. B. Hall and J. M. Sanders, US 3,907,321, 1975, to International Flavors and Fragrances, Inc.
89. J. B. Hall and J. M. Sanders, US Pat US 3,911,018, 1975, to International Flavors and Fragrances, Inc.
90. J. B. Hall and J. M. Sanders, US Pat 3,929,677, 1975, to International Flavors and Fragrances, Inc.
91. C. Nussbaumer, G. Fráter and P. Kraft, *Helv. Chim. Acta*, **1999**, *82*, 1016.
92. G. Fráter, J. A. Bajgrowicz and P. Kraft, *Tetrahedron*, **1998**, *54*, 7633.
93. L. A. Canova, US 4,018,842, 1977, to SCM Corp.
94. J. P. Bain, US 3,277,206, 1966, to The Glidden Co.
95. R. L. Webb, US 2,902,510, 1959, to The Glidden Co.
96. J. H. Blumenthal, US, 3,487,118, 1969, to International Flavors and Fragrances, Inc.
97. J. Ibareq and B. Lahourcade, FR 2,597,861, 1987, to Derives Resiniques et Terpeniques.
98. J. B. Hall, US 3,847,975, 1974, to International Flavors and Fragrances, Inc.
99. J. B. Hall and L. K. Lala, *J. Org. Chem.*, **1972**, *37*, 920.
100. H. R. Ansari and B. J. Jaggers, GB 1,254,198, 1970, to Bush Boake Allen.
101. R. L. Webb, US 3,038,431, 1962, to The Glidden Co.
102. J. B. Nicholas and H. R. Ansari, US 3,963,648, 1976, to Bush Boake Allen.
103. B. N. Jones, H. R. Ansari, B. N. Jaggers, and J. F. Janes, DE 2,255,119, 1972, to Bush Boake Allen.
104. Y. Matoubara and H. Yada, JP 7,413,158, 1974, to Yoshitomi Pharmaceutical Industries.
105. E. Demole, *Helv. Chim. Acta.*, **1964**, *47*, 319.
106. J. Dorsky and W. M. Easter, Jr, US 3,499,937, 1970, to Givaudan Corp.
107. G. K. Lange and K. Bauer, EP 19,845, 1981, to Haarmann and Reimer.
108. J. B. Hall and W. J. Wiegers, US 4,014,944, 1976, to IFF.
109. J. B. Hall and W. J. Wiegers, US 4,104,203, 1978, to IFF.
110. J. B. Hall and W. J. Wiegers, US 4,131,555, 1978, to IFF.
111. J. B. Hall and W. J. Wiegers, US 4,131,557, 1978, to IFF.
112. E. Demole, *Helv. Chim. Acta*, **1969**, *52*, 2065.
113. B. Arbusow, *Ber.*, **1935**, *68*, 1430.
114. J. B. Lewis and G. W. Hedrick, *J. Org. Chem.*, **1965**, *30*, 4271.
115. East German Patent 68936, 1969, to VEB Miltitz.
116. E. J. Brunke and E. Klein, DE 2,827,957, 1978, to Dragoco.
117. W. M. Easter and R. E. Naipawer, CH 629,461, 1977, to Givaudan.
118. R. E. Naipawer and W. M. Easter, US 4,052,341, 1976, to Givaudan Corp.
119. R. E. Naipawer, EP 203,528, 1985, to Givaudan SA.
120. R. E. Naipawer, US 4,696,766, 1986, to Givaudan.
121. K. -H. Schulte-Elte, B. Müller and H. Pamingle, EP 155,391, 1984, to Firmenich SA.
122. K. -H. Schulte-Elte and B. L. Müller, US 4,610,813, 1984, to Firmenich.
123. J. A. Bajgrowicz and G. Fráter, EP 801,049, 1997, to Givaudan-Roure.
124. J. A. Bajgrowicz, I. Frank, G. Fráter, M. Hennig, *Helvetica Chimica Acta*, **1998**, *81*(7), 1349–1358.
125. B. J. Kane, US 3,346,650, 1967, to The Glidden Co.
126. E. Goldstein, US 3,275,696, 1966, to Universal Oil Products Co.
127. R. B. Bates, E. S. Caldwell, and H. P. Klein, *J. Org. Chem.* **1969**, *34*, 2615.
128. A. F. Thomas, in Ed., J. Apsimon, *The Total Synthesis of Natural Products*, Vol. 2, John Wiley & Sons, Inc, New York, 1973, pp. 1–195.
129. H. Takaya, J. Ohta, R. Noyori, N. Sayo, H. Kumobayashi, and S. Akutagawa, US 4,739,084, 1988, to Takasago Perfumery Co.
130. H. Takaya, J. Ohta, R. Noyori, N. Sayo, H. Kumobayashi, and S. Akutagawa, US 4,739,085, 1988, to Takasago Perfumery Co.

131. S. Akutagawa and K. Tani in *Catalytic Asymmetric Synthesis*, Ed., I. Ojima, VCH, New York, 1993, p. 41.
132. R. S. DeSimone and P. S. Gradeff, US 4,029,709, 1977, to Rhodia, Inc.
133. P. S. Gradeff and G. Formica, *Tetrahedron Lett.*, **1976**, 4681.
134. M. Horner, M. Irrgang and A. Nissen, DE-OS 2,934,250, 1979, to BASF.
135. G. O. Schenck, G. Ohloff and E. Klein, DE 1 137 730, 1961, to Studiengesellschaft Kohle.
136. T. Yamamoto, H. Matsuda, Y. Utsumi, T. Hagiwara and T. Kanisawa, *Tet. Letts.*, **2002**, *43*, 9077.
137. H. Matsuda and T. Yamamoto, US 5,858,348, 1999.
138. R. Marbet and G. Saucy, DE 1,109,677, 1959, to Hoffmann La Roche.
139. H. Pommer, W. Reif, W. Pasedach and W. Hoffmann, DE-AS 1286019, 1967, to BASF.
140. GB 865,478, 1961, to Hoffmann-La Roche.
141. M. G. J. Beets and H. van Essen, GB 812,727, 1959, to Polak and Schwarz.
142. P. S. Gradeff, US 3,840,601, 1974, to Rhodia, Inc.
143. G. Ohloff and G. Schade, FR 1,355,944, 1962, to Studiengesellschaft Kohle.
144. L. Janitschke, W. Hoffman, L. Arnold, M. Strözel and H. J. Sheiper, US 4,431,844, 1984, to BASF.
145. G. Blume, R. Hopp and W. Sturm, DE 2,455,761, 1974, to Haarmann and Reimer.
146. K.-H. Schulte-Elte et al., *Liebigs Ann Chem*, **1975**, 484.
147. A. F. Morris, F. Näf and R. L. Snowden, *Perfum Flavor*, **1991**, *16*, 33.
148. G. Büchi and J. C. Vederas, *J. Am. Chem. Soc.*, **1972**, *94*, 9128.
149. K.-H. Schulte-Elte, B. L. Müller and G. Ohloff, *Helv. Chim. Acta*, **1973**, *56*, 310.
150. K. -H. Schulte-Elte, CH 563,951, 1972, to Firmenich.
151. C. Fehr and J. Galindo, EP 260,472, 1987, to Firmenich.
152. D. H. R. Barton, S. I. Parekh, D. K. Taylor and C.-L. Tse, *Tet. Letts.*, **1994**, *35*, 5801.
153. D. H. R. Barton, S. I. Parekh, D. K. Taylor and C.-L. Tse, US 5,463,089, 1994, to Quest International.
154. P. N. Davey, L. S. Payne and C.-L. Tse, EP 822,191, 1997, to Quest International.
155. P. N. Davey and C.-L. Tse, EP 1,171,433, 2000, to Quest International.
156. M. G. Bolster, Ph.D. Thesis, Wageningen Agricultural University, 2002, ISBN 90-5808-666-6.
157. C. Chauffat and A. Morris, *Perfumer and Flavorist*, **2004**, *29(2)*, 34.
158. G. Stork, A. W. Burgstahler, *J. Am. Chem. Soc.*, **1955**, *77*, 5068.
159. G. Lucius, *Angew Chem*, **1956**, *68*, 247.
160. G. Lucius, *Arch Pharm*, **1958**, *291*, 57.
161. G. Lucius, *Chem Ber*, **1960**, *93*, 2663.
162. A. Saito, H. Matsushita and H. Kaneko, *Chem Letts*, **1983**, 729.
163. G. Staiger and A. Macri, DE 3,240,054, 1982, to Consortium Elektrochemische Industrie GmbH.
164. G. Staiger and A. Macri, EP 107,857, 1985, to Consortium Elektrochemische Industrie GmbH.
165. T. Kawanobe, K. Kogami and M. Matsui, *Agric. Biol. Chem.*, **1986**, *50*, 1475.
166. T. Kawanobe and K. Kogami, EP 165,458, 1985, to Hasegawa.
167. P. F. Vlad, N. D. Ungur, and V. B. Perutskii, *Khim. Geterotslic. Soedin.*, **1990**, *26* 896.
168. K. -H. Schulte-Elte, R. L. Snowden, C. Tarchini, B. Bär and C. Vial, EP 403,945, 1989, to Firmenich SA.
169. R. L. Snowden and S. M. Linder, *Tet. Letts.*, **1991**, *32*, 4119.
170. R. L. Snowden, J.-C. Eichenberger, S. M. Linder, P. Sonnay, C. Vial and K.-H. Schulte-Elte, *J. Org. Chem.*, **1992**, *57*, 955.
171. R. L. Snowden, J. -C. Eichenberger, W. Giersch, W. Thommen and K. -H. Schulte-Elte, *Helv. Chim. Acta*, **1993**, *76*, 1608.
172. A. F. Barrero, E. J. Altarejos, E. J. Alvarez-Manzaneda, J. M. Ramos and S. Salido, *J. Org. Chem.*, **1996**, *61*, 2215.
173. N. L. J. M. Broekhof, J. J. Hofma, H. Renes and C. S. Sell, *Perf. Flavor.*, **1992**, *17*, 11–15.

174. EP Appl. 325,000 1988, to Quest Intl.
175. G. Wilke, B. Bogdanovič, P. Borner, H. Breil, P. Hardt, P. Heimbach, G. Hermann, H.-J. Kominsky, W. Keim, M. Kröner, H. Müller, E. W. Müller, W. Oberkirch, J. Schneider, J. Stedefeder, K. Tanaka, K. Weyer and G. Wilke, *Angew. Chem. Internat. Edn.*, **1963**, *2*(*3*), 105–115.
176. A. Eschenmoser, D. Felix and G. Ohloff, *Helv. Chim. Acta*, **1967**, *50*(2), 708–713 doi: 10.1002/hlca.19670500232.
177. H. Surburg and J. Panten, *Common Fragrance and Flavor Materials*, 5th Edn., Wiley-VCH, Weinheim, 2006, ISBN-10: 3-527-31315-X, ISBN-13: 978-3-527-31315-0.
178. *Ullmann's Encyclopedia of Industrial Chemistry*, Wiley, 2012, ISBN: 9783527306732, doi: 10.1002/14356007.

Chapter 7

The Design of New Fragrance Ingredients

HISTORICAL INTRODUCTION

Before the introduction of modern analytical methods and gas chromatography/mass spectrometry in particular, the perfume industry relied very much on secrecy for protection of the business. Nowadays, the composition of a perfume is relatively easily obtained by analysis, making patented technology and ingredients the soundest means of protection.

As described in Chapter 6, the second major technological innovation in perfumery was that of synthetic organic chemistry in the middle of the nineteenth century. Coumarin (**7.1**) was first synthesised in 1868, heliotropin (**7.2**) in 1869 and vanillin (**7.3**) in 1877. All of these were used in the classic fragrance Jicky which was created by Aimé Guerlin in 1889 (Figure 7.1) (1). These three chemicals all occur naturally and the synthetic material would nowadays be referred to as nature-identical. The advantages of using nature-identical substances in place of essential oils lie in security of supply, cost and purity. Production of essential oils and other natural products is at the mercy of adverse weather, earthquakes, tsunamis and the like. For example, patchouli production was seriously affected by the tsunami that devastated the Indonesian coast in 2004. Transport from the producer to the user can be affected by many factors such as wars, which is an issue for plants that grow only in certain geographic regions. Olibanum (frankincense) is an example of this. It grows in the horn of Africa which has been subject not only to drought but also to wars during the last decade. Production of natural extracts is labour intensive. For example, it takes about 7,000,000 jasmine flowers to produce 1 kg of oil and the flowers must be hand-picked in the early morning when the oil content is highest. Even flowers such as lavender, which can be harvested mechanically, still incur relatively high production costs and these are only likely to increase as competition for arable land for food production increases. Variations in

Chemistry and the Sense of Smell, First Edition. Charles S. Sell.
© 2014 John Wiley & Sons, Inc. Published 2014 by John Wiley & Sons, Inc.

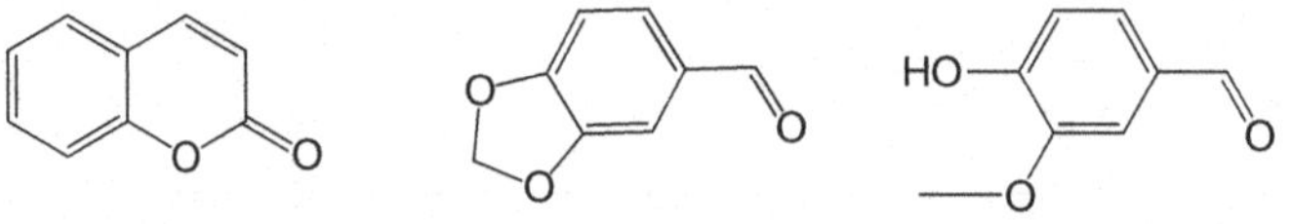

7.1   7.2   7.3

Figure 7.1   Nature-identical components of Jicky.

plant genetics, weather conditions, soil type and so on mean that material extracted from botanical sources inevitably varies in chemical composition. These variations in chemical composition lead to variations in odour quality and possibly also in terms of safety constraints. In contrast, a perfume ingredient that can be produced in a chemical factory can be produced with consistent odour quality with consistent high purity and in any geographical region. Some perfume ingredients with more complex structures, especially those requiring high enantiomeric purity for optimum performance, will always be produced more economically from natural sources but the majority of ingredients can be produced at much lower cost through synthetic organic chemistry.

Synthetic organic chemistry changed perfumery by providing nature-identical materials cheaply and with consistent odour quality and high purity. It also revolutionised perfumery in another way, by giving access to ingredients that have no counterpart in nature. In 1848, Bertagnini discovered that aliphatic aldehydes could be produced by pyrolysis of a mixture of calcium formate with the calcium salts of higher carboxylic acids. In 1921, the perfumer Ernest Beaux used aliphatic aldehydes to add a novel top note to an accord containing rose and jasmine oils and the famous Chanel No.5 was born. This landmark fragrance revolutionised perfumery and is still one of the top selling perfumes today. The dramatic success of this, and other perfumes using novel ingredients, led Beaux to remark that, 'One has to rely on chemists to find new aroma chemicals creating new, original notes. In perfumery, the future lies primarily in the hands of chemists'. When he said this, Beaux was thinking purely in terms of novelty of perfume notes. I believe that his assertion of the necessity for new ingredients is as true today as it was in 1921 but perhaps for different reasons.

THE NEED FOR NEW FRAGRANCE INGREDIENTS

Nowadays, with thousands of fragrance ingredients already available, one might question the need for further novel ingredients. However, the current pressures on the industry are such that the need for novel ingredients is greater than ever. Increasingly rigorous criteria for safety and biodegradability are resulting in the loss from the palette or restriction of use of traditional materials, including naturals. At the same time, the demands made of fragrance ingredients are becoming more stringent in terms of overall performance. As already mentioned in Chapter 6, any major fragrance company, or any company with the ambition of becoming one, needs to have a pipeline of novel ingredients for addition to its perfumery palette in order to deal with changing market conditions.

FACTORS DRIVING CHANGE IN THE FRAGRANCE INDUSTRY

There are many factors driving change in the fragrance industry. The major ones affecting ingredients are safety, environment, resources, performance requirements, market trends and scientific discovery. Each of these is discussed briefly below, and then they will be reflected in later sections of the chapter that discuss the selection criteria for novel ingredients and the process of molecular design.

Safety

Safety is, by far, the most important factor driving change in the fragrance industry of the early twenty-first century. Modern thinking covers the product from cradle to grave and so safety can be considered in three parts: safety during manufacture, safety in use and safety in the environment after use. Moral, financial and legal pressures combine to push fragrance companies to ensuring that they operate without causing harm. Even putting moral factors to one side, it would not be in the interests of any modern multinational company to cause harm to its employees, its customers or to the users of its products. For example, poor safety or environmental performance could result in financial loss through fines by government, law suits by injured parties or loss of business through adverse publicity from lobbyists.

Safety During Manufacture

Chapter 6 described in detail how modern process chemistry seeks always to improve the safety and environmental performance of both chemical manufacture and extraction of natural ingredients. Therefore, further discussion is unnecessary here.

Safety in Use

The Swiss alchemist and physician Theophrastus Bombastus von Hohenheim (1493–1541), better known as *Paracelsus*, was the first to formulate the basic principle of toxicology, 'Every substance is poisonous, the dose alone determines the poison'. One way of measuring toxicity is the use of the LD_{50} or dose (usually expressed in milligrams per kilogram bodyweight) required to cause death in 50% of tested animals. For ethical and other reasons, this test is only conducted infrequently but the historically obtained data in Table 7.1 can be used to show that typical fragrance ingredients such as farnesol (**7.7**), limonene (**7.8**), geraniol

Table 7.1   Acute Toxicity of Some Well-Known Chemicals

Compound	LD_{50} mg/kg	Structure
Sucrose (sugar)	33,000	**7.4**
Ethanol (alcohol)	13,000	**7.5**
Vitamin B1 (thiamine)	8,224	**7.6**
Farnesol	6,000	**7.7**
Limonene	4,400	**7.8**
Geraniol	3,600	**7.9**
Citronellol	3,450	**7.10**
Sodiuim chloride (salt)	3,000	—
Linalool	2,790	**7.11**
Oxalic acid (rhubarb)	375	**7.12**
Caffeine (coffee)	250	**7.13**
Digitoxin (foxglove)	200	**7.14**
Vitamin D2	42	**7.15**
Solanine (potatoes)	42	**7.16**
Theobromine (tea, chocolate)	26	**7.17**
Strychnine	5	**7.18**
Saxitoxin (red tide)	0.063	**7.19**
Tetrodotoxin (puffer fish = fugu)	0.01	**7.20**
Batrachotoxin (frogs)	0.002	**7.21**
Palytoxin (coral)	0.00015	**7.22**

(**7.9**), citronellol (**7.10**) and linalool (**7.11**) are relatively innocuous. They lie in the same region as vitamin B1 (**7.6**) and table salt and are much less toxic than chemicals such as caffeine (**7.13**) and theobromine (**7.17**) which are found in coffee, tea and chocolate. They are very much less toxic than the notorious poisons of red tide (**7.19**), frogs (**7.21**), coral (**7.22**) and the puffer fish (**7.20**) from which is obtained the sushi delicacy of fugu. Even a cursory inspection of the structures (Figure 7.2) of the compounds in Table 7.1 reveals that the structures of the fragrance ingredients are very dissimilar from those of the highly toxic compounds. The molecules of the fragrance ingredients are smaller and much less highly substituted than those of the potent toxins.

However, due care should always be taken and a perfume should not be ingested. Of course, perfume is not intended for ingestion but more for skin contact. Fine fragrance is applied directly to the skin, but even perfumes for use in laundry detergent and household cleaners are likely to come into contact with skin and so skin safety is a major concern of the fragrance industry. Therefore, the skin irritation and sensitisation potential of fragrance ingredients is assessed, and the use levels dependent on the application. A skin irritant is a substance that produces irritation immediately and whose action ceases when exposure ends, and a skin sensitiser is one that causes an allergic reaction to develop after previous exposure. Obviously, the maximum exposure levels to a possible allergen will be lower for a fine fragrance than for a perfume used in a household cleaner.

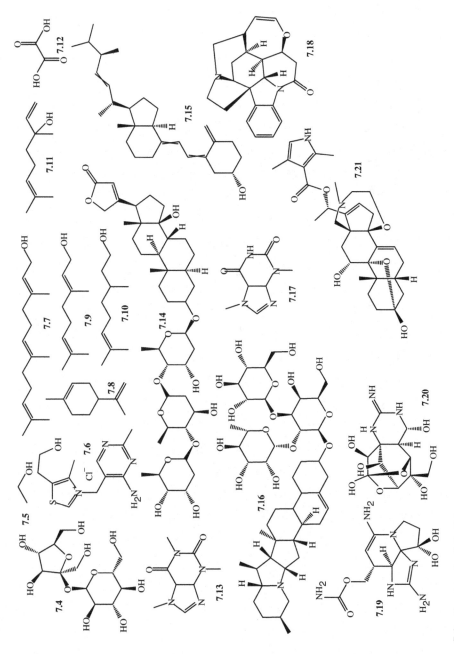

Figure 7.2 Structures for Table 7.1.

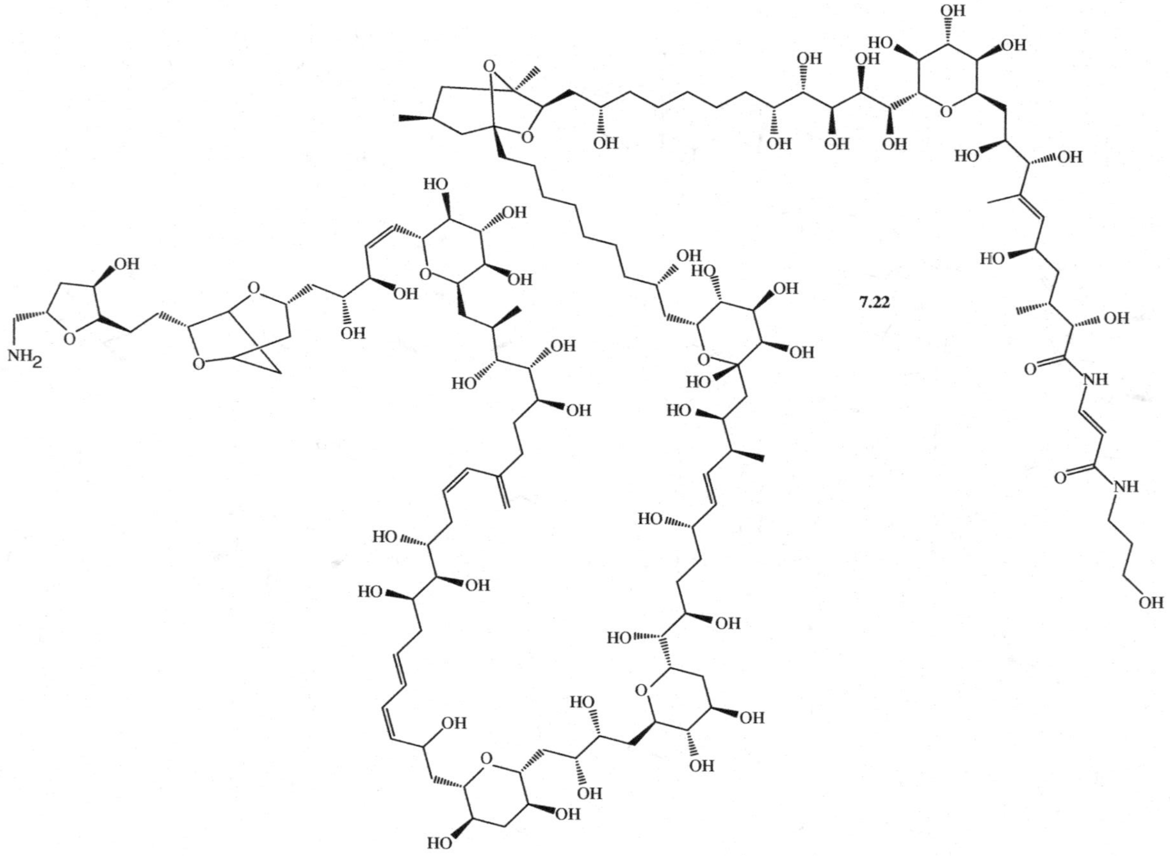

7.22

Figure 7.2 (*Continued*).

In the case of a functional product such as a hard surface cleaner containing hypochlorite bleach, the risk to skin is higher from the product than from the fragrance. Occasionally, some other adverse property of a fragrance ingredient is reported, and in those cases the safety in use will be assessed and appropriate measures taken.

The fragrance industry is, in part, self-regulating, and the safety of ingredients is evaluated by an independent body, the Research Institute for Fragrance Materials (RIFM). A scientific study by RIFM leads to recommendations which are passed to the International Fragrance Research Association (IFRA), who then publish use recommendations, including restrictions and bans where necessary, to the fragrance companies that have agreed to work within the IFRA guidelines. There are also legally imposed restrictions and bans, but these are usually less stringent than those imposed on themselves by the companies. In most countries, certainly in North America, Europe and Asia-Pacific, perfumes and perfume ingredients come under demanding chemicals regulations and these impose further constraints, relating, for example, to environmental toxicity and transport regulations. Regulations vary from one country to another and this presents an issue for companies operating across international boundaries. It might seem that the simplest solution would be to ensure that, for each aspect of safety, the most stringent of the various national conditions are applied internationally. However, the complexities of chemicals regulations are such that this is not always possible and the approach might not be practicable in any case. Thus, careful record keeping and good manufacturing practice are of vital importance to ensure that all appropriate legal demands are met.

The safety regulations laid down by governmental bodies or by the fragrance industry are based on chemical safety assessments, scientific hazard identification and risk assessment. However, there is at least one other factor in perfume safety, and that is public perception of threat. Evolutionary pressure for survival advantage has made humans generally very good at identifying hazards but poorer at evaluation of risks. Most people are happy to cross a road or drive a car despite the significant risk of a road traffic accident. When it comes to chemicals (or even more so to radiation), the general levels of familiarity and understanding are much lower and therefore fear of the unknown (and unseen?) tends to make people see only the hazard. There is also a common misconception nowadays that 'natural' is safe and 'synthetic' is not. A quick look at Table 7.1 will reveal that 'natural' is not synonymous with 'safe'. Scare stories are easily spread and can have a serious effect on the market. Therefore, consumer goods companies like to avoid anything that could attract such attention, even if science and logic demonstrate that the risk is so low as to be negligible.

An example of this arises from the Seventh Amendment to the European Cosmetics Directive. Dermatologists carried out a risk assessment for allergic reaction to perfume ingredients. They identified 26 substances that might correlate with allergic reactions and it became law for labels on cosmetics on sale in Europe to indicate whether any of the 26 substances was present at any level in the product. The 26 ingredients are shown in Table 7.2.

Table 7.2 The 26 Suspected Allergens of the Seventh Amendment to the EU Cosmetics Directive

1 Amylcinnamaldehyde (**7.23**)	14 Farnesol (**7.7**)
2 Amylcinnamyl alcohol (**7.24**)	15 Geraniol (**7.9**)
3 Anisyl alcohol (**7.25**)	16 Hexylcinnamaldehyde (**7.34**)
4 Benzyl alcohol (**7.26**)	17 Hydroxycitronellal (**7.35**)
5 Benzyl benzoate (**7.27**)	18 Isoeugenol (**7.36**)
6 Benzyl cinnamate (**7.28**)	19 Lilial (**7.37**)
7 Benzyl salicylate (**7.29**)	20 *d*-limonene (**7.8**)
8 Cinnamaldehyde (**7.30**)	21 Linalool (**7.11**)
9 Cinnamyl alcohol (**7.31**)	22 Lyral (**7.38**)
10 Citral (**7.32**)	23 methyl heptine carbonate (**7.39**)
11 Citronellol (**7.10**)	24 α-*iso*-Methylionone (**7.40**)
12 Coumarin (**7.1**)	25 Oakmoss extracts
13 Eugenol (**7.33**)	26 Treemoss extracts

The basic idea behind this regulation was that someone developing an allergy could be tested for sensitisation to the 26 substances in order to identify which, if any, was the culprit and then to avoid buying products containing that ingredient. However, the sudden appearance of chemical names on labels caused a similar, though less severe, response to that when E-numbers were allocated to food ingredients that had been approved as safe by European authorities and their inclusion on product labels became mandatory. One magazine article listed 24 of these suspect allergens (the two it failed to mention, oakmoss and treemoss extracts, are obviously natural extracts) and suggested that readers write to cosmetic manufacturers asking for these hazardous chemicals to be withdrawn from their products and replaced by safe natural oils. The oils that the writer suggested are shown in Table 7.3. The numbers following the name of each of the oils indicate which of the 26 suspected allergens has been found in samples of the oils. For example, Bergamot oil has been found to contain citral (**7.32**), citronellol (**7.10**), *d*-limonene (**7.8**) and linalool (**7.11**). These chemical components of the oils would therefore have to be labelled on any product containing them (Figure 7.3).

Furthermore, these oils also contain other suspect compounds. Methyleugenol (**7.41**) is another suspected skin sensitiser and has been found in frankincense, geranium, rosemary and ylang ylang oils. Estragole (**7.42**) is a suspected carcinogen and has been found in frankincense, jasmine, lavender, petitgrain and ylang ylang oils. Incensole acetate (**7.43**) is found in frankincense and it has been shown to act on TRPV3 channels in neurons in the brain and is psychoactive, being anxiolytic and inducing a feeling of calm (2). Lavender and ylang ylang contain benzyl acetate (**7.44**), which is a precursor for benzyl alcohol (**7.26**) and, therefore, oils containing it are likely to release the alcohol in use and so labelling as containing the alcohol is virtually a necessity. Terpinen-4-ol (**7.45**) is an irritant to the eyes, the skin and the respiratory system and requires warning labels. It is a major component of ti tree oil and has been found in frankincense, lavender, lemon, mandarin,

Table 7.3 Suggested Natural Oils as Replacements

Oil	Suspected allergens of Table 7.2
Bergamot	10, 11, 20, 21
Cedarwood (Atlas)	20
Chamomile	14, 15, 20, 21
Frankincense	11, 13, 20, 21
Geranium	10, 11, 13, 14, 15, 20, 21
Jasmine	4, 7, 13, 14, 15, 18, 21
Lavender (*angustifolia*)	8, 10, 11, 12, 13, 15, 20, 21
Lemon	10, 11, 15, 20, 21
Mandarin	9, 10, 12, 16, 23
Melissa	15
Myrrh	20
Neroli	10, 11, 14, 15, 20, 21
Palmarosa	10, 14, 15, 20, 21
Patchouli	20, 21
Peppermint	20, 21
Petitgrain	10, 11, 15, 20, 21
Rose (*centifolia*)	10, 11, 14, 15, 21
Rosemary	10, 11, 13, 15, 20, 21
Rosewood	5, 10, 15, 20, 21
Sandalwood (*S. Album*)	
Sweet orange oil	10, 11, 15, 20, 21
Ti tree	20, 21
Ylang ylang	5, 7, 10, 13, 14, 15, 18, 20, 21

neroli, petitgrain, rosemary, rosewood and orange oils. Ylang ylang has also been found to contain *cis*-anethole (**7.46**) and safrole (**7.47**), which are suspected carcinogens. Surprisingly for a writer in a magazine devoted to ecology, the article also failed to mention that rosewood and Indian sandalwood (*S. Album*) have been lost from the perfumers' palette because the trees producing them are now on the list of endangered species (Figure 7.3).

The story of this article serves to illustrate three main points. The first is that chemophobia, coupled with ignorance and the false belief that 'natural' is safe, means that even well-intentioned measures can lead to adverse consumer reaction. After such consumer reaction, attempts to explain the reality of the situation are often misunderstood as obfuscation, denial or excuse-making. It is, therefore, understandable that consumer goods companies would prefer to avoid labelling of ingredients. Similarly, they prefer to avoid inclusion of ingredients that are suspect in any way since analysis would quickly reveal their presence.

The second point relates to the issue of risk assessment and the fact that the law is not a science. To the chemist it seems illogical that tarragon oil is no longer

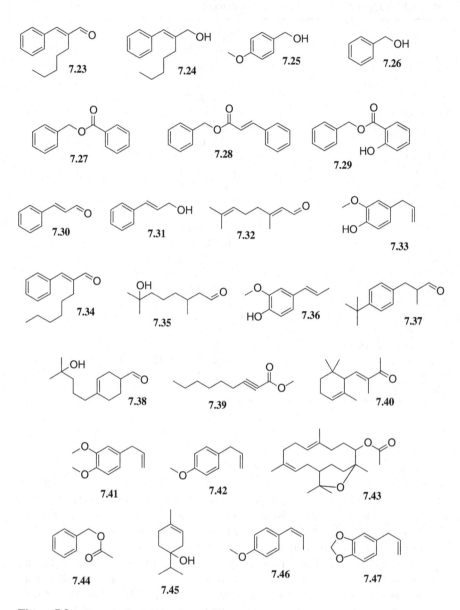

Figure 7.3   Structures for Tables 7.2 and related text.

used in perfumery because the estragole (**7.42**) it contains is a suspect carcinogen, yet tarragon herb, which also contains estragole (**7.42**), is allowed for use in food. This situation arises because perfume comes under chemicals regulations whereas flavours come under food regulations. Chemical regulations are based on chemical analysis, hazard identification and risk assessment. In food regulations, established safe use of a natural ingredient can ensure its continued use. Therefore when tarragon oil is considered for perfumery use, it is analysed, estragole (**7.42**) is identified as a component and the oil is not allowed because of the possible carcinogenicity of estragole (**7.42**). When tarragon is considered for use as a flavour, its safe historic use is easily demonstrated and therefore it is allowed.

Following on from the second point, a third arises. Chemical analysis and hazard identification shows that 'natural' is not synonymous with safe and, consequently, many natural oils are being lost from the perfumers' palette on safety grounds. This trend is likely to increase in the future, and so a new generation of ingredients, designed around safety parameters, is needed to replace those, both natural and synthetic, that have already been lost or will be lost in the coming decades.

Perhaps my account of the safety issues is rather simplistic, but I hope it serves to illustrate that safety in use is of utmost importance to the perfume industry, and that safety threats, whether real or not, are taken very seriously and provide an opportunity for novel ingredients.

Of course, increasing stringency in safety regulations is resulting in increasing costs for safety testing and raises the issue of animal testing. For both moral and financial reasons, the perfume industry has viewed animal testing of its ingredients as undesirable. It can be very expensive and deters all but the biggest companies from introducing new ingredients. From 2013, in Europe, animal testing of substances for use in cosmetics will be abolished, which creates an interesting situation. One law forbids it, whereas another in effect demands it. One solution offered by the industry is the use of *in vitro* testing, but governments have been slow to accept this. For the perfume industry, testing for skin sensitisation is imperative. *In vitro* tests have been developed thanks to our understanding of the mechanism of skin sensitisation. They are more acceptable on ethical grounds and much less expensive to conduct. An example is the Keratosens test developed by Givaudan (3,4), and the industry is co-operating with government regulators to evaluate this or similar tests to replace testing on animals or human volunteers. The test basically uses chemical reactivity of the material under test towards nucleophiles which mimic those in proteins. Skin sensitisation is the result of modification of proteins by electrophiles leading to formation of a new protein which is recognised as foreign by the immune system with consequent eliciting of an immune response.

In addition to *in vitro* testing strategies, to replace animal testing the fragrance industry has adopted a number of computer modelling approaches to aid in the prediction of possible chemical hazards. Such methods, based on quantitative structure activity relationship (QSAR) models, have a growing potential in predicting possible chemical hazards. Some models are now widely used not only by industry but also by government regulators such as the European Chemicals Agency (ECHA) and the US Environment Protection Agency (US EPA).

Both *in vitro* testing and QSAR modelling have significant potential, and the growth and development of these methods as well as their application is very encouraging.

Safety in the Environment

Environmental safety in the area around factories making fragrance ingredients was covered in the discussion of green chemistry in Chapter 6 as was the topic of sustainable agriculture, which also affects the environmental performance of ingredient production. The other environmental factor, and the one that involves ingredient design, is that of environmental fate of ingredients after use. Biodegradability and bioaccumulation are the key topics in this area. The trend in fragrance ingredients is very much towards chemicals that will degrade rapidly in the environment or that have more intense odours and therefore result in a lower environmental load because of lower use levels. Similarly, those that are prone to bioaccumulation are also to be avoided. The standard test for biodegradability of small molecules attempts to replicate what happens in a typical water treatment plant. A small amount of the test substrate is incubated with an aliquot of sewage sludge, and the rate of oxygen uptake and carbon dioxide production is measured. The test is somewhat problematic. The bacteria carry out more processes than digestion of the analyte and also incorporate some of the analyte's carbon into their biomass. This makes the measurement of the tiny amounts of oxygen consumed and carbon dioxide produced very difficult. The quality of the sewage sludge inoculum is, obviously, variable and can have a large effect on the results. Nevertheless, the test is useful, and comparison of rates of degradation of different molecules provides the basis for structure–activity work. Similarly, studies on piscine metabolism can give useful information regarding elimination of perfume ingredients from the bodies of fish and hence clues to design of chemicals that will not bioaccumulate.

Resources

As mentioned previously, some natural sources of perfume ingredients have been used at a greater rate than they can be replaced. This draws attention to the fact that a renewable resource is not necessarily a sustainable one. We are all aware that petrochemical reserves are inevitably dwindling and prices are rising in consequence. If the use of petroleum as fuel were to stop, the supply available for chemical manufacture would last very much longer but the price would rise because of the lower rate of production. Turpentine is currently a major feedstock for the perfume ingredient industry but most of that supply is crude sulfate turpentine (CST) which is a by-product of paper manufacture. Thanks to increased recycling of paper, the demand for new wood pulp is falling. If the paperless office were to become reality, then that demand would fall very significantly. Therefore, the supply of CST would be reduced and the price would rise sharply. Gum turpentine is more expensive than CST, and a shortage in the latter would inevitably lead to a price rise for gum turpentine. All crops grown for use as chemical feedstocks have to compete for land availability with food crops, and, therefore, with the rise in world population, that competition will increase and prices will rise. Ethical sourcing is now also a policy of major fragrance companies and this also tends to result in higher prices. All of these factors combine to make raw material sourcing a difficult issue and one that will be increasingly important in the future. When planning new synthesis routes, the fragrance chemist must take sustainable sourcing into consideration.

Performance Requirements

Commercial pressure and consumer demand drive the consumer goods companies, who are the customers of the fragrance industry, to search continually for more efficacious products. These new products often contain new active ingredients that interact with fragrance ingredients and therefore reduce the effective palette, thus creating the need for new fragrance ingredients. Even leaving such new active ingredients aside, there is a demand for fragrance ingredients that outperform existing ones in terms of price, stability and additional effects such as improved deodorancy.

Market Trends

Market trends can affect the demands made of perfume ingredients. Typical trends currently include the ageing population, the trend for 'natural' and an increased demand for products promoting health and well-being. However, experience tends to indicate that trends in fragrance are determined by novel ingredients rather than vice versa. Thus the opportunity for creation of new trends is more of a driver than a reactive response to market demands. This fact is enforced by the timescale of novel ingredient introduction, which is much longer than market trends arising from other directions. The nineteenth century trend for chypre, fougere and oriental fragrances created by perfumes such as Jicky and Shalimar has already been mentioned, as has the trend for aldehydic perfumes initiated by Chanel No.5 in 1921. The discovery of methyl dihydrojasmonate (**7.48**) led to a new fashion trend set by Eau Sauvage in 1966, and similarly dihydromyrcenol (**7.49**) allowed a new fashion in masculine freshness led by Drakkar Noir (1982) and Cool Water (1988) (Figure 7.4).

Scientific Discovery

Advances in chemistry and chemical technology often allow for serendipitous discovery of new perfume ingredients. The example of musk chemistry discussed

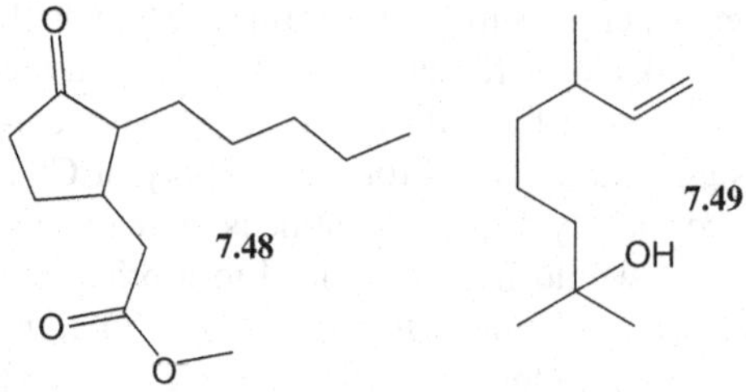

Figure 7.4 Two compounds that created new perfume trends.

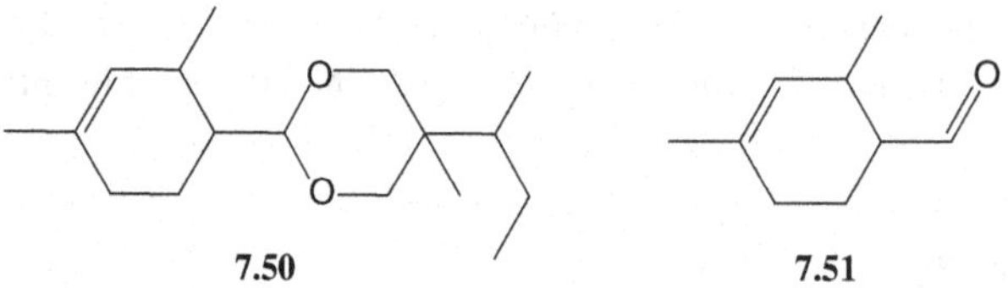

Figure 7.5 Karanal and Ligustral.

below shows how developments in synthetic methodology can be applied to the search for new routes to known ingredients and also to the discovery of new ones. Similarly, serendipity can also play a role in fragrance ingredient discovery by giving unexpected results in terms of odour properties. Karanal (**7.50**) is an example of this. The molecule was made in the hope of finding a long-lasting green note since it is prepared from Ligustral (**7.51**) which has a green grass odour. Many acetals also have green odours, and, so, by making an acetal of Ligustral (**7.51**) with a relatively large diol it was thought that the product might have a longer lived green character, thanks to its much higher molecular weight and thus higher boiling point. However, the acetal (**7.50**) proved to be the first of a new class of amber ingredients (5,6). This example also shows how essential it is for fragrance chemists to have experience of perfumery; otherwise, such chance findings could well be overlooked (Figure 7.5).

CRITERIA FOR NOVEL INGREDIENTS

During the course of a conversation on novel fragrance ingredients and what fragrance chemists can do for their internal customers, a perfumer said to me 'Just give me a lovely smell. If something smells good, I'll use it'. This line of thinking has led many chemists (and others), both in the industry and in academia, to try to understand the mechanism of olfaction in the hope of using that understanding to design more pleasant and more powerful odorants.

However, let us think again about the quotation. The perfumer in question has access to Rose Otto and Jasmine Absolute as ingredients, yet he doesn't use them.

Why not? Is it because he doesn't like their odour? Certainly not, he is in complete agreement with the rest of humanity that both oils have lovely smells. There are several good reasons why he does not use these ingredients in his creations and none of them has anything to do with odour. First, he creates perfumes for a functional product area. Neither of these natural ingredients would survive in the end products because these have a high pH and contain bleaching systems. Neither rose nor jasmine oil would survive long enough in the product for it to reach the supermarket shelf. In addition, neither oil could be supplied in sufficient quantity. The total annual world production for each is less than the volume of many of the briefs against which that perfumer works. Furthermore, the brief prices he has to work to are in the range $5–15 kg^{-1}, whereas rose and jasmine oils cost $3,000–5000 kg^{-1}. His colleagues who work in the cosmetics, toiletries and fine fragrance areas have another problem with rose and jasmine, that of safety. Both oils contain components that are restricted on safety grounds. Therefore, either the chemicals in question must be removed from the oil, or a restriction must be placed on the level of oil in the product. In Europe, the situation is about to become worse as the presence of several of the major components of rose oil will require labelling on the final product package. Similarly, jasmine contains a significant amount of benzyl acetate and, since this readily hydrolyses to benzyl alcohol, this will also lead to a requirement for labelling.

To be really contentious, one could ask (tongue in cheek) whether a nice odour is really important in perfumery. Quite a few of the materials we use have distinctly unpleasant odours. Examples such as scatole (**7.52**) and indole (**7.53**) spring to mind. Other materials such as thioterpineol (**7.54**) and phenylacetaldehyde (**7.55**) have unpleasant or irritating odours when concentrated and only become pleasant at high dilution. Furthermore, the highest volume materials used by the industry have little (e.g. ethanol (**7.5**), benzyl benzoate (**7.27**)) or no (e.g. dipropylene glycol (**7.56** and isomers)) odour (Figure 7.6).

This new line of argument raises an important question. If odour character is not the key consideration in the design of a novel fragrance ingredient, then what is? There are many parameters that are important in determining whether a material will make a successful fragrance ingredient. These can be grouped into five categories, each of which will be discussed below.

Often, the requirements made of a fragrance ingredient pull in opposite directions. For example, the desire for total stability in hypochlorite bleach drives us

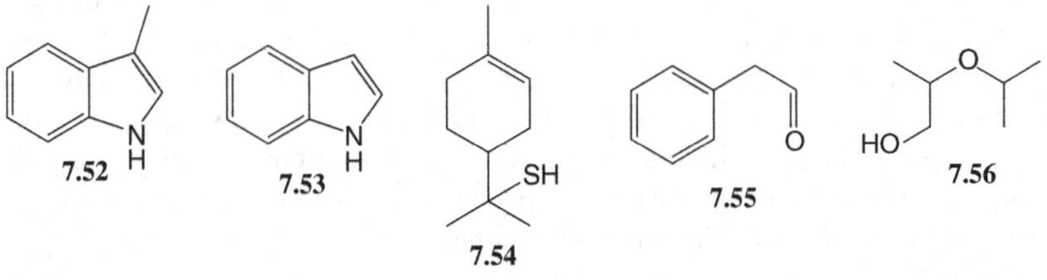

Figure 7.6 Unpleasant or odourless perfume ingredients.

towards very robust molecules with only the least reactive of functional groups present. However, when degrading compounds during sewage treatment, the bacterial enzymes can use only the same basic chemistry that we have in the laboratory and so molecules that survive treatment with hypochlorite are much less likely to succumb to bacterial oxidation. As shown in the discussion above, the demand for 'natural' often conflicts with the demand for safety. The demand from customers and consumers for high performance ingredients can easily conflict with that for low cost. No one molecule can be expected to satisfy all of the above criteria and so, in every case, we have to make a judgement based on an overall balance of all the parameters. The question is, 'Will this molecule be of sufficient value to repay the cost of developing it and launching it onto the market?' In order to make such a judgement, we need information on the performance of the material in each and every one of the categories. This means either testing or predicting each property and this involves a cost.

For some of the judgement criteria (such as safety, availability and performance), there are clear 'right' and 'wrong' answers, while for others, particularly odour properties, there are no wrong answers. We need all odour characters, including malodours such as that of indole (**7.53**), we need very volatile materials and poorly volatile, we need intense and gentle, long-lasting and fleeting and so on. Physical and chemical properties such as volatility or stability to strong base are relatively easy to predict but achieving a valuable balance of odour properties is difficult. So, could this be an argument for the shotgun approach to ingredient discovery? In evolution, nature makes all sorts of possibilities and then allows natural selection to sort out the winners. Could the same be true for fragrance ingredient discovery? In other words, should we just make lots of molecules, put them on the market and see which ones succeed? The barrier to such an approach is the cost of chemical process development and safety clearance for launch. Therefore, the number of candidates selected for development must be limited and the discovery chemist is obliged to design those with the greatest chance of success.

Odour

The odour of a material is important if it is to be a successful fragrance ingredient. Odour is usually described in terms of character, intensity and tenacity, and perfumers nowadays also refer to terms such as radiance, diffusion, trail (silage) or bloom. There is no 'right' or 'wrong' property for each of the first three, and the value of an ingredient depends essentially on its unique blend of the three. If in trying to make a muguet ingredient one produces a rose or a sandalwood, the discovery might be 'wrong' in terms of a commercial brief but it is not 'wrong' in perfumery, just different from what was expected, and it still offers an opportunity for the perfumer. This example is not unreasonable since, as is shown in Chapter 3, muguet is close to both rose and sandalwood in a map of odour space. Most fragrance chemists can tell of an experience when the odour of a novel ingredient was not what had been sought but actually turned out to be more valuable. The above example of Karanal is one such case.

The same holds true for intensity and tenacity. Each combination of character, intensity and tenacity is right for one application but wrong for another. The skill in this aspect of fragrance ingredient discovery is to be able to recognise the right application and evaluate it accurately.

As was evident from Chapter 2, odour is subjective in every respect, and the discovery chemist must always be aware of this. Judgements of a novel chemical's potential should always be made by a panel of experts, and each member of the panel should understand that his or her assessment of the odour is unique. In my experience, panels of perfumers work more effectively when all are confident of this because it eliminates the fear of expressing an opinion only to have it attacked by a colleague, especially if the colleague is more senior.

Characterisation of odour in terms of character, intensity and tenacity has been described in Chapter 3 and so repetition here is unnecessary. The terms radiance, trail, diffusion and bloom are becoming increasingly important in selection of ingredients for development, but the exact interpretation of each term varies from one source to another. All of them relate to the perceived intensity of a perfume or perfume ingredient when detected at a distance from the source. The commonest meaning of radiance is the ability of a perfume ingredient to fill a space when delivered from a point source such as a perfumer's blotter or an air freshener dispenser. However, some perfumers have a very different interpretation of the word; for example, one might see it as an indication of the role of an ingredient in an accord. Trail (silage in French) is commonly understood to mean just that. When someone wearing a fine fragrance walks through a building leaving a distinct impression of the fragrance in their wake, then the fragrance is said to have good trail. Bloom, like radiance, is more open to interpretation. An example of the commonest understanding of it would be the ability of the perfume in a bar of soap to fill the air in a bathroom. Low odour detection threshold (ODT) is probably the most important parameter as far as all of these properties is concerned. Vapour pressure is another important property. If the substance is too volatile, it will diffuse away in the air and rapidly be diluted below the detection threshold and so the molecules usually described as best in terms of radiance, bloom and trail tend to be towards the lower end of the vapour pressure range of fragrance ingredients. Blooming from a product such as soap will also depend on the affinity of the ingredient for the product base and how this changes when the product becomes wet.

Designing new ingredients with a required performance in terms of volatility is relatively easy. Structure–property relationships can give quite accurate predictions of boiling point and log P and, therefore, guide the chemist in producing odorants that are head, heart or base notes as required.

Predicting the odour character, intensity or threshold is a much more difficult task. One interesting paper discusses strategies for the design of drugs acting on G-protein coupled receptors (GPCRs) (7). However, the discovery chemist working in the pharmaceutical industry only has to target one GPCR at a time. His counterpart in the fragrance industry, when seeking a molecule with a specific set of odour properties, is trying to design a molecule that interacts to a specific level with an unspecified number of GPCRs, which is probably of the order of 400 but could be much greater if SNPs and CNVs lead to a large number of variants with different binding properties. (See Chapter 2 for details.) But even then, generation of an identical signal at the epithelium level does not necessarily guarantee the formation of an identical percept in the orbitofrontal cortex. Notwithstanding all of this, structure–odour relationships (SORs) do play a useful role in design of novel ingredients, as will become clear from Chapter 8. In addition to SOR models, the experience of the fragrance discovery chemist is important and must not be underestimated. The human brain seems to be better at spotting patterns and suggesting good candidate molecules than any of the mathematical models we have at present. Originality of approach is always interesting. For example, one publication shows how, rather than trying to design molecules to fit an olfactophore, combining structural fragments that are associated with different types of molecules eliciting the same percept can lead to useful new molecules. In that paper, the 2,2-dimethylpropan-1-ol-3-yl fragment, which is found in a number of muguet odorants, with hydrophobic fragments found commonly in other classes of muguet ingredients led to a variety of novel molecules eliciting a muguet odour (8).

Performance

Modern perfumery materials must perform well both in combination with other ingredients in perfume formulae and also in the range of products for which we are asked to provide fragrance. Product bases vary enormously in chemical properties. The pH can be anywhere from 1 to 14. There may be oxidants present, such as hypochlorite, peroxide, peracids and perborate. Thiols, glycerol, aluminium chloride and a host of other possible active ingredients all present specific issues as regards the chemical stability of fragrance ingredients. Furthermore, the fragrance must not have any deleterious effect of the product packaging; for example, fragrance ingredients can interfere with plasticiser systems to make plastic bottles either too plastic or too brittle.

Performance in Formulae

When perfume ingredients are mixed together, interactions occur at two levels. Simple chemical reactions can occur, such as the formation of hemiacetals when alcohols and aldehydes are mixed or Schiff's bases when methyl anthranilate and aldehydes are mixed. These reactions are reversible and serve a useful purpose in reducing volatility of top note materials but releasing them again slowly. Thus these reactions are reasonably predictable and advantageous. If the perfume, its container or the medium in which it is dissolved is not neutral, then other reactions, such as aldol condensations or transesterifications, can occur between ingredients or between perfume ingredients and other chemical species in the product. Oxygen is almost always present, and autoxidation is a particular problem for aldehydes. Autoxidation of aldehydes, of course, generates carboxylic acids and these can then catalyse a variety of other reactions. However, all of these chemical issues are of less importance than the sensory issues surrounding mixtures.

As we saw in Chapter 2, we cannot reliably and repeatedly predict the ultimate percept elicited by a single compound and prediction is even less possible for mixtures. Modification of percepts due to blending of inputs can occur at many levels, both in the receptors and in subsequent neuroprocessing, as already discussed in Chapter 2 and particularly in the section on mixtures. The only realistic way of determining how any new ingredient performs in perfume formulae is through experimentation by perfumers. This is time consuming and also costly in terms of materials, and so the perfumer, and his company, must be convinced that the expenditure of time and material is worth while. Selection of appropriate tests can improve the efficiency of such new product evaluation, and the experience of the perfumer in working with new ingredients is therefore of great importance.

Performance in Products

Chemical stability of new ingredients in consumer products is reasonably predictable. Soaps usually have a pH of about 8–9, that of laundry powders is likely to be 10–11 and of dishwash powders 13–14. Therefore, any molecule likely to undergo reaction with aqueous base is unlikely to show long-term stability in them. The latter two also have bleaching agents such as perborate in them and so susceptibility to oxidants would also give a prediction of instability. Anti-perspirants usually contain aluminium chloride or a similar active agent and therefore have a pH of about 3. Fabric conditioners have a similar low pH and, in some countries, lavatory cleaners are even more acidic. Sensitivity to acid hydrolysis or other acid-catalysed reactions, such as dehydration of tertiary alcohols, would then be indicators of poor stability in these media. Despite the predictability of chemical stability and other physical and chemical performance issues, it is normal practice to store novel ingredients in a representative range of (otherwise unperfumed) consumer products to determine performance properties. Storage conditions will usually include a range of temperatures from 0 °C to perhaps 50 °C, the higher temperatures giving an indication of what is likely to happen over longer storage times at lower temperatures. Storage conditions usually also include various humidity levels to represent different climates, and also ultraviolet light to indicate stability to sunlight. The products will be evaluated by a panel of perfumers at various times, for example, every 2 weeks up to a total of 12 weeks. Like testing in perfume formulae, this is costly in time and materials but the information generated is crucial for the evaluation of candidate ingredients and also in enabling perfumers to learn how to use each new ingredient.

Other Effects of Perfume Ingredients

Nowadays, the customer often expects fragrance ingredients to possess properties in addition to odour. Such additional benefits include deodorant properties such as malodour counteraction. Malodour counteraction can occur in various ways. One way in which fragrance ingredients can affect malodours is by chemical reaction with them. For instance, a malodorous amine could react with a fragrant aldehyde to produce a Schiff's base with much lower volatility, thereby reducing the concentration of the malodorant in the air. As discussed in Chapter 5, human body odour is the result of bacterial action on odourless chemicals produced by humans. Identification of the enzymes involved, therefore, opens up the possibility of designing fragrance ingredients that inhibit the enzymes or provide them with a more attractive substrate, releasing a pleasant odour rather than the customary malodour. Similarly, advances in understanding olfactory receptors could pave the way for the use of competitive agonism or antagonism of receptors responding to malodorants. Sufficient understanding of interactions during neuroprocessing is probably a little further away, but recent advances do give cause for optimism. Malodour counteraction is thus moving from purely swamping the malodour with a powerful pleasant odour to the state of a high-tech industry using biochemistry and molecular biology in the search for active ingredients.

Safety

Safety is the most important single parameter in fragrance ingredient design. If a material cannot be shown to be safe and registered as such by the competent authorities, it cannot be used. Thus EINECS/REACH listing is essential for use in Europe, ToSCA for the USA, MITI for Japan and so on. Modern fragrance houses actually exceed the legal requirements for safety testing and take a cradle-to-grave approach to their products. No responsible company would wish to harm its employees, its customers, the consumers or the environment. Consequently, serious attention will be given to the evaluation of all aspects of safety of a new ingredient before it is launched onto the market. For example, the starting materials, the synthesis process and the production plant used will all be subjected to a hazard and operability study, any skin sensitisation potential will be determined, as will aquatic toxicity and rate of degradation during sewage treatment.

Safety testing for skin sensitisation, carcinogenicity and environmental performance is expensive and so it is crucial for the discovery chemist to understand the basic mechanisms and to avoid structural features that would tend to lead to adverse biological properties. Otherwise (s)he would design too many substances that would fail at the most expensive hurdle. Skin sensitisation, mutagenicity and carcinogenicity all rely to a large extent on the ability of small molecules to modify protein structures, usually through action as electrophiles. For this reason, halogenated compounds have been ruled out of new odorant discovery for many decades. Similarly, structural features such as α,β-unsaturated aldehyde, ketone or ester functions are likely to create problems and so should be avoided, unless sufficient steric hindrance is present to prevent the molecule acting as a Michael acceptor. Multiple fused rings or a high degree of branching are features that are likely to reduce biodegradability through obstruction of the β-oxidation pathway. Unfortunately, these latter features are strongly associated with woody odorants and so a proper balance must be found between odour and biodegradability in that case.

The discovery chemist will use whatever synthetic methods that give the desired target molecule in the shortest possible time, since at the discovery stage the chemist's time is the largest cost factor. However, the discovery chemist must also be aware of the scope and limitations of process chemistry methods because any successful candidate molecule must be able to be produced eventually at suitable cost. In my experience, the novel molecules that show the fastest progress from laboratory to factory scale are those designed with production technology in mind.

Replacement of Threatened Ingredients

Safety is often seen as an obstacle to the introduction of novel fragrance ingredients because of the significant cost involved in worldwide registration. However, as discussed above, safety is also an important driver for the introduction of new ingredients because of the rate of loss of natural ingredients as a result of safety testing. In addition, there are also other reasons for loss of ingredients from the palette: for example, loss of Indian sandalwood oil due to over-harvesting, or loss of nitro musks due to poor environmental performance. The fragrance industry is small and so tends to rely on feedstocks from other industries. Changes in these other industries can therefore lead to loss of raw material for odorant manufacture. An example of this could be crude sulphate turpentine (CST) as described earlier.

Thus, for various reasons, fragrance ingredients can come under the threat of being lost from the palette. The usual response of commercial and perfumery sections of a fragrance company is to ask research to come up with a replacement. However, 'a replacement for molecule X' usually requires a molecule that elicits exactly the same odour as molecule X, costs no more than molecule X, performs identically to molecule X in perfume formulations and in consumer goods and does not have the undesirable properties of molecule X. Of course, no such molecule exists and such a research project is doomed to failure. While molecule X is still available, perfumers will not accept an alternative with different properties because this will change the fragrances containing molecule X. The perfumers are not to blame because the customers will not accept the altered fragrance and, equally, they are merely reflecting what the consumer wants. Therefore, alternatives that have different odour properties tend only to succeed when molecule X is no longer available. It takes a brave fragrance house to try to change fragrance fashion while molecule X is still being used in competitors' fragrances.

If researchers accept a brief for direct replacement of a threatened ingredient, they are condemning themselves to failure. If they do not accept it, they risk being charged with being lazy, uncooperative or incompetent. A more realistic approach would be to allow chemists the creative freedom to design novel molecules that will be safe in use, have a good environmental profile, be inexpensive and so on, and then work with perfumers to use these to drive new fragrance directions and establish new associations between odour and effect in the mind of the consumer. For example, there is an association between lemon scent and freshness. If, because of the skin sensitisation potential of citral (**7.32**), lemon oil were to come under threat in some application where it is used to indicate freshness, then an alternative note such as white floral might be acceptable as a potential alternative to lemon as a signal of freshness.

Availability and Cost

The cost of a new fragrance ingredient is determined by the cost of the raw materials, the operating cost of the process and the capital cost of the plant. The contribution of the raw material cost contains elements of both intrinsic cost of the starting materials and also the efficiency of the process. A low-yielding process wastes starting material and, therefore, leads to increased product cost. Starting materials and reagents that are lost as waste also add to the product price through disposal costs. Some processes are more expensive than others in terms of labour costs and/or capital costs of the plant required.

Starting material availability is also an issue, and security of supply of the product will depend directly on that of the starting materials. Launch of a new fragrance ingredient is something of a lottery. The ingredient will be offered to perfumers for use in formulae, and they will not be able to experiment with it unless they have sufficient quantities available. There is therefore a minimum production level for a successful launch. But what will the rate of volume growth be? This is a very difficult question to answer as there are a number of unpredictable factors involved. These include the rate of incorporation into formulae by perfumers, the level of incorporation into formulae, the success of formulae containing the new ingredient when these are submitted against briefs and the size of the briefs. The production unit must, therefore, be able to respond to rapid increases in production volume.

The aims of development must, therefore, be to make the target material available at a cost that provides value for money and with the ability to produce it at anything from the kilogram to tonne scale at short notice.

These issues are the domain of the process R&D chemists but a good discovery chemist will bear them all in mind in order to avoid designing materials that will fail in development or, even worse, after launch.

THE PROCESS OF DEVELOPING NOVEL INGREDIENTS

Fragrance chemistry R&D is intrinsically longer term in nature, and commercial success invariably suffers if reactive and short-term projects are allowed to derail longer term projects that are genuinely capable of delivering sound results. In my opinion, the most effective way to deal effectively with loss of ingredients from the palette is to have a steady pipeline of novel ingredients that have good performance, are capable of clearing regulatory hurdles and can be made at an appropriate cost, and then for perfumers to learn how to use the new generation ingredients and accept the loss of the old ones. If one odour facet is lost from the palette, then perfumers must adapt and use new odours to create new fashions in fragrance.

Novel odorant design is a notoriously difficult field of research for all the reasons described above. It currently costs about $500,000 to develop, register and launch a new material. Novel ingredients must offer an advantage over naturals and existing synthetic materials; otherwise, they will not be accepted when introduced. The advantage could be in safety, biodegradability, performance, cost or any combination of these. Cost efficiency is crucial, and any ingredient that increases the cost of a perfume formulation without providing an advantage worth the additional cost will not survive. Therefore, because of the time taken to build up tonnage of a new ingredient, the payback period is necessarily long. Consequently, only a company with a fragrance compounding business can afford to introduce new ingredients, recovering the development cost of the ingredient through sales of perfume containing it. Even so, the high cost means that only the largest fragrance companies can afford to introduce new ingredients. Chemical companies outside the fragrance industry cannot recover the development and registration costs of new fragrance molecules in an acceptable time. One way round this difficulty would be to form an alliance with a fragrance company, but the latter would always insist on exclusivity.

New fragrance ingredient discovery is somewhat akin to a lottery. To be absolutely certain of winning the lottery, one must either buy all of the tickets or buy the winning ticket. To buy all of the tickets almost invariably means spending more than the prize, and the second approach fails because the identity of the winning ticket is not known until the draw. Similarly, with perfume ingredients design, to synthesise and evaluate every possible molecule would cost more than the fragrance industry could ever earn, and the identity of the block-buster ingredient would become clear only after it has been launched and tested in use by perfumers. So, discovery chemistry in the fragrance industry has relied on various methods between the two extremes. Random screening, for example of substances from other industries or of newly discovered natural materials, can play a role and is a way of finding unexpected breakthrough materials. However, the success rate is very low. For all of the reasons outlined in Chapters 2 and 7, rational design remains an imprecise science. So fragrance chemists rely heavily on structure–activity relationships and SORs. The activities that are subject of these relationships include safety, biodegradability, deodorant activity, effect on plastics, stability in consumer goods, release from customer bases such as soap and so on. The many published examples of SORs have been reviewed, and particularly useful reviews are those of Rossiter (9), Fráter et al. (10) and Kraft et al. (11). SORs are important but other structure–activity relationships are even more so, especially those relating to safety in use and biodegradability. These tend not to be published but remain proprietary tools of the companies involved in ingredient design.

Cost of Testing

Having synthesised a library of candidate molecules, it is essential to test them for odour, performance, safety and so on, and this testing invariably costs more than the initial synthesis. The use of predictive models is, therefore, important in order to prevent waste of physical and financial resource through evaluation of substances that fail during the screening process. The cost of different tests varies enormously, and so each fragrance house has developed a testing protocol to minimise the overall cost, usually by carrying out the tests in ascending order of cost.

The test for odour character is the fastest and cheapest of all. One sniff and you know whether it is rose, jasmine, musk or whatever. Furthermore, the cost of producing the samples for smelling is falling as the result of advances in methodology for robot-assisted and parallel synthesis. To evaluate the behaviour of a material when combined with other ingredients requires the expertise of a perfumer and a great deal of laboratory work. The chemical performance in various products can be estimated reasonably accurately. Physical delivery from product media is more difficult to predict and so storage tests are usually carried out, and these will also confirm any predictions of chemical instability. Sensory tests such as underarm malodour counteraction are much more expensive because they involve the use of panels of trained assessors. However, all of the other testing costs are dwarfed by the costs of safety testing. On average, more than half of the total development cost of a new ingredient will have been spent on safety testing. This is the area where predictive tools stand to save most money. However, it is also the area where there is most resistance to using prediction as an alternative to testing.

Initial evaluation of odour character of the pure substance takes only minutes of sample preparation and minutes of evaluation time. Even using a panel of perfumers, the cost is only tens of dollars per sample. Evaluation of tenacity is only slightly more costly, but accurate estimation of perceived intensity is more expensive since the use of an olfactometer is required. Testing of olfactive stability requires longer term (e.g. several months) storage of a wide variety of perfumed products, followed by evaluation by a perfumery panel. The costs associated with

Table 7.4 Priority Order for Developing Predictors

1	Safety in use
2	Safety in the environment
3	Chemical process costs
4	Performance in formulae
5	Performance in products
6	Underarm malodour counteraction
7	Other malodour counteraction
8	Odour – intensity
9	Odour – tenacity
10	Odour – character

this run into hundreds of dollars per sample. *In vitro* and *in vivo* malodour counteraction testing costs are in thousands of dollars per sample, especially for *in vivo* testing of underarm malodour counteractancy, where the bill is more likely to be in tens of thousands of dollars. A chemical process development evaluation is necessary in order to determine likely manufacturing costs with a requisite degree of confidence. These vary with the complexity of the synthetic process and typical figures are likely to be in hundreds of thousands of dollars, second only to the cost of the toxicological testing needed to satisfy regulatory bodies. A priority order for development of predictive models, based on the cost of relevant testing, might therefore be something like that shown in Table 7.4.

In order to provide a good predictive model, we need data on each step of the process being modelled as well as a good library of structures with known activity levels. The more steps there are in a process, the more difficult it is to model and the more the data needed. So, for example, stability to an aqueous acid will always be easier to model than a biological process such as skin sensitisation where transport properties, reactivity of small molecules towards biological macromolecules and the various steps of the immune response all come into play. The most complex property of all is odour character, where transport phenomena, nasal chemistry, the combinatorial nature of odorant recognition and the complex neuroprocessing from the receptors to the orbito-frontal cortex are all involved. Understanding of biodegradation and toxicological mechanisms are helping fragrance chemists to improve their predictive models and hence to design molecules with improved safety and environmental profiles. Discovery chemists continue to improve structure/odour models, and fragrance process chemistry is benefitting from advances in "green" chemistry in general. Modelling plays a significant role nowadays and is likely to continue to do so for the foreseeable future.

At some point during development, the fragrance company will assess the patentability of a new ingredient. If it cannot be protected by a patent, then the development and registration costs are at risk because any competitor could use the ingredient also. One possible alternative means of protecting new ingredients is through the use of synthetic technology that is either patented or is inaccessible to competitors. If competitors cannot manufacture the ingredient at a competitive cost, then it is protected for the owner of the technology. However patent protection, especially if patenting of the composition of matter is possible, is the preferred method of securing selectivity to enable return of development costs. The ideal patent would include composition of matter, use in perfumery and method of manufacture.

New ingredients are released only to the company's own perfumers initially. This secures maximum creative advantage for the company's perfumers. Patents mostly only have a 20-year lifetime, and so the company that develops a successful new ingredient knows that after that time competitors and third parties will be free to use the ingredient also. Therefore, in order to gain maximum advantage from the discovery, it will use that 20-year period to improve the manufacturing route, so that when the patent expires it will be difficult for any third party to compete effectively. It is also possible to take out derivative patents claiming specific advantages of the ingredient that came to light only after granting of the original patent. Another commonly employed approach is to slowly release the ingredient for third-party sales. Initially, sales will probably be to a company such as a large soap manufacturer that buys perfume ingredients for in-house compounding but does not compete with fragrance houses for perfume sales. Eventually, the ingredient will be offered for sale on the open market in order to build up a sales history with other companies, including direct competitors, and thus discourage them from investing in development of their own manufacturing process for it. The end result is a somewhat incestuous industry with competitors buying and selling ingredients from each other.

EXAMPLES OF PROGRESS IN INGREDIENT DESIGN

The roles of coumarin (**7.1**), heliotropin (**7.2**) and vanillin (**7.3**) in Jicky and the ensuing fashion in fougere and chypre fragrances, of methyl dihydrojasmonate (**7.48**), in Eau Sauvage (a perfumery colleague once remarked to me that the Paris Metro smelt of methyl dihydrojasmonate (**7.48**) thanks to the popularity of Eau Sauvage and derivative perfumes) and of dihydromyrcenol (**7.49**) with the masculine freshness fashion initiated by Drakkar Noir and Cool Water have already been mentioned. These are not exceptions; novel fragrance ingredients have played a key role in perfumery over the last century.

Geraniol (**7.9**) and citronellol (**7.10**) are available at low cost from either turpentine or petrochemical feedstocks, and 2-phenylethanol (**7.57**) is a by-product of the styrene monomer/propylene oxide (SMPO) process as described in Chapter 6. Thus, simple but effective rose accords can be made using inexpensive nature-identical ingredients (Figure 7.7). However, the fruity top note of *Rosa damascena*, the main variety used in production of rose oils, is due to the damascone family of terpenoids. Nature-identical damascones are expensive, so

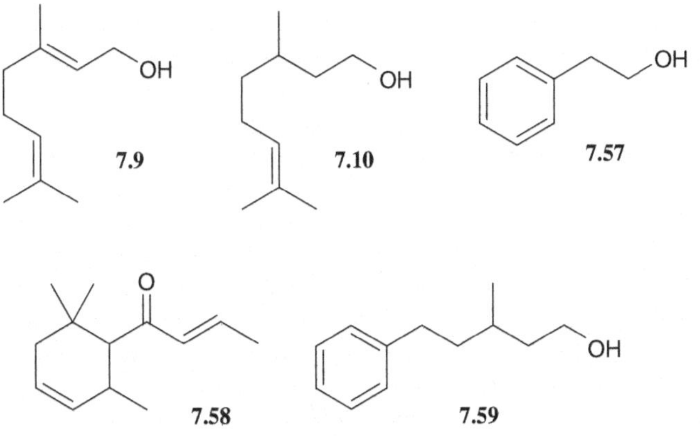

Figure 7.7   Ingredients of rose accords.

the introduction of analogues such as δ-damascone (**7.58**) was welcome since they allow better rose accords to be produced at a low cost and in volumes that could never be achieved through the natural oils. The technique of using isosteres (different structural fragments occupying a similar volume and shape of space) for analogue synthesis is illustrated by the design of the alcohol (**7.59**) first introduced by Naarden International (now part of Givaudan) under the name Mefrosol and later by IFF under the name Phenoxanol. In this example, the benzene ring emulates the isobutenyl group of citronellol (**7.10**) and produces a molecule eliciting a similar odour to that elicited by citronellol (**7.10**).

No essential oil of muguet (lily of the valley) (*Convalaria majalis*) has ever been produced, yet the odour of this flower is of value for its freshness and association with springtime. Dihydrofarnesol (**7.60**) is one of the main chemical components of the natural odour but it does not give the whole impression of the scent (Figure 7.8). Hydroxycitronellal (**7.35**) and Lilial (**7.37**) are early examples of success in the search for muguet ingredients and both have become important fragrance ingredients, the former considered to be the synthetic ingredient eliciting an odour closest to that of the flower. However, both suffer from the chemical instability of the aldehyde function and there are some issues regarding safety of use. The search for improved muguet ingredients has therefore been the subject of a great deal of research in the fragrance industry, and alcohols have been investigated in order to avoid the stability and safety issues associated with aldehydes. Mayol (**7.61**) and the tetrahydropyran derivative (**7.62**) known as *Florol* and *Florosa*, respectively, are examples of stable and skin-safe ingredients that are used successfully to add muguet notes, as for example the use of the latter in J'Adore and Fragil. Tetrahydropyran (**7.62**) has a softer character whereas the carbocyclic analogue Rossitol (**7.63**), although a skin-safe alcohol, captures more of the sparkling freshness

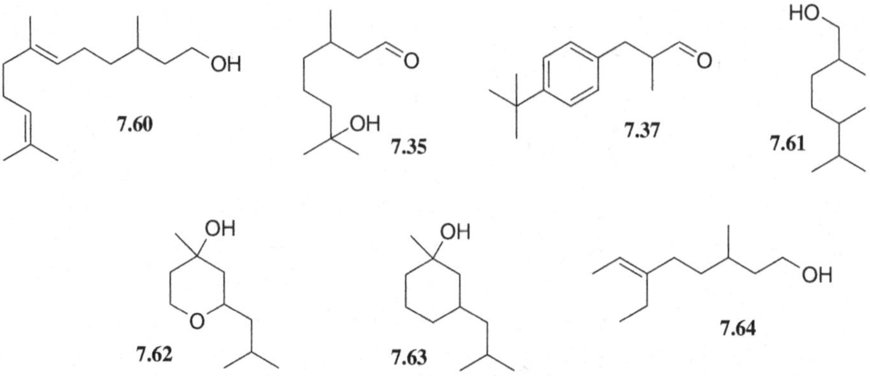

Figure 7.8   Development of muguet ingredients.

commonly associated with the aldehydic materials. It is used to good effect in perfumes such as Miracle. Molecular modelling and the olfactophore approach have
led to the introduction of Super Muguet (**7.64**) onto the palette, and this is another
ingredient of increasing significance.

As mentioned earlier, it is often the case that different requirements impose
conflicting constraints on the fragrance discovery chemist. Perhaps the most frus-
trating is the fact that the structural requirements for a material to elicit a woody
odour (i.e. a compact structure with multiple, usually fused, rings and a high degree
of chain branching and methyl group substitution) are exactly those that tend to
reduce the rate of biodegradation. It is, therefore, a welcome discovery when a
totally different structural class produces a molecule eliciting a woody odour. An
example is the discovery of Agarbois (**7.65**) which elicits an odour reminiscent of
agarwood (also known as *aloes*, *oud* or *eagle wood*) and vetiver (Figure 7.9). Agarbois (**7.65**) has been used to enhance agarwood notes in perfumes such as John
Varvatos.

The musk odour area provides a good example of how new molecule design
has served to produce an evolving family of odorants seeking always to produce
a better balance of properties and to overcome issues as they become apparent in
the changing social and commercial climate. Some examples of musk odorants
are used to illustrate this in Table 7.5. In the past, natural musk was extracted
from the anal glands of the musk deer (*Moschus moschiferus* and related species).
This necessitated killing the animal, and, so, for ethical reasons, natural musk is

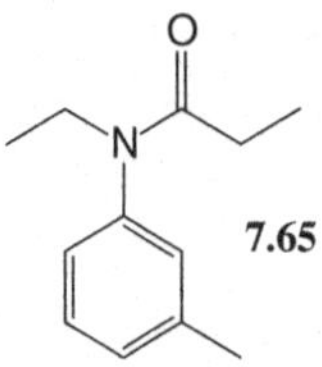

Figure 7.9   Agarbois.

Table 7.5   Evolution of Musk Odorants

No.	Structure	Environmental performance	Odour detected threshold	Cost	Year
7.66		good	4.5 ng/L	$$$$	
7.67		poor	0.1 ng/L	$	**1894**
7.68		good	2.1 ng/L	$$$	**1927**
7.69		poor	0.9 ng/L	$	**1965**
7.70		good	0.1 ng/L	$$	**2001**

no longer in use. Natural musk was also very expensive and so the serendipitous discovery by Baur that some of the nitro-aromatics he synthesised as potential explosives elicited musk odours was a very welcome one. The chemical respon-sible for the musk odour in the natural material is muscone (**7.66**), which has an ODT of 4.5 ng/L. Musk ketone (**7.67**) is a typical nitro-musk. It was discovered in 1894 and has a much lower ODT (0.1 ng/L) than muscone (**7.66**). Its cost is very much lower than that of muscone (7.66), which made it an immediate success, especially in view of its low ODT. However, nowadays, its poor environmental performance is a major obstacle and has led to its demise. In 1927, Ruzicka's discovery of a technique to produce macrocyclic compounds made synthetic musks such as cyclopentadecanolide (**7.68**) available. The cost then was high, but not so much as that of the natural. Macrocyclic esters and ketones have good environmental performance. However, in the case of cyclopentadecanolide (**7.68**), the ODT is 20 times that of Musk Ketone (**7.67**). The discovery of the polycyclic musks led to compounds such as Galaxolide (**7.69**), which was discovered in 1965. Its ODT is better than that of cyclopentadecanolide (**7.68**) and its cost is low, factors that made it a very successful fragrance ingredient. However, its rate of biodegradaton is not good and so the search for better alternatives continues. Nirvanolide (**7.70**) is typical of the latest generation of musks. Its ODT is equal to that of Musk Ketone and its environmental performance is good. It is less costly than the original macrocyclic musks but a lower price would be desirable. Thus, scanning Table 7.5 from top to bottom, one can see how the balance between ODT, environmental performance and cost has continued to improve from the late nineteenth century to the present day, and a recent publication on musk design shows that the search for an even better overall performance continues (12)

CONCLUSIONS

The fragrance discovery chemist must have a wide knowledge not only of synthetic organic chemistry on laboratory scale and of SAR methods but also of process chemistry, the biochemistry of skin sensitisation, mutagenicity and carcinogenicity and the biochemistry of factors affecting environmental fate such as bacterial and piscine metabolism as well as a basic understanding of olfaction and sensory science. In the field of odorant design, there is no substitute for the experience of a fragrance chemist. The human brain seems to be particularly good at holding large amounts of disparate and fuzzy data and making good guesses based on some equally fuzzy logic. I am convinced that, as long as we wish to design new fragrance ingredients, there will be a need for creative and experienced fragrance chemists.

REFERENCES

1. M. Edwards, *Perfume Legends*, HM Editions, Levallois, 1996. ISBN 0 646 27794 4.
2. A. Moussaieff, N. Rimmerman, T. Bregman, A. Straiker, C. C. Felder, S. Shoham, Y. Kashman, S. M. Huang, H. Lee, E. Shohami, K. Mackie, M. J. Catarina, J. M. Walker, E. Fride, and R. Mechoulam, *Faseb Journal*, **2008**, *22(8)*, 3024–3034. doi: 10.1096/fj07-101865.
3. A. Natsch and H. Gfeller, *Tox. Sci.*, **2008**, *106(2)*, 464–478.
4. A. Natsch, R. Emter, and G. Ellis, *Tox. Sci.*, **2008**, *107(1)*, 106–121.
5. C. P. Newman, K. J. Rossiter, and C. S. Sell, E. P. 276998 A2 1988.
6. C. S. Sell, *Chem. Ind.*, **1990**, *16*, 516–20.
7. T. Klabunde and G. Hessler, *ChemBioChem*, **2002**, *3(10)*, 928–944.
8. K. J. Rossiter, in Flavours and Fragrances, Proceedings of the 1997 RSC/SCI International Conference on Flavours and Fragrances, K. A. D. Swift, Ed., Royal Soc. Chem., Cambridge, 1997, pp 21–35.

9. K. J. Rossiter, *Chem. Rev.*, **1996**, *96*, 3201–3240.

10. G. Fráter, J. A. Bajgrowicz and P. Kraft, *Tet.*, **1998**, *54(27)*, 7633–7703.

11. P. Kraft, J. A. Bajgrowicz, C. Denis, and G. Fráter, *Angew. Chem. Int. Edn.*, **2000**, *39(17)*, 2980–3010.

12. Y. Zou, H. Mouhib, W. Stahl, A. Goeke, Q. Wang, and P. Kraft, *Chem. Eur. J.*, **2012**, *18*, 7010–7015 doi: 10.1002/chem.201200882.

Chapter 8

The Relationship Between Molecular Structure and Odour

OVERVIEW

The earliest debate on the relationship between structure and odour was between the schools of Aristotle and Democritus/Epicurus in Greece in the fourth century BC. Aristotle argued that all matter was continuous and the substances of the universe were composed of varying proportions of four elements: earth, air, fire and water. Democritus (and subsequently Epicurus) argued that matter was discontinuous and composed of minute indivisible particles called *atoms* (from the Greek ατομος meaning indivisible). In order to account for odour, Democritus argued that atoms of the odorous substance travelled through the air and into the nose. Substances such as vinegar, he postulated, had sharp or pointed atoms which irritated the nose, whereas those of sweet smelling substances had soft rounded atoms which produced a much gentler interaction. Following the logic of Democritus, Titus Lucretius Carus proposed in 50 BC. that each odorant had a unique structure (1). In order to fit his views to odour, Aristotle had to argue that odorous substances radiated their odour to the nose. It is interesting that Aristotle's flawed cosmology prevailed, was adopted by the Renaissance in Western Europe to the detriment of the science of the medieval monasteries (2) and persists today in such forms as astrology and terms such as 'essential oil'.

As mentioned in Chapter 7, perfume relied on natural extracts up to the late nineteenth century when synthetic organic chemistry began to provide ingredients which are single, pure chemical entities with known molecular structures. Chemists were thus able to produce series of molecules with small sequential modifications to the molecular structure. It then became apparent that the perceived odour depended

Chemistry and the Sense of Smell, First Edition. Charles S. Sell.
© 2014 John Wiley & Sons, Inc. Published 2014 by John Wiley & Sons, Inc.

to a degree on molecular structure, and the search for patterns began. However, from the outset, and still today, the patterns were not always easy to analyse and debates raged. The reasons for the inability to develop structure/odour relationships (SORs) that are consistently precise and accurate are obvious from Chapters 2 and 3. Over the last three decades, huge advances in our understanding of olfaction have enabled us to account for the limits to our ability to develop perfect SORs. Some of these developments have led to Nobel Prizes for olfaction scientists, as will be discussed in Chapter 9.

This chapter will review the techniques and limitations of structure/activity correlation and particularly those that relate to odour. Theories of olfaction based on SORs have been a feature of fragrance chemistry since the nineteenth century and some of these will also be discussed.

TECHNIQUES OF STRUCTURE/ACTIVITY CORRELATION

Since odorants must reach the receptors in the nose through the air, it follows that they must be volatile. Therefore, the vast majority of odorants will have a molecular weight below 300 Da and will be relatively hydrophobic. Any molecules with highly polar groups, or a large number of polar groups, will experience inter-molecular electrostatic forces which would reduce their vapour pressure below the required limit. These requirements will, therefore, be included in any structure/odour model.

SARs are based on the assumption that the molecular structure of any chemical will determine all of its physical, chemical and biological properties. This assumption is generally held but has come under challenge recently. As far as odour is concerned, this biological property is not consistently, accurately and precisely predictable from molecular structure, and evidence to demonstrate this has been reviewed by the author (3). In the same year, Jansen and Schoen challenged the assumption at an even deeper level, by showing that no chemical properties of a proposed structure are accurately and consistently predictable (4).

Nonetheless, there must be some link between the molecular structure and the physical, chemical and biological properties of a substance. When looking at a set of molecular structures, some of which have a desired property and others do not, virtually any chemist will start to draw SARs in his or her mind. In some cases, apparently simple relationships are immediately obvious. For example, substances with a camphoraceous odour such as camphor (**8.1**), borneol (**8.2**) and 1.8-cineole (**8.3**) immediately suggest a hydrophobic ellipsoidal shape, and Amoore's camphor model is based on this. His suggested dimensions are 9.5Å for the long axis and 7.5Å for the short one (5). Synthesis of new molecules fulfilling the criteria then can be used to test the hypothesis. Thus the fact that *t*-butyl acetate (**8.4**) elicits a camphoraceous odour rather than the fruity character associated with *n*-butyl acetate (**8.5**) would support Amoore's model. Similarly, visual inspection of the natural sandalwood odorants α-santalol (**8.6**) and β-santalol (**8.7**) and the synthetic sandalwood odorants Radjanol (**8.8**), Brahmanol (**8.9**) and Sandalore (**8.10**) immediately suggests a model with a hydrophobic ball of specific size and a projection with an alcoholic function at a set distance from the centre of the ball (Figure 8.1).

More rigorous models fall into three main categories: those based on mechanistic lines involving physical or chemical parameters; those based on statistics; and those based on molecular modelling.

A typical chemical modelling technique would be Hansch analysis which uses regression analysis to correlate electronic, steric and hydrophobic properties with the biological activity in question. By investigating a limited set of parameters (usually based on a guess as to a mode of action), such approaches can bias the result by identifying a contributing parameter but giving it the appearance of being the dominant one.

Principal component analysis (PCA) is a statistical technique that reduces a multi-dimensional input (physical properties of molecules) to two or three dimensions which then aids in correlation with the biological activity. However, the principal component might be a complex blend of various parameters. The paper by Lavine et al. is a good example of how statistical approaches to SOR are carried out (6). They used a neural network and also PCA to investigate a set of 147 indanes and tetralins which either elicit a musk odour or are structurally similar to those that do. They included a total of 1344 different molecular descriptors for each structure in their data set. Their conclusions are more realistic than some in the literature in that they point out that it is a balance between 45 different descriptors rather than any single one or two of them that is important in determining whether the molecules they studied will elicit musk odour. Descriptors, such as log P, relating to transport phenomena did not seem to be important. However, as the authors point out, all of the molecules in the set under investigation have similar transport properties and, therefore, their analyses would not identify these

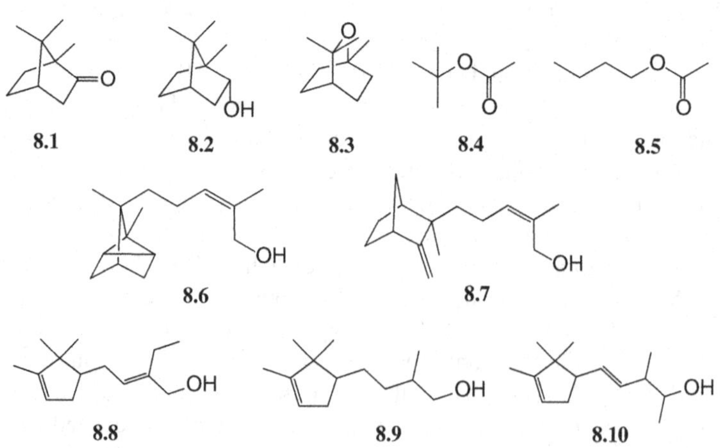

Figure 8.1   Camphor and sandalwood.

as discriminating between musks and non-musks. The descriptors that were most significant related to the shape of the hydrophobic part of the molecules. This is consistent with the concept of molecules binding into a specific pocket in an olfactory receptor.

Molecular modelling approaches such as COMFA (comparative molecular field analysis) usually attempt to model the complete stereo-electronic profile of the odorants in question and identify those features associated with the desired activity. The olfactophore approach uses both active odorants and inactive molecules with structures similar to those of the active analogues to build a three-dimensional model showing regions in space where electron density should be high or low and where hydrophobic bulk should be present or absent. If the input relates to receptor binding rather than odour, such a model represents an impression of the ligand binding pocket. A good example of the use of an odour-based olfactophore is the design of new marine odorants by Kraft and Eichenberger (7).

A comparison of Hansch analysis, PCA and COMFA in the correlation of structure with fruity odours provides a useful introduction to the three techniques and shows that, in this instance at least, they give similar results (8). The approaches used in structure/odour modelling have been comprehensively reviewed by Rossiter (9), Fráter et al. (10) and Kraft et al. (11), and a review of structure activity modelling techniques in general can be found in the book by Livingstone (12).

LIMITATIONS OF STRUCTURE/ACTIVITY CORRELATION

Structure/activity analysis is a useful tool but, like all tools, it should be used with due care and attention. There are a number of pitfalls into which the unwary researcher can fall.

Data Selection

Data selection can introduce a bias into the work. The larger and more varied the data in the starting data set, the less likely it will be to bias results. Certainly, any data set used must include both active substances and structurally similar ones lacking the desired activity.

The work of Lavine et al. (6) referred to above does give useful information about musk odorants but it cannot give a total picture because the data set contained only indane and tetralin musks such as Tonalid (**8.11**) and Phantolide (**8.12**) (Figure 8.2). Any model based on polycyclic musks such as these and Galaxolide (**8.13**) would give information about that family of musk odorants but would not present a total picture of the musk odour since it would exclude macrocyclic molecules such as hexadecanolide (**8.14**) and Musk MC4 (**8.15**) and nitro-aromatics such as Musk Ketone (**8.16**) and Musk Xylene (**8.17**). If either of these categories were unknown at the time, the model would certainly be of no use in predicting them as potential novel ingredients.

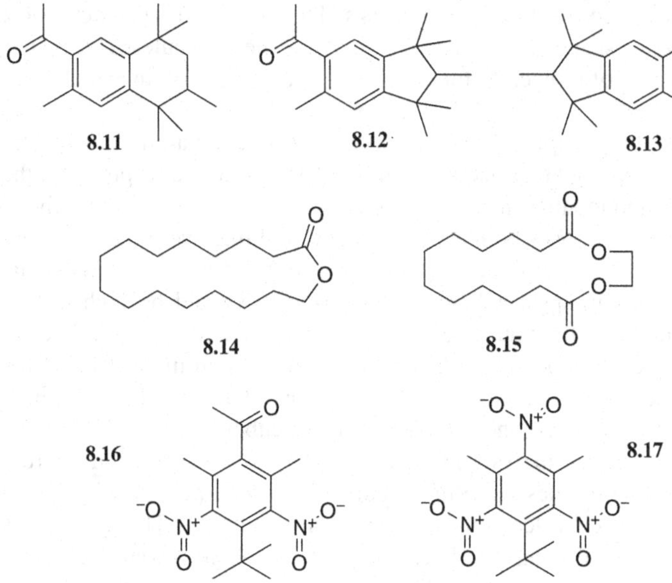

Figure 8.2   Some musks.

Similarly, once a lead structure has been identified for an odour area, chemists will synthesise many analogues of it, and use of these in the data set could bias the result of a structure/activity study. For example, if there were 100 polycyclic musks and 100 macrocyclic musks in existence but a structure activity study were to be carried out using all of the polycyclic musks and only 5 of the macrocyclic musks, the result would be distorted compared with the genuine result using all 200 molecules.

Conformation of Odorants

One problem facing all molecular modellers seeking to understand the interaction of small molecules with proteins is that of ligand conformation. Molecular mechanics calculations can predict the conformations with lowest energy, but at ambient temperature many higher energy states will also be occupied and the modeller must allow for all reasonable possibilities. The protein to which the small molecule binds will also be conformationally mobile and the final configuration of the complex will depend on all the energetics of both molecules, the binding energy of each point to point interaction and the energy gain from displacement of water from the binding pocket of the receptor and from around the ligand. Even in a relatively simple clathrate compound, hydrocarbon chains are compressed by the cage, as is demonstrated by the work of Ajami and Rebek (13). The effects were found to occur on the NMR (nuclear magnetic resonance) timescale and so could be studied effectively. Also using NMR, Frederick et al. were able to study a receptor protein, calmodulin, as it bound to ligands and used their results to estimate entropy changes during binding (14).

The fact that the conformation of a ligand when docked into its cognate receptor is not the minimum energy conformation is nicely demonstrated by the X-ray crystal structure of the bioactive jasmonate plant hormone (3R,7S)-jasmonyl-L-leucine (**8.18**) docked into the receptor coronatine insensitive 1 (COI1). The jasmonic acid fragment of the ligand is held in a conformation which is certainly not that which would be obtained by MM2 energy minimisation (15). Another example of the effect of a protein on ligand conformation is that of the pheromone-binding protein of the silk moth *Bombyx mori*. As was mentioned in Chapter 2, this protein has been shown to bind 1-iodohexadecane (**8.19**) and 2-isobutyl-3-methoxypyrazine (**8.20**) in addition to the pheromone bombykol (**8.21**) (16). The configuration of bombykol (**8.21**) in the ligand binding pocket is far from that of a structure minimised by molecular mechanics. The 1-iodohexane (**8.19**) adopts a similar configuration when it is bound to the protein as that adopted by bound bombykol (**8.21**). In order for the pyrazine (**8.20**) to bind, two molecules must occupy the site simultaneously. Similarly, as also mentioned in Chapter 2, the free fatty acid receptor GPR40 binds oleic acid (**8.22**), linoleic acid (**8.23**) and the synthetic compound GW9508 (**8.24**), all with low nanomolar potency (17,18). It is therefore reasonable to believe that the configurations of the two fatty acids in the binding site is as shown in Figure 8.3 and not the corresponding minimum energy conformations shown at the bottom of Figure 8.3.

Correlation and Causality

It must always be remembered that structure/activity correlations are statistical and do not necessarily represent mechanistic reality, especially in the field of olfaction. This will be discussed in more detail below. Correlation between two properties must not be assumed to imply a causal relationship. The two properties could be correlated simply because they are both results of a common cause, or the correlation could be purely chance. A somewhat tongue-in-cheek report has shown that there is a correlation between the performance of the Welsh international rugby team over a season and the chances of a Pope dying in the same year (19). The ratio of points scored by the Welsh team to those scored against them shows a statistically significant correlation with Papal deaths over the period 1880–2008. To imply a causal relationship would clearly be ridiculous, especially as none of the Popes during the period studied were Welsh and probably none were interested in rugby. Similarly, there is a strong correlation between *per capita* chocolate consumption in a country and *per capita* rate of gaining Nobel Prizes (20). Eating a large quantity of chocolate will not guarantee anyone a Nobel Prize.

The more mechanistic steps involved between a molecule and the observed activity, the less meaningful is any correlation between the two. As is obvious from Chapter 2, there are very many steps between an odorant entering the nasal cavity and an odour percept forming in the consciousness. Therefore, SORs should be

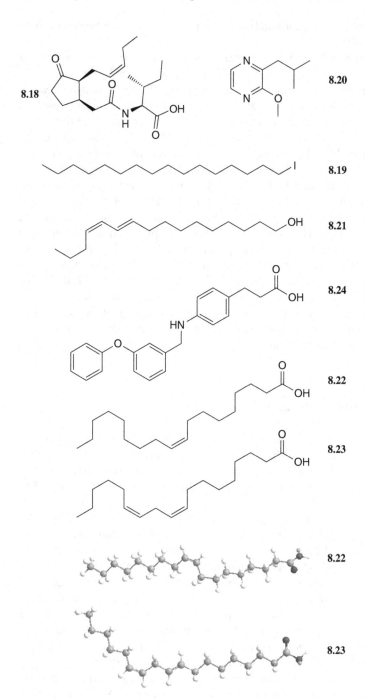

Figure 8.3 Some ligands and conformations.

regarded only as useful tools for the design of fragrance ingredients and not as indications of mechanism.

OBSTACLES TO STRUCTURE/ODOUR CORRELATION

Structure/odour correlation suffers from all of the pitfalls and difficulties of structure/activity correlation in general, but has many additional obstacles of which the researcher should be aware.

Organoleptic Purity

The issue of organoleptic purity is often overlooked or underestimated by those unaccustomed to working with fragrance chemicals in laboratory practice. It is no use claiming that samples used for evaluation were >95% pure, or even >99% pure. There are plenty of examples where minute traces of an impurity completely distort the odour of a sample. For example, a mixture of the alcohols (**8.25**) and (**8.26**) was found to have a powerful muguet odour due to traces of the isomeric aldehyde (**8.27**) even though (**8.27**) was present at such a low level as to be barely detectable by gas chromatography (Figure 8.4) (21,22). Another striking example of the issue is that of 3-methylthiobutanal. The (*R*)-enantiomer (**8.28**) has such a low odour threshold that even a 99.9% enantiomerically pure sample of the odourless (*S*)-enantiomer (**8.29**) would smell on GC smelling because of the 0.02 ng of (**8.28**) present in it, unless smelt from a chiral GC column (23). Traces of highly odorous volatile molecules containing sulfur or nitrogen atoms are notorious in fragrance chemistry for their tendency to dominate the odour of any substance containing them. Thus, for example, one is inclined to speculate that odour descriptions given (24) for amino acids such as cysteine, methionine and proline, all of which should be too poorly volatile to elicit odours, are actually due to trace impurities or decomposition products.

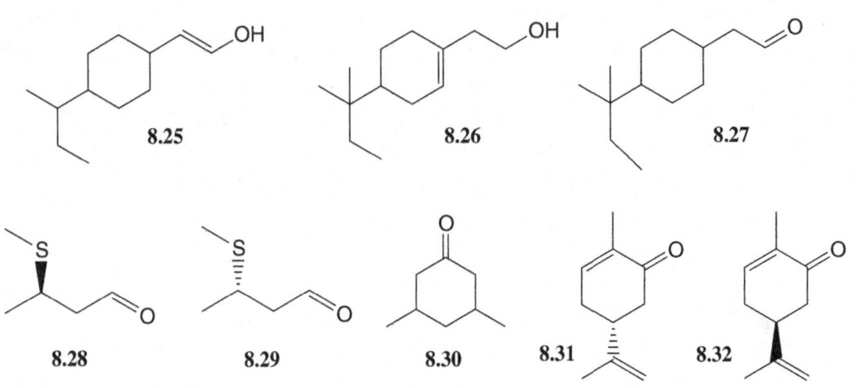

Figure 8.4   Organoleptic purity.

The debate on odour differences (or not, as the case may be) between enantiomers provides a number of examples of careful and thoughtful consideration of organoleptic purity. In the early twentieth century, von Braun and co-workers prepared all three isomers of 3,5-dimethylcyclohexanone (**8.30**) and, to ensure that any differences were not due to impurities, they used a single starting material and identical reagents in preparing the three disatereomers (25–27). To establish the difference in odour between *l*-carvone (**8.31**) and *d*-carvone (**8.32**), Friedman and Miller interconverted the enantiomers in both directions, thus confirming the difference between the spearmint character of (**8.31**) and the caraway character of (**8.32**) (28). Further examples are reviewed in a paper by the author (29).

Odour Data

Only large fragrance companies have significant databases of molecular structures with associated odour descriptors which are generated using a consistent methodology. Researchers in other institutions have to compile their data from diverse literature sources and thus they obtain sets based on different terms of reference and, consequently, inconsistent input. But even when the data set is derived using a single methodology and odour terminology, there are inherent issues of which researchers should be aware and in some cases, about which nothing can be done. An example of how the method of data collection will affect the result is that of anosmia to androstenone (**8.33**) (Figure 8.5). In a review of published results, it was found that measurements of the percentage of anosmia to androstenone (**8.33**) in the population varied from 1.8 to 75 (30).

Descriptors are Associative

As discussed in Chapter 3, senses other than smell have reference points associated with measurable physical properties. For instance, colour vision is dependent on the frequency/wavelength of electromagnetic radiation (light); in hearing, pitch depends on the frequency/wavelength of sound waves in air; touch depends on pressure; and salt taste depends on concentration of sodium chloride. Smell has no fixed reference points and all descriptions are associative. The problems of measuring odour have already been discussed in Chapter 3. We can only describe odours by reference to other odours. So, when we say that a molecule has a jasmine odour, what we really mean is that it elicits a percept similar

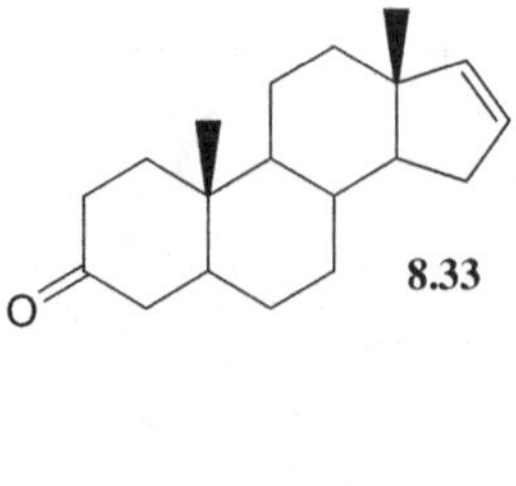

8.33

Figure 8.5 Androstenone.

to that elicited by the mixture of volatiles released from jasmine flowers. The headspace around a jasmine flower is a mixture of hundreds of discrete chemical substances. A simple accord of benzyl acetate (**8.34**), hexylcinnamic aldehyde (**8.35**) and indole (**8.36**) also elicits a jasmine odour as does an accord of isobornyl acetate (**8.37**), Jasmacyclene (**8.38**), 2-cyclopentylidenecyclopentanol (**8.39**) and methyl dihydrojasmonate (**8.40**), none of which is found in natural jasmine (Figure 8.6). Therefore, it is clear that, although the term jasmine might be useful in describing the odour percept, it is of limited use in correlating odour with structure. Furthermore, a single chemical entity, jasmone (**8.41**) for example, might also be described as having a jasmine odour, even though it elicits one facet of jasmine rather than the odour of the entire oil. Similarly, as described in Chapter 3, odour classifications and maps chart out our semantic descriptors but these are not necessarily meaningful in terms of linking odour percept with chemical structure.

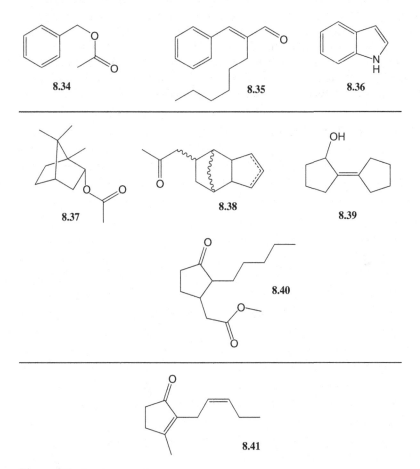

Figure 8.6   Jasmine accords.

Concentration

Odour descriptors are concentration dependent. All fragrance chemists are aware of the fact that sulfur-containing odorants such as the passionfruit oxathiane (**8.42**) and the grapefruit thiol (**8.43**) often elicit pleasant or tolerable odours at low concentration but are foul at higher concentration (Figure 8.7). However, this phenomenon is not limited to sulfurous odorants. For example, Laing et al. measured odour descriptors at various concentrations for five odorants, namely, heptanal (**8.44**), methyl heptanoate (**8.45**), 2-octanone (**8.46**), heptanoic acid (**8.47**) and 1-heptanol (**8.48**) (31). Taking the liberty of treating 2-octanone (**8.46**) as 1-methylheptanal, these molecules all share a seven-membered carbon chain with an oxygenated function on the terminal carbon, all have different functional groups and all elicit different odours. In every case, the odour descriptors allocated by a panel varied with concentration of the odorant. Only 2-octanone (**8.46**) had descriptors that persisted at all concentrations tested (from detection threshold up to 729 times detection threshold), and even then minor aspects of the description did vary with concentration. So, which descriptor should be used for structure/odour correlation? In this case, the concentration dependence of the odour is known, but in most published work an odour description is reported but without mention of concentration dependence and often without even an indication of the concentration at which the descriptor was obtained.

Perireceptor Chemistry

As discussed in Chapter 2, the olfactory mucus is not an inert medium but contains various enzymes that are capable of metabolising odorants in a timescale which

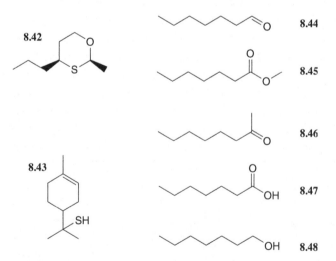

Figure 8.7   Concentration dependence of odour character.

means that the signal generated at the epithelium could well be a composite of the odorant and its metabolites, rather than just the pure odorant. Again, referring back to Chapter 2, Schilling and co-workers have established beyond doubt that, when a subject describes the nor-sesquiterpene ketone (**8.49**) as eliciting a raspberry percept, this percept is actually the result of detecting the metabolite (**8.50**) (32–36) (Figure 8.8). If a researcher is unaware of the biotransformation, the structure (**8.49**) would be fed into an SOR as eliciting a raspberry percept and this would distort the SOR considerably. Comparison of the real raspberry odorant (**8.50**) with raspberry ketone (**8.51**) would give a much better idea of structural requirements, in this case an OH function and a ketone function at a distinct distance from each other. It is often a matter of speculation as to why some hydrocarbons should elicit intense odours when they can form only weak hydrogen bonds. Perhaps the answer lies in oxidative metabolism.

Just a cursory inspection of the examples in Figure 2.7 will show the potential magnitude of this issue. Even if the researcher considers the possibility of enzymic activity in the mucus, establishment of nasal metabolism in any given instance is difficult and, consequently, in the vast majority of available odour descriptors the possibility is ignored. It is therefore likely that many SORs are founded on incorrect data input.

In addition to reaction with enzymes in the mucus, odorants might well be capable of reacting with the receptors themselves. Dods et al. have reported reaction between lysine residues in a membrane protein and ester functions in the lipids forming the membrane (37). If this reaction is possible, then reaction of aldehydes with lysine residues and thiols with cysteine or cysteine must also be possible. Modification of protein structures in this way could well affect their ability to recognise odorants, thus effectively antagonising them, or could stabilise an active state of the receptor, thus effectively rendering them constitutively active. Either set of conditions would then have an effect on odour perception and confuse any SOR based on the resultant data.

The Nature of Olfaction

The largest obstacle to devising consistently accurate and precise odour descriptions is the mechanism of olfaction itself. I have often joked to discovery chemists

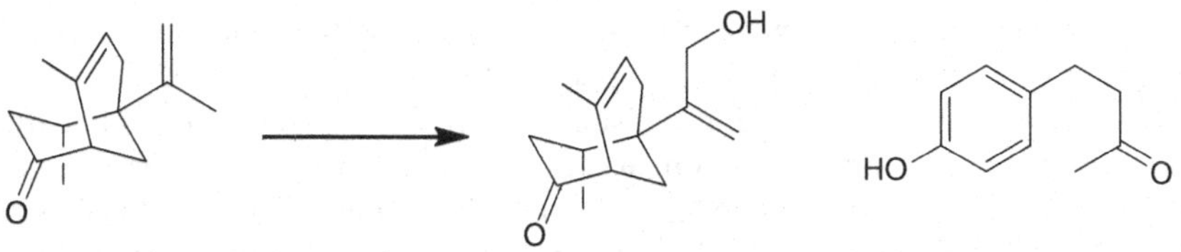

Figure 8.8 Nasal chemistry and structure/odour relationship.

in the pharmaceutical industry that they have an easy task. They only have to target one protein at a time. Fragrance discovery chemists have to simultaneously target an unspecified and variable number of receptors in the region possibly of 400, but if SNPs and CNVs are included, this number could be 10 times that, or more. This may sound facetious but it is the overall pattern of activation and not the activation of any one receptor that generates the signal which is eventually interpreted as an odour percept. Even in a relatively simple organism, the fruit fly *Drosophila melanogaster*, which has only 60 different types of olfactory receptors, it requires simultaneous activation of two different receptors to detect carbon dioxide, a very simple odorant (38). Humans can distinguish an essentially infinite variety of odours and this can only be achieved using recognition of complex patterns. Furthermore, any odorant could act as an agonist on some receptors and as an antagonist on others and it could also have an effect on the second messenger system. These parameters add another layer of complexity, as does the fact that the chemotopic map of the olfactory epithelium and olfactory bulb is lost in the piriform cortex (39). The multitudinous interactions during neuroprocessing, including input from other senses and from non-sensory sources, complicate the picture even further. As mentioned in Chapter 2, Gottfried and co-workers used functional MRI (fMRI) to show how neuroprocessing can alter the signal coming from the receptors, leading them to conclude that, 'It is widely presumed that odour quality is a direct outcome of odorant structure, but human studies indicate that molecular knowledge of an odorant is not always sufficient to predict odour quality. Indeed, the same olfactory input may generate different odour percepts depending on prior learning and experience' (40).

The Subjectivity of Odour

The subjectivity of odour has plagued SOR research throughout its history. Different people can give quite different descriptions of odour character of the same molecule, and, equally, the thresholds of detection and recognition can vary enormously from one person to another. Even when detailed experimental protocols are rigorously applied in collection of odour data, the results will vary from one individual to another. It has been claimed that this phenomenon has been used by lazy or inefficient chemists to excuse imprecision in their SOR results. However, subjectivity is an intrinsic part of odour perception as made clear in the relevant section of Chapter 2. Each of us has a unique perception of odour, irrespective of the terminology we are trained to apply to it. Therefore, any SOR is either a statistical average across a group of people or is applicable to only one individual.

To illustrate the danger of using statistical averages, I usually point out that the average family in the United Kingdom at present has 1.8 children but I have never met anyone with 1.8 children. A concrete example in the field of olfaction is that of the odour detection threshold (ODT) of the enantiomers of geosmin (Figure 8.9). In a study, which is a model of how such sensory evaluations should be carried out, Ernst Polak found that the average threshold of the naturally occurring (−)-geosmin

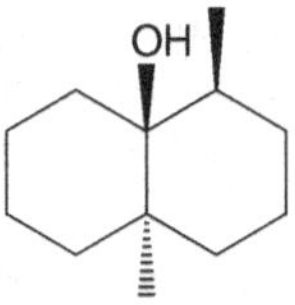

Figure 8.9   Geosmin enantiomers.

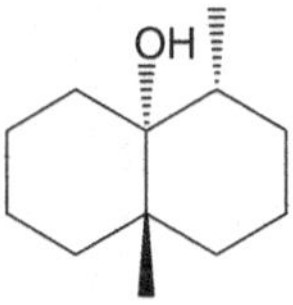

(**8.52**) had an ODT of 0.0095 ppb in water, whereas that of the (+)-enantiomer (**8.53**) was 0.078 ppb in water. (41) This means that, for the 'average person', the natural enantiomer can be detected at just under one-tenth of the concentration at which the unnatural one becomes odourless. This headline result from the abstract of Polak's paper is almost invariably quoted as if it implied a molecular property of the enantiomers. However, the detail of Polak's experimental work shows that some people are more sensitive to the unnatural enantiomer and that there is quite a distribution of values across the spectrum. Very few "average people" exist in terms of this data set. Similarly, Polak showed that ODTs for *l*-carvone (**8.31**) varied from 0.07 to 2,000 ppb from water depending on the subject, and those for *d*-carvone (**8.32**) from 1–4,000 ppb from water. The *l*-carvone/*d*-carvone ratio varied from 28 to 230 (42). More significantly, in both cases the ratio of ODTs depends on the observer and not on the chirality of the molecule. The example given in Chapter 2 of the use of aversive learning to enable discrimination between enantiomers is evidence of the fact that odour percepts are malleable and a property of the observer rather than the observed molecule (43). People who were previously unable to distinguish between the odours of two enantiomers learnt to do so after exposure to one enantiomer had been paired with an electric shock.

However, odour properties must depend to some extent on the physical and chemical properties of molecules and so predictive models are possible, albeit on a statistical basis only. The range of detection thresholds of the geosmin enantiomers is limited, and a model might be able to place the predicted threshold within the correct range, even though it cannot predict the precise value for any individual observer. Thus, for example, Abraham et al. were able to develop an algorithm for prediction of ODT for a set of 353 molecules (44). It is based on solubility parameters, suggesting the importance of transport phenomena in odorant detection.

PROBLEMS IN STRUCTURE ODOUR CORRELATION

There have always been puzzling observations that have intrigued chemists looking for simple relationships between structure and odour and that have served to frustrate attempts to develop a mechanistic understanding based on SORs. For example, sometimes similar chemical structures elicit quite different odour percepts, whilst in other instances dissimilar structures can elicit almost identical percepts. Sometimes a small structural change in a molecule produces a large change in odour, whereas in other cases gross structural changes produce little change in odour. Sometimes the chemical functional group in a molecule is important, for example, the ester group and the fruity odour, whereas at other times the shape of the molecule is more important, for example, for the camphor odour. Absolute stereochemistry of a molecule sometimes affects its odour and sometimes does not. Some of these problem observations are illustrated in the following examples.

Figure 8.10 shows two sets of three compounds. In the first set, there are α-methylcinnamaldehyde (**8.54**), Lilial (**8.55**) and Florosa (**8.56**). The first two are similar in that they are both substituted 3-phenylpropionaldehydes yet they elicit very different odours. α-Methylcinnamaldehyde (**8.54**) produces a cinnamon odour whereas, Lilial (**8.55**) is a classic muguet (lily of the valley) odorant. Florosa (**8.56**) has an odour similar to Lilial (**8.55**) although it has no structural features in common. The second set is even more striking. As its name suggests, cuminaldehyde (**8.57**) elicits an odour that is distinctly reminiscent of cumin yet benzaldehyde (**8.58**) and hydrogen cyanide (**8.59**) both elicit an almond odour. Cuminaldehyde (**8.57**) is a simple isopropylated analogue of benzaldehyde (**8.58**) and so the two are very similar and both are very different in chemical and physical properties from hydrogen cyanide (**8.59**). The difference in physical and chemical properties between benzaldehyde (**8.58**) and hydrogen cyanide (**8.59**) is so great that theories to account for their similarity in odour include a proposal that HCN tetramerises in the receptor. Since benzaldehyde (**8.58**) and hydrogen cyanide (**8.59**) almost invariably occur together in nature, it is more likely that they activate different patterns at the receptor level but that both patterns are learnt to be interpreted as reminiscent of the same natural odour.

The effect of stereochemistry is equally puzzling and unpredictable. The enantiomers of wine lactone (**8.60**)–(**8.67**) (Figure 8.11) show how absolute stereochemistry can have a dramatic effect on odour threshold, in this case even more

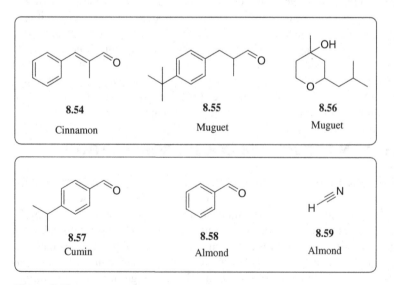

Figure 8.10   Puzzles in structure/odour correlation.

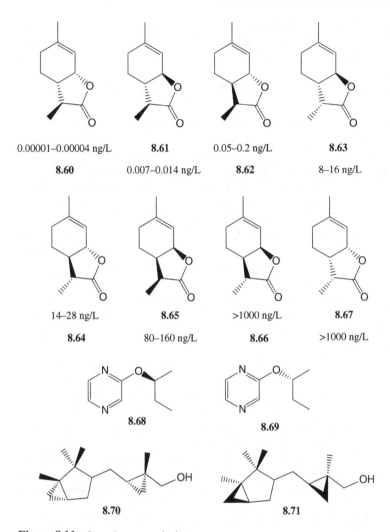

Figure 8.11 Stereoisomers and odour.

so than cis/trans isomerism around a ring, the threshold of (**8.60**) being more than eight orders of magnitude lower than its enantiomer (**8.66**) (45). On the other hand, the enantiomers (**8.68**) and (**8.69**) of 2-isobutoxypyrazine have identical detection thresholds (46). These enantiomers also elicit identical odour qualities, whereas the enantiomer (**8.70**) of the potent sandalwood ingredient Javanol (**8.71**) has a muguet odour (47). Examples of all possible combinations of effects of stereoisomerism on odour character and/or threshold can be found in the review by Sell (29).

These and similar puzzles have provided much scope for speculation and debate in the past and have caused a great deal of frustration. The reason for this is, of course, that the facts do not fit any of the theories advanced on the basis of

SOR results. Following the advances of the last few decades, we now understand why simple and perfect SORs do not exist and why such apparently impossible observations should not cause any surprise.

SUCCESS IN STRUCTURE/ODOUR CORRELATION

Notwithstanding the conclusion in the preceding paragraph, SORs have proved very useful in the design of novel fragrance ingredients. Many of the successful models and the techniques employed to design them have been reviewed by various authors including Boelens (48), Rossiter (9), Fráter et al. (10) and Kraft et al. (11).

The most consistently accurate structure/odour model is Amoore's camphor model (5). As mentioned previously, the model indicates that hydrophobic molecules with an ellipsoidal shape having a long axis of 9.5Å and a short axis of 7.5Å will possess a camphoraceous odor, and compounds (**8.1**)–(**8.4**) are examples of some that meet the criteria. Another example of a simple but quite effective model is Boelens' model for jasmine odorants (48). Shown in Figure 8.12, this model proposes that, in order to possess a jasmine odor, a molecule should contain a central carbon atom surrounded by a strongly polar group, a weakly polar group and an alkyl chain. Two examples of structures fitting the model are jasmone (**8.41**) and hexylcinnamic aldehyde (**8.35**) and these are shown *per se* in Figure 8.12 and also fitted over Boelens' model.

Another example of a simple SOR based essentially on visual inspection of structures is that of a series of aliphatic esters with fruity odours (49,50). It was found that steric congestion around the ester function of open chain aliphatic esters reduced the degree of fruitiness in the odours, as perceived by a trained sensory panel. Using the data from these papers and computer molecular modelling, Rossiter was able to quantify these effects and obtain a more precise model (8). It is not surprising that steric accessibility affects recognition of ester groups and so these results were in line with expectations of any contemporary fragrance chemist. However, the early data set did show up one of the factors of which those working in SOR should be aware, that of semantics. As explained in Chapter 3, there are no physical references for odour, and classification systems are derived from factors unrelated to odour. Thus, although steric congestion reduces the perception of fruitiness in simple aliphatic esters, it was found to increase it for a group of cyclohexyl esters. So for example, the fruitiness of compounds (**8.73**)–(**8.78**) diminishes in the order shown, whereas that of compounds (**8.79**)–(**8.82**) diminishes in the order as shown even though that is the opposite order in terms of steric hindrance to hydrogen-bond formation around the ester function. The answer to this apparent contradiction lies in semantics. The sensory panel scores in Figure 8.13 relate to fruitiness, but this was obtained by summing the scores given for individual fruits. When the data is examined in more detail, it is seen that the fruitiness of compounds (**8.73**)–(**8.78**) is pear whereas that of compounds (**8.79**)–(**8.82**) is apple. We relate the terms apple and pear because they are similar fruits, but in terms of the olfactory system these are distinct percepts and different epithelial activation patterns are associated with each.

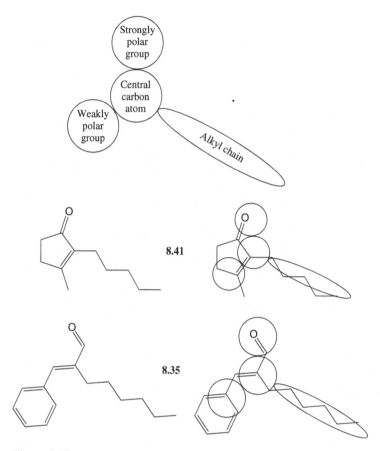

Figure 8.12   Boelens' jasmine model.

Computer-aided molecular modelling has advanced considerably over the last few decades and now allows the construction of olfactophores, the olfactory equivalent of medicinal chemistry's pharmacophores. These are three-dimensional models derived from the structures of active odorants with a desired target odour and also the structures of analogous substances which have either a different odour or no odour. The olfactophore shows volumes of space where steric bulk is required and also those where bulk should be absent. They also indicate where there should be regions of electron density or regions where a partial positive charge is needed, and they give the location and direction of hydrogen-bond donors and acceptors. Figure 8.14 shows an olfactophore for aliphatic esters with a fruity odour. One typical ester is docked into the olfactophore, showing how it fills the required volume of steric bulk, avoids the region just above the ethereal oxygen atom of the ester group where steric bulk is undesirable and provides the necessary hydrogen-bond acceptors (indicated by cones). The review by Kraft et al. gives a number of other examples of olfactophores (11).

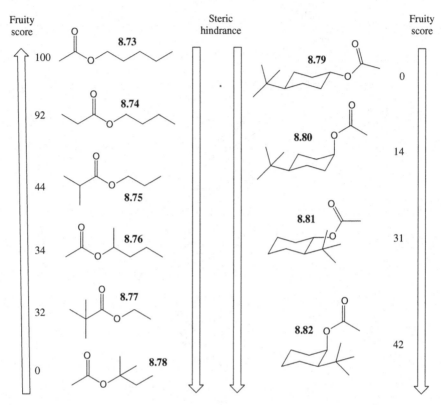

Figure 8.13   Steric effects and fruitiness of esters.

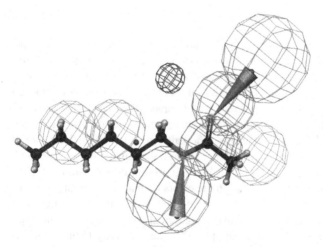

Figure 8.14   A fruity olfactophore (By kind permission of Dr Philip Kraft.)

The advantage of olfactophores is that a fragrance chemist can look at them and visualise novel molecules that might match the requirements. Thus they can be used in two ways in molecular design. They can be used directly to inspire research ideas and also as tools for what has become known as *in silico* screening. In this latter technique, potential candidates for synthesis are screened against a computer model to assess the probability of their meeting the requirements of the target. Olfactophores and related techniques have proved of considerable value in the design of novel ingredients, and examples include the highly successful fragrance ingredients Azurone (**8.83**) (51,7), Belambre (**8.84**) (52,53), Javanol (**8.71**) (54,47) and Rossitol (**8.85**) (55,56) (Figure 8.15).

The musk model of Lavine et al. (6), which has already been described, is a good example of a study using PCA and shows how the technique can identify which parameters are important and to what extent. It also clearly demonstrates how the link between structure and odour is dependent on many molecular parameters and not just one or two. Such models can be used for *in silico* screening but are not very helpful in suggesting novel structures to the discovery chemist.

As explained in Chapter 2, volatility and polarity/polarisability are important in molecular recognition, and this places constraints on odorants. Thus models will always focus on molecules with a maximum of 18–20 carbon atoms in the molecule, equivalent to a molecular weight of about 300 Da, and with a log $P_{oct/water}$ in the region of 2–5. Conformational flexibility is another important factor in SOR models and, not surprisingly, many of the more successful models involve relatively conformationally constrained molecules. Thus, there are more examples of reasonably good SORs in odour areas such as musk and camphor than in odour areas such as muguet (lily of the valley) which are associated with conformationally flexible molecules.

A good example of this, and one that also illustrates the problem of data distortion through following leads, is that of ambergris. For many years, Ohloff's triaxial rule (57) dominated ambergris chemistry and gave accurate predictions of whether candidate molecules would elicit the prized ambergris odour. Ohloff's

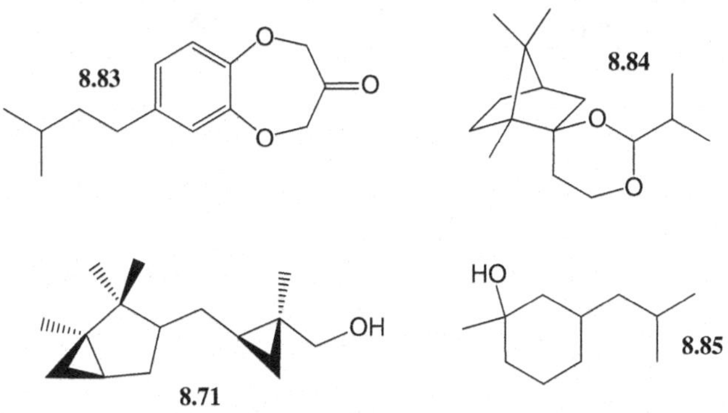

Figure 8.15 Successful fragrance ingredients designed using molecular modelling.

model proposes that, in order to possess an ambergris odour, a molecule should have a *trans*-decalin structure with three axial substituents in a 1,2,4-relationship, as shown in Figure 8.16, and that one of these should be an oxygen function (outline structure **8.86**). This model was based on analogues of the natural compound (**8.87**) found in ambergris produced by the sperm whale *Physeter macrocephalus*. The serendipitous discovery of Karanal (**8.88**) as a potent ambergris odorant (58,59) invalidated the model. Since then, ambergris models have mostly been based on olfactophores and these have been used to help design such ingredients as Belambre (**8.84**) (52,53) and Ambermax (**8.89**) (60). Another ambergris model is that of Bersuker et al., which is based on their electron topological theory of odour (61). They propose that ambergris odorants have two hydrogen atoms and one oxygen atom located at the corners of a triangle, the dimensions of which are shown in Figure 8.16. They also require that all three atoms make a significant contribution to the lowest unoccupied molecular orbital (LUMO) of the molecule, that the coefficients of both of these hydrogen atoms should coincide, that the negative charge on the oxygen atom should be between 0.24 and 0.31 of that of an electron, that the charge on H_i should be negative and that the charge density over the triangle should be -0.1 e/Å^2.

OLFACTION THEORIES

In 1870, Ogle identified a yellow pigment in the nose and suggested that it interacted vibrationally with odorants to produce heat, which then triggered the perception of smell (62). A later development of this theme was that the colour was due to

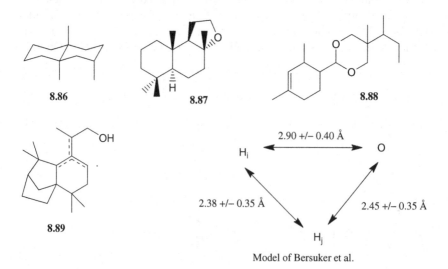

Figure 8.16   Ambergris structure/odour relationships.

carotenoids bound to the receptors and that these interacted with odorants to create a semi-conductor current which activated the sensory neurons (63). Although administration of carotenoids alleviated anosmia in one set of trials (64), there is no evidence that carotenoids are associated with the olfactory receptor proteins and all of the evidence concerning GPCRs indicates a different mechanism of action, as is clear from Chapter 2. Another theory was that odorants diffuse across the membrane of the olfactory sensory neurons (OSNs) leaving an ion pore in their wake (65). Since the odorants and membrane are both hydrophobic, this is a difficult concept to accept. The electron topological approach proposes an interaction between the frontier orbitals of the odorant with those of the receptor. Such interactions are part of recognition between odorant and receptor but his is only a part of the process. The electron topological approach is essentially SOR-based and will be further discussed below. The chromatographic theory of Mozell is also an SOR-based approach and will also be included below, along with the two best known SOR-based theories, namely, the shape and vibration theories.

Olfaction Theories Based on Structure Odour Correlation

The availability of sets of molecular structures coupled to odour descriptions inevitably tempted chemists and others to propose theories of olfaction based on them. Mechanistic theory formulation based on structure/activity relationships is an intellectually hazardous activity at the best of times and is particularly so in the case of olfaction where the number of mechanistic steps from the structure to the odour percept is so huge. Proponents of theories have tended to cite only those examples that fit their theories and ignore examples that go against them. Some of the various theories will be discussed briefly below. The two most prominent of the SOR-based theories are the shape theory of Amoore and the vibration theory of Dyson, Wright and Turin. Even a cursory inspection of Figure 8.10 will reveal that the structures with similar odours in each of the groups of three odorants have very much less in common in terms of shape, molecular vibrations, functional group, solubility properties and so on, than one of them does with the odorant that elicits a distinctly different odour. For example, Lilial (**8.55**) has much more in common in terms of physical and chemical properties with α-methylcinnamaldehyde (**8.54**) than with Florosa (**8.56**), even though it is Lilial (**8.55**) and Florosa (**8.56**) that smell alike. The example of cuminaldehyde (**8.57**), benzaldehyde (**8.58**) and hydrogen cyanide (**8.59**) is even more striking because the last two have very little in common physically or chemically yet have quite similar odours. The old adage 'the exception proves the rule' is widely misunderstood because, in this context the verb 'prove' has the meaning of 'test' rather than 'establish'. The examples of Figure 8.10 challenge the 'rules' of all of the SOR-based theories and certainly do not validate them. In any case, there are too many exceptions to all the 'rules' for any of them to be taken seriously. The description of the process of olfaction in Chapter 2 and the many examples of receptive ranges of receptors in that chapter are further proof that the simplistic SOR-derived theories are woefully inadequate.

The fact that the various accords shown in Figure 8.6 all elicit similar percepts is further evidence that simplistic theories relating molecular structure to odour are at best inadequate.

Of course, some of the SOR-based olfaction theories do have an element of truth in them. For example, the shape of a molecule, or more precisely the variety of shapes that it can adopt, do influence recognition by receptor proteins and, similarly, the diffusion processes in the nasal air space and between that and the mucus of the olfactory epithelium do affect the activation pattern across the epithelium, or at least the temporal aspects of it.

I and at least three other fragrance chemists known to me have set up experiments concerning the predictive ability of models derived from SOR-based olfaction theories and compared them with the predictive ability of experienced fragrance chemists. All of us found that the predictions of the experienced chemists tend to have an accuracy of 70–80%, whereas those from theoretical models tend to be closer to random. For me, this serves to illustrate the power of the human brain in pattern recognition. I do not fully understand what it is that we recognise, but clearly there is something (probably a combination of many molecular parameters) that is recognisable and gives a degree of predictability.

False Assumptions

All SOR-derived theories of olfaction make some basic assumptions that we now know to be false. The assumptions are never stated, and it is possible that proponents of the various olfactory theories are unaware that they are making them. Four false assumptions are described briefly below.

False Assumption 1: Odour is a Molecular Property   The tuning of the olfactory receptors clearly indicates that molecules activate them depending on the physical and chemical properties and not on odour. It is the activation of a pattern of receptors and not any individual one that leads to formation of an olfactory percept. The section on subjectivity in Chapter 2 shows clearly how genetic variation means that each of us uses a different repertoire of receptors leading to a different epithelial activation pattern for any given molecule. The resultant signals are then processed through neural networks that are also unique to each individual and they receive inputs that vary not only with the individual's experience but also with non-sensory input. The ultimate odour percept is synthesised using all of these inputs and is unique to the individual. Thus we each live in our own unique sensory universe and, although we may learn to apply the same semantic terms to odour percepts, the actual percept is likely to be different. Just as the ODT of any pure chemical entity varies over a range of values across the population, so it is likely that qualitative percepts also vary over a range. This would allow for sufficient commonality for the construction of SORs but the percept must be seen as the result of both the molecular structure of the odorant and the anatomy, biochemistry, genetics, physiology and history of the person perceiving it.

False Assumption 2: The Odorant Delivered to the Nose is the Substance Detected by the Receptors The section on nasal metabolism in Chapter 2 shows how we know that, in some cases at least, the volatile chemical entering the nose is metabolised by nasal enzymes and the odour percept is dependent on the metabolites as well as the molecules of the original substance that reach a receptor before meeting a metabolic enzyme. It is likely that this is a common phenomenon and, therefore, odour descriptions based on the starting structure will be due to metabolites. Therefore, the data used in SOR studies is questionable, and in some cases erroneous.

False Assumption 3: Olfactory Receptors are Tuned to Odour Experimental evidence now shows that olfactory receptors can, and many do, show a remarkable breadth of tuning. It is obvious from the receptive ranges of the more narrowly tuned receptors that they recognise the stereo-electronic properties of molecules and not their odour. Even with broadly tuned receptors such as OR1D2, which is tuned to phenylpropionaldehyde and related structures and responds to ligands that elicit a very diverse range of odour types (30); OR52D1 (66) or OR17-40 (67), investigation of their receptive ranges shows that the tuning is not to odour but to stereo-electronic properties of the ligands. As a further example, the majority of carvone-responding receptors respond to both enantiomers and are, therefore, not tuned to either spearmint or caraway. Many further examples can be found in chapter 2.

False Assumption 4: Odorant–Receptor Interaction is the Only Significant Event in the Process of Olfaction Nasal metabolism is known to occur and clearly will be a significant obstacle as far as this assumption is concerned. If a metabolite is detected by the receptor, then an SOR based on the original odorant will be wrong even if the odorant–receptor interaction is the most significant step of the overall process. SOR-based olfaction theories assume that, once a specific receptor has been activated, the formation of a given percept in the orbitofrontal cortex is inevitable. Some might even replace the phrase 'specific receptor' by 'specific set of receptors'. However, the evidence from neuroscience is that the input signal is used to construct a mental image. The process of synthesis depends on the brain and is affected by inputs from other senses and also from non-sensory inputs as described in Chapter 2. Receptor activation is only the start of a long and complex process in the brain.

SOME SOR-BASED ODOUR THEORIES

In ancient Greece, the school of Democritus and Epicurus argued that odorous atoms interacted with the nose to create odour. They suggested that atoms with a smooth surface were responsible for sweet smells because they brushed softly against the nose, whereas atoms with sharp edges or projections irritated the nose and produced odours such as that of vinegar. This theory was proposed in the absence of any serious physical evidence and is clearly untenable in the light of current knowledge.

In 1970, Mozell proposed a chromatographic model of olfaction based on measurements of the activity gradients across the olfactory mucosa during sniffing of various odorants and the relative behaviour of the same odorants on gas chromatography (68). It is not surprising that he found a correlation at that level since chromatographic effects must play a part in at least the temporal aspects of the pattern of receptor activation across the epithelium. The exact significance of this effect has yet to be clarified in terms of subsequent neuroprocessing.

The Electron Topological Theory

This theory, developed by the Moldavian team of Bersuker, Vlad and Gorbachov, proposes that contact is made between the frontier orbitals of the odorant and those of the receptor and that the latter will be activated if certain criteria are met by the odorant. An example of their results is shown in Figure 8.16 and relates to the ambergris odour (61). The figure shows how two hydrogen atoms and one oxygen atom are involved and the spatial relationship between them. The accompanying text describes the contributions that the molecular orbitals of each of the atoms should make to the molecular orbitals. The model predicts that any molecule satisfying these criteria will possess an ambergris odour. The team also developed similar triangular models for other odours such as musk (69). The ambergris model was later refined by Gorbachov and Rossiter (70). These models might prove useful in molecular design but more recent research on GPCRs, including X-ray crystal structure determinations, has shown that the receptor completely envelops the ligand rather than contacting it at only three points.

Amoore's Shape/Anosmia Theory

The lock and key of pharmacological action was first stated by Emil Fischer (71), and it has served medicinal chemistry well in the century since then. The concept is that biological macromolecules such as enzymes and receptors have a pocket into which ligands fit, analogous to a key fitting into a lock, the correct shape of ligand activating the macromolecule just as the correct key opens the lock. In 1920, Ružička adapted this to odorants by arguing that the character of an odorant is determined by its shape, with the functional group causing only variations in the character (72). Forty years later, Beets recognised that heteroatoms played a greater role and suggested that the functional group serves to determine the affinity of an odorant for the receptor and that the shape profile determines the odour (73,74). We now know that the olfactory ligands are conformationally flexible as are many of the odorants that bind to them and so the 'lock and key' model of odorant/receptor binding must be modified to a 'flexible key in an array of flexible locks' model.

In the late 1950s and early 1960s, John Amoore built on the lock-and-key model of olfaction. He investigated the odour descriptions of many hundreds of molecules and noticed that seven key terms made frequent and prominent appearances. These were camphoraceous, musky, floral, minty, ethereal, pungent and putrid. He considered the possibility that these might be 'primary odours' analogous to the three 'primary colours' (red, blue and green) of vision. He, therefore, looked at the structures of odorants falling into these categories and proposed receptor pocket shapes for five of them, with pungent being associated with molecules carrying a (partial) positive charge and putrid with a negative one. Examples of his receptor shapes are the ellipsoidal shape into which the camphoraceous molecules (**8.1**)–(**8.4**) of Figure 8.1 would fit and a flat oval dish into which the musks of Figure 8.2 would fit. He later abandoned the classification based on semantics and postulated that the 'primary' odours could be identified by anosmia. The argument was that specific anosmias indicated missing receptors and therefore a primary odour characteristic. He collected a list of about 30 anosmias. As discussed in the section on anosmia in Chapter 2, there is a very much larger number of reported anosmias, and research into receptor genetics and genotype/phenotype correlations show that the relationship between genotype and anosmia is complex and not dependent on single genes. Of course, such studies were not possible in the days when Amoore was developing his theory. His approach is detailed in various papers and a book that he wrote on the subject (75–79). It is clear that stereo-electronic effects must be a part of molecular recognition but olfaction is more complex than the simple theories imply, as is evident from Chapter 2.

Vibration Theory

Dyson proposed a theory of odour in 1938 in which odorant molecules are recognized by their vibrational frequencies in the infrared spectrum (80). He had no suggestions for a biological mechanism for such recognition but one was proposed by Turin in 1996 (81). Turin proposed that the receptors contain two electrodes separated by a gap into which an odorant molecule could fit. If an electron could tunnel through the odorant, then a current could pass from one electrode to the other. In tunnelling, the electron would lose an amount of energy and if this matched the energy gained by vibrational excitation of the odorant molecule, then the electrical circuit would be completed. Theoretical calculations show that such a mechanism is plausible in terms of the physics involved (82). However, there is no evidence for the existence of such electrodes in olfactory receptors.

The Shape Versus Vibration Debate

The proponents of the shape and vibration theories have held a long-running, and at times very acrimonious, debate. The vibration theory was championed by Wright in the 1950s and 1960s at the time when Amoore was most active regarding the shape theory. Both Dyson's and Amoore's theories rely on some basic assumptions which we now know to be false, as discussed above. The debate, therefore, serves to remind us that in scientific research it is important to recognise any basic assumptions that have been made and to test them experimentally to determine their validity. To be fair to Dyson and Amoore, it must be said that, at that time, they did not have the tools to test most of those basic assumptions. The receptors were not available until Buck and Axel's landmark discovery in 1991 (83), and neuroscience equally has made enormous progress in the last two decades. However, it would seem that, when the debate was at its most heated, Ernst Polak was almost alone in recognising that factors such as the diversity of odorants and odour descriptors, and the subjectivity of odour, brought into question the assumption that odour is a fundamental property of molecules (84). In the light of current knowledge, it is easy to see that the shape versus vibration debate is irrelevant, but even at the time there were so many known experimental exceptions to both theories that researchers should have been warned to look again and seek a better explanation. As already discussed, the two sets of molecules in Figure 8.10 are examples showing that both theories are incorrect. Molecules (**8.54**) and (**8.55**) are much closer to each other in both shape and infrared absorption frequencies than either is to (**8.56**), yet it is (**8.55**) and (**8.56**) that elicit similar odours. Similarly, (**8.57**) and (**8.58**) are close in shape and spectra but (**8.58**) elicits a similar odour to that elicited by (**8.59**) and not that elicited by (**8.57**).

Because of the issues in describing odour (see Chapter 3) and its subjectivity (see Chapter 3), the use of odour descriptors can be used in different ways. For example, how different is different? Two rose odorants might have different aspects but are clearly distinguishable from each other and from peppermint. So one person might choose to claim that the two rose odorants have different odours, while another will say that they are both rose and not peppermint. The data can, therefore, be interpreted to suit a theory, and this is one factor that kept the debate alive. Few researchers went to the lengths that von Braun and his co-workers (25,85,27) or Friedman and Miller (28) did in ensuring organoleptic purity and, therefore, no doubt, trace contaminants also added fuel to the flames.

Isotopes and Enantiomers

The issue of isotopes and enantiomers is of particular relevance to the shape versus vibration debate because supporters of the shape theory claimed that odour differences between enantiomers disproved the vibration theory (enantiomers have identical infrared spectra) while supporters of the vibration theory claimed that any difference between odours of isotopic analogues would disprove the shape theory. Organoleptic purity is of considerable importance in these arguments, but even without that issue the reality is more complex than proponents of either theory admit. It is, therefore, worthwhile looking at both topics in a little more detail.

Some factors affect both the enantiomer and isotope issues. For example, both assume very narrow tuning of olfactory receptors, but if we consider evolutionary pressure, it is clear that olfactory receptors will not all be narrowly tuned. Most GPCRs other than ORs have evolved to recognise one hormone or neurotransmitter only. So, for example, it is important that an adrenergic receptor with a specific function should recognise only adrenalin, or a related hormone, and it will have evolved to do exactly that. However, ORs have evolved to deal with an infinite variety of volatile materials and so have evolved to be more flexible, and thus many respond to both enantiomers of their cognate ligands. It would, therefore, be reasonable to expect that breadth of tuning would be such that isotopic substitution would also have little or no effect. Enzymes can be quite specific in action and so metabolism by enzymes in the nasal mucosa could well be affected by enantiomeric or isotopic substitutions. If considering odour descriptions given by humans or behavioural effects of intact, living animals, this might give the unwary experimenter the mistaken impression that the difference was due to olfactory receptors. Concentration dependence of odour character is another warning that olfaction is not so simple as either theory would suggest. Such change in odour character with concentration implies that some receptors are more selective than others and respond to low odorant concentrations, but as the concentration increases, more receptor types respond to the odorant. This ability to recognise non-ideal ligands and to show concentration dependence is not really consistent with either the shape or vibration theories.

Isotopes

The critical feature concerning isotopes is whether isotopic substitution changes the shape or other properties of a molecule. The proponents of the vibration theory argue that isotopic substitution has no effect on the shape of a molecule, but this is not the case. For example, the C–D bond effectively has a slightly smaller van der Waals radius than the C–H bond, so perdeuteration of a molecule will make it smaller, giving rise to the so-called steric isotope effect. An example of this is the different rate of diffusion of protium and deuterium isotopologues from a simple clathrate compound (86). Moreover, substitution of hydrogen by deuterium will affect the polarisation and polarisability of neighbouring functions, such as a ketone. If the hydrogen or deuterium atom in question is involved in enzymic reactions such as oxidation by a nasal P450, isotope effects could be very significant. My own laboratory experience is of a 30-fold difference in reaction rate because of deuteration (87). D_2O has a higher boiling point (101.42 °C) than H_2O (100.00 °C) and a higher melting point (3.82 °C as opposed to 0.00 °C) and deuteration is known to affect binding to receptors (88). There is a report of *D. melanogaster* being able to distinguish between isotopologues (89); however, the result should be treated with caution since the assay is behavioural and not at the receptor level, and other groups have shown that *D. melanogaster* uses combinatorial coding for identification of odorants (90,91). In any case, the considerable differences between fruit fly and human olfaction (see Chapter 2) means that such results cannot necessarily be taken to imply a mechanism in humans. The effect of isotopic substitution in human olfaction has been investigated by Keller and Vosshall using a well-constructed sensory experiment, but no evidence was found to support the vibration theory (92).

Enantiomers

As with isotopes, because of the flexibility and breadth of tuning of olfactory receptor proteins, it is not surprising that many respond to both enantiomers of odorants. As shown by the examples in Chapter 2, most of the carvone responding receptors respond to both enantiomers. Friedman and Miller showed that the enantiomers do have different odour characters (28), but Leitereg et al. showed that this was neither clear-cut nor consistent across a group of subjects (93). Similarly, Polak et al. showed that the ratio of *d*-carvone (**8.31**)/*l*-carvone (**8.32**) ODTs varied from 28 to 230 in a group of subjects they studied (42). The answer is that a pattern of receptors is activated, some responding to one enantiomer and others to both and the brain makes a judgement about odour based on a synthesis of this input together with other factors. Since the pattern and the other factors are unique to individual subjects, the percepts vary. Ignoring the complexity of reality in an attempt to prove or disprove the shape theory merely serves to hide the truth from all. In some cases, absolute stereochemistry does affect odour properties, and in others it does not. A review of the subject shows examples of all possible combinations of effect/no effect on character and/or threshold and there is no predictable pattern (29).

CONCLUSIONS

Contrary to ideas from the early history of the subject, SOR is neither a straightforward task nor a precise science. The number, nature and complexity of the steps involved between a molecule entering the nasal cavity and the formation of an odour percept in the orbitofrontal cortex of the brain are such that there is no simple direct correlation. SORs are not consistently precise and accurate tools but are statistical in nature and provide only a general guidance for the discovery chemist. This is something that fragrance chemists, patent examiners and senior executives in fragrance companies have to learn. Similarly, those who dabble on the fringes of fragrance chemistry and the science of olfaction must learn that SORs cannot either prove or disprove any theories of olfaction.

REFERENCES

1. *De Rerum Natura (On the Nature of Things)* A. M. Esolen, translator, Johns Hopkins University Press, Baltimore, 1995.
2. J. Hannam, *God's Philosophers: How the Medieval World Laid the Foundations of Modern Science*, Icon Books, London 2009, ISBN 978-184831-070-4.
3. C. S. Sell, *Angew. Chem. Int. Edn.*, **2006**, *45(38)*, 6254–6261 doi: 10.1002/anie.200600782.
4. M. Jansen and J. C. Schoen, *Angew. Chem. Int. Edn.*, **2006**, *128*, 1346–1352.

5. J. E. Amoore, *Molecular Basis of Odor*, Charles C. Thomas, Springfield, Illinois, 1970 Library of Congress Card No. 70-97521.
6. B. K. Lavine, C. White, N. Mirjankar, C. M. Sundling, C. M. Beneman, *Chem. Senses*, **2012**, *37*, 723–736 doi: 10.1093/chemse/bjs058.
7. P. Kraft and W. Eichenberger, *Eur. J. Org. Chem.*, **2003**, *19*, 3735–3743 doi: 10.1002/ejoc.200300174.
8. K. J. Rossiter, *Perfumer and Flavorist*, 1996, *21(2)*, 33–36, 38-40, 42-4, 46.
9. K. J. Rossiter, *Chem. Rev.*, **1996**; *96*, 3201–3240.
10. G. Fráter, J. A. Bajgrowicz, and P. Kraft, *Tetrahedron*, **1998**; *54(27)*, 7633–7703.
11. P. Kraft, J. A. Bajgrowicz, C. Denis, and G. Fráter, *Angew. Chem. Int. Edn.*, **2000**, *39(17)*, 2980–3010.
12. D. Livingstone, *Data Analysis*, Oxford Science Publishers, Oxford, 1995.
13. D. Ajami and J. Rebek, *Nature Chem.*, **2009**, *1*, 87–90 doi: 10.1038/NCHEM.111.
14. K. K. Frederick, M. S. Marlow, K. G. Valentine, and A. J. Wand, *Nature*, **2007**, *448*, 325–330 doi: 10.1038/nature.05959.
15. L. B. Sheard, X. Tan, H. Mao, J. Withers, G. B. Nissan, T. R. Hinds, Y. Kobayashi, F.-F. Hsu, M. Sharon, J. Browse, S. Y. He, Y. Rizo, G. A. Howe, and N. Zheng, *Nature*, **2010**, *468*, 400–407.
16. C. Lautenschlager, W. S. Leal, and J. Clardy, *Structure*, **2007**, *15(9)*, 1148–1154 doi: 10.1016/j.str.2007.07.013.
17. I.G. Tikhonova, C. S. Sum, S. Neumann, C. J. Thomas, B. M. Raaka, S. Costanzi, and M. C. Gershengorn, *J. Med. Chem.*, **2007**, *50*, 2981–2989.
18. D. M. Garrido, D. F. Corbett, K. A. Dwornik, A. S. Goetz, T. R. Littleton, S. C. McKeown, W. Y. Mills, T. L. Smalley Jr., C. P. Briscoe, and A. J. Peat, *Bioorg. Med. Chem. Letts.*, **2006**, *16*, 1840–1845.
19. G. C. Payne, R. E. Payne, and D.M. Farewell, *BMJ*, **2008**, *337(a2768)*, 1435–1437.
20. F. H. Messerli, *New Engl. J. Med.*, **2012**; *367*, 1562–1564 doi: 10.1056/NEJMon1211064.
21. C. S. Sell, *Perfumer and Flavorist*, **2000**, *25*, 67–73.
22. D. Munro, EP 1029845 (A1), 2000, to Quest International.
23. B. Weber and A. Mosandl, *Z. Lebensm.-Unters.-Forsch*, **1997**, *A204*, 194–197.
24. M. Laska, *Chem. Senses*, **2010**, *35*, 279–287.
25. J. von Braun and W. Kaiser, *Ber. Dtsch. Chem. Ges.*, **1923**, *56B*, 2268.
26. J. von Braun and W. Hänsel, *Ber. Dtsch. Chem. Ges.*, **1926**, *59B*, 1999.
27. J. von Braun and E. Anton, *Ber. Dtsch. Chem. Ges.*, **1927**, *60B*, 2438.
28. L. Friedman and J. G. Miller, *Science*, **1971**, *172*, 1044–1046.
29. C. S. Sell, *Chem. Biodiv.*, **2004**, *1*, 1899–1920.
30. A. Triller, E. A. Boulden, A. Churchill, H. Hatt, J. Englund, M. Spehr, and C. S. Sell, *Chem. Biodiv.*, **2008**, *5*, 862–886.
31. D. G. Laing, P. K. Legha, A. L. Jinks, and I. Hutchinson, *Chem. Senses*, **2003**, *28*, 57–69.
32. B. Schilling, R. Kaiser, A. Natsch, and M. Gautschi, *Chemoecology*, **2010**, *20*, 135–147 doi: 101007/s00049-009-0035-5 and references cited therein.
33. B. Schilling, CH2005/000412, to Givaudan.
34. B. Schilling, WO 2006/007751, to Givaudan.
35. B. Schilling, WO 2006/007752, to Givaudan.
36. B. Schilling, T. Granier, G. Fráter, A. Hanhart, WO 2008 116338, to Givaudan.
37. R. H. Dods, J. A. Mosely, and J. M. Sanderson, *Org. Biomol. Chem.*, **2012**, *10*, 5371–5378 doi: 10.1039/c2ob07113d.
38. W. D. Jones, P. Cayirlioglu, I. G. Kadow, and L. B. Vosshall, *Nature*, **2007**, *445*, 86–90.
39. D. D. Stettler and R. Axel, *Neuron*, **2009**, *63*, 854–864 doi: 10.1016/j.neuron.2009.09.005.
40. W. Li, E. Luxenberg, T. Parrish, and J. A. Gottfried, *Neuron*, **2006**, *52(6)*, 1097–1108.
41. E. H. Polak and J. Provasi, *Chem. Senses*, **1992**, *17*, 23–26.
42. E. H. Polak, A. M. Fombon, C. Tilquin, and P. H. Punter, *Behav. Brain Research*, **1989**, *31*, 199.
43. W. Li, J. D. Howard, T. B. Parrish, and J. A. Gottfried, *Science*, **2008**, *319*, 1842–1845.
44. M. H. Abraham, R. Sánchez-Moreno, J. E. Cometto-Muñiz, and W. S. Cain, *Chem. Senses*, **2012**, *37*, 207–218 doi: 10.1093/chemse/bjr094.

45. H. Guth, *Helv. Chim. Acta*, **1996**, *79(6)*, 1559–1571 doi: 10.1002/hlca.19960790606.
46. M. Masuda and S. Mihara, *Agric. Biol. Chem.*, **1989**, *53*, 3367.
47. J. A. Bajgrowicz, I. Frank, and G. Fráter, *Helv. Chim. Acta*, **1998**, *81*, 1349–1358 doi: 10.1002/hlca.19980810545.
48. H. Boelens, *Cosmet. Perfum.*, **1974**, *89*, 70–78.
49. C. S. Sell, *Seifen, Öle, Fette, Wachse*, **1986**, *112(8)*, 267–70.
50. C. S. Sell, *Dev. Food Sci.*, **1988**, *18 (Flavors Fragrances)*, 777–795.
51. P. Kraft, EP 1136481, 2001 assigned to Givaudan.
52. J. A. Bajgrowicz, EP 761664, 1997 assigned to Givaudan-Roure.
53. J. A. Bajgrowicz and I. Frank, *Tetrahedron: Asymmetry*, **2001**, *12(14)*, 2049–2057.
54. J. A. Bajgrowicz and G. Fráter, EP 801049, 1997, assigned to Givaudan-Roure.
55. K. J. Rossiter, WO 9847842, 1998, assigned to Quest Int.
56. K. J. Rossiter, *Chimia*, **2001**, *55(5)*, 388–399.
57. G. Ohloff, Relationship between odor sensation and stereochemistry of decalin ring compounds. in *Gustation and Olfaction*, Eds, G. Ohloff and A. F. Thomas, Academic Press, London, New York, 1971, pp. 178–183.
58. C. S. Sell, *Chem. Ind.*, **1990**, *16*, 516–520.
59. C. P. Newman, K. J. Rossiter, and C. S. Sell, EP 0 276 998, 1987, assigned to Quest Intl.
60. J. A. Bajgrowicz and I. Frank, WO 2007030963, 2007, assigned to Givaudan.
61. I.B. Bersuker, A. S. Dimoglo, M. Yu. Gorbachov, M. Koltsa, and P. F. Vlad, *New J. Chem.* **1985**, *9(3)*, 211–218.
62. W. Ogle, *Med.-Chir. Trans.* (**1870**), *53*, 263–290.
63. B. Rosenberg, T. N. Misra, and R. Switzer, *Nature*, **1968**, *217*, 423–427.
64. M. H. Briggs and R. B. Duncan, *Nature*, **1961**, *191*, 1310–1311.
65. J. T. Davies and F. H. Taylor, *Biol. Bull. Marine Lab, Woods Hole*, **1959**, *117*, 222–238.
66. G. Sanz, C. Schlegel, J.-C. Pernollet, and L. Briand *Chem. Senses*, **2005**, *30*, 69–80, doi: 10.1093/chemse/bji002.
67. O. Baud, S. Etter, M. Spreafico, L. Bordoli, T. Schwede, H. Vogel, and H. Pick, *Biochem.*, **2011**, *50*, 843–853 doi: 10.1021/bi1017396.
68. M. M. Mozell, *J. Gen. Physiol.*, **1970**, *56*, 46–63.
69. B. Bersuker, A. S. Dimoglo, M. Y. Gorbachov, and P. F. Vlad, *New J. Chem.* **1991**, *15(5)*, 307–320.
70. M. Y. Gorbachov and K. J. Rossiter, *Chem. Senses*, **1999**, *24*, 171–178.
71. E. Fischer, *Ber. Dtsch. Chem. Ges.*, **1894**, *27*, 2985.
72. L. Ružička, *Chem. Zeit.*, **1920**, *44*, 93–94.
73. M. G. J. Beets, *Amer. Perfumer Ess. Oil Rev.*, **1961**, *76(6)*, 54.
74. M. G. J. Beets, *Amer. Perfumer*, **1961**, *76(10)*, 12.
75. J. E. Amoore, *Perfum. Ess. Oil Record* , **1953**, *43*, 321–330.
76. J. E. Amoore, *Nature*, **1963**, *198*, 271–272.
77. J. E. Amoore, *Nature*, **1963**, *199*, 912–913.
78. J. E. Amoore, *Nature*, **1967**, *214*, 1095–1098.
79. J. E. Amoore, *Molecular Basis of Odour*, Charles C. Thomas, Springfield, IL, 1970.
80. G. M. Dyson, *Chem. Ind.*, **1938**, *57*, 647–651.
81. L. Turin, *Chem. Senses*, **1996**, *21*, 773–791.
82. J. C. Brookes, F. Hartoutsiou, A. P. Horsfield, and A. M. Stoneham, *Phys. Rev. Lett.* **2007**, *98*, 038101 doi: 10.1103/PhysRevLett.98.038101.
83. L. B. Buck and R. Axel, *Cell*, **1991**, *65*, 175–187.
84. E. H. Polak, *J. Theor. Biol.*, **1973**, *40*, 469–484.
85. J. von Braun and W. Hänsel, *Ber. Dtsch. Chem. Ges.*, **1926**, *59B*, 1999.
86. J. S. Mugridge, R. G. Bergman, and K. N. Raymond, *Angew. Chem. Int. Edn.*, **2010**, *49*, 3635–3637 doi: 10.1002/anie.200906569.
87. B. T. Golding, C. S. Sell, and P. J. Sellars, *Chem. Comm.*, **1977**, *20*, 693–694.
88. V. L. Schramm, *Curr. Opin. Chem. Biol.*, **2007**, *11*, 529–536.
89. M. I. Franco, L. Turin, A. Mershin, and E. M. Skoulakis, *PNAS*, **2011**, *108(9)*, 3797–3802 doi: 10.1073/pnas.1012293108 doi: 10.1073%2Fpnas.1012293108.

90. M. de Bruijne, K. Foster, and J. R. Carlson, *Neuron*, **2001**, *30*, 537–552.

91. E. A. Hallem and J. R. Carlson, *Cell*, **2006**, *125*, 143–160.

92. A. Keller and L. B. Vosshall, *Nature Neuroscience*, **2004**, *7(4)*, 337–338.

93. T. J. Leitereg, D. G. Guadagni, J. Harris, T. R. Mon, and R. Teranishi, *J. Agric. Food Chem.*, **1971**, *19*, 785.

Chapter 9

Intellectual Challenges in Fragrance Chemistry and the Future

CHALLENGES OVERCOME

When the University of Plymouth proposed to run a course in perfumery, combining the science and business aspects of the subject, the press, both local and national, ran articles containing comments along the lines of 'You don't need a degree to sell scent in a department store'. The university changed the name of its proposed course which now runs as a successful distance-learning one. To me, the attitude of the press indicated a degree of ignorance of the subject. It prompted me to think about the intellectual challenges of fragrance science and where better to start than with a look at the list of Nobel Laureates. The content of their Nobel lectures shows that many Nobel Laureates have worked on subjects of relevance to fragrance science. Some had more direct references than others to odorants and odour perception in their prize-winning lectures (Figure 9.1). In my view, admittedly a biased one, nine Nobel Prizes awarded for pure chemistry and another eight for work on G-coupled protein receptors (GPCRs) are all of relevance to odorants and olfaction.

In 1905, Adolf von Baeyer was awarded the Nobel Prize for his work including ring strain in cyclic terpenoids such as camphor (**9.1**), which is an odorant in both nature and perfumery. His work also contributed to later developments in the understanding of terpenoids and carbocation chemistry, subjects of great importance to the perfume ingredients industry, as is evident from Chapter 6. Otto Wallach's Prize in 1910 was of similar pioneering value in that he studied the various monoterpenoid hydrocarbons and showed how they differ from each other in their structures. The definitive work on carbocation chemistry, adding to all the results of von Bayer and Wallach and therefore accounting for much of terpenoid

Chemistry and the Sense of Smell, First Edition. Charles S. Sell.
© 2014 John Wiley & Sons, Inc. Published 2014 by John Wiley & Sons, Inc.

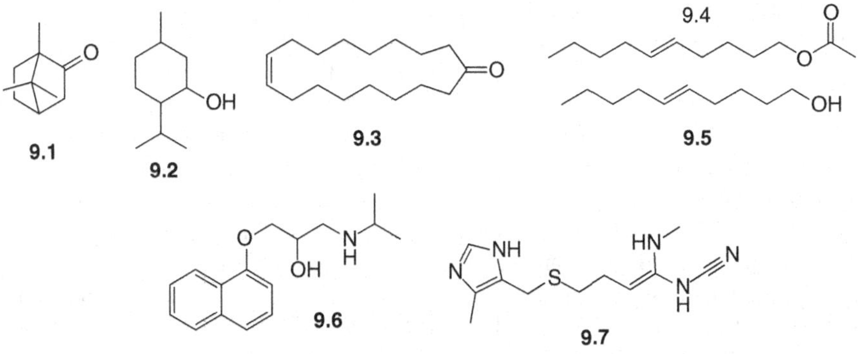

Figure 9.1 Some compounds associated with Nobel Prizes.

chemistry, was that which won the 1994 prize for George Olah. Sir Robert Robinson was awarded the 1947 Prize for elucidation of the isoprene rule which led to a much deeper understanding of this important family of natural products, many members of which are key odorants. The annulation reaction devised by Robinson is also of great value in odorant synthesis. In 1939, Leopold Ruzicka was awarded the Prize for his work on polymethylenes and higher terpenes, and his lecture mentioned many odorants. His work on polymethylenes included the development of the high-dilution technique for the synthesis of macrocyclic compounds, a methodology of great significance for the macrocyclic musks. Sir Derek Barton and Odd Hassel were awarded the Prize in 1967 for their work on conformational analysis and the effect of the shape of rings and fused rings on the chemistry of the molecules containing them. Although they did not mention odorants in their lectures, the basic concepts that they described are of considerable use in fragrance chemistry, especially in the terpenoid field. Barton also worked on a number of topics directly of significance for the fragrance industry. Elias Corey developed the technique of retro-synthetic analysis, and this is another topic of relevance to the synthesis of odorants. Like Barton, he also published work directly related to odorants. William Knowles and Ryoji Noyori won the Prize in 2001 for work on asymmetric catalysis, the prime example cited in Noyori's lecture being that of menthol (**9.2**). His chemistry had a significant effect on the world menthol market. The 2005 Chemistry Prize went to Richard Schrock, Robert Grubbs and Yves Chauvin for their work on olefin metathesis. This is another technique that is increasingly used in the fragrance ingredients industry. In their lectures in Stockholm, both Schrock and Grubbs described the use of olefin metathesis in preparing macrocyclic structures, Schrock citing the musk civetone (**9.3**) as an example, and Grubbs also describing the use of olefin metathesis in the synthesis of the mixture of (**9.4**) and (**9.5**), which is a pheromone of the peach twig borer (*Anarsia lineatella*).

Of the eight Nobel Prizes awarded for work on GPCRs, only one is directly related to olfactory receptor (OR) proteins, that of Linda Buck and Richard Axel (2004) for the discovery of the olfactory receptor proteins and their work on the organisation of the olfactory system. Nonetheless, olfactory receptors are part of the GPCR family and the very significant work that has been done on other members of the family is of vital importance in our understanding of the mechanism of action of the olfactory receptors, as was described in Chapter 2. In 1971, Earl Sutherland received the Nobel Prize for elucidation of the mode of action of hormones via second messengers, a mechanism paralleled exactly in olfactory sensory neurons. In 1977, Roger Guillemin and Andrew Schally shared the Prize for work on peptidic hormones in the brain. Work on prostaglandins won the Prize for Sune Bergström, Bengt Samuelson and John Vane in 1982, and in 1988, James Black, Gertrude Elion and George Hitchings won it for their work on the principles of drug action and the discovery of propranolol (**9.6**) and cimetidine (**9.7**). In 1994, Alfred Gilman and Martin Rodbell were awarded the Prize for work on the roles of G-proteins, and Arvid Carlson, Paul Greengard and Eric Kandel in 2000 for their study of the role of dopamine receptors in Parkinson's disease. The 2012 Nobel Prize in Chemistry was awarded jointly to Robert Lefkowitz and Brian Kobilka for studies of GPCRs. Their enormous contribution to the elucidation of the mechanism of action of the receptors is clear from the citations in Chapter 2.

The discoveries of other Nobel Prize winners have also been of importance in fragrance research, though their application is less directly relevant. The list would include synthetic methods that are used daily in fragrance chemistry such as the palladium-catalysed cross-coupling reaction of Richard Heck, Ei-ichi Negishi and Akira Suzuki (2010), chirally catalysed oxidation (Barry Sharpless, 2001), catalytic hydrogenation (Paul Sabatier, 1912) and the reactions bearing the names of their discoverers, namely Otto Diels and Kurt Alder (1950) and Victor Grignard (1912). Robert Woodward's work on synthesis (1965) was directed mostly at porphyrins and corrins and so is less pertinent to fragrance chemistry than that of Corey but still has an influence on the thinking of fragrance chemists. The 1952 Noble Prize was awarded to Archer Martin and Richard Synge for their work on chromatography, an analytical technique that has transformed perfumery and the study of natural volatile oils. The work of Richard Kuhn (1938) and Paul Karrer (1937) on carotenoids has a bearing on the ionone and damascene families of odorants. The application of the green fluorescent protein to molecular biology and biochemistry has been used in the study of olfaction and won the Nobel for Osamu Shimomura, Martin Chalfie and Roger Tsien in 2008.

Research in olfaction can be hampered by many obstacles, most of which are common to all scientific research. Those that are more specific to the study of odour and odorants are usually the result of underestimating the complexity of olfaction. For example, in Chapter 8 we have seen the pitfall of assuming that structure/odour relationship patterns tell us about the receptive range of olfactory receptors, and in Chapter 2 we saw that a specific anosmia does not indicate a missing receptor for a primary odour. Indeed, from the previous chapters in this book, it is clear how olfaction theories based on overly simplified hypotheses, issues regarding olfactory purity and issues related to odour measurement in the light of the subjectivity of olfaction have all led to much wasted research effort. Nonetheless, very significant advances have been made over the last few decades and there are many examples of intellectual challenges that have been met successfully. All of these leave no doubt that fragrance science is a serious subject, full of significant intellectual challenge.

CHALLENGES FOR THE FUTURE

There are still plenty of challenges for future research in olfaction, and some of them are discussed briefly below.

Challenges in Olfaction

The enormous strides in the last two decades have taught us a great deal about how the sense of smell functions, but they have also made it increasingly clear that smell is a very complex sense and that we still have a great deal more to learn about it. No one has yet managed to crystallise an olfactory receptor protein, and so we do not have such detailed and accurate structural information about ORs that we do for some other class A GPCRs. This remains a major challenge and its solution will give us much more precise information about the binding sites and receptive ranges of ORs. Similarly, crystal structures of ligand-bound ORs complexed to $G_{\alpha olf}$ will provide improved insight into the mechanism of second messenger activation. We need to learn more about the variations in ORs and the occurrence and significance of allelic variation, SNPs and copy number variations. The largest challenge is, doubtlessly, in improving our understanding of how glomerular activation patterns are translated into conscious odour percepts.

Challenges in Application

Health and well-being are becoming more significant issues for olfactory science. We are beginning to see how odours produced by the body can be used as indicators of disease. The smell of acetone (**9.8**) on the breath of diabetics or the odour of acetophenone (**9.9**) from those suffering from phenylketonuria has been known for over a century (Figure 9.2). However, we are now aware that many disease states are indicated by specific odorants, and novel diagnostic techniques could well be developed, following the example of those in Chapter 4. The use of dogs to smell cancer in patients is increasing. For instance, dogs can distinguish between the urine of healthy subjects from that of sufferers from bladder cancer. In many cases, there does not seem to be a single distinctive marker chemical as there is with diabetes and phenylketonuria, so the dogs are probably identifying characteristic patterns of metabolite composition. The challenge is therefore to identify what it is that the dog smells and to find a suitable clinical technique based on this. Loss of sense of smell is increasingly recognised as an indicator for various diseases such as Parkinson's and Alzheimer's. The degree and nature of smell loss seem also to offer a means of distinguishing between different forms of these diseases. Moreover, smell loss seems to be an early indicator, often occurring long before other clinical symptoms appear. The clinical significance is therefore high and will be even more so

9.8   **9.9**   **Figure 9.2** Odorants as indicators of disease.

when effective treatments are developed. Improvements in clinical methodology for assessing smell loss are therefore likely to be important. The use of odour as an adjunct to various medical treatments is a growing topic in popular practice. The effects are probably mediated by the higher brain rather than through direct pharmaceutical action, and this needs to be explored further. Mood effects of fragrance are used in marketing, but perhaps they could be used also to complement other medical treatments, for instance in mental health.

Challenges in Process Chemistry

The significant challenges for chemical synthesis are nowadays related to sustainability. For process chemistry, therefore, the key issues are safe and clean manufacturing technology and use of sustainable starting materials. The relatively small size of the fragrance industry will always present economic obstacles in terms of return on R&D investment. Large volume products will be targets for larger chemical companies that can invest more in process technology, often spreading development costs across different products. Therefore, the fragrance industry must adopt a tactical approach and seek versatile technology that can be used at the scale at which it operates. Clearly, catalytic processes, including biotechnology, will become increasingly important. Atom utilisation is a useful tool in evaluating the environmental impact of a process, but often the energy costs are also important and these contribute to environmental impact, albeit outside the fragrance ingredients factory. Tools such as Roger Sheldon's E-factor are therefore useful since they can include an allowance for energy consumption. Many fragrance ingredient processes run without solvents, and this represents an advantage in both reactor utilisation and in energy terms since there is no requirement for solvent removal and recovery. Such processes will doubtless gain in significance. The use of water as solvent has been advocated in recent years, but this tends to introduce problems in terms of wastewater contaminated by organic materials. The solutions to this issue usually involve energy consumption and, so, avoidance of solvents is preferable. Stoichiometric reduction–oxidation (redox) processes represent an environmental issue, and the simple biotech response using whole organisms is far from ideal. Clean chemical redox methods and alternatives to cofactors in biotech redox systems are worthwhile research targets.

Currently, synthetic chemists are trained in the techniques of mentally breaking down target molecules into small fragments. In order to take advantage of sustainable precursors, chemists must learn to think in terms of connections between feedstocks and targets. Thus we should adapt Corey's above-mentioned technique of retrosynthetic analysis to one that works backwards from the target structure to sustainable small building blocks.

Most fragrance ingredients are synthetically accessible from either natural or petrochemical starting materials. The balance, as shown in Chapter 6, can change over time and so the industry must be able to adapt to such changes. A common misconception is that renewable is synonymous with sustainable. A number of renewable materials have been lost, the Indian sandalwood oil being an example. Current starting materials based on turpentine derivatives rely on crude sulphate turpentine (CST), but this is a by-product of paper manufacture. Demand for new paper production is falling as recycling of paper and electronic information systems rise in significance. CST is therefore becoming more expensive and, if we were to find an alternative to softwood (e.g. tropical grasses) for paper manufacture, then CST-derived fragrance ingredients might well become less sustainable than their petrochemical equivalents. CST is not the only by-product from other industries that the fragrance industry has come to rely on as a starting material for synthesis. Changes in those other industries can therefore have a significant effect on the fragrance industry. Consequently, process chemists must work together with procurement specialists to ensure the sustainability of perfume ingredients. Safety, health and environment (SHE) issues and other pressures might lead to an increasing emphasis on homochiral products, but the cost/benefit ratio must always be better than that of the racemate. Awareness of the homochiral pool of natural starting materials is therefore important.

Challenges in Odorant Design

Discovery chemistry continues to be essential for the fragrance industry in order to produce a new generation of ingredients with improved safety in use and environmental fate profiles. Natural oils can be lost for reasons of sustainability (e.g. Indian sandalwood and rosewood) and alternatives need to be found if that odour aspect is to be secured for use. Rational design in terms of odour remains a challenge because of the issue of subjectivity, and therefore continued improvement in structure/odour correlation is needed. Another challenge for the discovery chemist is to find new methods of delivering a fragrance effect. There have been a number of developments in terms of pro-fragrances such as Givaudan's Tonkarose (**9.10**) (Figure 9.3). On exposure to long-range ultraviolet light such as sunlight, the *E*-isomer (**9.10**) isomerises to the *Z*-isomer (**9.11**), which then undergoes a cyclic elimination reaction to give a fragrant accord which is a combination of coumarin (**9.12**) and 9-decenol (**9.13**). Improved understanding of olfaction in topics such as nasal metabolism and receptor technology will open up new avenues for research in fragrance ingredient discovery. This whole area on novel effect delivery offers great scope for creativity in the future.

The relatively slow volume growth of new products and the fact that they must compete against current materials mean that processes and products must be able

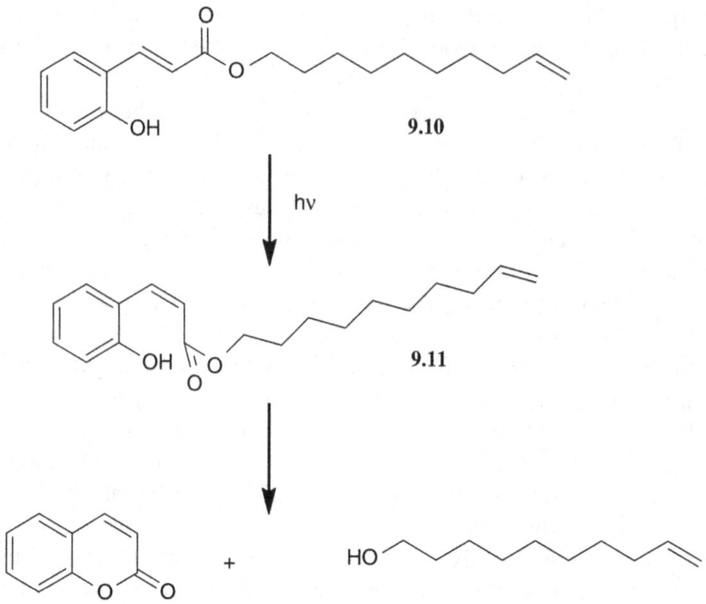

Figure 9.3   Tonkarose as an example of a pro-fragrance.

to enter with low volume at a cost per kilogram which is not too far from that anticipated for the eventual production volume.

WHAT WILL BE NEEDED TO MEET THESE CHALLENGES OF THE FUTURE?

Fragrance science has always been intrinsically multidisciplinary, and the requirement for multidisciplinary teams will only increase since no one discipline can cover all of the necessary expertise base. People with different academic backgrounds look at things in different ways and therefore contribute different insights to subjects and problems. For example, a molecular biologist might see aldehydes as just another type of compound, but an organic chemist will see them as reactive species; similarly, an organic chemist will see thiols as very different from alcohols, which a biologist might not. A chemist will not see metal ions as just ionic species but as redox reagents. Therefore a chemist looking at interactions of an odorant with an OR and its mucosal environment will think in terms of reactive chemistry as well as molecular recognition. Similarly, a chemist is likely to take too simple and mechanical a view of problems in biology, and so the biologist's input is necessary to point out the complexity of natural systems.

Researchers working in disciplinary silos will tend to turn inwards intellectually and lose sight of the total picture. The greatest breakthroughs will come at interfaces between disciplines and so are most likely to be found by multi-disciplinary teams. However, multi-disciplinary work creates a problem for editors and reviewers (referees) of learned journals. It is difficult to find a reviewer who is an expert in everything and, if a large number of disciplines is involved, then a large number of reviewers would be needed but this increases complexity and would be likely to introduce delays, and, if different reviewers disagree on the merits of a proposed publication, how is the editor to judge between them?

Remembering Albert Einstein's comment that 'The most beautiful thing we can experience is the mysterious. It is the source of all true art and science', I hope it is evident from this conclusion and from all that has preceded it in the book that the science of olfaction still has enough mystery and intellectual challenge to inspire many future generations of scientists.

Glossary

Adenylyl cyclase III (AC III)	A membrane-bound enzyme that converts adenosine triphosphate (ATP) to cyclic adenosine monophosphate (cAMP).
Accessory olfactory bulb (AOB)	The part of the olfactory bulb that accepts signals from olfactory sensory neurons located in the vomeronasal organ.
Acellular	A system that does not include living or intact cells.
Agonist	A ligand that binds to a receptor protein and stabilises it in its active state.
Allelle	A version of a gene. Different members of a species may carry different versions of the same gene and, since each individual carries two copies of each gene, these could be different alleles.
Allosteric modulator	A substance that changes the response of a receptor to its cognate ligand by binding at a site other than the site at which the ligand binds. (See also *Positive allosteric modulator* and *Negative allosteric modulator*.)
Allosteric binding site	A binding site in a receptor protein other than that at which an agonist binds.
Amino terminus	The end of a peptide chain that contains a free α-amino group. In mammalian olfactory receptors (ORs), this is the tail of the protein that lies outside the cell membrane. In insect ORs, it is the tail that lies inside the cell membrane.
Anosmia	The inability to smell. General anosmia is the inability to smell anything; specific anosmia is the inability to smell one (or more) specific compounds despite normal ability to smell others.

Chemistry and the Sense of Smell, First Edition. Charles S. Sell.
© 2014 John Wiley & Sons, Inc. Published 2014 by John Wiley & Sons, Inc.

Antagonist	A substance that, when co-administered to a receptor with an agonist of the receptor, produces a decrease in the observed response of the receptor/reporter system relative to that produced by the agonist alone. An antagonist could be a neutral agonist, an inverse agonist or a substance that acts elsewhere on the second messenger or reporter systems. The term is often used to imply an inverse agonist. (See also *Neutral antagonist* and *Inverse agonist*.)
Axon	An extension of a neuron through which it sends signals to another neuron via a synapse.
Basal activity	The level of signal generated by a receptor in the absence of any ligand.
Cladogram	A cluster tree analysis or dendrogram of a family of receptors showing the degree of difference in sequence between the receptors of that family.
Cognate ligand	The corresponding ligand for a receptor.
Conserved residue	An amino acid that is likely to be found at a given position in a protein sequence. The chance of finding a particular amino acid at a particular position is described as the degree, level or percentage conservation.
Carboxyl terminus	The end of a peptide chain that contains a free carboxylate group. In mammalian olfactory receptors (ORs) this is the tail of the protein that lies inside the cell membrane. In insect ORs, it is the tail that lies outside the cell membrane.
Cell culture	A medium containing cells in which the cells are kept alive and possibly can reproduce.
Cell line	Cultured cells that can be maintained indefinitely in a suitable medium.
Central	A term used in in neuroscience to refer to the central nervous system and brain.
Cognate ligand	The ligand that activates a given receptor.
Constitutively active	A receptor protein that has been modified so that it remains in the active state.
Copy number variation (CNV)	Organisms can carry more than the usual two copies of a gene. This term relates to variations in the numbers of copies carried by different individuals.
Cyclic AMP (cAMP)	Cyclic adenosine monophosphate, a nucleotide used as a chemical messenger in cells.
Cyclic nucleotide gated channel (CNG channel)	A membrane protein that allows ions to pass through the cell membrane when activated by a cyclic nucleotide such as cAMP.

Cytochrome	A family of oxidative enzymes characterised by their being deeply coloured.
Cytosol	The material constituting the internal content of a cell.
Dendrite	An extension of a neuron through which it receives signals from another neuron via a synapse.
Deorphanisation	The process of identifying a ligand for an orphan receptor.
2-Deoxyglucose (2-DG) imaging	A technique using radio-labelled 2-deoxy-glucose to identify areas of high activity in the brain.
Dizygotic twin	An individual whose twin does not share an identical genome.
Efficacy	The kinetics of a ligand in its interaction with a receptor and hence its effect on the size of the resultant signal.
Electroencaphalography (EEG)	A technique in which electrodes are placed on the exterior of the head to detect electrical signals in the brain.
Expression (gene expression)	The process of producing the protein encoded by a gene.
Extracellular loop (ECL, EL)	A loop between two of the transmembrane helices and which lies outside the cell membrane.
Extracellular surface (ECS)	The part of the receptor that lies outside the cell membrane and, in olfactory sensory neurons, is exposed to the olfactory mucus.
Functional expression	The production of a receptor into a cell in such a state that the receptor can be activated by its cognate ligand.
Functional magnetic resonance imaging (fMRI)	A form of nuclear magnetic resonance spectroscopy which uses signals from protons (essentially those in water) to produce images of different body parts and to measure changes in them due to, for example, activation of neural circuits.
GDP	Guanosine diphosphate, a nucleotide used as a chemical messenger in cells.
Gene	A fragment of nucleic acid that encodes a protein.
Genome	The entirety of an orgsanism's genes.
Genotype	Characterised by genetic make-up.
Glomerulus	A region of the olfactory bulb that receives signals from olfactory sensory neurons expressing the same type of olfactory receptor.
GMP	Guanosine monophosphate, a nucleotide used as a chemical messenger in cells.

GPCR	A membrane-bound receptor protein that is coupled to a G-protein through which it activates a second messenger system.
G-protein	A protein that binds a guanine-derived nucleotide such as GDP or GTP.
Grüneberg ganglion (GG)	A sensory organ found in the nose of rodents.
Homologue	A receptor that is structurally similar to another and therefore is found in the same branch of the cladogram.
Homology	The overall level of conservation of amino acids between different proteins. The percentage homology indicates the number of instances when the same amino acid is found at the same position as a percentage of the total number of amino acids in the protein.
Intracellular loop (ICL, IL)	A loop between two of the transmembrane helices and which lies inside the cell membrane.
Intracellular surface (ICS)	The part of the receptor that lies inside the cell membrane and interacts with the G-protein; in olfactory sensory neurons, this is G_{olf}.
Inverse agonist	A ligand that binds to a receptor protein and stabilises it in an inactive form thus suppressing the basal level of activity.
Ligand	A substance that binds to a receptor.
Ligand binding pocket (LBP)	The location in a receptor where the ligand binds.
Neutral antagonist	An antagonist that binds to the receptor and does not affect is basal firing rate.
Negative allosteric modulator (NAM)	A substance that binds at an allosteric site and decreases the sensitivity of the receptor to its ligands.
Interneuron	A type of neuron found in the olfactory bulb.
Inverse agonist	A ligand that binds to a receptor, stabilises it in its inactive state and reduces the basal firing rate.
Jacobsen's organ	See vomeronasal organ.
Lipocalin	A family of proteins that binds small molecules, usually for the purpose of elimination via the bloodstream.
Major olfactory bulb (MOB)	The part of the olfactory bulb that accepts signals from olfactory sensory neurons located in the olfactory epithelium.
Major urinary proteins (MUPs)	A family of proteins belonging to the lipocalin family which constitute the major protein content of murine urine.

Mitral cell	A type of neuron found in the olfactory bulb.
Monozygotic twin	An individual whose twin shares an identical genome.
Mutagenesis	The replacement of one amino acid in a protein sequence by another one. This technique is used to identify those amino acids that are vital for ligand binding and/or activation of a receptor. It is often used to support predictions from modelling.
Mutant	An individual whose genome has been altered relative to the normal genome of the species.
Negative modulator	A substance that, when co-administered to a receptor with an agonist of the receptor, produces a decrease in the observed response of the receptor/reporter system compared to that produced by the agonist alone.
Negative allosteric modulator	A substance that, when co-administered to a receptor with an agonist of the receptor, produces a decrease in the observed response of the receptor compared to that produced by the agonist alone, by binding to site on the receptor other than that to which the agonist binds.
Neuron	A cell that transmits electrical impulses in the nervous system or brain.
Nonspecific antagonist	A substance that, when co-administered to a receptor with an agonist of the receptor, produces a decrease in the observed response of the receptor/reporter system compared to that produced by the agonist alone and which is independent of the nature of the receptor: in other words, an antagonist that acts on the second messenger cascade or the reporter system used to detect the signal, rather than on the receptor.
Normosmia	Normal ability to smell.
Odour binding protein (OBP)	Proteins belonging to the lipocalin family which are found in the nasal mucus and bind odorant molecules.
Olfactory bulb (OB)	The part of the brain that accepts signals from olfactory sensory neurons.
Olfactory cortex	Cortical brain regions in which olfactory signals are processed. The term is sometimes used to refer specifically to the piriform cortex.
Olfactory epithelium (OE)	A patch of tissue in the nose which contains olfactory sensory neurons.
Olfactory receptor protein (OR)	A protein belonging to the group of class A GPCRs which is expressed in olfactory sensory neurons and

	detects odorants from nasal air via the olfactory mucus.
Olfactory sensory neuron (OSN)	A neuron that recognises an odorant and, on doing so, sends a signal to the olfactory bulb.
Oocyte	Female reproductive cell (egg).
Optogenetics	A technique in which light can be used to initiate or terminate signals in neurons and can therefore be used to study neuroprocessing.
Orphan receptor	A receptor for which the structure is known but for which no ligand has yet been identified.
Orthologues	Genes that have diverged in different species from a common ancestral gene, along with the divergence of the species, and which share a high level of homology with each other and have the same function. Since they have been conserved across species, it is expected that their ligands are of general significance and will therefore be similar.
Orthonasal olfaction	Smelling by intake of air through the nose.
Orthosteric binding site	The site at which the agonist binds.
Paralogues	Genes in a single species that have arisen by duplication and divergence within the species. Since they have diverged within a species, it is expected that, despite a high level of homology, they could show different receptive ranges.
Partial agonist	A ligand that binds to a receptor but causes only weak activation.
Peripheral	A term used in in neuroscience to refer to structures or activities at the edge of the nervous system, foe example, in olfactory sensory neurons.
Phenotype	Characterised by activity or physical or chemical properties.
Piriform cortex	A part of the brain that is important in the processing of olfactory signals.
Positron emission tomography (PET)	An imaging technique that uses γ-ray emissions from a radionuclide to build a 3-D image. By tagging the radionuclide to fluorodeoxyglucose (FDG), a glucose analogue, the rate of metabolic activity can be studied.
Positive modulator	A substance that, when co-administered to a receptor with an agonist of the receptor, produces an increase in the observed response of the receptor/reporter system compared to that produced by the agonist alone.
Positive allosteric modulator (PAM)	A substance that, when co-administered to a receptor with an agonist of the receptor, produces an

increase in the observed response of the receptor compared to that produced by the agonist alone, by binding to a site on the receptor other than that to which the agonist binds.

Potency	The degree of affinity of a ligand for a receptor and hence its effect on EC_{50}.
Projects	The sending of an axon from a neuron to the dendrite of another neuron.
Pseudogene	A gene that does not code for a full-length functional protein.
Pyramidal cell	A type of neuron found in the piriform cortex.
Reporter gene	A gene for an enzyme or other protein that is incorporated into a cell in order to indicate when a specific piece of cell chemistry has been activated, for example, firefly luciferase which is activated by cAMP and therefore can be used as an indicator of activation of the olfactory second messenger cascade.
Retronasal olfaction	Smelling by passing air from the mouth and airways across the olfactory epithelium during exhalation.
Second messenger	A chemical that is produced by action of a receptor protein and which serves to initiate membrane depolarisation in a sensory neuron and hence a nerve impulse.
Sensory neuron	A neuron that detects an external stimulus.
Septal organ (SO)	A sensory organ found in the nose of rodents.
Sequence	The sequence of nucleotides in a nucleic acid or amino acids in a protein.
Single nucleotide polymorphism (SNP)	Difference between two genes characterised by only one nucleotide being different.
Site-directed mutagenesis	A technique to modify a gene by exchanging one specific nucleotide for another.
Specific antagonist	A substance that, when co-administered to a receptor with an agonist of the receptor, produces a decrease in the observed response of the receptor/reporter system relative to that produced by the agonist alone and which is dependent on the nature of the receptor: in other words, an antagonist that acts on the receptor rather than other elements of the second messenger cascade or the reporter system used to detect the signal.
Synapse	The electrical connection between the axon of one neuron and the dendrite of another.
Trace amine activated receptors (TAARs)	A family of receptors thath occur in the nasal epithelia and which respond to small amines.

Transgenic	Having a genome that has been altered by the artificial insertion or deletion of one or more genes.
Transient receptor potential channels (TRP)	A family of transmembrane proteins that allow ions to pass through the cell membrane.
Tufted cell	A type of neuron found in the olfactory bulb.
Vomeronasal organ (VNO) (also known as Jacobsen's organ)	An organ found in the nose of many mammalian species and which detects and responds to volatile materials in inhaled air. It is thought to have a major role in recognition of mammalian semiochemicals. It sends signals to the accessory olfactory bulb.
Vomeronasal receptors (VNRs)	A family of receptor proteins that are located in the voneronasal organ.
Xenobiotic	A substance that is not produced by an organism and is not therefore a normal part of its chemistry but has entered the organism from the environment.

Index

Printed in the United States
By Bookmasters